Basal Ganglia and Thalamus
in Health and Movement Disorders

Basal Ganglia and Thalamus in Health and Movement Disorders

Edited by

Kristy Kultas-Ilinsky

and

Igor A. Ilinsky

The University of Iowa
Iowa City, Iowa

Kluwer Academic/Plenum Publishers
New York, Boston, Dordrecht, London, Moscow

Library of Congress Cataloging-in-Publication Data

Basal ganglia and thalamus in health and movement disorders/edited by Kristy
Kultas-Ilinsky and Igor A. Ilinsky.
    p. cm.
    Papers presented at an international workshop held in Moscow, Russia, 29–31 May 2000.
    Includes bibliographical references and index.
    ISBN 0-306-46543-4
      1. Movement disorders—Congresses. 2. Basal ganglia—Congresses 3.
Thalamus—Congresses. I. Kultas-Ilinsky, Kristy. II. Ilinsky, Igor A.
    [DNLM]: 1. Movement Disorders—physiopathology—Congresses. 2. Movement
Disorders—therapy—Congresses. 3. Basal Ganglia—physiology—Congresses. 4. Basal
Ganglia—physiopathology—Congresses. 5. Thalamus—physiology—Congresses 6.
Thalamus—physiopathology—Congresses. WL 390 B297 2001]
    RC376.5 .B37 2001
    616.7'07—dc21

                                                          00-067119

ISBN 0-306-46543-4

©2001 Kluwer Academic/Plenum Publishers, New York
233 Spring Street, New York, N.Y. 10013

http://www.wkap.nl/

10 9 8 7 6 5 4 3 2 1

A C.I.P. record for this book is available from the Library of Congress.

Printed in the United States of America

This book is dedicated
in memory of Agnes Kultas

Contributors

SERGEY V. AFANAS'EV • Sechenov Institute of Evolutionary Physiology and Biochemistry, St. Petersburg, Russia.

MARJORIE E. ANDERSON • Departments of Rehabilitation Medicine, Physiology and Biophysics, and Program in Neurobiology and Behavior, University of Washington, Seattle, USA.

CLAIRE ARDOUIN • Department of Neurosciences, Joseph Fourrier University, Grenoble, France.

PATRIZIA AVONI • Institute of Clinical Neurology, University of Bologna Medical School, Bologna, Italy.

HAZAR AWAD • Department of Pharmacology, Emory University School of Medicine, Atlanta, USA.

TIPU AZIZ • Department of Neurosurgery, Radcliffe Infirmary, Oxford, U.K.

PETER BAIN • Department of Neurology and Imperial College of Medicin, Charing Cross Hospital, London, U.K.

ALIM-LOUIS BENABID • Department of Neurosurgery and Neuroscience, Joseph Fourier University, Grenoble, France.

ABDEL-HAMID BENAZZOUZ • Department o f Neuroscience, Joseph Fourier University, Grenoble, France.

DEBRA A. BERGSTROM • Experimental Therapeutics Branch, National Institute of Neurological Disorders and Stroke, National Institutes of Health, Bethesda, USA.

MARK D. BEVAN • MRC Anatomical Neuropharmacology Unit, Department of Pharmacology, Oxford, U.K.

J. PAUL BOLAM • MRC Anatomical Neuropharmacology Unit, Department of Pharmacology, Oxford, U.K.

RABIA BOUALI-BENAZZOUZ • Department o f Neuroscience, Joseph Fourier University, Grenoble, France.

STEFANIA R. BRADLEY • Department of Pharmacology, Emory University School of Medicine, Atlanta, USA.

SORIN BREIT • Department of Neurosciences, Joseph Fourier University, Grenoble, France.

JOHN A. BUFORD • Division of Physical Therapy, The Ohio State University, Columbus, USA.

NANCY N. BYL • Department of Physical Therapy, University of California at San Francisco, San Francisco, USA.

ELENA CAPUTO • Department of Neurosciences, Joseph Fourier University, Grenoble, France.

STEPHAN CHABARDES • Department of Neurosurgery, Joseph Fourier University, Grenoble, France.

JIN WOO CHANG • Department of Neurosurgery and Brain Research Institute, Yonsei University College of Medicine, Seoul, Korea.

JONG HEE CHANG • Department of Neurosurgery, Yonsei University College of Medicine, Seoul, Korea.

ALI CHARARA • Division of Neuroscience, Yerkes Regional Primate Center, and Department of Neurology, Emory University, Atlanta, USA.

MARIE-FRANÇOISE CHESSELET • Department of Neurology, University of California at Los Angeles, Los Angeles, USA.

SANG SUP CHUNG • Department of Neurosurgery and Brain Research Institute, Yonsei University College of Medicine, Seoul, Korea.

P. JEFFREY CONN • Department of Pharmacology, Emory University School of Medicine, Atlanta, USA.

PIETRO CORTELLI • Institute of Clinical Neurology, University of Bologna Medical School, Bologna, Italy.

KAREN D. DAVIS • Institute of Medical Science and Department of Surgery, University of Toronto, Toronto Western Research Institute and Toronto Western Hospital, Toronto, Canada.

PHILIPPE DE DEURWAERDÈRE • Department of Neurology, University of California at Los Angeles, Los Angeles, USA.

MAHLON R. DELONG • Department of Neurology, Emory University, Atlanta, USA

JONATHAN O. DOSTROVSKY • Department of Physiology, University of Toronto, Toronto, Canada.

CHRISTOPHER M. ELDER • Department of Neurology, Emory University School of Medicine, Atlanta, USA.

MICHEL FILION • Unite de Recherche en Neuroscience, Centre Hospitalier de l'Universite Laval, Sainte-Foy, Canada.

VALERIE FRAIX • Department of Neurosciences, Joseph Fourier University, Grenoble, France.

LAUREN E. FREEMAN • Experimental Therapeutics Branch, National Institute of Neurological Disorders and Stroke, National Institutes of Health, Bethesda, USA.

AURÉLIE FUNKIEWIEZ • Department of Neurosciences, Joseph Fourier University, Grenoble, France.

IRA M. GARONZIK • Department of Neurosurgery, Johns Hopkins University, Baltimore, USA.

DONGMING GAO • Department of Physiology, Jinzhou Medical College, Jinzhou, P.R. China.

RALPH GREGORY • Department of Neurology, Radcliffe Infirmary, Oxford, U.K.

MARK HALLETT • Human Motor Control Section, National Institute of Neurological Disorders and Stroke, Bethesda, USA.

JESSE E. HANSON • Division of Neuroscience, Yerkes Regional Primate Center, and Department of Neurology, Emory University, Atlanta, USA.

SHERWIN E. HUA • Department of Neurosurgery, Johns Hopkins University, Baltimore, USA.

GEORGE W. HUBERT • Division of Neuroscience, Yerkes Regional Primate Center, and Department of Neurology, Emory University, Atlanta, USA.

WILLIAM D. HUTCHISON • Departments of Physiology and Surgery, University of Toronto, and Toronto Western Hospital, Toronto, Canada.

IGOR A. ILINSKY • Department of Anatomy and Cell Biology, University of Iowa College of Medicine, Iowa City, USA.

MASAHIKO INASE • Department of Physiology, Kinki University School of Medicine, Osaka-Sayama, Osaka, Japan.

DANIEL JEANMONOD • Department of Functional Neurosurgery, University Hospital Zurich, Zurich, Switzerland.

CAROLE JOINT • Department of Neurosurgery, Radcliffe Infirmary, Oxford, U.K.

GLENDA L. KEATING • Department of Neurology, Emory University School of Medicine, Atlanta, USA.

ADNAN KOUSIE • Department of Neurosciences, Joseph Fourier University, Grenoble, France.

PAUL KRACK • Neurology Department, University of Kiel, Kiel, Germany; and Department of Neurosciences, Joseph Fourier University, Grenoble, France.

KRISTY KULTAS-ILINSKY • Department of Anatomy and Cell Biology, University of Iowa College of Medicine, Iowa City, USA.

MASAAKI KUWAJIMA • Division of Neuroscience, Yerkes Regional Primate Center, and Department of Neurology, Emory University, Atlanta, USA.

BAE HWAN LEE • Medical Research Center and Brain Research Institute, Yonsei University College of Medicine, Seoul, Korea.

JUNG–I. LEE • Department of Neurosurgery, Johns Hopkins University, Baltimore, USA.

MYUNG SIK LEE • Department of Neurology, Yonsei University College of Medicine, Seoul, Korea.

FREDERICK A. LENZ • Department of Neurosurgery, Johns Hopkins University, Baltimore, USA.

TATIANA A. LEONTOVICH • Brain Research Institute of the Russian Academy of Medical Sciences, Moscow, Russia.

ANDRES M. LOZANO • Institute of Medical Science and Department of Surgery, University of Toronto, Toronto Western Research Institute and Toronto Western Hospital, Toronto, Canada.

ELIO LUGARESI • Institute of Clinical Neurology, University of Bologna Medical School, Bologna, Italy.

ROBERT MACKEL • The Rockefeller University, New York, USA.

MICHEL MAGNIN • Department of Functional Neurosurgery, University Hospital Zurich, Zurich, Switzerland.

MICHAEL J. MARINO • Department of Pharmacology, Emory University School of Medicine, Atlanta, USA.

ARPESH MEHTA • Department of Neurology, University of California at Los Angeles, Los Angeles, USA.

GREGORY F. MOLNAR • Institute of Medical Science, University of Toronto, Toronto, Canada.

LANCE R. MOLNAR • Experimental Therapeutics Branch, National Institute of Neurological Disorders and Stroke, National Institutes of Health, Bethesda, USA.

PASQUALE MONTAGNA • Institute of Clinical Neurology, University of Bologna Medical School, Bologna, Italy.

ANNE MOREL • Department of Functional Neurosurgery, University Hospital Zurich, Zurich, Switzerland.

ELENA MORO • Department of Neurosciences, Joseph Fourier University, Grenoble, France.

ZHONGGE NI • Department of Neurosciences, Joseph Fourier University, Grenoble, France.

CHIHIRO OHYE • Functional and Gamma Knife Surgery Center, Hidaka Hospital, Takasaki, Japan.

ALEXANDR A. ORLOV • Sechenov Institute of Evolutionary Physiology and Biochemistry, St. Petersburg, Russia.

YONG GOU PARK • Department of Neurosurgery, Medical Research Center and Brain Research Institute, Yonsei University College of Medicine, Seoul, Korea.

SIMON PARKIN • Departmennt of Neurology, Radcliffe Infirmary and Imperial College School of Medicine, Charing Cross Hospital, London, U.K.

MICHAEL G. PCHELIN • Pavlov Institute of Physiology, Russian Academy of Sciences, St.-Petersburg, Russia.

JOHN B. PENNEY • Department of Neurology, Massachusetts General Hospital and Harvard Medical School, Boston, USA. Deceased.

ANDREW PILLIAR • Department of Physiology, University of Toronto, Toronto, Canada.

PIERRE POLLAK • Department of Neurology, Joseph Fourier University, Grenoble, France.

SVETLANA RAEVA • Institute of Chemical Physics, Russian Academy of Sciences and Burdenko Neurosurgery Institute, Moscow, Russia.

SERGEY P. ROMANOV • Pavlov Institute of Physiology, Russian Academy of Sciences, St.-Petersburg, Russia.

SUSAN T. ROUSE • Department of Pharmacology, Emory University School of Medicine, Atlanta, USA.

MARK RUFFO • Program in Neurobiology and Behavior, University of Washington, Seattle, USA.

DAVID N. RUSKIN • Experimental Therapeutics Branch, National Institute of Neurological Disorders and Stroke, National Institutes of Health, Bethesda, USA.

DAVID B. RYE • Department of Neurology, Emory University School of Medicine, Atlanta, USA.

RICHARD SCOTT • Department of Neuropsychology, Radcliffe Infirmary, Oxford, U.K.

VLADIMIR SHABALOV • Burdenko Institute of Neurosurgery, Moscow, Russia.

KSENIA SHAPOVALOVA • Pavlov Institute of Physiology of the Russian Academy of Sciences, St. Petersburg, Russia.

TOHRU SHIBAZAKI • Functional and Gamma Knife Surgery Center, Hidaka Hospital, Takasaki, Japan.

YOLAND SMITH • Division of Neuroscience, Yerkes Regional Primate Center, and Department of Neurology, Emory University, Atlanta, USA.

JOHN F. STEIN • Department of Physiology, University of Oxford, Oxford, U.K.

BORIS F. TOLKUNOV • Sechenov Institute of Evolutionary Physiology and Biochemistry, St. Petersburg, Russia.

MICHAEL V. UGRUMOV • Institute of Developmental Biology of the Russian Academy of Sciences, and Institute of Normal Physiology of the Russian Academy of Medical Sciences, Moscow, Russia.

JERROLD L. VITEK • Department of Neurology, Emory University School of Medicine, Atlanta, USA.

JUDITH R. WALTERS • Experimental Therapeutics Branch, National Institute of Neurological Disorders and Stroke, National Institutes of Health, Bethesda, USA.

TOMAS WICHMANN • Department of Neurology, Emory University, Atlanta, USA

MARION WITTMANN • Department of Pharmacology, Emory University School of Medicine, Atlanta, USA.

ANNE B. YOUNG • Department of Neurology, Massachusetts General Hospital and Harvard Medical School, Boston, USA.

Acknowledgments

Both the International Workshop "Basal Ganglia and Thalamus in Health and Movement Disorders" and this publication were made possible by the generous support of several corporations, including Eli Lilly, Allergan, Solvay, Medtronic, SmithKline Beecham, Integrated Surgical Systems, BrainLab, Radionics, and FHC. To all of them the editors express their sincere gratitude. Special thanks are due to the University of Iowa and the Council on Science and Technology of the Russian Federation for their support, and to the Burdenko Institute of Neurosurgery for hosting the meeting and providing the organizational infrastructure.

Finally, this publication would not have been possible without the dedication of Eric Jones, who assisted in editing and formatting the text, and Paul Reimann, who worked on the final illustrations for the volume.

Preface

This volume is comprised of the majority of lecture presentations and a few select posters presented at the International Workshop, *"Basal Ganglia and Thalamus in Health and Movement Disorders,"* held in Moscow, Russia, on May 29-31, 2000. The International Committee responsible for organizing this workshop included Alexander Konovalov, Director, Burdenko Institute of Neurosurgery of the Russian Academy of Medical Sciences, Mahlon DeLong, Chair, Department of Neurology, Emory University, Atlanta, USA, Alim Louis Benabid, Chief, Neurosurgery Service, University of Joseph Fourrier, Grenoble, France, and the two undersigned.

The workshop was conceived out of a desire to provide a forum for discussions of both basal ganglia- and motor thalamus-related issues by bringing together basic scientists and clinicians representing different disciplines, research directions, and philosophies. The primary goals were to encourage an exchange of information and ideas in an informal environment, to stimulate integration of the data from different disciplines, and to identify controversial issues and the most essential questions to be addressed in future research.

The choice of a workshop venue was determined by two factors. One was a desire to promote international cooperation and exchange, particularly with Russian scientists, who have made remarkable contributions to basal ganglia and thalamus research in the past, but who have in the last decade been somewhat isolated from mainstream developments in the field. Another important factor in the choice of venue was the enthusiasm and generosity of the workshop's host, Dr. A. Konovalov, who offered the state-of-the-art conference facilities of the newly renovated Burdenko Neurosurgical Institute for this event.

Recent years have witnessed remarkable advances in our understanding of the functional and neurochemical organization of movement control systems, as well as the introduction of new pharmacological and surgical treatments for movement disorders, including high frequency stimulation (HFS) of deep brain structures. However, many questions remain about the specific changes treatments cause in target structures, the exact mechanisms by which the new treatments work, and the overall contribution of individual basal ganglia and thalamic nuclei to pathological symptoms of movement disorders. During the three days of the workshop, participants discussed various aspects of the organization of the basal ganglia and motor thalamic circuits, potential mechanisms responsible for pathological manifestations in movement disorders, and the therapeutic potential of pharmacological targeting of specific parts of the basal ganglia-thalamic circuits.

One session of special interest took the form of a clinical round table discussion focusing mainly on parkinsonism, dystonia, and some related conditions. The topics addressed in it included the following: mechanisms of tremorogenesis and other manifestations of movement disorders; mechanisms responsible for the therapeutic effects of HFS; the choice and identification of targets for lesioning and HFS; the efficacy of these treatments in different targets; comparison of clinical outcomes of lesioning and HFS in the medial globus pallidus, subthalamic nucleus, and motor thalamus; and causes and mechanisms of potential complications. Unfortunately, it was not possible to include in the book all the lively exchanges and debates of this informal session, but the appendix and a few of the chapters in the final section highlight some of the themes and divergent approaches discussed.

The primary challenge for the editors of this volume was to group the chapters so that there would be some continuity and logical succession of topics. This turned out to be an unattainable goal. Nonetheless, individual chapters have been organized into sections that contain underlying themes and unifying motifs, although this may not always be apparent from chapter titles.

Due to time constraints, the scientific program of the workshop did not afford comprehensive coverage of all aspects of basic and clinical research on the basal ganglia and thalamus. Hence this book does not aspire to provide comprehensive coverage of the field. Yet, a patient reader will be rewarded by thoughtful analyses of the current status in each field of research represented, and by a variety of unorthodox and often contradictory ideas and approaches in explaining normal and pathological interactions in the basal ganglia and motor thalamus. This, from our standpoint, is the most valuable feature of the volume.

Igor A. Ilinsky, M.D., Ph.D.
Iowa City

Kristy Kultas-Ilinsky, Ph.D.
Iowa City

Contents

PART VI. NEURONAL ACTIVITY IN MOVEMENT DISORDERS

PART VII. MECHANISMS AND EFFICIENCY OF NOVEL TREATMENTS FOR MOVEMENT DISORDERS

PART I. BASAL GANGLIA CIRCUITRY IN MOVEMENT DISORDERS: HISTORICAL PERSPECTIVE AND OVERVIEW OF THE CURRENT STATUS OF THE FIELD

WHAT WE KNOW AND WHAT WE HAVE
LEFT TO LEARN

ANNE B. YOUNG AND JOHN B. PENNEY[*]

INTRODUCTION

When John Penney and I went to join the faculty in the Neurology Department at the University of Michigan in 1978, we started a movement disorders clinic and established a laboratory to study the neurotransmitter pathways involved in the control of movements. At the time, there existed a disconnect between neurophysiologists and neuropharmacologists. For the most part, the motor physiologists paid little attention to the transmitters used at various synapses but focused instead on the electrical responses of individual neurons (with intracellular recording) and populations of neurons (with extracellular recordings). Other neuroscientists focused on the classical methods of neuropathology and anatomy. Many of these individuals were basic scientists and neurologists. The classic physiologists and anatomists emphasized the cortico-striatal-pallido-thalamo-cortical circuits (Kemp and Powell, 1971) but paid little attention to the dopamine nigrostriatal pathway. Conversely, the neuropharmacologists paid only rudimentary attention to the anatomy and physiology but rather examined the effects of drugs on various biochemical and behavioral responses. This group was composed primarily of basic scientists and psychiatrists. They focused on the nigrostriatal-nigral circuit (McGeer and McGeer, 1976). Jack and I saw an opportunity, by combining experimental neuropathologic and neuropharmacologic viewpoints, to lend a new approach to studying the functional neuroanatomy of motor pathways.

After several years observing firsthand the complexity of basal ganglia disorders, we developed a model of basal ganglia functional anatomy (Penney and Young, 1983). We wanted to provide a framework for testing hypotheses about the function of various components of the circuits, their transmitters, and their regulation (and modification) after specific lesions (Penney and Young, 1983). To approach the problem, we developed a method of film-based, autoradiographic analysis that allowed us to examine both the pharmacology and anatomy of neurotransmitter pathways (Penney et al., 1981). Previous receptor autoradiographic techniques involved the use of emulsion-coated slides to observe the anatomy of transmitter receptors (Herkenham and Pert, 1980; Kuhar, 1981). As soon as tritium-sensitive film became available, we modified the method for autoradiographic

---

[*] ANNE B. YOUNG • Department of Neurology, Massachusetts General Hospital and Harvard Medical School, Boston, MA 02114. JOHN B. PENNEY • Department of Neurology, Massachusetts General Hospital and Harvard Medical School, Boston, MA 02114. Deceased.

studies of ^{14}C-deoxyglucose metabolism to tritium-labeled ligand binding to various receptors. The method allowed for calculation of B_{max} and K_D of ligands in discrete areas of brain (Penney et al, 1981). Thus, we could use this method on both animal and human brain to study the changes of receptor number and affinity in animal models of disease and in human postmortem brain tissue.

We agreed with several previous authorities that the basal ganglia were critical for reinforcing particular behavioral sets while suppressing unwanted behaviors. We hypothesized that this selection (and suppression) process took place primarily at the level of the striatum, which was a key locus in a cortico-striatal-pallidal-thalamo-cortical feedback loop (Penney and Young, 1983). Based on our interpretation of the predominantly pharmacological literature, we assumed that dopamine was inhibitory onto striatal neurons (see Albin et al., 1989b). Based on physiological data, we assumed that the subthalamic nucleus was inhibitory onto pallidal and striatal neurons (Rouzarie-Dubois et al., 1983; Hammond et al., 1983).

If our model were correct, we concluded that striatal lesions should result in GABAergic denervation of both entopeduncular nucleus (EPN) (counterpart of medial globus pallidus, MGP, in human), globus pallidus (GP) (counterpart of lateral globus pallidus, LGP, in human) and the substantia nigra pars reticulata (SNR). Such denervation was expected to produce postsynaptic GABA receptor supersensitivity in these same structures. This indeed appeared to be the case (Pan et al., 1984).

Furthermore, we hypothesized that if dopamine were inhibitory on striatum, then postsynaptic GABA receptors in EPN, SNR, and GP should be downregulated after lesions of the nigrostriatal pathway. Surprisingly, we discovered that this was not the case. As predicted, GABA receptors in GP were downregulated, but, paradoxically, GABA receptors in EPN and SNR were upregulated after lesions of the dopaminergic input to striatum (Pan et al., 1985). Furthermore, in early Huntington's disease and animal models, we and others found the same thing (Mitchell et al., 1985; Albin et al., 1990; Reiner et al., 1988; Sapp et al., 1995). These observations suggested that dopamine had dual effects on striatal neurons: excitatory on those that project to EPN and SNR and inhibitory on those that project to GP.

At about this time, additional data also became available suggesting that the STN was not inhibitory but excitatory (Nakanishi et al., 1987) and likely glutamatergic (Smith and Parent, 1988; Albin et al., 1989a).

THE 'CLASSIC' MODEL

These findings, along with complementary data from other laboratories, prompted us to modify our model (Penney and Young, 1986). Our new model predicted two circuits through the basal ganglia: (1) a reinforcing cortico-striatal-medial pallido-thalamo-cortical circuit, and (2) a suppressing cortico-striatal-lateral pallidal-subthalamo-medial pallidal-thalamo-cortical circuit. We suggested that dopamine was excitatory on the former and inhibitory on the latter. We also predicted that in Parkinson's disease both the subthalamic nucleus and the medial globus pallidus would be overactive but that in Huntington's disease, the subthalamic nucleus would be underactive.

Over the next several years, further work substantiated the contrasting effects of dopamine on the different pathways and also indicated that the excitatory dopamine effects appeared to be mediated by dopamine D1 receptors and the inhibitory effects by dopamine D2 receptors (see Albin et al., 1989b). D1 receptors were primarily expressed in GABA/substance P/dynorphin medium spiny striatal neurons projecting to the MGP and SNR, and D2 receptors were primarily expressed in GABA/enkephalin medium spiny striatal neurons projecting to the LGP. Additional studies confirmed that the subthalamic nucleus was excitatory and glutamatergic (Smith and Parent, 1988; Albin et al., 1989a; Feger et al., 1989).

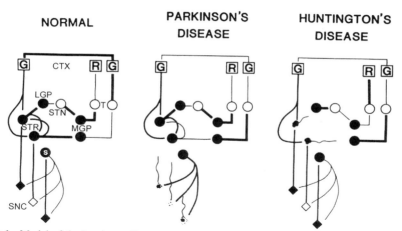

Figure 1. Model of the basal ganglia, as proposed by Penney and Young in 1986. The model proposed the existence of two distinct pathways out of the striatum (STR), one of which enhanced the intended movements (Grasp, "G"), while inhibiting unwanted postural mechanisms (Release, "R"). These two pathways were differentially affected in Parkinson's and Huntington's diseases, suggesting overactivity of the STN and MGP in parkinsonism, and underactivity of the STN in Huntington's disease.

McKenzie et al. (1989) first termed the two different striatal circuits "direct" and "indirect" pathways in their studies of turning behavior in rodents. DeLong and colleagues followed up on this terminology in their studies of subthalamic and pallidal activity in MPTP primates. They (DeLong, 1990) confirmed the excessive activity of STN and MGP neurons in primate MPTP models of Parkinson's disease as predicted by earlier hypotheses (Penney and Young, 1986; Young et al., 1989; Albin et al., 1989b; McKenzie et al., 1989). Furthermore, elegant studies by DeLong and his group showed that lesions of the subthalamic nucleus ameliorated parkinsonian symptoms without inducing dyskinesias (Bergman et al., 1990).

The "model" became extremely popular. It appeared to account for many features of basal ganglia pathology and it was generally adapted by the neuroscience community. The model predicted that lesions of the globus pallidus and subthalamic nucleus should be beneficial in Parkinson's disease. Stereotaxic surgery had been carried out for decades before the advent of dopamine agonists and other effective therapies (Leksell, 1951; Laitinen, 1972). In the early days, however, the locus of the lesions was identified using crude techniques. In the early 1990s, several groups began operating on Parkinson's patients using special imaging techniques (Laitinen et al., 1992; Bakay et al., 1992). Small radio frequency lesions could be placed accurately in almost any region of the brain. The results of pallidotomy in Parkinson's disease were particularly striking and, in certain aspects, surprising. Not only did MGP lesions improve tremor, rigidity, and fine motor incoordination, but they also drastically improved dyskinesias (Fine et al., 2000; Eskandar et al., 2000). These findings were certainly not predicted or explained by the "classic model" of basal ganglia anatomy. Despite the fact that MGP activity appeared to be decreased during dyskinesias, in subthalamic lesions, and in Huntington's disease, further ablation of the nucleus appeared to improve dyskinetic movements, ballism, and chorea. Neither did the model explain tremor, acute oculogyric crises, dystonia, or biphasic dyskinesias. It also didn't acknowledge the known connections of the MGP and SNR with the brainstem locomotor regions and pedunculo-pontine nucleus.

CHALLENGES AND REVISIONS TO THE MODEL

Basal ganglia aficionados, however, continued to challenge the validity of the classic model, noting exceptions to the predicted responses of particular pathways to drugs, lesions, and disease (Graybiel et al., 1994; Chesselet and Delfs, 1996; Obeso et al., 1997; Levy et al., 1997; Parent and Cicchetti, F., 1998; Ruskin et al., 1999). DeLong and colleagues pointed out that observations in the diseased basal ganglia were unlikely to reflect directly on the function of the normal basal ganglia. Perhaps in disease, they speculated, no output from MGP was better than abnormal or disordered output.

Additional information about the anatomy of different pathways, their relative influence, and their pharmacology were being discovered. The notion that the subthalamic nucleus was just part of a side loop was changed. Indeed, inputs from cortex to subthalamic nucleus and thence to medial globus pallidus arrived at the pallidum before the inputs from cortex to striatum to pallidum did. The striatal circuit became a side pathway of the cortico-subthalamo-pallidal-thalamo-cortical loop. Furthermore, the subthalamic nucleus not only gives off excitatory projections to MGP and SNR but also to LGP and striatum (Parent and Cicchitti, 1998). In turn, the LGP not only projects to STN but also directly to MGP (Kincaid et al., 1991). There also appear to be dopamine influences on STN activity.

The output of MGP and SNR to thalamus does not synapse only on thalamo-cortical relay cells but also onto a complex web of GABAergic interneurons (Ilinsky and Kultas-Ilinsky, this volume). Thalamus not only projects directly to various areas of cerebral cortex but also to striatum, where it appears to preferentially excite cholinergic interneurons but also the medium spiny neurons of the direct pathway (Sidibe and Smith, 1996, 1999). In contrast, cerebral cortex preferentially effects medium spiny neurons of the indirect pathway (Parthasarathy and Graybiel, 1997).

The brainstem connections to the basal ganglia are also very rich and complex (Rye et al., 1987). At the midbrain pontine junction in the area of the pedunculopontine nucleus (PPN) there are actually several clusters of neurons, one which is the PPN proper and the other which is the midbrain extrapyramidal area (MEA). Nearby, there are also the dopaminergic neurons of A8. The PPN neurons are cholinergic, and they also contain nitric oxide synthase, corticotrophin releasing factor, and substance P. They have very diffuse afferents (MGP, SNr, locus coeruleus, dorsal raphe, the orexin cells of the hypothalamus (for latter pathway see Kilduff and Peyron, 2000)) and efferents (SNR, MGP, VM, VL, reticular thalamus, and the bulbospinal inhibitory zone). Their activity is very state-dependent. The MEA neurons are glutamatergic and also express parvalbumin. They receive projections from limbic/associative and sensorimotor cortices, MGP, and SNR and have major projections to SNC, STN, and the bulbospinal inhibitory zone (Bevan and Bolam, 1995).

The PPN is very involved in arousal, cardiorespiratory function, nociception, sleep, and eye movements. The MEA influences the reticulospinal and cerebellar-related motor pathways. Both the MEA and PPN converge onto the glycinergic inhibitory neurons of the bulbospinal inhibitory zone and thereby play a role in atonia during sleep. The PPN at the same time facilitates rapid eye movement sleep.

The pathways from basal ganglia to thalamus are also much more complex than previously appreciated in the classic model. MGP and SNR inputs not only contact thalamocortical relay neurons directly but also local circuit GABAergic interneurons (LCN) in the same region. Furthermore, the cortex sends excitatory afferents to both thalamocortical relay neurons and to LCN. The LCN neurons, in turn, send inhibitory inputs to the thalamocortical relay neurons. In addition, reticulothalamic GABAergic pathways also inhibit both thalamocortical relay neurons and LCN and the thalamocortical neurons send excitatory inputs to the reticulothalamic neurons.

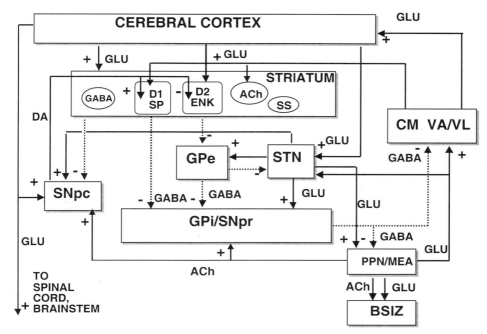

Figure 2. Updated model of the basal ganglia. This incorporates recent data on the connections of the structures, and includes the PPN and BSIZ (bulbospinal inhibitory zone, which inhibits movement during sleep). Despite these improvements, this type of model neglects the functional, temporal and plastic interactions of the neural structures.

More detailed electrophysiological and pharmacological studies suggest that the cellular response to a single neurotransmitter can vary greatly depending on the 'state' of the cell (Hernandez-Lopez et al., 1997; Plenz and Kitai, 1998; Galarraga et al., 1999). For instance, medium spiny neurons in a slightly hyperpolarized state 'downstate' may have no response or an inhibitory response to dopamine input but subsequently have a robust excitatory response when the medium spiny neuron is in the 'upstate'. Thus depending on the timing and strength of various inputs, the response of the cell to other inputs may be drastically different. When long-term plasticity is also taken into account, responses in a lesioned or diseased circuit may be very different from their responses in the intact state. The details of these phenomena and their functional significance is just now being characterized.

Furthermore, the plasticity of the system had been underestimated and such plasticity clearly has to be taken into account. The responses of individual cells to altered inputs appear to be constantly modified (receptor desensitization, supersensitivity, redistribution). Data in the rat, for instance, indicate that loss of dopamine input to the striatum alters the number and subunit constitution of NMDA receptor subunits in the cellular membrane. Stimulation by dopamine agonists normalizes these receptor changes (Dunah et al., 1999). Thus one transmitter can influence the makeup of other membranous receptors not only by phosphorylation but also by changing the number and composition of those receptors. These plastic changes have been observed not only in effects of dopamine lesions on striatal neurons but also on pallidal and thalamic neurons (Chesselet et al., and Kultas-Ilinsky and Ilinsky, this volume). Presumably, these effects are mediated via second messenger systems that are linked directly to enzyme function but also via changes in transcriptional regulation

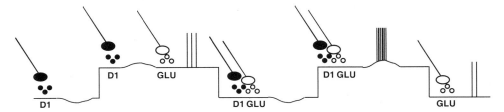

Figure 3. The interaction of up and down states of striatal neurons with dopamine and glutamate. The oscillating membrane potential of a medium spiny neuron is illustrated. In hyperpolarized states, D1 activation is stabilizing, and occludes the excitatory effect of glutamate. At depolarized potentials, D1 and glutamate are both independently marginally excitatory, but co-activation produces repetitive bursting.

of additional pathway components. Furthermore, neurotransmitter receptors at the surface of the membrane are connected to complex intracellular scaffolds that influence these secondary pathways and are also tightly regulated by cellular events (O'Brien et al., 1998; Kennedy, 1998).

SUMMARY

Future models will have to be displayed in four dimensions in order to reflect even our current stage of knowledge. The models must incorporate the new pathways discovered to influence motor function such as the LGP to MGP inputs and the PPN and MEA influences. They will need to take into account plastic changes in the connections themselves via sprouting and modification of local circuits. The models must also incorporate plastic changes in receptor composition, number, and distribution. It is very likely that the circuits subserving abnormal movements in primates and humans with specific lesions are very different than those subserving normal movements in intact subjects. Defining and displaying these changes in new models will necessitate the creation of dynamic circuits that continually change in response to perturbations. With new methods of modeling, such attributes can be accounted for and these evolving models can be tested and challenged in the future. In order to achieve these goals, anatomic, genetic, pharmacologic, and physiologic tools will need to be applied and integrated to provide a fuller understanding of normal and abnormal motor function.

ACKNOWLEDGMENTS: This chapter was particularly difficult for me because Jack and I thought about these issues together all the time, and he always referenced the papers and I usually wrote the text and did the figures. This time I bungled along on my own. Fortunately, David G. Standaert, M.D., Ph.D. was kind enough to help me when I got desperate. Supported by USPHS grant NS38372.

REFERENCES

Albin, R.L., Aldridge, J.W., Young, A.B., and Gilman, S., 1989a, Feline subthalamic nucleus neurons contain glutamate-like but not GABA-like or glycine-like immunoreactivity, *Brain Res.* 491:185.

Albin, R.L., Young, A.B., and Penney, J.B., 1989b, The functional anatomy of basal ganglia disorders, *Trends Neurosci.* 12:366.

Albin, R.L., Young, A.B., Penney, J.B., Handelin, B., Balfour, R., Anderson, K.D., Markel, D.S., Tourtelotte W.W., and Reiner, A, 1990, Abnormalities of striatal projection neurons and N-methyl-D-aspartate receptors in presymptomatic Huntington's disease, *N. Engl. J. Med.* 322:1293.

Bakay, R.A., DeLong, M.R., and Vitek, J.L., 1992, Posteroventral pallidotomy for Parkinson's disease, *J. Neurosurg.* 77:487.

Bergman, H., Wichman, T., and DeLong, M.R., 1990, Reversal of experimental parkinsonism by lesions of the subthalamic nucleus, *Science,* 249:1436.

Bevan, M. D., and Bolam, J.P., 1995, Cholinergic, GABAergic, and glutamate-enriched inputs from the mesopontine tegmentum to the subthalamic nucleus in the rat, *J. Neurosci.* 15:7105.

Chesselet, M.-F., and Delfs, J.M., 1996, Basal ganglia and movement disorders: an update, *Trends Neurosci.* 19:417.

DeLong, M.R., 1990, Primate models of movement disorders of basal ganglia origin, *Trends Neurosci.* 13:281.

Dunah, A.W., Wang, Y.H., Yasuda, R.P., Kameyama, K., Huganir, R.L., Wolfe, B.B., and Standaert, D.G., 2000, Alterations in subunit expression, composition and phosphorylation of striatal NMDA glutamate receptors in the rat 6-OHDA models of Parkinson's disease, *Mol. Pharmacol.* 57:242.

Eskandar, E.N., Shinobu, L.A., Penney, J.B., Cosgrove, G.R., and Counihan, T.J., 2000, Stereotactic pallidotomy performed without using microelectrode guidance in patients with Parkinson's disease: surgical technique and 2-year results, *J. Neurosurg.* 92:375.

Feger, J., Vezole, I., Renwart, N., and Robledo, P., 1989, The rat subthalamic nucleus: electrophysiological and behavioral data, in: *Neural Mechanisms in Disorders of Movement*, A.R. Crossman and M.A. Sambrook, eds., John Libbey, London.

Fine, J., Duff, J. Chen, R., Chen, R., Chir, B., Hutchison, W., Lozano, A.M., and Lang, A.E., 2000, Long-term follow-up of unilateral pallidotomy in advanced Parkinson's disease, *N. Engl. J. Med.* 342:1708.

Galarraga, E., Hernandez-Lopez, S., Reyes, A., Miranda, I., Bermudez-Rattoni, F., Vilchis, C., and Bargas, J., 1999, Cholinergic modulation of neostriatal output: a functional antagonism between different types of muscarinic receptors, *J. Neurosci.* 19:3629.

Graybiel, A.M., Aosaki, T., Flaherty, A., and Kimura, M., 1994, The basal ganglia and adaptive motor control, *Science,* 265:1826.

Hammond, C., Rouzaire-Dubois, B., Feger, J., Jackson, A., and Crossman, A.R., 1983, Anatomical and electrophysiological studies on the reciprocal projections between the subthalamic nucleus and nucleus tegmenti pedunculopontinus in the rat, *Neuroscience,* 9:41.

Herkenham, M., and Pert, C.B., 1980, In vitro autoradiography of opiate receptors in rat brain suggests loci of 'opiatergic' pathways, *Proc. Natl. Acad. Sci. USA* 77:5532.

Hernandez-Lopez, S., Bargas, J., Surmeier, D.J., Ryes, A., and Galarraga, E., 1997, D1 receptor activation enhances evoked discharge in neostriatal medium spiny neurons by modulating an L-type Ca^{2+} conductance, *J. Neurosci.* 17:3334.

Kemp, J.M., and Powell, T.P.S., 1971, The connections of the striatum and globus pallidus: synthesis and speculation, *Phil. Trans. R. Soc. (London Ser B)* 262:441.

Kennedy, M.B., 1998, Signal transduction molecules at the glutamatergic postsynaptic membrane, *Brain Res. Rev.* 26:243.

Kilduff, T.S., and Peyron, C., 2000, The hypocretin/orexin ligand-receptor system: implications for sleep and sleep disorders, *Trends Neurosci.* 23:359.

Kincaid, A.E., Penney, J.B. Jr., Young, A.B., and Newman, S.W., 1991, Evidence for a projection from the globus pallidus to the entopeduncular nucleus in the rat, *Neurosci. Lett.* 128:121.

Kuhar, M.J., 1978, Histochemical localization of opiate receptors and opioid peptides, *Fed. Proc.* 37:153.

Laitinen, L., 1972, Surgical treatment, past and present, in Parkinson's disease, *Acta. Neurol. Scand. Suppl.* 51:43.

Laitinen, L.V., Bergenheim, A.T., and Hariz, M.I., 1992, Leksell's posteroventral pallidotomy in the treatment of Parkinson's disease, *J. Neurosurg.* 76:53.

Leksell, L., 1951, The stereotaxic method and radiosurgery of the brain, *Acta. Chir. Scand.* 102:316.

Levy, R., Hazerati, L.-N., and Herrero, M.-T., et al., 1997, Re-evaluation of the functional anatomy of the basal ganglia in normal and Parkinsonian states, *Neuroscience,* 76:335.

McGeer, P.L., and McGeer, E.G., 1976, The GABA system and function of the basal ganglia: Huntington's disease, in: *GABA in Nervous System Function: Huntington's Disease,* E. Roberts, T.N. Chase, and D.B. Tower, Raven Press, New York.

McKenzie, J.S., Shafton, A.D., and Stewart, C.A., 1989, Striatal output pathways involved in mechanisms of rotation in rats, in: *Neural Mechanisms in Disorders of Movement*, A.R. Crossman and M.A. Sambrook, eds., John Libbey, London.

Mitchell, I.J., Jackson, A., Sambrook, M.A., and Crossman, A.R., 1985, Common neural mechanisms in experimental chorea and hemiballismus in the monkey. Evidence from 2-deoxyglucose autoradiography, *Brain Res.* 339:346.

Nakanishi, H., Kita, H., and Kitai, S.T., 1987, Intracellular study of rat substantia nigra pars reticulata neurons in an in vitro slice preparation: electrical membrane properties and response characteristics to subthalamic stimulation, *Brain Res.* 437:45.

Obeso, J.A., Rodriguez, M.C., and DeLong M.R., 1997, Basal ganglia pathophysiology. A critical review, *Adv. Neurol.* 74:3.

O'Brien, R. J., Lau, L-F., and Huganir, R.L., 1998, Molecular mechanisms of glutamate receptor clustering at excitatory synapses, *Curr. Opinion in Neurobiol.* 8:364.

Pan, H. S., Penney, J.B., and Young, A.B., 1984, Characterization of benzodiazepine receptor changes in substantia nigra, globus pallidus and entopeduncular nucleus after striatal lesions, *J. Pharmacol. Exp. Ther.* 45:768.

Pan, H.S., Penney, J.B., and Young, A.B., 1985, γ-Aminobutyric acid and benzodiazepine receptor changes induced by unilateral 6-hydroxydopamine lesions of the medial forebrain bundle, *J. Neurochem.* 334:215.

Parent, A., and Cicchitti, F., 1998, The current model of basal ganglia organization under scrutiny, *Mov. Disord.* 13:199.

Parthasarathy, H.B., and Graybiel, A.M., 1997, Cortically driven immediate-early gene expression reflects modular influence of sensorimotor cortex on identified striatal neurons in the squirrel monkey, *J. Neurosci.* 17:2477.

Penney, J.B., Pan, H.S., Young, A.B., Frey, K.A., and Dauth, G.W., 1981, Quantitative autoradiography of 3H-muscimol binding in rat brain, *Science,* 214:1036.

Penney, J.B., and Young, A.B., 1983, Speculations on the functional anatomy of basal ganglia disorders, *Annu. Rev. Neurosci.* 6:73.

Penney, J.B. Jr., and Young, A.B., 1986, Striatal inhomogeneities and basal ganglia function, *Mov. Dis.* 1:3.

Plenz, D., and Kitai, S.T., 1998, Up and down states in striatal medium spiny neurons simultaneously recorded with spontaneous activity in fast-spiking interneurons studied in cortex-striatum-substantia nigra organotypic cultures, *J. Neurosci.* 18:266.

Reiner, A., Albin, R.L., Anderson, K.D., D'Amato, C.J., Penney, J.B., and Young, A.B., 1988, Differential loss of striatal projection neurons in Huntington's disease, *Proc. Natl. Acad. Sci. USA* 85:5733.

Rouzaire-Dubois, B., Scarnati, E., Hammond, C., Crossman, A.R., and Shibazaki, T., 1983, Microiontophoretic studies on the nature of the neurotransmitter in the subthalamo-entopeduncular pathway of the rat, *Brain Res.* 271:11.

Ruskin, D.N., Bergstrom, D.A., Kaneoke, Y., Patel, B.N., Twery, M.J., and Walters, J.R., 1999, Multisecond oscillations in firing rate in the basal ganglia: robust modulation by dopamine receptor activation and anesthesia, *J. Neurophysiol.* 81:2046.

Rye, D. B., Saper, C.B., Lee, H.J., and Wainer, B.H., 1987, Pedunculopontine tegmental nucleus of the rat: cytoarchitecture, cytochemistry, and some extrapyramidal connections of the mesopontine tegmentum, *J. Comp. Neurol.* 259:483.

Sapp, E., Ge, P., Aizawa, H., Bird, E., Penney, J., Vonsattel, J.P., and DiFiglia, M., 1995, Evidence for a preferential loss of enkephalin immunoreactivity in the external globus pallidus in low grade Huntington's disease using high resolution image analysis, *Neuroscience,* 65:397.

Sidibe, M., and Smith, Y., 1996, Differential synaptic innervation of striatofugal neurones projecting to the internal or external segments of the globus pallidus by thalamic afferents in the squirrel monkey, *J. Comp. Neurol.* 365:445.

Sidibe, M., and Smith, Y., 1999, Thalamic inputs to striatal interneurons in monkeys: synaptic organization and co-localization of calcium binding proteins, *Neuroscience,* 89:1189.

Smith, Y., and Parent, A., 1988, Neurons of the subthalamic nucleus in primates display glutamate but not GABA immunoreactivity, *Brain Res.* 453:353.

Young, A.B., Albin, R.L., and Penney, J.B., 1989, Neuropharmacology of basal ganglia functions: relationship to pathophysiology of movement disorders, in: *Neural Mechanisms in Disorders of Movement,* A.R. Crossman and M.A. Sambrook, eds., John Libbey, London.

BASAL GANGLIA CIRCUITS IN MOVEMENT AND MOVEMENT DISORDERS

THOMAS WICHMANN AND MAHLON R. DELONG*

INTRODUCTION

Research over the past decades has lead to major insights into the structure and function of the basal ganglia and to the development of circuit models of these nuclei and of their role in the pathophysiology of movement disorders (Albin et al., 1989; DeLong, 1990; Albin, 1995; Brooks, 1995; Chesselet and Delfs, 1996; Wichmann and DeLong, 1996). Although these models have had significant heuristic value, especially in guiding new research into surgical and pharmacologic treatments for movement disorders, it was clear even at the time of their inception that they were oversimplified, and would need almost continuous updating. This brief overview will focus on the principal features of these circuit models, especially with regard to their application as models to explain the pathophysiology of movement disorders.

NORMAL ANATOMY AND FUNCTION OF THE BASAL GANGLIA

The basal ganglia are components of larger segregated circuits, that involve cerebral cortex and thalamus (Alexander et al., 1990). Each circuit's designation reflects the presumed function of its cortical areas of origin and termination, i.e. motor, oculomotor, associative, and limbic. Each circuit originates in specific cortical areas, passes through separate portions of the basal ganglia and thalamus, and projects back onto the original frontal cortical area. In each of these circuits, the striatum serves as the principal input stage of the basal ganglia, and GPi and SNr as output stations. This anatomic organization is consistent with the clinical evidence for motor and non-motor functions of the basal ganglia and the development of cognitive and emotional/behavioral disturbances in diseases of the basal ganglia (see below).

Of the aforementioned circuits, the motor circuit has particular clinical relevance for understanding the pathophysiology of movement disorders. This circuit originates in pre- and postcentral sensorimotor fields, which project to the putamen, either directly, or via the

---
* THOMAS WICHMANN AND MAHLON R. DELONG • Department of Neurology, Emory University, Suite 6000, WMRB, Atlanta, GA 30030.

11

intercalated centromedian nucleus (CM) of the thalamus (Kemp and Powell, 1971; Wilson et al., 1983; Smith and Parent, 1986; Dube et al., 1988; Nakano et al., 1990; Sadikot et al., 1988, 1992). As shown in the left half of Figure 1, putamenal output reaches GPi/SNr via two pathways, a direct, monosynaptic route, and an indirect, polysynaptic, route that passes through the external pallidal segment (GPe) to reach GPi either directly or via the subthalamic nucleus (STN) (Hazrati et al., 1990; Parent and Hazrati, 1995). The putamenal source neurons for the direct and indirect pathways differ anatomically in that neurons that receive input from CM appear to predominantly give rise to the monosynaptic pathway, while neurons that receive input from cortex, project to GPi/SNr via the polysnaptic route (for discussion, see Sadikot et al., 1992; Sidibe and Smith, 1986). Although the main transmitter of all striatal output neurons is GABA, an additional difference between the source neurons for the direct and indirect pathways is that the direct pathway neurons in the striatum contain the neuropeptide substance P, while the source neurons of the indirect pathway carry the neuropeptides enkephalin and dynorphin.

In addition to the cortico-striatal pathway, the cortico-subthalamic pathway (Hartman - von Monakow et al., 1978; Kita, 1994; Nambu et al., 1996) is another important cortico-basal ganglionic route which may influence basal ganglia activity at all levels, even at the striatal level (Smith and Parent, 1986, Smith et al., 1990). The potential importance of this pathway is underscored by the fact that neuronal responses to sensorimotor examination in GPe and GPi are greatly reduced after lesions of the STN (Hamada and DeLong, 1992). Basal ganglia output is directed towards the thalamic ventral anterior, ventral lateral, and intralaminar nuclei (VA, VL, CM/Pf, respectively)(Schell and Strick, 1984; Goldman Rakic and Porrino, 1985; Strick,

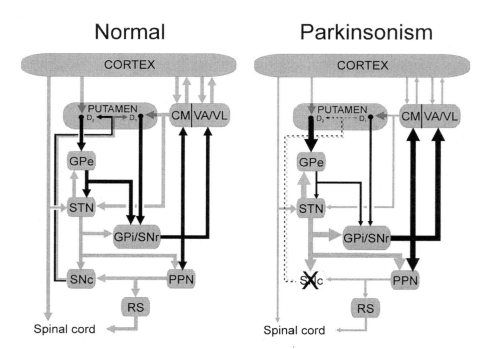

Figure 1. Normal anatomy of the basal ganglia and changes in the basal ganglia-thalamocortical circuitry in parkinsonism. Putamen and STN receive cortical input, which, via the basal ganglia output nuclei GPi and SNr, is transferred to thalamus and cortex. In parkinsonism, the dopaminergic SNc degenerates, resulting in increased activity along the 'indirect' pathways, and reduced activity along the 'direct' pathway, leading to increased basal ganglia output. RS: reticulospinal system. For other abbreviations see text.

1985; Nambu et al., 1988; Hoover and Strick, 1993; Jinnai et al., 1993; Midleton and Strick, 1994; Inase and Tanji, 1995; Sidibe et al., 1997), and to the brainstem, in particular to portions of the pedunculopontine nucleus (PPN) which may serve to connect the basal ganglia to spinal centers (Rye et al., 1988; Lavoie and Parent, 1994; Bevan and Bolam, 1995; Jaffer et al., 1995; Steininger et al., 1997). In addition, output from the SNr reaches the superior colliculus; this phylogenetically old connection may be relevant in the control of orienting behavior and eye movements (Anderson and Yoshida, 1977; Deniau et al., 1978; Hikosaka and Wurtz, 1983; Parent et a., 1983; Wurtz and Hikosaka, 1986; Lynch et al., 1994). With the exception of the excitatory (glutamatergic) efferents of the STN, intrinsic and output connections of the basal ganglia are inhibitory (GABAergic).

The activity along the intrinsic basal ganglia circuitry is modulated by changes in the release of striatal dopamine from terminals of the nigrostriatal projection. It appears that, at the striatal level, dopamine facilitates the transmission over the direct pathway and inhibits the transmission over the indirect pathway via dopamine D_1 and D_2 receptors, respectively (Gerfen, 1995). The overall effect of striatal dopamine release is to reduce basal ganglia output, leading to disinhibition of thalamocortical projection neurons.

By virtue of being part of the aforementioned cortico-subcortical re-entrant loops that terminate in the frontal lobes, the basal ganglia may have a major impact on cortical function, and, thus, on the control of behavior. Both GPi and SNr output neurons exhibit a high tonic discharge rate in the intact animal (DeLong, 1971; DeLong et al., 1983; Rodriguez et al., 1997; Wichmann et al., 1999). Modulation of this discharge by alteration in phasic and tonic activity over multiple afferent pathways is thought to play a role in the control of voluntary movements, as well as in the development of involuntary movements. Although details of the basal ganglia mechanisms involved in the control of voluntary movements are still not clear, it is thought that motor commands generated at the cortical level, are transmitted to the putamen both directly and via the CM. Phasic activation of the direct striatopallidal pathway will then result in reduction of tonic inhibitory basal ganglia output, resulting in disinhibition of thalamocortical neurons, and facilitation of movement. By contrast, phasic activation of the indirect pathway leads to increased basal ganglia output (Kita, 1994) and to suppression of movement.

This combination of information traveling via the direct and the indirect pathways of the motor circuit may serve basic motor control functions such as scaling or focusing of movements (Alexander et al., 1990; Mink and Thach, 1991; Mink, 1996, Wenger et al., 1999). Scaling or termination of movements would be achieved if in an orderly temporal sequence striatal output would first inhibit GPi/SNr neurons via the direct pathway (facilitating a movement in progress), followed by disinhibition of the same GPi/SNr neuron via the indirect pathway (terminating the movement). By contrast, focusing would be achieved if separate target populations of neurons in GPi/SNr would receive simultaneous input via the direct and indirect pathways in a center-surround (facilitating/inhibiting) manner (Mink and Thach, 1991; Mink, 1996, Wenger et al., 1999). Increased activity along the direct pathway would lead to inhibition of some GPi/SNr neurons, allowing intended movements to proceed, while increased activity along the indirect pathway would activate other GPi/SNr neurons, acting to inhibit unintended movements. Similar models have been proposed for the generation of saccades in the oculomotor circuit (Hikosaka et al., 1993). Direct anatomical support for either of these functions is lacking, because it is uncertain whether the direct and indirect pathways (emanating from neurons that are concerned with the same movement) converge on the same, or on separate neurons in GPi/SNr (Bolam and Smith, 1992; Hazrati and Parent, 1992; Bevan et al., 1994; Parent and Hazrati, 1995), and thus, whether focusing or scaling would be anatomically possible. Furthermore, it is uncertain whether corticostriatal neurons indeed carry information that would be relevant for either focusing or scaling (see above).

The lack of effect of STN lesions on voluntary movement are difficult to reconcile with either hypothesis. Such lesions induce dyskinesias (hemiballism, see below), but voluntary movements can still be carried out. Lack of focusing would be expected to result in inappropriate activation of antagonistic muscle groups (dystonia), and lack of scaling would be expected to result in hypo-or hypermetric movements. Both effects are not seen after STN lesions.

In general, movement-related changes in discharge occur too late in the basal ganglia to influence the initiation of movement. However, such changes in discharge could still influence the amplitude or limit the overall extent of ongoing movements (Wichmann et al., 1994; Jaeger et al., 1995; Nambu et al., 2000). Conceivably, neurons with shorter onset latencies or with preparatory activity may indeed play such a role (Jinnai et al., 1993; Kubota and Hamada, 1979; Crutcher and Alexander, 1988; Alexander and Crutcher, 1989; Apicella et al., 1991; Anderson et al., 1992; Schultz and Romo, 1992; Jaeger et al., 1993). Recent PET studies have reported that basal ganglia activity is modulated in relation to low-level parameters of movement, such as force or movement speed (Dettmers et al., 1995; Turner et al., 1998) supporting a scaling function of the basal ganglia.

The basal ganglia may also serve more global functions such as the planning, initiation, sequencing and execution of movements (Marsden and Obeso, 1994; Martin et al., 1994). Most recently, an involvement of these structures in the performance of learned movements, and in motor learning itself has also been proposed (Graybiel, 1995; Kimura, 1995; Kimura et al., 1996; Watanabe and Kimura, 1998). For instance, both dopaminergic nigrostriatal neurons and tonically active neurons in the striatum have been shown to develop transient responses to sensory conditioning stimuli during behavioral training in classical conditioning tasks (Aosaki et al., 1994, 1994; Graybiel, 1995), and shifts in the response properties of striatal output neurons during a procedural motor learning task have been demonstrated (Jog et al., 1999).

A major problem with all schemes that attribute a significant indispensable motor function to the basal ganglia is the fact that lesions of the basal ganglia output nuclei do not lead to obvious motor deficits in humans or experimental animals. Most studies have found either no effect or only subtle short-lived effects on skilled fine movements after such lesions (DeLong and Georgeopolous, 1981; Aziz et al., 1991; Laitinen, 1992; Baron et al., 1996; but see also Mink and Thach, 1991; Wenger et al., 1999). Given the paucity of motor side-effects in animals and humans with lesions in the pallidal or thalamic motor regions, one would conclude that basal ganglia output does not play a significant role in the initiation or execution of most movements (Marsden and Obeso, 1994). Of course, the motor functions of the basal ganglia could be compensated for by other areas of the circuitry, but the striking lack of even immediate motor side-effects after lesions of GPi argues against this possibility, because reorganization and changes in synaptic strength in other areas would require time to develop. A complicating factor in the interpretation of the apparent lack of motor side effects associated with pallidal lesions in patients with movement disorders is that the organization of the motor pathways in these patients differs from that in normal individuals because of the movement disorder that predated the lesions (see Brooks, 1991; Calne and Snow, 1993; Eidelberg et al., 1994; Turner et al., 1997). For instance, patients with longstanding parkinsonism most certainly compensate for the progressive loss of dopamine by a variety of mechanisms, including the recruitment of cortical and subcortical circuits that are normally not involved in movement control (Turner et al., 1997). In these patients, lesions of GPi may fail to produce a deficit because of compensatory mechanisms are at work that no longer require basal ganglia input. It can also not be ruled out with certainty that lesions of the basal ganglia may, in fact, impair some specific functions in motor control that are not addressed in most studies, such as impaired procedural motor learning, or responses to novel conditions (Marsden and Obeso, 1994).

MOVEMENT DISORDERS

Movement disorders represent a spectrum ranging from hypokinetic disorders, such as Parkinson's disease, to hyperkinetic disorders such as Huntington's disease. Specific alterations in the activity along various pathways in the aforementioned model of basal ganglia circuitry have been proposed to underlie the various movement disorders.

Pathophysiologic Models of the Major Movement Disorders

Parkinson's Disease. In Parkinson's disease (Figure 1), the source neurons of the dopaminergic nigrostriatal projection degenerate, resulting in loss of dopamine in the striatum. According to the above-mentioned model, this will result in increased activity along the indirect pathway, and reduced activity along the direct pathway. Both effects together will lead to increased excitation of GPi and SNr neurons, and to greater inhibition of thalamocortical cells, and, therefore reduced excitation of cortex, clinically manifest in the development of the cardinal parkinsonian signs of akinesia and bradykinesia (Wichmann and DeLong, 1996). In addition, descending basal ganglia output to the PPN may play a role in the development of parkinsonism. The PPN region was shown to be metabolically overactive in parkinsonian animals, consistent with a major increase of input to this region (Palombo et al., 1990; Mitchell et al., 1989) and it has been shown that PPN inactivation of this nucleus alone is sufficient to induce a form of akinesia in experimental animals (Kojima et al., 1997; Munro-Davies et al., 1999) although it is not certain how this syndrome relates to parkinsonism.

The finding that increased basal ganglia output is a major pathophysiologic step in the development of parkinsonian motor signs has provided a rationale for attempts to reduce this output pharmacologically and surgically. The demonstration that lesions of the STN in MPTP-treated primates reverses all of the cardinal signs of parkinsonism by reducing GPi activity has contributed to these efforts (Bergman et al., 1990; Aziz et al., 1991). Stereotactic lesioning of the motor portion of GPi (GPi pallidotomy), which has been reintroduced in human patients, has been shown to be effective against all major parkinsonian motor signs (Laitinen, 1992; Baron et al., 1996; Dogali et al., 1995; Lozano et al., 1995; 1997; Lang et al., 1997; Starr et al., 1998). PET studies have shown that frontal motor areas whose metabolic activity was reduced in the parkinsonianm state were again active following pallidotomy (Dogali et al., 1995, Cebalos-Bauman et al., 1994).

The latest addition to the neurosurgical armamentarium used in the treatment of parkinsonism is deep brain stimulation (DBS). DBS of the STN and of Gpi has been shown to be highly effective against parkinsonism. Although the exact mechanism of action remains uncertain, the clinical and neuroimaging effect of DBS closely mimics that of ablation, suggesting a net inhibitory action (Benazzouz et al., 1995), perhaps induced by depolarization block, although other mechanisms may also be at play, such as activation of inhibitory pathways, a jamming of abnormal activity along output pathways, and use-induced depletion of neurotransmitters (see below).

Hyperkinetic disorders. In hyperkinetic disorders basal ganglia output is thought to be reduced, resulting in disinhibition of thalamocortical systems and dyskinesias (Albin et al., 1989; DeLong, 1990). This is best explored for hemiballism (Figure 2), a disorder that follows

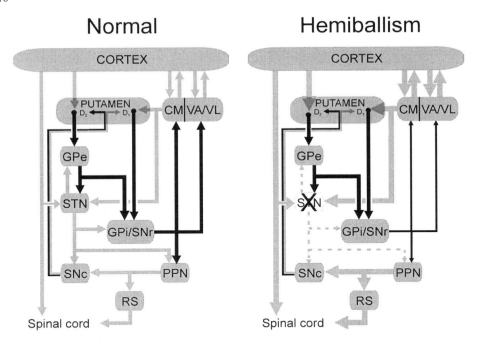

Figure 2. In hemiballism, a lesion of the STN results in reduced 'driving' of the basal ganglia output nuclei, and disinhibition of thalamocortical neurons. For abbreviations see Figure 1.

discrete lesions of the STN which result in reduced activity in GPi in both experimental primates and in humans (Hamada and DeLong, 1992). The mechanisms underlying chorea in Huntington's disease are thought to be similar to those in hemiballism in that degeneration of striatal neurons projecting to GPe (indirect pathway) leads to disinhibition of GPe, followed by increased inhibition of the STN and thus reduced output from GPi (Albin, 1995; Albin et al., 1990). Thus, whereas in hemiballism there is a distinct lesion in the STN, in Huntington's disease there is a functional lesion of this nucleus. Drug-induced dyskinesias may also result from a similar reduction in STN and GPi activity. Support for the validity of these models comes from direct recording of neuronal activity (Miller and DeLong, 1987; Filion and Tremblay, 1991; Vitek et al., 1993; Papa et al., 1999) as well as metabolic studies in primates, and a number of PET studies investigating cortical and subcortical metabolism in humans with movement disorders (Brooks, 1995; 1993).

Dystonia. This is a disorder characterized by slower, more sustained abnormal movements and postures with co-contraction of agonist-antagonist muscle groups, and overflow phenomena. Pathophysiologically it may be a hybrid disorder with both hypo- and hyperkinetic features. Multiple causes of dystonia have been identified, ranging from genetic disorders to focal lesions of the basal ganglia or other structures to disorders of dopamine metabolism. As a generalization, however, it seems that dystonia appears to result in most cases from changes after partial lesions of the basal ganglia or with changes in striatal dopamine supply.

The lack of suitable animal models and the likely heterogeneity of the human disease have impeded our understanding of the pathophysiologic changes in the basal ganglia that lead to dystonia. Based on intraoperative recording in a small number of human patients undergoing neurosurgical procedures for treatment of dystonia, appears that the activity along both the direct and indirect pathways is increased in dystonia (Figure 3). Supporting this concept, recent recording studies in dystonic patients undergoing pallidotomy revealed low average discharge rates in both pallidal segments (Cardoso et al., 1995; Lenz et al., 1998; Vitek et al., 1999). The reduction of discharge in GPe in these dystonic patients attest to increased activity along the indirect pathway, which by itself would have led to increased GPi discharge. The fact that discharge rates in GPi were actually reduced, argues therefore in favor of additional overactivity along the direct pathway. However, because in both dystonia and ballismus output from GPi is reduced, it would appear that factors other than rate are playing a significant role. Most likely, a major part of the pathophysiology of dystonia is abnormally increased synchronization of the activity of basal ganglia output neurons, which is not accounted for by the model (see below, and Vitek et al., 1999). Changes suggestive of a reorganization of the activity of the basal ganglia-thalamocortical circuits in dystonia have also been shown outside of the basal ganglia. For instance, at the level of the thalamus, abnormal receptive field have been described in the thalamic nucleus Vc in dystonic patient undergoing thalamotomy (Lenz and Byl, 1999; Lenz et al., 1999) and, at the cortical level, a degredation of the discrete cortical representation of individual body parts has been demonstrated in dystonic patients (Bara-Jimenez et al., 1998; Hallet and Toro, 1996; Ikoma et al., 1996).

Critique of the Pathophysiologic Models

The pathophysiologic models mentioned above may account for changes in discharge rates and global metabolic activity in the basal ganglia, but fail to explain several key findings in animals and humans with such disorders. These shortcomings of the traditional models are best

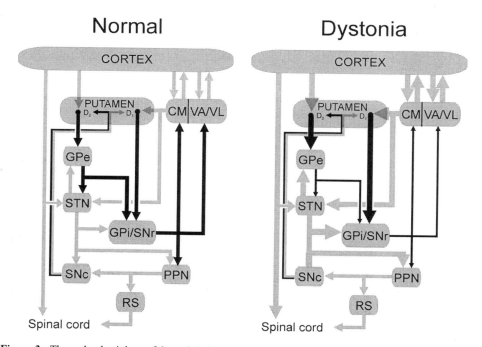

Figure 3. The pathophysiology of dystonia is less well understood than that of other movement disorders. It is thought that dystonia results form concomitant overactivity along both, the 'direct' and 'indirect' pathways. For abbreviations see Figure 1.

explored with regard to the pathophysiology of parkinsonism and will be elaborated for this disease in particular. However, similar considerations also hold for the other movement disorders mentioned above.

The most serious shortcoming of the aforementioned scheme of parkinsonian pathophysiology is, perhaps, that changes in the spontaneous discharge patterns, and the responses of basal ganglia neurons to external stimuli is significantly different in the parkinsonian state when compared to the normal state (Miller and DeLong, 1987; Bergman et al., 1994; Filion and Tremblay, 1991; Filion et al., 1988; Vitek et al., 1990; Raz at al., 1996). Thus, the neuronal responses to passive limb manipulations in STN, GPi and thalamus are increased (Miller and DeLong, 1987; Bergman et al., 1994; Filion et al., 1988; Vitek et al., 1990), suggesting an increased gain by the subcortical portions of the circuit as compared to the normal state. In addition to the changes in the responses to peripheral inputs, there is also a marked increase in the synchronization of discharge between neurons in the basal ganglia under parkinsonian conditions, as demonstrated in primates through cross correlation studies of the neuronal discharge in GPi and STN (Bergman et al., 1994). This is in contrast to the virtual absence of synchronized discharge of these neurons in normal monkeys (Wichmann et al., 1994). In addition, a very prominent abnormality in the pattern of basal ganglia discharge in parkinsonian primates and humans is an increase in the proportion of cells in STN, GPi and SNr with oscillatory discharge characteristics (Miller and DeLong, 1987; Bergman et al., 1994; Filion and Tremblay, 1991; Vitek et al., 1993; Vitek et al., 1994; Wichmann et al., 1996). Increased oscillatory activity and synchronization have also been demonstrated in the basal ganglia-receiving areas of cortex in parkinsonian individuals (Meufield et al., 1988; Zijlmans et al., 1998; Primavera and Novello, 1992; Soikkeli et al., 1991; Jelles et al., 1995). In patients with unilateral parkinsonian tremor, EMG and contralateral EEG were found to be coherent at the tremor frequency (or its first harmonic), particularly over cortical motor areas (Hellwig et al., 2000). Dopaminergic therapy has been shown to desynchronize cortical activity in parkinsonian subjects (Boulton and Marjerrison, 1972; Marjerrison et al., 1972; Brown and Marsden, 1999; Wang at al., 1999).

The mechanisms underlying increased synchronization and oscillations in the basal ganglia-thalamocortical circuitry are not clear. Attempts have been made to link these changes to the prominent dopaminergic deficit in the striatum. For instance, it has been shown that the predominant oscillation frequency in many basal ganglia neurons is influenced by the presence or absence of dopamine. Thus, low frequency oscillations appear to be enhanced in the presence of dopamine receptor agonists (Ruskin et al., 1999), but are less common in the dopamine-depleted, parkinsonian state (at least in primates, unpublished observation). On the other hand, oscillatory discharge with predominant frequencies > 3 Hz are present in only a small percentage of neurons in GPe, STN, GPi and SNr in the normal state, but are strikingly enhanced in the parkinsonian state (Wichmann et al., 1999; Bergman et al., 1994; Karmon and Bergman, 1993; Hutchison et al., 1997; Hurtado et al., 1999; Bergman et al., 1998).

The most obvious consequence of increased synchronized oscillatory discharge in the basal ganglia-thalamo-cortical loops may be tremor (Bergman et al., 1994; 1998; Hallet, 1998; Rothwell, 1998; Hua et al., 1998; Elble, 1996; Nini et al., 1995). For instance, parkinsonian African green monkeys show prominent 5 Hz tremor, along with striking 5-Hz oscillations in STN and GPi (Bergman et al., 1994). Oscillatory discharge at higher dominant frequencies (8-15 Hz) is seen in the basal ganglia of parkinsonian Rhesus monkeys, which typically do not exhibit tremor (see Wichmann et al., 1999; Bergman et al., 1998). It seems obvious that strong oscillations in the basal ganglia could be very disruptive with regard to the transfer and processing of information in these nuclei, and in the basal ganglia-thalamocortical circuitry as a whole. Via their widespread influence on frontal cortex, synchronized oscillatory basal ganglia activity may adversely influence cortical activity in a large part of frontal cortex, and could therefore (in a rather non-specific manner) contribute to cortical dysfunction which may underlie parkinsonian signs such as akinesia or bradykinesia. The dramatic changes in PET activation patterns seen in parkinsonian subjects may indicate not only reduced excitation via the thalamocortical pathway, as predicted by the traditional circuit model, but may indicate a functional impairment of cortical activation through subcortical "noise" (i.e., oscillatory activity, synchronization and other abnormal neuronal discharge patterns), which would perhaps induce plastic changes in cortex (Hallet, 1995; 1999). The development of parkinsonian motor abnormalities such as akinesia or bradykinesia may to some extent depend on the inefficiencies of partial redistribution of premotor activities at the cortical level, induced by relatively non-specific subcortical "noise."

The view that the pathophysiology of movement disorders may be far more complicated than suggested by the simple rate model introduced above is further supported by the observation that pallidotomy and DBS, relatively non-specific procedures, not only ameliorate (as expected) *parkinsonian* abnormalities, but also most of the major *hyperkinetic* syndromes, including drug-induced dyskinesias (Lozano et al., 1995; Johansson et al., 1997; Samuel et al., 1998) hemiballismus (Vitek et al., 1999), and dystonia (Lenz et al., 1997; 1998; Vitek et al., 1999) which, according to the earlier pathophysiologic scheme should be worsened by lesions of GPi. Furthermore, these hyperkinetic disorders can also be treated with lesions of VA/VL thalamus (without producing parkinsonism). These findings suggest that, rather than counteracting specific pathophysiologic changes, pallidotomy, thalamotomy, and DBS may act by removing abnormal disruptive signals or reducing the amount of "noise" in cortical motor areas resulting from abnormal basal ganglia discharge, which secondarily may render previously dysfunctional cortical areas again functional.

Another significant area of discussion with regard to the models introduced above relates to the role of GPe in the development of movement disorders. For instance, on the basis of metabolic and biochemical studies, it has been questioned whether reduced GPe activity is indeed important in the development of parkinsonism as stated by the traditional models (Brooks, 1995; Soghomonian et al., 1994; Herrero et al., 1996). In addition, the view that GPe activity is increased in drug-induced dyskinesias, as postulated by the traditional model has been challenged by studies such as those by Bedard et al. in which excitotoxic lesions of GPe did not resolve levodopa-induced dyskinesias in parkinsonian primates (Blanchet et al., 1994).

Finally, while the anatomy and function of the intrinsic basal ganglia circuitry is known in considerable detail, the anatomy and function of several key portions of the circuit models outside of the basal ganglia are far less explored. Among others, this includes the interaction of the basal ganglia with brainstem nuclei such as the PPN, as well as input and output connections between the basal ganglia and the intralaminar thalamic nuclei CM and Pf. Particularly the connection with CM and Pf may form potentially important (positive) feedback circuits that may exaggerate any change in basal ganglia output, and may therefore have an important role in the pathophysiology of movement disorders. Furthermore, the processing of basal ganglia output at the thalamic level is not well understood. Finally, the characteristics of cortical output to the basal ganglia and the influence of thalamocortical reentrant pathway neurons remains uncertain.

CONCLUSIONS

There has been very significant progress in the understanding of basal ganglia anatomy and physiology over the last years, but the functions of these nuclei, most likely related to the functions of cortical areas they are connected to, remain unclear. The current models of basal ganglia function have been of tremendous value in stimulating basal ganglia research and providing a rationale for large-scale neurosurgical treatment efforts directed at movement disorders. At the same time, the scientific shortcomings of these models have become increasingly obvious, particularly with regard to the fact that the models are predominantly based on anatomic data, and do not take into account the multiple dynamic changes that take place in the basal ganglia in individuals with movement disorders. It is also likely that an increased understanding of the relationship between the basal ganglia and related areas of cortex will greatly help to better understand the normal function of the basal ganglia and the pathophysiology of movement disorders.

REFERENCES

Albin, R.L., 1995, The pathophysiology of chorea/ballism and parkinsonism, *Parkinsonism & Relat. Disord.,* 1:3.

Albin, R.L., Reiner, A., Anderson, K.D., Penney, J.B., and Young, A.B., 1990, Striatal and nigral neuron subpopulations in rigid Huntington's disease: implications for the functional anatomy of chorea and rigidity-akinesia, *Ann.Neurol.* 27:357.

Albin, R.L., Young, A.B., and Penney, J.B., 1989, The functional anatomy of basal ganglia disorders, *Trends Neurosci.* 12:366.

Alexander, G.E., and Crutcher, M.D., 1989, Coding in spatial rather than joint coordinates of putamen and motor cortex preparatory activity preceding planned limb movements, in: *Neural mechanisms in disorders of movement,* M. A. Sambrook and A. R. Crossman, ed., Blackwell, London.

Alexander, G.E., Crutcher, M.D., and DeLong, M.R., 1990, Basal ganglia-thalamocortical circuits: parallel substrates for motor, oculomotor, `prefrontal' and `limbic' functions, *Prog.Brain.Res.* 85:119.

Anderson, M., and Yoshida, M., 1977, Electrophysiological evidence for branching nigral projections to the thalamus and the superior colliculus, *Brain Res.* 137:361.

Anderson, M., Inase, M., Buford, J., and Turner,R., 1992, Movement and preparatory activity of neurons in pallidal-receiving areas of the monkey thalamus, *Role of Cerebellum and Basal Ganglia in Voluntary Movement,* 39.

Aosaki, T., Kimura, M., and Graybiel, A.M., 1995, Temporal and spatial characteristics of tonically active neurons of the primate striatum, *J. Neurophysiol.* 73:1234.

Aosaki, T., Tsubokawa, H, Watanabe, K, Graybiel, A.M., and Kimura, M., 1994, Responses of tonically active neurons in the primate's striatum undergo systematic changes during behavioral sensory-motor conditioning, *J. Neurosci.* 14:3969.

Apicella, P., Scarnati, E., and Schultz, W., 1991, Tonically discharging neurons of monkey striatum respond to preparatory and rewarding stimuli, *Exp. Brain Res.* 84:672.

Aziz, T.Z., Peggs, D., Sambrook, M.A., and Crossman, A.R., 1991, Lesion of the subthalamic nucleus for the alleviation of 1-methyl-4-phenyl-1,2,3,6-tetrahydropyridine (MPTP)-induced parkinsonism in the primate, *Mov. Disord.* 6:288.

Bara-Jimenez, W., Catalan, M.J., Hallett, M., and Gerloff, C., 1998, Abnormal somatosensory homunculus in dystonia of the hand, *Ann. Neurol.* 44:828.

Baron, M.S., Vitek, J.L., Bakay, R.A.E., Green, J., Kaneoke, Y., Hashimoto, T., Turner, R.S., Woodard, J.L., Cole, S.A., McDonald, W.M., and DeLong, M.R., 1996, Treatment of advanced Parkinson's disease by GPi pallidotomy: 1 year pilot-study results, *Ann. Neurol.* 40:355.

Benazzouz, A., Piallat, B., Pollak, P., and Benabid, A.L., 1995, Responses of substantia nigra pars reticulata and globus pallidus complex to high frequency stimulation of the subthalamic nucleus in rats: electrophysiological data, *Neurosci. Lett.* 189:77.

Bergman, H., Feingold, A., Nini, A., Raz, A., Slovin, H., Abeles, M., and Vaadia, E., 1998, Physiological aspects of information processing in the basal ganglia of normal and parkinsonian primates, *Trends in Neurosci.* 21:32.

Bergman, H., Raz, A., Feingold, A., Nini, A., Nelken, I., Hansel, D., Ben-Pazi, H., and Reches, A., 1998, Physiology of MPTP tremor, *Mov. Disord.* 13, Suppl. 3:29.

Bergman, H., Wichmann, T., and DeLong, M.R., 1990, Reversal of experimental parkinsonism by lesions of the subthalamic nucleus, *Science,* 249:1436.

Bergman, H., Wichmann, T., Karmon, B., and DeLong, M.R., 1994, The primate subthalamic nucleus. II. Neuronal activity in the MPTP model of parkinsonism, *J. Neurophysiol.* 72:507.

Bevan, M.D., and Bolam, J.P., 1995, Cholinergic, GABAergic, and glutamate-enriched inputs from the mesopontine tegmentum to the subthalamic nucleus in the rat, *J. Neurosci.* 15:7105.

Bevan, M.D., Bolam, J.P., and Crossman, A.R., 1994, Convergent synaptic input from the neostriatum and the subthalamus onto identified nigrothalamic neurons in the rat, *Eur. J. Neurosci.* 6:320.

Blanchet, P.J., Boucher, R., and Bedard, P.J., 1994, Excitotoxic lateral pallidotomy does not relieve L-dopa-induced dyskinesia in MPTP parkinsonian monkeys, *Brain Res.* 650:32.

Bolam, J.P., and Smith, Y., 1992, The striatum and the globus pallidus send convergent synaptic inputs onto single cells in the entopeduncular nucleus of the rat: a double anterograde labelling study combined with postembedding immunocytochemistry for GABA, *J. Comp. Neurol.* 321:456.

Boulton, A.A., and Marjerrison, G.L., 1972, Effect of L-dopa therapy on urinary p-tyramine excretion and EEG changes in Parkinson's disease, *Nature,* 236:76.

Boussaoud, D., and Kermadi, I, 1997, The primate striatum: neuronal activity in relation to spatial attention versus motor preparation, *Eur. J. Neurosci.* 9:2152.

Brooks, D.J., 1991, Detection of preclinical Parkinson's disease with PET, *Neurology,* 41 (suppl. 2):24.

Brooks, D.J., 1993, Functional imaging in relation to parkinsonian syndromes, *J. Neurol. Sci.* 115:1.

Brooks, D.J., 1995, The role of the basal ganglia in motor control: contributions from PET, *J. Neurol. Sci.* 128:1.

Brown, P., and Marsden, C.D., 1999, Bradykinesia and impairment of EEG desynchronization in Parkinson's disease, *Mov. Disord.* 14:423.

Calne, D.,and Snow, B.J., 1993, PET imaging in Parkinsonism, *Adv. Neurol.* 60:484.

Cardoso, F., Jankovic, J., Grossman, R.G., and Hamilton, W.J., 1995 Outcome after stereotactic thalamotomy for dystonia and hemiballismus, *Neurosurgery,* 36:501.

Ceballos-Bauman, A.O., Obeso, J.A., Vitek, J.L., DeLong, M.R., Bakay, R., Linaasoro, G., and Brooks, D.J., 1994, Restoration of thalamocortical activity after posteroventrolateral pallidotomy in Parkinson's disease, *Lancet,* 344:814.

Chesselet, M. F., and Delfs, J.M., 1996, Basal ganglia and movement disorders: an update, *Trends Neurosci.* 19:417.

Crutcher, M.D., and Alexander, G.E., 1988, Supplementary motor area (SMA): Coding of both preparatory and movement-related neural activity in spatial rather than joint coordinates, *Soc. Neurosci. Abstr.* 14:342.

DeLong, M. R., 1990, Primate models of movement disorders of basal ganglia origin, *Trends Neurosci.* 13:281.

DeLong, M.R., Activity of pallidal neurons during movement, *J. Neurophysiol.* 34:414 (1971).

DeLong, M.R., and Georgopoulos, A.P., 1981, Motor functions of the basal ganglia, in: *Handbook of Physiology. The Nervous System. Motor Control. Sect. 1, Vol. II, Pt. 2,* J. M. Brookhart, V. B. Mountcastle, V. B. Brooks and S. R. Geiger, ed., American Physiological Society, Bethesda.

DeLong, M.R., Crutcher, M.D., and Georgopoulos, A.P., 1983, Relations between movement and single cell discharge in the substantia nigra of the behaving monkey, *J. Neurosci.* 3:1599.

Deniau, J.M., Hammond, C., Riszk, A., and Feger, J., 1978, Electrophysiological properties of identified output neurons of the rat substantia nigra (pars compacta and pars reticulata): evidences for the existence of branched neurons, *Exp. Brain Res.* 32:409.

Dettmers, C., Fink, J.R., Lemon, R.N., Stephan, K.M., Passingham, R.E., Silbersweig, D., Holmes, A., Ridding, M.C., Brooks, D.J., and Frackowiak, R.S., 1995, Relation between cerebral activity and force in the motor areas of the human brain, *J. Neurophysiol.* 74:802.

Dogali, M., Fazzini, E., Kolodny, E., Eidelberg, D., Sterio, D., Devinsky, O., and Beric, A., 1995, Stereotactic ventral pallidotomy for Parkinson's disease, *Neurology,* 45:753.

Dube, L., Smith, A.D., and Bolam, J.P., 1988, Identification of synaptic terminals of thalamic or cortical origin in contact with distinct medium-size spiny neurons in the rat neostriatum, *J. Comp. Neurol.* 267:455.

Eidelberg, D., Moeller, J.R., Dhawan, V., Spetsieris, P., Takikawa, S., Ishikawa, T., Chaly, T., Robeson, W., Margouleff, D., Przedborski, S., and Fahn, S., 1994, The metabolic topography of Parkinsonis, *J. Cereb. Blood Flow Metab.* 14:783.

Elble, R.J., 1996, Central mechanisms of tremor, *J. Clin. Neurophysiol.* 13:133.

Filion, M., and Tremblay, L., 1991, Abnormal spontaneous activity of globus pallidus neurons in monkeys with MPTP-induced parkinsonism, *Brain Res.* 547:142.

Filion, M., Tremblay, L., and Bedard, P.J., 1988, Abnormal influences of passive limb movement on the activity of globus pallidus neurons in parkinsonian monkeys, *Brain Res.* 444:165.

Gerfen, C.R., 1995, Dopamine receptor function in the basal ganglia, *Clin. Neuropharmacol.* 18:S162.

Goldman-Rakic, P.S., and Porrino, L.J., 1985, The primate mediodorsal (MD) nucleus and its projection to the frontal lobe, *J. Comp. Neurol.* 242:535

Graybiel, A.M., 1995, Building action repertoires: memory and learning functions of the basal ganglia, *Current Opinion in Neurobiology* 5:733.

Hallett, M. and Toro, C., 1996, Dystonia and the supplementary sensorimotor area, *Adv. Neurol.* 70:471.

Hallett, M., 1995, The plastic brain [editorial] [see comments], *Ann. Neurol.* 38:4.

Hallett, M., 1998, Overview of human tremor physiology, *Mov. Disord.* 13:43.

Hallett, M., 1999, Motor cortex plasticity, *Electroencephalogr. & Clin. Neurophysiol. Supplement* 50:85.

Hamada, I., and DeLong, M.R., 1992, Excitotoxic acid lesions of the primate subthalamic nucleus result in reduced pallidal neuronal activity during active holding, *J. Neurophysiol.* 68:1859.

Hartmann-von Monakow, K., Akert, K., and Kunzle, H., 1978, Projections of the precentral motor cortex and other cortical areas of the frontal lobe to the subthalamic nucleus in the monkey, *Exp. Brain Res.*

    33:395.

Hazrati, L., and Parent, A., 1992, Convergence of subthalamic and striatal efferents at pallidal level in primates: an anterograde double-labeling study with biocytin and PHA-L, *Brain Res.* 569:336.

Hazrati, L.N., Parent, A., Mitchell, S., and Haber, S.N., 1990, Evidence for interconnections between the two segments of the globus pallidus in primates: a PHA-L anterograde tracing study, *Brain Res.* 533:171.

Hellwig, B., Haussler, S., Lauk, M., Guschlbauer, B., Koster, B., Kristeva-Feige, R., Timmer, J., and Lucking, C.H., 2000, Tremor-correlated cortical activity detected by electroencephalography, *Clin. Neurophysiol.* 111:806.

Herrero, M.T., Levy, R., Ruberg, M., Javoy-Agid, F., Luquin, M.R., Agid, Y., Hirsch, E.C., and Obeso, J.A., 1996, Glutamic acid decarboxylase mRNA expression in medial and lateral pallidal neurons in the MPTP-treated monkeys and patients with Parkinson's disease, *Adv. Neurol.* 69:209.

Hikosaka, O., Matsumara, M., Kojima, J., and Gardiner, T.W., 1993, Role of basal ganglia in initiation and suppression of saccadic eye movements, in: *Role of the cerebellum and basal ganglia in voluntary movement,* N. Mano, I. Hamada and M. R. DeLong, ed., Elsevier, Amsterdam.

Hikosaka, O.and Wurtz, R.H., 1983, Visual and oculomotor functions of monkey substantia nigra pars reticulata. IV. Relation of substantia nigra to superior colliculus, *J. Neurophysiol.* 49:1285.

Hoover, J.E. and Strick, P.L., 1993, Multiple output channels in the basal ganglia, *Science,* 259:819.

Hua, S., Reich, S.G., Zirh, A.T., Perry, V., Dougherty, P.M., and Lenz, F.A., 1998, The role of the thalamus and basal ganglia in parkinsonian tremor, *Mov. Disord.* 13 Suppl 3:40.

Hurtado, J.M., Gray, C.M., Tamas, L.B., and Sigvardt, K.A., 1999, Dynamics of tremor-related oscillations in the human globus pallidus: a single case study, *Proce. Nat. Acad. Sci. USA* 96:1674.

Hutchison, W.D., Lozano, A.M., Tasker, R.R., Lang, A.E., and Dostrovsky, J.O., 1997, Identification and characterization of neurons with tremor-frequency activity in human globus pallidus, *Exp. Brain Res.* 113:557.

Ikoma, K., Samii, A., Mercuri, B., Wassermann, E.M., and Hallett, M., 1996, Abnormal cortical motor excitability in dystonia, *Neurology,* 46:1371.

Ilinsky, I.A., Jouandet, M.L., and Goldman-Rakic, P.S., 1985, Organization of the nigrothalamocortical system in the rhesus monkey, *J. Comp. Neurol.* 236:315.

Inase, M., and Tanji, J., 1995, Thalamic distribution of projection neurons to the primary motor cortex relative to afferent terminal fields from the globus pallidus in the macaque monkey, *J. Comp. Neurol.* 353:415.

Jaeger, D., Gilman, S., and Aldridge, J.W., 1993, Primate basal ganglia activity in a precued reaching task: preparation for movement, *Exp. Brain Res.* 95:51.

Jaeger, D., Gilman, S., and Aldridge, J.W., 1995, Neuronal activity in the striatum and pallidum of primates related to the execution of externally cued reaching movements, *Brain Res.* 694:111.

Jaffer, A., Van der Spuy, G.D., Russell, V.A., Mintz, M., and Taljaard, J.J., 1995, Activation of the subthalamic nucleus and pedunculopontine tegmentum: does it affect dopamine levels in the substantia nigra, nucleus accumbens and striatum? *Neurodegeneration,* 4:139.

Jelles, B., Achtereekte, H.A., Slaets, J.P., and Stam, C.J., 1995, Specific patterns of cortical dysfunction in dementia and Parkinson's disease demonstrated by the acceleration spectrum entropy of the EEG, *Clin. Electroencephalogr.* 26:188.

Jinnai, K., Nambu, A., Yoshida, S., and Tanibuchi, I., 1993, The two separate neuron circuits through the basal ganglia concerning the preparatory or execution proceses of motor control, in: *Role of the Cerebellum and Basal Ganglia in Voluntary Movement,* N. Mamo, I. Hamada and M. R. DeLong, ed., Elsevier Science.

Jog, M.S., Kubota, Y., Connolly, C.I., Hillegaart, V., and Graybiel, A.M., 1999, Building neural representations of habits, *Science,* 286:1745.

Johansson, F., Malm, J., Nordh, E., and Hariz, M., 1997, Usefulness of pallidotomy in advanced Parkinson's disease, *J. Neurol., Neurosurg. Psychiatr.* 62:125.

Karmon, B., and Bergman, H., 1993, Detection of neuronal periodic oscillations in the basal ganglia of normal and parkinsonian monkeys, *Israeli J. Med. Sci.* 29:570.

Kemp, J. M., and Powell, T.P.S., 1971, The connections of the striatum and globus pallidus: synthesis and speculation, *Phil. Trans. R. Soc. Lond.* 262:441.

Kimura, M., 1995, Role of basal ganglia in behavioral learning, *Neurosci. Res.* 22:353.

Kimura, M., Kato, M., Shimazaki, H., Watanabe, K., and Matsumoto, N., 1996, Neural information transferred from the putamen to the globus pallidus during learned movement in the monkey, *J. Neurophysiol.* 76:3771.

Kita, H., 1994, Physiology of two disynaptic pathways from the sensorimotor cortex to the basal ganglia output nuclei, in: *The Basal Ganglia IV. New ideas and data on structure and function,* G. Percheron, J. S. McKenzie and J. Feger, ed., Plenum Press, New York and London.

Kojima, J., Yamaji, Y., Matsumura, M., Nambu, A., Inase, M., Tokuno, H., Takada, M., and Imai, H., 1997, Excitotoxic lesions of the pedunculopontine tegmental nucleus produce contralateral hemiparkinsonism in the monkey, *Neurosci. Lett.* 226:111.

Kubota, K., and Hamada, I., 1979, Preparatory activity of monkey pyramidal tract neurons related to quick movement onset during visual tracking performance, *Brain Res.* 168:435.

Laitinen, L.V., Bergenheim, A.T., and Hariz, M.I., 1992, Leksell's posteroventral pallidotomy in the treatment of Parkinson's disease, *J. Neurosurg.* 76:53.

Lang, A., Lozano, A., Montgomery, E., Duff, J., Tasker, R., and Hutchinson, W., 1997, Posteroventral medial pallidotomy in advanced Parkinson's disease, *New Engl. J. Med.* 337:1036.

Lavoie, B., and Parent, A., 1994, Pedunculopontine nucleus in the squirrel monkey: projections to the basal ganglia as revealed by anterograde tract-tracing methods, *J. Comp. Neurol.* 344:210.

Lenz, F.A., and Byl, N.N., 1999, Reorganization in the cutaneous core of the human thalamic principal somatic sensory nucleus (Ventral caudal) in patients with dystonia, *J. Neurophysiol.* 82:3204.

Lenz, F.A., Jaeger, C.J., Seike, M.S., Lin, Y.C., Reich, S.G., DeLong, M.R., and Vitek, J.L., 1999, Thalamic single neuron activity in patients with dystonia: dystonia-related activity and somatic sensory reorganization, *J. Neurophysiol.* 82:2372.

Lenz, F.A., Suarez, J.I., Metman, L.V., Reich, S.G., Karp, B.I., Hallett, M., Rowland, L.H., and Dougherty, P.M., 1998, Pallidal activity during dystonia: somatosensory reorganisation and changes with severity, *J. Neurol., Neurosurg. Psychiatr.* 65:767.

Lozano, A.M., Kumar, R., Gross, R.E., Giladi, N., Hutchison, W.D., Dostrovsky, J.O., and Lang, A.E., 1997, Globus pallidus internus pallidotomy for generalized dystonia, *Mov. Disord.* 12:865.

Lozano, A.M., Lang, A.E., Galvez-Jimenez, N., Miyasaki, J., Duff, J., Hutchinson, W.D., and Dostrovsky, J.O., 1995, Effect of GPi pallidotomy on motor function in Parkinson's disease, *Lancet,* 346:1383.

Lozano, A.M., Lang, A.E., Hutchinson, W.D., and Dostrovsky, J.O., 1997, Microelectrode recording-guided posteroventral pallidotomy in patients with Parkinson's disease, *Adv. In Neurol.* 74:167.

Lynch, J.C., Hoover, J.E., and Strick, P.L., 1994, Input to the primate frontal eye field from the substantia nigra, superior colliculus, and dentate nucleus demonstrated by transneuronal transport, *Exp. Brain Res.* 100:181.

Marjerrison, G., Boulton, A.A., and Rajput, A.H., 1972, EEG and urinary non-catecholic amine changes during L-dopa therapy of Parkinson's disease, *Diseas. Nerv. Syst.* 33:164.

Marsden, C.D., and Obeso, J.A., 1994, The functions of the basal ganglia and the paradox of stereotaxic surgery in Parkinson's disease, *Brain,* 117:877.

Martin, K.E., Phillips, J.G., Iansek, R., and Bradshaw, J.L., 1994, Inaccuracy and instability of sequential movements in Parkinson's disease, *Exp. Brain Res.* 102:131.

Middleton, F.A., and Strick, P.L., 1994, Anatomical evidence for cerebellar and basal ganglia involvement in higher cognitive function, *Science,* 266:458.

Miller, W.C., and DeLong, M.R., 1987, Altered tonic activity of neurons in the globus pallidus and subthalamic nucleus in the primate MPTP model of parkinsonism., in: *The Basal Ganglia II,* M. B. Carpenter and A. Jayaraman, eds., Plenum Press, New York.

Mink, J.W., 1996, The basal ganglia: focused selection and inhibition of competing motor programs, *Progr. Neurobiol.* 50:381.

Mink, J.W., and Thach, W.T., 1991, Basal ganglia motor control. III. Pallidal ablation: normal reaction time, muscle cocontraction, and slow movement, *J. Neurophysiol.* 65:330.

Mitchell, I.J., Clarke, C.E., Boyce, S., Robertson, R.G., Peggs, D., Sambrook, M., and Crossman, A.R., 1989, Neural mechanisms underlying parkinsonian symptoms based upon regional uptake of 2-deoxyglucose in monkeys exposed to 1-methyl-4-phenyl-1,2,3,6-tetrahydropyridine, *Neuroscience,* 32:213.

Munro-Davies, L.E., Winter, J., Aziz, T.Z., and Stein, J.F., 1999, The role of the pedunculopontine region in basal-ganglia mechanisms of akinesia, *Exp. Brain Res.* 129:511.

Nakano, K., Hasegawa, Y., Tokushige, A., Nakagawa, S., Kayahara, T., and Mizuno, N., 1990, Topographical projections from the thalamus, subthalamic nucleus and pedunculopontine tegmental nucleus to the striatum in the Japanese monkey, Macaca fuscata, *Brain Res.* 537:54.

Nambu, A., Takada, M., Inase, M., and Tokuno, H., 1996, Dual somatotopical representations in the primate subthalamic nucleus: evidence for ordered but reversed body-map transformations from the primary motor cortex and the supplementary motor area, *J. Neurosci.* 16:2671.

Nambu, A., Tokuno, H., Hamada, I., Kita, H., Imanishi, M., Akazawa, T., Ikeuchi, Y., and Hasegawa, N., 2000, Excitatory cortical inputs to pallidal neurons via the subthalamic nucleus in the monkey, *J. Neurophysiol.* 84:289.

Nambu, A., Yoshida, S., and Jinnai, K., 1988, Projection on the motor cortex of thalamic neurons with pallidal input in the monkey, *Exp. Brain Res.* 71:658.

Neufeld, M.Y., Inzelberg, R., and Korczyn, A.D., 1988, EEG in demented and non-demented parkinsonian patients, *Acta Neurol. Scandin.* 78:1.

Nini, A., Feingold, A., Slovin, H., and Bergman, H., 1995, Neurons in the globus pallidus do not show correlated activity in the normal monkey, but phase-locked oscillations appear in the MPTP model of parkinsonism, *J. Neurophysiol.* 74:1800.

Palombo, E., Porrino, L.J., Bankiewicz, K.S., Crane, A.M., Sokoloff, L., and Kopin, I.J., 1990, Local cerebral glucose utilization in monkeys with hemiparkinsonism induced by intracarotid infusion of the neurotoxin MPTP, *J. Neurosci.* 10:860.

Papa, S.M., Desimone, R., Fiorani, M., and Oldfield, E.H., 1999, Internal globus pallidus discharge is nearly suppressed during levodopa-induced dyskinesias, *Ann. Neurol* 46:732.

Parent, A., and Hazrati, L.N., 1995, Functional anatomy of the basal ganglia. I. The cortico-basal ganglia-thalamo-cortical loop, *Brain Res. Rev.* 20:91.

Parent, A., Mackey, A., Smith, Y., and Boucher, R., 1983, The output organization of the substantia nigra in primate as revealed by a retrograde double labeling method, *Brain Res. Bull.* 10:529.

Primavera, A., and Novello, P., 1992, Quantitative electroencephalography in Parkinson's disease, dementia, depression and normal aging, *Neuropsychobiol.* 25:102.

Raz, A., Feingold, A., Zelanskaya, V., and Bergman, H., 1996, Neuronal synchronization of tonically active neurons in the striatum of normal and parkinsonian primates, *J. Neurophysiol.* in press.

Rodriguez, M.C., Gorospe, A., Mozo, A., Guridi, J., Ramos, E., Linazasoro, G., Obeso, J.A., Chockkan, V., Mewes, K., Vitek, J., and DeLong, M., 1997, Characteristics of neuronal activity in the subthalamic nucleus (STN) and substantia nigra pars reticulata (SNr) in Parkinson's disease (PD), *Soc. Neurosci. Abstr.* 23:470.

Rothwell J.C., 1998, Physiology and anatomy of possible oscillators in the central nervous system, *Mov. Disord.* 13 Suppl 3:24.

Ruskin, D.N., Bergstrom, D.A., Kaneoke, Y., Patel, B.N., Twery, M.J., and Walters, J.R., 1999, Multisecond oscillations in firing rate in the basal ganglia: robust modulation by dopamine receptor activation and anesthesia, *J. Neurophysiol.* 81:2046.

Rye, D.B., Lee, H.J., Saper, C.B., and Wainer, B.H., 1988, Medullary and spinal efferents of the pedunculopontine tegmental nucleus and adjacent mesopontine tegmentum in the rat, *J. Comp. Neurol.* 269:315.

Sadikot, A.F., Parent, A., and Francois, C., 1992, Efferent connections of the centromedian and parafascicular thalamic nuclei in the squirrel monkey: a PHA-L study of subcortical projections, *J. Comp. Neurol.* 315:137.

Sadikot, A.F., Parent, A., Smith, Y., and Bolam, J.P., 1992, Efferent conncetions of the centromedian and parafascicular thalamic nuclei in the squirrell monkey: a light and electron microscopic study of the thalamostriatal projection in relation to striatal heterogeneity, *J. Comp. Neurol.* 320:228.

Samuel, M., Caputo, E., Brooks, D.J., Schrag, A., Scaravilli, T., Branston, N.M., Rothwell, J.C., Marsden, C.D., Thomas, D.G., Lees, A.J., and Quinn, N.P., 1998, A study of medial pallidotomy for Parkinson's disease: clinical outcome, MRI location and complications, *Brain,* 121:59.

Schell, G.R. and Strick, P.L., 1984, The origin of thalamic inputs to the arcuate premotor and supplementary motor areas, *J. Neurosci.* 4:539.

Schultz, W., 1998, The phasic reward signal of primate dopamine neurons, *Adva. in Pharmacol.* 42:686.

Schultz, W., and Romo, R., 1992, Role of primate basal ganglia and frontal cortex in the internal generation of movement. I. Preparatory activity in the anterior striatum, *Exp. Brain Res.* 91:363.

Sidibe, M., and Smith, Y., 1996, Synaptic interactions between thalamic afferents and striatal interneurons in monkeys, *Soc. Neurosci. Abstr.* 22:411.

Sidibe, M., Bevan, M.D., Bolam, J.P., and Smith, Y., 1997, Efferent connections of the internal globus pallidus in the squirrel monkey: I. Topography and synaptic organization of the pallidothalamic projection, *J. Comp. Neurol.* 382:323.

Smith, Y., and Parent, A., 1986, Differential connections of caudate nucleus and putamen in the squirrel monkey (Saimiri Sciureus), *Neuroscience,* 18:347.

Smith, Y., Hazrati, L.N., and Parent, A., 1990, Efferent projections of the subthalamic nucleus in the squirrel monkey as studied by PHA-L anterograde tracing method, *J. Comp. Neurol.* 294:306.

Soghomonian, J.J., Pedneault, S., Audet, G., and Parent, A., 1994, Increased glutamate decarboxylase mRNA levels in the striatum and pallidum of MPTP-treated primates, *J. Neurosci.* 14:6256.

Soikkeli, R., Partanen, J., Soininen, H., Paakkonen, A., and Riekkinen, P. Sr., 1991, Slowing of EEG in Parkinson's disease, *Electroencephalogr. Clin. Neurophysiol.* 79:159.

Starr, P.A., Vitek, J.L., and Bakay, R.A.E., 1998, Pallidotomy: Clinical Results, in: *Neurosurgical treatment of movement disorders,* I. M. Germano, ed., American Association of Neurological Surgeons, Park Ridge, Ill..

Steininger, T.L., Wainer, B.H., and Rye, D.B., 1997, Ultrastructural study of cholinergic and noncholinergic neurons in the pars compacta of the rat pedunculopontine tegmental nucleus, *J. Comp. Neurol.* 382:285.

Strick, P.L., 1985, How do the basal ganglia and cerebellum gain access to the cortical motor areas? *Behav. Brain Res.* 18:107.

Turner, R.S., Grafton, S.T., Votaw, J.R., Delong, M.R., and Hoffman, J.M., 1998, Motor subcircuits mediating the control of movement velocity: a PET study, *J. Neurophysiol.* 80:2162.

Turner, R.S., Hoffmann, J.M., Grafton, S.T., Bakay, R.A.E., Vitek, J.L., and DeLong, M.R., 1997, The reorganization of movement rate-related cerebral activation following pallidotomy for Parkinson's disease, *Soc. Neurosci. Abstr.* 23:543.

Vitek, J.L., Ashe, J., DeLong, M.R., and Alexander, G.E., 1990, Altered somatosensory response properties of neurons in the 'motor' thalamus of MPTP treated parkinsonian monkeys, *Soc. Neurosci. Abstr.* 16:425.

Vitek, J.L., Ashe, J., DeLong, M.R., and Alexander, G.E., 1990, Altered somatosensory response properties of neurons in the `motor' thalamus of MPTP treated parkinsonian monkeys, *Soc. Neurosci. Abstr.* 16:425.

Vitek, J.L., Ashe, J., DeLong, M.R., and Alexander, G.E., 1994, Physiologic properties and somatotopic organization of the primate motor thalamus, *J. Neurophysiol.* 71:1498.

Vitek, J.L., Chockkan, V., Zhang, J.Y., Kaneoke, Y., Evatt, M., DeLong, M.R., Triche, S., Mewes, K., Hashimoto, T., and Bakay, R.A.E., 1999, Neuronal activity in the basal ganglia in patients with generalized dystonia and hemiballismus, *Ann. Neurol.* 46:22 .

Vitek, J.L., Kaneoke, Y., Turner, R., Baron, M., Bakay, R.A.E., and DeLong, M., 1993, Neuronal activity in the internal (GPi) and external (GPe) segments of the globus pallidus (GP) of parkinsonian patients is similar to that in the MPTP-treated primate model of parkinsonism, *Soc. Neurosci. Abstr.* 19:1584.

Wang, H.C., Lees, A.J., and Brown, P., 1999, Impairment of EEG desynchronisation before and during movement and its relation to bradykinesia in Parkinson's disease, *J. Neurol. Neurosurg. Psychiatr.* 66:442.

Watanabe, K., and Kimura, M., 1998, Dopamine receptor-mediated mechanisms involved in the expression of learned activity of primate striatal neurons, *J. Neurophysiol.* 79:2568.

Wenger, K.K., Musch, K.L., and Mink, J.W., 1999, Impaired reaching and grasping after focal inactivation of globus pallidus pars interna in the monkey, *J. Neurophysiol.* 82:2049.

Wichmann, T., and DeLong, M.R., 1996, Functional and pathophysiological models of the basal ganglia, *Current Opinion Neurobiol.* 6:751.

Wichmann, T., Bergman, H., and DeLong, M.R., 1994, The primate subthalamic nucleus. I. Functional properties in intact animals, *J. Neurophysiol.* 72:494.

Wichmann, T., Bergman, H., and DeLong, M.R., 1996, Comparison of the effects of experimental parkinsonism on neuronal discharge in motor and non-motor portions of the basal ganglia output nuclei in primates, *Soc. Neurosci. Abstr.* 22:415.

Wichmann,, T., Bergman, H., Starr, P.A., Subramanian, T., Watts, R.L., and DeLong, M.R., 1999, Comparison of MPTP-induced changes in spontaneous neuronal discharge in the internal pallidal segment and in the substantia nigra pars reticulata in primates, *Exp. Brain Res.* 125:397.

Wilson, C.J., Chang, H.T., and Kitai, S.T., 1983, Origins of post synaptic potentials evoked in spiny neostriatal projection neurons by thalamic stimulation in the rat, *Exp Brain Res* 51:217.

Wurtz, R.H.. and Hikosaka, O., 1986, Role of the basal ganglia in the initiation of saccadic eye movements, *Progre. Brain Res.* 64:175.

Zijlmans, J.C., Pasman, J.W., Horstink, M.W., Stegeman, D.F., Van't Hof, M.A., Poortvliet, D.J., Notermans, S.L., and Jonkman, E.J., 1998, EEG findings in patients with vascular parkinsonism, *Acta Neurol. Scandin.* 98:243.

PART II. ANATOMICAL AND FUNCTIONAL ORGANIZATION OF BASAL GANGLIA AND MOTOR THALAMIC CIRCUITS

MICROCIRCUITS OF THE STRIATUM

J. PAUL BOLAM AND MARK D. BEVAN[*]

INTRODUCTION

The striatum is the major division of the basal ganglia, a group of subcortical nuclei involved in a variety of processes including motor, associative, cognitive and mnemonic functions. The dorsal division of the basal ganglia consists of the striatum, the globus pallidus (GP, external segment of the globus pallidus, or GPe, in primates), entopeduncular nucleus (EP, internal segment of globus pallidus, or GPi, in primates), the subthalamic nucleus (STN), and the substantia nigra (SN). The SN is divided into two main parts, the dorsal pars compacta (SNC), in which the dopaminergic nigrostriatal neurons reside, and the more ventral pars reticulata (SNR). The ventral division of the basal ganglia, which is primarily associated with limbic functions, consists of the ventral striatum or nucleus accumbens, ventral pallidum, and ventral tegemental area.

The major input to the basal ganglia is derived from the cortex; virtually the whole of the cortical mantle projects onto the basal ganglia in a highly topographical manner. The main point of entry of this cortical information to the basal ganglia is the striatum, although there are significant cortical projections to the STN. The corticostriatal projection imparts functionality on to the striatum and consequently other divisions of the basal ganglia. The functional organization is such that cortical information carried by the corticostriatal projection is processed within the striatum, integrated with the many other inputs to the basal ganglia (e.g. intralaminar thalamic nuclei, amygdala, hippocampus, dorsal raphe), which primarily innervate the striatum, and then the "processed information" is transmitted to the output nuclei of the basal ganglia, the EP or GPi, and the SNR (Albin et al., 1989; DeLong, 1990; Smith et al., 1998). The basal ganglia then influences behavior by these output nuclei projecting to the ventral thalamus and then back to the cortex or by projecting to subcortical "premotor" regions including the superior colliculus, the pedunculopontine nucleus, or the reticular formation (Figure 1) (see Albin et al., 1989; DeLong, 1990; Bolam and Bennett, 1995; Gerfen and Wilson, 1996; Smith et al., 1998; Bolam et al., 2000, for recent reviews).

The transmission of cortical information through the basal ganglia occurs through two routes, the so-called "direct" and "indirect" pathways (Albin et al., 1989; DeLong, 1990). In the *direct pathway* corticostriatal information is transmitted directly from the striatum to the output nuclei. In the *indirect pathway* corticostriatal information is transmitted indirectly to

[*] J. PAUL BOLAM AND MARK D. BEVAN • MRC Anatomical Neuropharmacology Unit, Department of Pharmacology, Mansfield Road, Oxford, U.K.

the output nuclei via the complex network interconnecting the GP and STN (Shink et al., 1996) (Figure 1). Data from a variety of disciplines, but particularly the pioneering work of Deniau and colleagues (Chevalier and Deniau, 1990), Hikosaka and Wurtz (1983a,b,c,d), Albin et al. (1989), and DeLong (1990) has shown that the output signal of the basal ganglia under resting conditions is one of inhibition, and that there is a loss of inhibition or disinhibition during a basal ganglia associated behaviour. This is brought about by the neurochemical nature of neurons in the pathways and their basal activity. Striatal projection neurons are GABAergic and quiescent under resting conditions, while basal ganglia output neurons are also GABAergic but have a high discharge rate, tonically inhibiting the targets of the basal ganglia, i.e. neurons in the ventral thalamus or subcortical premotor regions (Kawaguchi, 1993; Wilson, 1993). When the system is activated by the firing of corticostriatal glutamatergic neurons, striatal neurons discharge, which in turn causes inhibition of basal ganglia output neurons in the SNR and EP. This reduction in firing of basal ganglia output neurons leads to a release from inhibition, or disinhibition, of neurons in the targets of the basal ganglia and is associated with "basal ganglia behaviour". Activation of the indirect pathway(s) leads to the opposite effect, i.e. increased firing of output neurons and increased inhibition of basal ganglia targets. It is possible that tonic activity of STN neurons is the driving force for the resting activity in basal ganglia output nuclei (Nakanishi et al., 1987; Bevan and Wilson, 1999). During basal ganglia associated behaviors, the output is a complex spatiotemporal pattern of increased and decreased firing i.e. inhibition and disinhibition, and it has been suggested that while the direct pathway 'selects' a behaviour, the indirect pathway acts to attenuate or terminate a basal ganglia movement or to suppress unwanted sequences of movement (Mink and Thach, 1993).

There are many feedback pathways within the basal ganglia, the major one of which is the dopaminergic projection from the SNC to the striatum. This projection modulates the flow of cortical information through the basal ganglia. Loss of these neurons in Parkinson's disease leads to an imbalance of the flow of cortical information through the basal ganglia in favor of the indirect pathway and hence the akinetic behaviour associated with this disorder (Albin et al., 1989; DeLong, 1990).

In the following sections we will briefly review the microcircuitry of the striatum with particular emphasis on two classes of neurons, the spiny projection neurons and GABA interneurons. For detailed discussions of the anatomical and functional organization of the basal ganglia see reviews by Gerfen and Wilson (1996) and Smith et al. (1998).

STRIATAL NEURONS

The striatum contains both projection neurons and several populations of interneurons (Bolam and Bennett, 1995; Kawaguchi et al., 1995; Kawaguchi, 1997). The major type of projection neuron is the medium size densely spiny neuron (spiny neuron), which accounts for 90-95% of the total population of striatal neurons (Kemp and Powell, 1971b). These neurons are of medium size and their secondary dendrites are densely laden with spines. They utilize GABA as their major neurotransmitter and are subdivided into two major populations on the basis of their projection region, pattern of axonal collateralisation, and neurochemical content (for reviews see Smith and Bolam, 1990; Bolam and Bennett, 1995; Gerfen and Wilson, 1996; Kawaguchi, 1997; Smith et al., 1998). One sub-population projects preferentially, but not exclusively, to the output nuclei of the basal ganglia and expresses, in addition to GABA, the neuropeptides substance P and dynorphin and the D1 subtype of dopamine receptors. The second subpopulation projects almost exclusively to the GP and expresses enkephalin and the D2 subtype of dopamine receptors. They give rise to extensive local axon collaterals, one of the major synaptic targets of which are other spiny neurons (Pickel et al., 1980; Wilson and Groves, 1980; Aronin et al., 1981; Somogyi et al.,

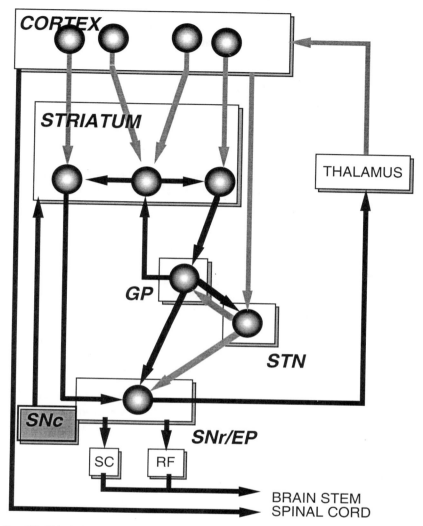

Figure 1. Simplified block diagram of the circuitry of the basal ganglia. Inhibitory projections are shown in black, excitatory projections in grey. Cortical information that reaches the striatum is conveyed to the basal ganglia output structures (EP/SNr, entopeduncular nucleus/substantia nigra pars reticulata) via two pathways, a direct inhibitory projection from the striatum to EP/SNr and an indirect pathway, which involves an inhibitory projection from the striatum to globus pallidus (GP), an inhibitory projection from the GP to the subthalamic nucleus (STN) and to the output nuclei, and an excitatory projection from the STN to EP/SNr. The information is then transmitted back to the cerebral cortex via the thalamus or conveyed to various brain stem structures including the superior colliculus (SC) and the parvicellular reticular formation (RF). Dopaminergic neurons of the SNc provide a massive feedback projection to the striatum and modulate the flow of cortical information. A proportion of GP neurons also feedback to the striatum where they innervate interneurons which also receive cortical input. Cortical information also reaches the basal ganglia via the corticosubthalamic projection.

et al., 1981; DiFiglia et al., 1982; Somogyi et al., 1982; Bolam et al., 1983b; Bouyer et al., 1984a; Bolam and Izzo, 1988; Pickel et al., 1992; Yung et al., 1996). Spiny neurons located in the matrix division of the striatum express the calcium binding protein calbindin, whereas those located in the patches or striosomes do not (Gerfen and Wilson, 1996).

Electron microscopic analysis of spiny neurons reveals that they receive afferent synaptic input from terminals that form symmetrical and asymmetrical synaptic specializations. Terminals forming asymmetric synapses are primarily localized at the heads of dendritic spines, whereas those forming symmetrical synapses are more sparse and are located on the necks of spines, on dendritic shafts, and perikarya. Anterograde degeneration and tracing studies have demonstrated that corticostriatal terminals form a large proportion of the asymmetrical synapses located on the heads of dendritic spines of the spiny projection neurons (Kemp and Powell, 1971a,c; Frotscher et al., 1981; Somogyi et al., 1981; Dubé et al., 1988; Hersch et al., 1995). Wilson and colleagues (Kincaid et al., 1998) have demonstrated that an individual cortical neuron makes very few synaptic contacts with an individual striatal neuron, that there is a high degree of convergence of corticostriatal neurons onto individual striatal neurons but that close neighbours do not share common cortical inputs. The activation of corticostriatal neurons leads to the release of glutamate, activation of both AMPA and NMDA receptors that are localized almost exclusively within the synapse (Bernard et al., 1997; Bernard and Bolam, 1998), which leads in turn to depolarization of the neuron (Wilson, 1993; Kita, 1996); a volley of action potentials follows if there is sufficient convergent excitatory input to an individual spiny neuron (Wilson, 1993; Stern et al., 1997, 1998). Spiny neurons also receive excitatory input from terminals derived from the intralaminar thalamic nuclei (Dubé et al., 1988; Xu et al., 1991; Sidibé and Smith, 1996), and there is some indication, in primates at least, that there is a preferential innervation of neurons of the indirect pathway by thalamic afferents (Sidibé and Smith, 1996).

The excitatory cortical input to spiny neurons is modulated by the many other inputs to spiny neurons, including those from extrinsic sources and from local interneurons (see Smith and Bolam, 1990; Bolam and Bennett, 1995; Kawaguchi, 1997). Of particular importance is the input from the dopaminergic terminals derived from the SNC, which degenerate in Parkinson's disease. These terminals form symmetrical synaptic contacts, mainly with the necks of dendritic spines of spiny projection neurons (Bouyer et al., 1984b; Freund et al., 1984; Smith et al., 1994; Hanley and Bolam, 1997). The heads of spines that receive the dopaminergic input invariably receive input from terminals forming asymmetrical synapses (Freund et al., 1984), which are generally derived from the cortex (Bouyer et al., 1984b; Smith et al., 1994). This anatomical arrangement is ideally suited for the dopamine released from the nigrostriatal terminal, which is likely to act upon dopamine receptors localized both within the synapse and at extrasynaptic sites (Yung et al., 1995) to very selectively modulate the response to the excitatory input at the head of the spine. Other inputs to spiny neurons, e.g. cholinergic input, exhibit a similar anatomical organization (Bolam and Izzo, 1988; Pickel and Chan, 1990), and GABAergic terminals are also observed in contact with the necks of spines (Bolam et al., 1985). Synaptic input from GABA interneurons and from other spiny neurons occurs primarily in the proximal regions of the neurons.

The major classes of interneurons in the striatum include cholinergic interneurons, a population of neurons that express neuropeptide Y, somatostatin, and nitric oxide synthase (NOS), GABAergic interneurons which consist of those that express parvalbumin (PV) (see below) and those that express calretinin, as well as various minor populations of neurons that express neuropeptides (VIP, CCK, SP) (Bolam and Bennett, 1995; Kawaguchi et al., 1995; Gerfen and Wilson, 1996; Kawaguchi, 1997; Bolam et al., 2000). Although interneurons account for a small proportion of striatal neurons, they possess dense axonal arbors by which they profoundly influence the processing and flow of cortical information through the striatum.

GABA INTERNEURONS IN THE STRIATUM

GABAergic interneurons were originally identified on the basis of the selective uptake of exogenous [³H]GABA (Iversen and Schon, 1973; Bolam et al., 1983a) and glutamate decarboxylase immunocytochemistry (Ribak et al., 1979; Bolam et al., 1985; Kita and Kitai, 1988). There are several sub-types (Kubota et al., 1993; Kawaguchi, 1997), however the largest populations are characterized by the presence of PV (Cowan et al., 1990; Kita et al., 1990; Kita, 1993). They account for only a small proportion of the total population of striatal neurons, possess smooth, spine-free dendrites, have the ability to fire at high rates (Kawaguchi, 1993), and are interconnected by gap junctions (Kita, 1993; Koos and Tepper, 1999). Electron microscopic analysis has revealed that these cells receive synaptic input from many terminals that form asymmetric synaptic specializations (Bolam et al., 1983a; 1985; Kita et al., 1990), and anterograde tracing studies have demonstrated that one of their major inputs is from the cortex (Lapper et al., 1992; Bennett and Bolam, 1994). As with the cortical input to spiny neurons, the input to GABA interneurons is associated with AMPA receptors localized within the synaptic specialization (Bernard et al., 1997). It is interesting to note that following cortical stimulation these neurons increase expression of Fos over a larger area of striatum than do spiny neurons (Parthasarathy and Graybiel, 1997). Furthermore, these neurons may subserve some role in the integration of cortical afferents from different functional territories, as preliminary double anterograde tracing studies suggest that an individual neuron may receive input from both motor and sensory regions of cortex (Hanley et al., 1999; and unpublished observations). Indeed, in individual sections up to 25% of all PV-positive interneurons were found to be apposed by boutons derived from both motor and sensory regions of cortex. Furthermore, up to 50% of those interneurons that were identified as receiving cortical input were apposed by terminals from both the motor and sensory regions of cortex.

The major synaptic target of the GABA interneurons are spiny output neurons (Kita et al., 1990; Bennett and Bolam, 1994). Individual spiny neurons receive a basket-like innervation around their perikarya from parvalbumin-positive terminals, the majority of which are likely to be derived from the parvalbumin-positive GABA interneurons. On the basis of this anatomical organization and the failure to detect inhibitory signals mediated by the collaterals of spiny neurons (Jaeger et al., 1994) despite the presence of synapses (Wilson and Groves, 1980; Somogyi et al., 1981; Yung et al., 1996), it has been proposed that the GABA interneurons are the principal mediators of lateral inhibition in the striatum and provide a feed-forward inhibition of cortical information to spiny neurons (Pennartz and Kitai, 1991; Jaeger et al., 1994; Plenz and Kitai, 1998). Indeed, paired recordings *in vitro* have shown that GABA interneurons produce large unitary IPSPs in spiny neurons (Koos and Tepper, 1999). Although the precise role of this inhibition is unknown, there are several possibilities. First, the effect may simply be to shunt coincident cortical excitation and limit the duration of excitation (Pennartz and Kitai, 1991; Plenz and Kitai, 1998). Second, depending on their pattern of connectivity, the interneurons may underlie surround inhibition, thereby focusing cortical excitation (Parthasarathy and Graybiel, 1997). Third, by their dense local axonal arbors they might synchronize sub- and suprathreshold activity of groups of neighboring spiny output neurons (Plenz and Kitai, 1998; Koos and Tepper, 1999). It is thus evident that the PV-positive GABA interneurons, despite their relatively low numbers, are in a position to powerfully control the activity of spiny neurons and hence the output of the striatum.

NEURONS OF THE GLOBUS PALLIDUS

It has recently been demonstrated that the main synaptic target of spiny neurons that give rise to the indirect pathway, GABAergic neurons of the GP, are in a position to powerfully influence activity of the microcircuits of the striatum (Bevan et al., 1998). Neurons of the GP project to the STN, the output nuclei of the basal ganglia, and the SNC. The results of tracing studies at the electron microscopic level, combined with post-embedding immunolabeling for glutamate and GABA, suggest that *individual* neurons of the globus pallidus innervate the STN and output structures of the basal ganglia (see Bolam et al., 1993; Smith et al., 1998). Tracing and physiological studies have also indicated that the globus pallidus, in addition, provides a feedback to the striatum (for references see Bevan et al., 1998). Single cell filling studies have confirmed these suggestions (Kita and Kitai, 1994; Bevan et al., 1998; Sato et al., 2000). In the rat, all pallidal neurons give rise to local axon collaterals within the GP and collateral projections to the STN, EP, and SN. About a quarter of pallidal neurons also give rise to collateral projections to the striatum (Kita and Kitai, 1994; Bevan et al., 1998). On average, pallidostriatal neurons give rise to approximately 790 boutons within the striatum. From the known number of neurons in the striatum and globus pallidus (2.79×10^6 and 4.6×10^4 respectively) (Oorschot, 1996) and the proportion of pallidal neurons giving rise to striatal collaterals, it can be calculated that on average a striatal neuron will receive input from 3.3 pallidal boutons in total. It is unlikely that such a small number of synapses in a projection can impart significant information on, or significantly affect, the function of the striatum. However, the combination of single-cell filling with immunolabelling for sub-populations of striatal neurons has revealed that pallidostriatal axons *selectively* innervate striatal interneurons (Bevan et al., 1998). Up to 66% (mean±sd: 43.7±17.8) of the striatal terminals of an individual pallidostriatal neuron make contact with PV-positive GABA interneurons. On average, an individual PV neuron receives 7 boutons from an individual pallidal axon, and these make contact primarily in the proximal regions of the neurons. A quantitative model of the connectivity between pallidal neurons and GABA interneurons (Bolam et al., 2000), suggests that each GABA interneuron receives input form seven pallidal neurons that give rise to a total of approximately 49 synaptic boutons. In addition, 3-32% of terminals of a single pallidal neuron make contact with NOS interneurons. The synaptic target of the remainder of the pallidal boutons is at present unknown.

Thus, despite the relatively small number of neurons and boutons that comprise the pallidostriatal pathway, pallidostriatal neurons are in a position to powerfully control the activity of the striatum by selective innervation of GABA interneurons, which in turn control the activity of the output neurons of the striatum. Since GP neurons receive monosynaptic and/or rapid disynaptic activation (via the STN) (Tremblay and Filion, 1989; Ryan and Clark, 1991; Kita, 1992; Naito and Kita, 1994; Plenz and Kitai, 1998) following cortical activation, they are well placed to modulate the cortical activation of GABA interneurons (Pennartz and Kitai, 1991; Plenz and Kitai, 1998) through shunting of coincidental cortical excitatory postsynaptic potentials and/or phase-lock or prevent action potential generation (c.f. Pennartz and Kitai, 1991; Cobb et al., 1995). The same pallidostriatal neurons that innervate PV-positive GABA interneurons also provide a major input to NOS interneurons, which themselves are likely to regulate striatal activity through the release of GABA (Kubota et al., 1993), nitric oxide (Hanbauer et al., 1992; Guevara-Guzman et al., 1994; Lonart and Johnson, 1994; Stewart et al., 1996), and neuropeptides (Radke et al., 1993).

CONCLUSIONS

Given the current knowledge of striatal microcircuitry, it is possible to speculate on the roles that GP neurons may play during the cortical activation that is associated with basal ganglia-mediated behaviour. The essential feature of the microcircuit is that GABAergic neurons of the GP selectively innervate striatal GABAergic interneurons which, in turn, selectively innervate the spiny output neurons. Thus, GP neurons are in a position to control the activity of selected groups of GABA interneurons, which will control the activity of selected groups of spiny output neurons.

Under resting conditions the spiny projection neurons and GABA interneurons are hyperpolarized and thus inactive. In contrast, GP neurons are tonically active due to intrinsic membrane properties and excitatory synaptic drive from the STN. Thus, under resting conditions, GABA interneurons may be further inhibited by tonic, GABAergic synaptic input from the GP. This would ensure that only the most synchronous and powerful patterns of cortical input would lead to activation of striatal GABA interneurons.

During basal ganglia associated behavior, activation of corticostriatal neurons will lead to co-activation of spiny neurons and GABA interneurons by virtue of their direct cortical input, and activation of GP neurons directly from the corticopallidal pathway and/or indirectly via the corticosubthalamic pathway. The role of GP neurons will thus be dependent on the timing of their activity relative to the activation of striatal neurons and the spatial arrangement of their axonal arbors within the striatum. If we assume that there is simultaneous activation of spiny neurons, GABA interneurons, and neurons in the GP, then the pallidostriatal projection may limit cortical activation of GABA interneurons by shunting the effects of excitatory cortical potentials. If there is not precise co-activation of striatal and GP neurons but activation of GP neurons follows that of striatal neurons, then the activity of GP neurons may be important in terminating cortical activation of interneurons. Thus, GP neurons may lead to the disinhibition of spiny neurons. Activation of GP neurons may have more complex roles. As pallidostriatal neurons also innervate all other nuclei of the basal ganglia, they may play a role in synchronizing oscillatory activity in the striatum relative to downstream basal ganglia nuclei. Similarly, since their activity is ultimately influenced by spiny neurons, whose activity they indirectly affect within the striatum, they may synchronize oscillatory activity in the striatum-GP-striatum loop.

Alternatively or in addition to these possible roles in the temporal patterning of activity, GP neurons may play a role in the spatial patterning of the activity in the microcircuits. Activated GP neurons may selectively target groups of GABA interneurons leading to selective disinhibition of groups of spiny neurons and thus selection of appropriate motor programs. Whatever the roles of the GP neurons, it is clear that the GABA interneurons play a critical role in the striatum-GP-striatum loop, and the precise identity of their targets is thus crucial for understanding the functional role of this microcircuit.

It is thus apparent that neurons of the GP not only play a role in the expression of the activity of the indirect network by disinhibiting STN and basal ganglia output neurons but also play a role in patterning the activity of populations of striatal interneurons and hence spiny output neurons.

ACKNOWLEDGMENTS: The authors' work described in this review was supported by the Medical Research Council UK; the Wellcome Trust (MDB) and The European Community (BIOMED 2 Project: BMH4-CT-97-2215). The authors would like to thank Caroline Francis, Liz Norman and Paul Jays for technical assistance.

REFERENCES

Albin, R.L., Young, A.B., and Penney, J.B., 1989, The functional anatomy of basal ganglia disorders, *Trends Neurosci.* 12:366.

Aronin, N., DiFiglia, M., Liotta, A.S., and Martin, J.B., 1981, Ultrastructural localization and biochemical features of immunoreactive leu-enkephalin in monkey dorsal horn, *J. Neurosci.* 1:561.

Bennett, B.D., and Bolam, J.P., 1994, Synaptic input and output of parvalbumin-immunoreactive neurones in the neostriatum of the rat, *Neurosci.* 62:707.

Bernard, V., and Bolam, J.P., 1998, Subcellular and subsynaptic distribution of the NR1 subunit of the NMDA receptor in the neostriatum and globus pallidus of the rat: co-localization at synapses with the GluR2/3 subunit of the AMPA receptor, *Eur. J. Neurosci.* 10:3721.

Bernard, V., Somogyi, P., and Bolam, J.P., 1997, Cellular, subcellular, and subsynaptic distribution of AMPA-type glutamate receptor subunits in the neostriatum of the rat, *J. Neurosci.* 17:819.

Bevan, M.D., Booth, P.A.C., Eaton, S.A., and Bolam, J.P., 1998, Selective innervation of neostriatal interneurons by a subclass of neuron in the globus pallidus of the rat, *J. Neurosci.* 18:9438.

Bevan, M.D., and Wilson, C.J., 1999, Mechanisms underlying spontaneous oscillation and rhythmic firing in rat subthalamic neurons, *J. Neurosci.* 19:7617.

Bolam, J.P. and Bennett, B., 1995, The microcircuitry of the neostriatum, in: *Molecular and cellular mechanisms of neostriatal functions*, M. Ariano, and D.J. Surmeier, eds., R.G. Landes Company, Austin.

Bolam, J.P., Booth, P.A.C., Hanley, J.J., and Bevan, M.D., 2000, Synaptic organization of the basal ganglia, *J. Anatomy*, 196:527.

Bolam, J.P., Clarke, D.J., Smith, A.D., and Somogyi, P., 1983a, A type of aspiny neuron in the rat neostriatum accumulates (^{3}H) γ-aminobutyric acid: combination of Golgi-staining, autoradiography and electron microscopy, *J. Comp. Neurol.* 213:121.

Bolam, J.P., and Izzo, P.N., 1988, The postsynaptic targets of substance P-immunoreactive terminals in the rat neostriatum with particular reference to identified spiny striatonigral neurons, *Expl. Brain Res.* 70:361.

Bolam, J.P., Powell, J.F., Wu, J.-Y., and Smith, A.D., 1985, Glutamate decarboxylase-immunoreactive structures in the rat neostriatum. A correlated light and electron microscopic study including a combination of Golgi-impregnation with immunocytochemistry, *J. Comp. Neurol.* 237:1.

Bolam, J.P., Smith, Y., Ingham, C.A., von-Krosigk, M., and Smith, A.D., 1993, Convergence of synaptic terminals from the striatum and the globus pallidus onto single neurones in the substantia nigra and the entopeduncular nucleus, in: *Chemical Signaling in the Basal Ganglia. Progress in Brain Research 99*, G.W. Arbuthnott, and P.C. Emson, eds., Elsevier, Oxford.

Bolam, J.P., Somogyi, P., Takagi, H., Fodor, I., and Smith, A.D., 1983b, Localization of substance P-like immmunoreactivity in neurons and nerve terminals in the neostriatum of the rat: a correlated light and electron microscopic study, *J. Neurocytol.* 12:325.

Bouyer, J.J., Miller, R.J., and Pickel. V.M., 1984a, Ultrastructural relation between cortical efferents and terminals containing enkephalin-like immunoreactivity in rat neostriatum, *Reg. Peptides*, 8:105.

Bouyer, J.J., Park, D.H., Joh, T.H., and Pickel, V.M., 1984b, Chemical and structural analysis of the relation between cortical inputs and tyrosine hydroxylase-containing terminals in rat neostriatum, *Brain Res.* 302:267.

Chevalier, G., and Deniau, J.M., 1990, Disinhibition as a basic process in the expression of striatal functions, *Trends Neurosci.* 13:277.

Cobb, S.R., Buhl, E.H., Halasy, K., Paulsen, O., and Somogyi, P., 1995, Synchronization of neuronal activity in hippocampus by individual GABAergic interneurons, *Nature,* 378:75.

Cowan, R.L., Wilson, C.J., Emson, P.C., and Heizmann, C.W., 1990, Parvalbumin-containing GABAergic interneurons in the rat neostriatum, *J. Comp. Neurol.* 302:197.

DeLong, M.R., 1990, Primate models of movement disorders of basal ganglia origin, *Trends Neurosci.* 13:281.

DiFiglia, M., Aronin, N., and Martin, J.B., 1982, Light and electron microscopic localization of immunoreactive leu-enkephalin in the monkey basal ganglia, *J. Neurosci.* 2:303.

Dubé, L., Smith, A.D., and Bolam, J.P., 1988, Identification of synaptic terminals of thalamic or cortical origin in contact with distinct medium size spiny neurons in the rat neostriatum, *J. Comp. Neurol.* 267:455.

Freund, T.F., Powell, J., and Smith, A.D., 1984, Tyrosine hydroxylase-immunoreactive boutons in synaptic contact with identified striatonigral neurons, with particular reference to dendritic spines, *Neurosci.* 13:1189.

Frotscher, M., Rinne, U., Hassler, R., and Wagner, A., 1981, Termination of cortical afferents on identified neurons in the caudate nucleus of the cat. A combined Golgi-EM degeneration study, *Exp. Brain Res.* 41:329.

Gerfen, C.R., and Wilson, C.J., 1996, The basal ganglia, in: *Handbook of Chemical Neuroanatomy, Integrated systems of the CNS, Part III*, A. Björklund, T. Hökfelt, and L. Swanson eds., Elsevier Science, Amsterdam.

Guevara-Guzman, R., Emson, P.C., and Kendrick, K.M., 1994, Modulation of in vivo striatal transmitter release by nitric oxide and cyclic GMP, *J. Neurochem.* 62:807.

Hanbauer, I., Wink, D., Osawa, Y., Edelman, G.M. and Gally, J.A., 1992, Role of nitric oxide in NMDA-evoked release of [^{3}H]-dopamine from striatal slices, *Neuroreport,* 3:409.

Hanley, J.J., and Bolam, J.P., 1997, Synaptology of the nigrostriatal projection in relation to the compartmental organization of the neostriatum in the rat, *Neurosci.* 81:353.

Hanley, J.J., Deniau, J.M., and Bolam, J.P., 1999, Convergence of distinct cortical regions onto GABAergic parvalbumin-positive interneurons in the neostriatum, *Soc Neurosci. Abs.* 25:1923.

Hersch, S.M., Ciliax, B.J., Gutekunst, C.A., Rees, H.D., Heilman, C.J., Yung, K.K.L., Bolam, J.P., Ince, E., Yi, H., and Levey, A.I., 1995, Electron microscopic analysis of D1 and D2 dopamine receptor proteins in the dorsal striatum and their synaptic relationships with motor corticostriatal afferents, *J. Neurosci.* 15:5222.

Hikosaka, O., and Wurtz, R.H., 1983a, Visual and oculomotor functions of monkey substantia nigra pars reticulata. I. Relation of visual and auditory responses to saccades, *J. Neurophysiol.* 49:1230.

Hikosaka, O., and Wurtz, R.H., 1983b, Visual and oculomotor functions of monkey substantia nigra pars reticulata. II. Visual responses related to fixation of gaze, *J . Neurophysiol.* 49:1254.

Hikosaka, O., and Wurtz, R.H., 1983c, Visual and oculomotor functions of monkey substantia nigra pars reticulata. III. Memory-contingent visual and saccade responses, *J . Neurophysiol.* 49:1268.

Hikosaka, O., and Wurtz, R.H., 1983d, Visual and oculomotor functions of monkey substantia nigra pars reticulata. IV. Relation of substantia nigra to superior colliculus, *J . Neurophysiol.* 49:1285.

Iversen, L.L., and Schon, F.E., 1973, The use of autoradiographic techniques for the identification and mapping of transmitter-specific neurones in CNS, in: New Concepts in Neurotransmitter Regulation, Plenum Press , New York-London.

Izzo, P.N., and Bolam, J.P., 1988, Cholinergic synaptic input to different parts of spiny striatonigral neurons in the rat neostriatum, *J. Comp. Neurol.* 269:219.

Jaeger, D., Kita, H., and Wilson, C.J., 1994, Surround inhibition among projection neurons is weak or nonexistent in the rat neostriatum, *J. Neurophysiol.* 72:2555.

Kawaguchi, Y., 1993, Physiological, morphological, and histochemical characterization of three classes of interneurons in rat neostriatum, *J. Neurosci.* 13:4908.

Kawaguchi, Y., 1997, Neostriatal cell subtypes and their functional roles, Neurosci. Res. 27:1.

Kawaguchi, Y., Wilson, C.J., Augood, S.J., and Emson, P.C., 1995, Striatal interneurones: chemical, physiological and morphological characterization, *Trends Neurosci.* 18:527.

Kemp, J.M., and Powell, T.P.S., 1971a, The site of termination of afferent fibres in the caudate nucleus, *Phil. Trans. R. Soc. Lond.* B262:413.

Kemp, J.M., and Powell, T.P.S., 1971b, The structure of the caudate nucleus of the cat: light and electron microscopy, *Phil. Trans. R. Soc. Lond.* B262:383.

Kemp, J.M., and Powell, T.P.S., 1971c, The termination of fibres from the cerebral cortex and thalamus upon dendritic spines in the caudate nucleus: a study with the Golgi method, *Phil. Trans. R. Soc. Lond.* B262:429.

Kincaid, A.E., Zheng, T., and Wilson, C.J., 1998, Connectivity and convergence of single corticostriatal axons, *J. Neurosci.* 18:4722.

Kita, H., 1992, Responses of globus pallidus neurons to cortical stimulation: intracellular study in the rat, *Brain Res.* 589:84.

Kita, H., 1993, GABAergic circuits of the striatum, in: *Chemical Signaling in the Basal Ganglia. Progress in Brain Research 99,* G.W. Arbuthnott, and P.C. Emson, eds., Elsevier, Oxford.

Kita, H., 1996, Glutamatergic and GABAergic postsynaptic responses of striatal spiny neurons to intrastriatal and cortical stimulation recorded in slice preparations, *Neurosci.* 70:925.

Kita, H., and Kitai, S.T., 1988, Glutamate decarboxylase immunoreactive neurons in rat neostriatum: their morphological types and populations, *Brain Res.* 447:346.

Kita, H., and Kitai, S.T., 1994, The morphology of globus pallidus projection neurons in the rat: an intracellular staining study, *Brain Res.* 636:308.

Kita, H., Kosaka, T., and Heizmann, C.W., 1990, Parvalbumin-immunoreactive neurons in the rat neostriatum: a light and electron microscopic study, *Brain Res.* 536:1.

Koos, T., and Tepper, J.M., 1999, Inhibitory control of neostriatal projection neurons by GABAergic interneurons, *Nature Neurosci.* 2:467.

Kubota, Y., Mikawa, S., and Kawaguchi, Y., 1993, Neostriatal GABAergic interneurones contain NOS, calretinin or parvalbumin, *Neuroreport,* 5:205.

Lapper, S.R., Smith, Y., Sadikot, A.F., Parent, A., and Bolam, J.P., 1992, Cortical input to parvalbumin-immunoreactive neurones in the putamen of the squirrel monkey, *Brain Res.* 580:15.

Lonart, G., and Johnson, K.M., 1994, Inhibitory effects of nitric oxide on the uptake of [^{3}H]dopamine and [^{3}H]glutamate by striatal synaptosomes, *J. Neurochem.* 63:2108.

Mink, J.W., and Thach, W.T., 1993, Basal ganglia intrinsic circuits and their role in behavior, *Current Opinion Neurobiol.* 3:950.

Naito, A., and Kita, H., 1994, The cortico-pallidal projection in the rat: an anterograde tracing study with biotinylated dextran amine, *Brain Res.* 653:251.

Nakanishi, H., Kita, H., and Kitai, S.T., 1987, Intracellular study of rat substantia nigra pars reticulata neurons in an in vitro slice preparation: electrical membrane properties and response characteristics to subthalamic stimulation, *Brain Res.* 437:45.

Oorschot, D.E., 1996, Total number of neurons in the neostriatal, pallidal, subthalamic, and substantia nigral nuclei of the rat basal ganglia: a stereological study using the cavalieri and optical disector methods, *J. Comp. Neurol.* 366:580.

Parthasarathy, H.B., and Graybiel, A.M., 1997, Cortically driven immediate-early gene expression reflects modular influence of sensorimotor cortex on identified striatal neurons in the squirrel monkey *J. Neurosci.* 17:2477.

Pennartz, C.M., and Kitai, S.T., 1991, Hippocampal inputs to identified neurons in an in vitro slice preparation of the rat nucleus accumbens: evidence for feed-forward inhibition, *J. Neurosci.* 11:2838.

Pickel, V.M., and Chan, J., 1990, Spiny neurons lacking choline acetyltransferase immunoreactivity are major targets of cholinergic and catecholaminergic terminals in rat striatum, *J. Neurosci. Res.* 25:263.

Pickel, V.M., Chan, J., and Sesack, S.R., 1992, Cellular basis for interactions between catecholaminergic afferents and neurons containing leu-enkephalin-like immunoreactivity in rat caudate-putamen nuclei, *J. Neurosci. Res.* 31:212.

Pickel, V.M., Sumal, K.K., Beckley, S.C., Miller, R.J., and Reis, D.J., 1980, Immunocytochemical localization of enkephalin in the neostriatum of rat brain: a light and electron microscopic study, *J. Comp. Neurol.* 189:721.

Plenz, D. and Kitai, S.T., 1998, Up and down states in striatal medium spiny neurons simultaneously recorded with spontaneous activity in fast-spiking interneurons studied in cortex-striatum-substantia nigra organotypic cultures, *J. Neurosci.* 18:266.

Radke, J.M., Spyraki, C., and Thermos, K., 1993 Neuronal release of somatostatin in the rat striatum - an in vivo microdialysis study, *Neurosci.* 54:493.

Ribak, C.E., Vaughn, J.E., and Roberts, E., 1979, The GABA neurons and their axon terminals in rat corpus striatum as demonstrated by GAD immunocytochemistry, *J. Comp. Neurol.* 187:261.

Ryan, L.J., and Clark, K.B., 1991, The role of the subthalamic nucleus in the response of globus pallidus neurons to stimulation of the prelimbic and agranular frontal cortices in rats, *Expl. Brain Res.* 86:641.

Sato, F., Lavallée, P., Lévesque, M., and Parent, A., 2000, Single-axon tracing study of neurons of the external segment of the globus pallidus in primate, *J. Comp. Neurol.* 417:17.

Shink, E., Bevan, M.D., Bolam, J.P., and Smith, Y., 1996 The subthalamic nucleus and the external pallidum: two tightly interconnected structures that control the output of the basal ganglia in the monkey, *Neurosci.* 73:335.

Sidibe, M., and Smith, Y., 1996, Differential synaptic innervation of striatofugal neurones projecting to the internal or external segments of the globus pallidus by thalamic afferents in the squirrel monkey, *J. Comp. Neurol.* 365:445.

Smith, A.D., and Bolam, J.P., 1990, The neural network of the basal ganglia as revealed by the study of synaptic connections of identified neurones, *Trends Neurosci.* 13:259.

Smith, Y., Bennett, B.D., Bolam, J.P., Parent, A., and Sadikot, A.F., 1994, Synaptic relationships between dopaminergic afferents and cortical or thalamic input in the sensorimotor territory of the striatum in monkey, *J. Comp. Neurol.* 344:1.

Smith, Y., Bevan, M.D., Shink, E., and Bolam, J.P., 1998 Microcircuitry of the direct and indirect pathways of the basal ganglia, *Neurosci.* 86:353.

Somogyi, P., Bolam, J.P., and Smith, A.D., 1981, Monosynaptic cortical input and local axon collaterals of identified striatonigral neurons. A light and electron microscopic study using the Golgi-peroxidase transport-degeneration procedure, *J. Comp. Neurol.* 195:567.

Somogyi, P., Priestley, J.V., Cuello, A.C., Smith, A.D., and Takagi, H., 1982, Synaptic connections of enkephalin-immunoreactive nerve terminals in the neostriatum: a correlated light and electron microscopic study, J. Neurocytol. 11:779.

Stern, E.A., Jaeger, D., and Wilson, C.J., 1998, Membrane potential synchrony of simultaneously recorded striatal spiny neurons *in vivo, Nature,* 394:475.

Stern, E.A., Kincaid, A.E., and Wilson, C.J., 1997, Spontaneous subthreshold membrane potential fluctuations and action potential variability of rat corticostriatal and striatal neurons in vivo, *J. Neurophysiol.* 77:1697.

Stewart, T.L., Michel, A.D., Black, M.D., and Humphrey, P.P.A., 1996, Evidence that nitric oxide causes calcium-independent release of [^{3}H]dopamine from rat striatum in vitro, *J. Neurochem.* 66:131.

Tremblay, L., and Filion, M., 1989, Responses of pallidal neurons to striatal stimulation in intact waking monkeys, *Brain Res.* 498:1.

Wilson, C.J., 1993, The generation of natural firing patterns in neostriatal neurons, in: *Chemical Signaling in the Basal Ganglia. Progress in Brain Research 99,* G.W. Arbuthnott, and P.C. Emson, eds., Elsevier, Oxford.

Wilson, C.J., and Groves, P.M., 1980, Fine structure and synaptic connections of the common spiny neuron of the rat neostriatum: a study employing intracellular injection of horseradish peroxidase, *J. Comp. Neurol.* 194:599.

Xu, Z.C., Wilson, C.J., and Emson, P.C., 1991, Restoration of thalamostriatal projections in rat neostriatal grafts: An electron microscopic analysis, *J. Comp. Neurol.* 303:22.

Yung, K.K.L., Bolam, J.P., Smith, A.D., Hersch, S.M., Ciliax, B.J., and Levey, A.I., 1995, Ultrastructural localization of D1 and D2 dopamine receptors in the basal ganglia of the rat by light and electron microscopic immunocytochemistry, *Neurosci.* 65:709.

Yung, K.K.L., Smith, A.D., Levey, A.I., and Bolam, J.P., 1996, Synaptic connections between spiny neurons of the direct and indirect pathways in the neostriatum of the rat: evidence from dopamine receptor and neuropeptide immunostaining, *Eur. J. Neurosci.* 8:861.

LOCAL AND EFFERENT NEURONS AND INTRINSIC MODULAR ARRANGEMENT OF THE HUMAN STRIATUM

TATIANA A. LEONTOVICH[*]

INTRODUCTION

The goal of this article is to give a short overview of the structural organization of human striatum as revealed in our studies. We will focus specifically on the three aspects of this organization that appear in the title: local circuit neurons, efferent neurons, and intrinsic modules. Although the format of a book chapter excludes the possibility of presenting detailed data, the extreme complexity of the structural organization of human striatum, a complexity exceeding that of any previous models, should become obvious from the description provided below.

LOCAL CIRCUIT (SHORT AXON) NEURONS

Golgi studies of human short axon striatal neurons (Ramon y Cajal, 1911; Leontovich, 1954; Eder et al., 1980; Braak and Braak, 1982; Graveland et al., 1985; Yelnik et al., 1991) include descriptions of a total of about five types of cells, with a different number of types described in each study. Approximately the same five to six cell types have been identified in the monkey striatum. We have studied 21 cases of Golgi-impregnated striatum (see Leontovich, 1969, 1978 for procedure), obtained from patients at autopsy within 6 hours of death, who had not suffered from neurological or psychiatric diseases (9 adults and 12 children's brains). The neurons were drawn using camera-lucida from serial 120 – 130 μm thick sections mounted in canadian balsam. Quantitative analysis of morphological features of neurons was done using nine different morphometric parameters, that were measured from cell drawings (after Leontovich, 1973, 1975, 1978); the data were subjected to statistical analyses.

In both caudate nucleus and putamen we identified total 18 species of short axon neurons (including those described earlier), with each type being present in at least several cases. The neurons were identified as short axon cells based on the branching pattern of their axons. These bifurcated repeatedly into branches of equal thickness until terminating

[*] TATIANA A. LEONTOVICH ● Brain Research Institute of the Russian Academy of Medical Sciences, Pereulok Obukha 5, Moscow 103064, Russia.

as thin beaded branches mainly in close proximity to somata. Each of these cell species displayed a unique set of structural features and parameter values that differed significantly (confirmed by Mann-Whitney nonparametric test) from all other striatal cells including medium spiny neurons. The specific differences were in the soma size and shape, branching pattern, number and character of dendrites (their arrangement, curving, length etc.), existence of spines, etc. The parameter "cell branching" (mean branching of non-cut dendrites multiplied by their total number) was considered as the main discriminating factor. Based on this parameter, all short axon neurons were combined into two groups: sparsely branched and densely branched cells, with the discriminative value of the cell branching around 25. The branching specificity of dendritic tree is critical for the mode of processing of afferent impulses by a neuron and therefore has important functional implications. This grouping of the short axon cells into two subpopulations is in accord with our grouping of long axon neurons of the mammalian brain into sparsely branched and densely branched populations (Leontovich, 1975, 1978), which closely correlates, although does not exactly coincide, with isodendritic or allodendritic and idiodendritic long axon neurons of Ramon-Moliner and Nauta (1966), respectively.

Species of short axon neurons within each of the two subpopulations displayed some common characteristics (Figure 1). Sparsely branched neurons were small or medium sized, had only a few relatively long and straight dendrites with long interbranch segments, and always displayed some spines. Densely branched neurons distributed within a much wider range of sizes (from tiny to giant) and possessed numerous dendrites (except the bipolar cell species), which were short, curved, and free of spines, but displayed quite a few varicosities. These two groups of short axon neurons were found to be significantly different, not only in the cell branching, but also in the maximal and relative radiuses of their dendritic fields and the lengths of their dendritic segments. The values for all these three parameters were higher in sparsely branched cell group.

A detailed description of each of the 18 cell species is currently in preparation for publication. Here we will focus only on some cell species from sparsely and densely branched cell groups not described earlier to provide a general idea of the structural specificity of these subpopulations.

The population of sparsely branched neurons included 10 species. Five of these are illustrated in Figure 1A. These included: (1) tripolar ordinary cells – medium-sized with three straight, long, and extremely poorly branched dendrites; (2) miniature spine-rich cells with several, wavy sparsely branched dendrites loaded with spines (although these cells resembled long axon medium spiny neurons, the density of spines was lower in them and five quantitative parameters were significantly different); (3) double bouquet cells - small-sized with two short, richly branched dendrites; (4) twisted cells – small-sized with 2 –3 dendrites circling around or next to the cell body; (5) bipolar long-dendritic neurons - medium or larger sized with two bunches of long, straight dendrites.

Densely branched neurons included 8 cell species. Four of them are illustrated in Figure 1B. These are: (1) tufted cells - small or medium-sized, displaying several dendrites with long primary segments ending with brushy branches; (2) bipolar densely branched cells – small-sized with two short, richly branched dendrites; (3) large multidendritic cells with extremely numerous (up to 21 in our material) straight, long, and thin dendrites; (4) large, fanlike cells with a few thick dendrites starting from one end of the soma and leaving the other bald end for an axon.

These results demonstrate an extremely rich local circuit neuron population in the human striatum, which is in line with its complex intrinsic compartmentalization and heterogeneous functions including motor as well as other integrative and cognitive aspects.

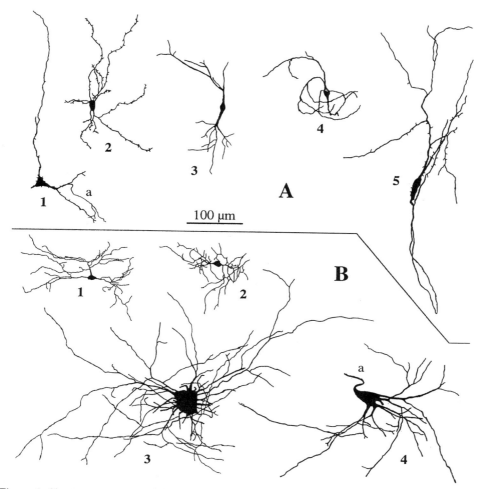

Figure 1. Short axon neurons of the human striatum. (A) Some of the sparsely branched cell species: 1.Tripolar, ordinary. 2. Miniature, spine-rich. 3. Double bouquet. 4. Twisted. 5. Bipolar long-dendritic. (B) Some of the densely branched cell species: 1. Tufted. 2. Bipolar, densely branched. 3. Large, multidendritic. 4. Large, fanlike. Golgi. Camera lucida drawings. a – axon.

NONSPINY EFFERENT (LONG AXON) NEURONS AND THEIR PROJECTIONS

The structure of the main, long axon neurons of striatum – medium spiny cells – has repeatedly been described in mammals and in man, and their long axon nature and projection sites have been analyzed in numerous experimental studies. We shall focus here on other species of efferent striatal neurons, which we named "nonspiny long axon neurons", in contrast to the term "spiny neurons", used for the major medium-sized long axon striatal cell population.

Structure of Nonspiny Long Axon Neurons

Large cells with very long dendrites, different from medium spiny neurons and considered by most authors as long axon cells, have been observed by all investigators studying human striatum with the Golgi method (Ramon y Cajal, 1911; Leontovich, 1954; Eder et al., 1980; Braak and Braak, 1982; Graveland et al., 1985; Yelnik et al., 1991).

We have identified four species of "nonspiny" neurons, which based on their axon type, belong to the population of long axon cells. According to the cell branching parameter (under 25), they belong to the sparsely branched cell category. These cell species are illustrated in Figure 2A and include: (1) medium smooth-dendritic cells; (2) large reticular spiny dendritic cells; (3) large reticular smooth-dendritic cells; (4) cactus-like cells. Our term "reticular" refers to the resemblance of these two cell species to neurons of the reticular formation (Leontovich and Zhukova, 1963; Leontovich, 1975, 1978). In Mann-Whitney test these four cell species differed significantly from each other as well as from medium spiny neurons and all the above-mentioned short axon cells. Except for very specific small cactus-like neurons, the neurons of this group were distinguished from other striatal neurons by their robust appearance. Large reticular cells were mostly multipolar, with several very long, straight, thick and sparsely branched dendrites. The large reticular *smooth-dendritic cells* lacked spines, were very large, and had the longest and thickest dendrites compared to all other striatal cells, whereas large reticular *spiny-dendritic* cells had some spines, shorter dendrites, and were somewhat smaller than the smooth dendritic reticular neurons. Thick, straight axons of these neurons issued some thinner collaterals inside their dendritic field. It seems that due to their long dendrites, these cells should be considered as striatal elements, capable of collecting and integrating influences from a large number of striatal neurons to pass their output to somewhere by their main axons and to neighboring cells by their axon collaterals. Such "intrinsic" integrative function of these cells correlates with primarily symmetrical synapses found on their surface (on the surface of "spiny II cells" of Di Figlia and Carey, 1986, which correspond to our large reticular spiny-dendritic neurons). Symmetrical synapses have been considered to be typical of intrinsic striatal connections, whereas asymmetrical synapses – typical of extrinsic afferents (Hassler et al., 1977). In the above-mentioned Golgi studies, large striatal cells with long, thick dendrites, corresponding to our large reticular cells, were described either as spine free or as possessing spines. It seems that in these studies either one or another large reticular smooth-dendritic and spiny-dendritic cell species was found but not both together.

Projection Sites of Nonspiny Long Axon Neurons

Modern schemes of basal ganglia connections ignore direct strio-cortical pathways. Our previous finding of cortical projections of large reticular neurons in the cat striatum (Mukhina et al., 1986) prompted us to explore such a pathway in the human. For this task we used applications of DiI, retro- and anterograde fluorescent tracer, either to subcortical white matter and stratum subcallosum (to obtain retrograde labeling of presumed strio-cortical cells) or to putamen (to obtain anterograde labeling of postulated strio-insular fibers) of human brains obtained at autopsy (5 cases). The sites of DiI application were sufficiently far from the striatal tissue that stain diffusion into it was not an issue. After 1–2 years of exposure, striatal nuclei were cut on a Vibratome into serial 120 μm thick sections, mounted in glycerol, and examined under fluorescent Zeiss microscope.

Large reticular neurons with their long, sparsely branched, thick dendrites were retrogradely labeled in putamen and caudate nucleus after DiI application to the subcortical white matter dorsal to the putamen (Figure 2B, #1) and anterior to the caudate nucleus, respectively (Figure 2B, # 2). Similar labeled neurons were found in caudate nucleus after DiI application to the stratum subcallosum. We identified about total of 50 labeled cells in our material. In both striatal nuclei, the majority of these cells were spine free, but a few bore some spines. Thus, both species of large reticular cells, spiny- and smooth-dendritic, were labeled. Several medium-sized cells with rather long and straight, spine-free, sparsely branched dendrites were also labeled in the putamen after DiI application to the subcortical white matter. We defined these as medium, smooth-dendritic, long axon cells (Figure 2A, #

3). Although we only found several labeled cells of this type in the human striatum, we suspect that these are also cortex-projecting neurons, since they are sparse in Golgi material as well. Neither medium spiny, long-axon neurons nor any short axon cells were labeled in the examined cases. DiI application to putamen resulted in labeling of scattered thick axons, characteristic of large reticular neurons, coursing in the horizontal plane in the

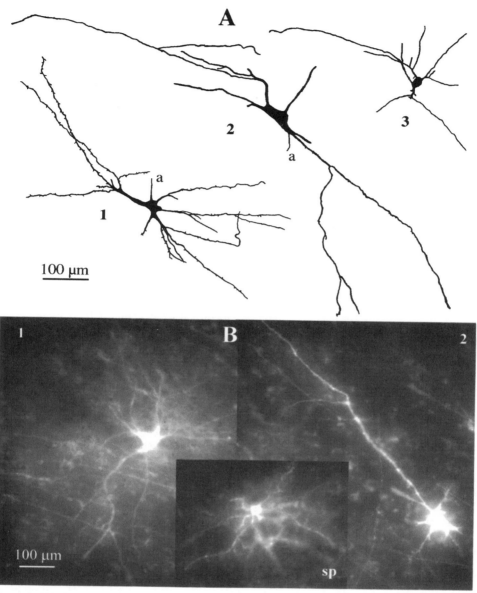

Figure 2. (A) Nonspiny, long axon neurons of human striatum: 1. Large reticular spiny dendritic. 2. Large reticular smoothdendritic. 3. Medium smooth-dendritic. Golgi. Camera lucida drawings. a – axon. (B) Large reticular neurons of putamen (1) and caudate nucleus (2) retrogradely labeled after DiI application to subcortical white matter. Sp. – medium spiny cell (for comparison), retrogradely labeled after DiI application to pallidum. Digital camera.

putamen toward c.externa. Some of these fibers continued their straight course, crossing c.externa, claustrum, and c.extrema at right angles and entering insula. Other fibers joined c.externa and coursed within it dorsally to enter subcortical white matter. These thick fibers never formed bundles but followed a parallel course at an equal distance from one another.

Thus, the results of these experiments provide evidence that both species of large reticular neurons of the human striatum (and perhaps medium smooth-dendritic cells as well) project to the cortex. These data are in accord with the neocortical projections of "large reticular cells" (Mukhina et al., 1986), "large cells with long thick dendrites" (Dimova and Usunoff, 1989), or just "large cells" (Jayaraman, 1980; Maiskyi, 1983) of the cat striatum, revealed by retrograde labeling with horseradish peroxidase. Moreover, Ermolenko et al. (1976) traced a direct caudate-frontal lobe pathway in the monkey with Nauta technique. These studies identified very widely dispersed cortical targets of these cells (in the monkey - from caput of the caudate nucleus to fields 1, 3, 4, 6, 8, 9, 11, 32 and orbital cortex, and in the cat, in sum, to practically all neocortical areas), suggesting involvement of the large reticular striatal neurons not only in motor control, but also (or even primarily?) in other integrative and/or cognitive functions of the striatum.

MODULAR ARRANGEMENT OF STRIATUM AND ITS RELATIONSHIP TO STRIOSOMES

To study the cells and myeloarchitecture of striatum, blocks of the adult human dorsal striatum (10 cases) were cut in different planes into 20 μm thick serial sections (after paraffin embedding), or into 40 μm thick serial Vibratome sections, stained with Luxol-Fast-Blue (Kluver and Barrera, 1953). For striosome definition, some sections from the Vibratome series were immunohistochemically stained for calbindin (4 cases), which is one of the most reliable and widely used markers for striosome-matrix compartments and valid also for human dorsal striatum (Holt et al., 1997). To reveal the relationship of cortical fibers to striatal moduli, DiI was applied to the subcortical white matter in three cases.

Modular Arrangement of Striatum

The brilliant concept of histochemical compartmentalization of mammalian striatum into striosomes ("patches") and matrix contrasts with scanty data on its structural compartments. The search for specific structural units within the monkey striatum revealed "small-celled or dense-celled islands" scattered in a more loosely packed matrix and encircled by cell-poor "rings" in Nissl-stained sections (Goldman-Rakic, 1982; Selemon et al., 1994), and "oval or irregular elongated shapes" in fiber-stained sections (Quinn and Graybiel, 1994). In the dog's caudate nucleus (Leontovich, 1982) and in the dog's and human putamen (Leontovich, 1994), we have described specific structural moduli consisting of small-celled cores and fibrous shells, as seen in Luxol-Fast-Blue stained sections.

Our microscopic analysis of the human striatum, stained for fibers and cells, revealed its very complex structural organization. At first glance the striatal tissue appears to be a homogeneous mixture of radial fiber bundles (RFB) pointing toward pallidum and piercing through a small-celled striatal tissue with scattered single or clustered large cells. But upon more careful examination of serial sections in different planes, our attention was drawn to a peculiar trajectory of RFBs. At certain points RFBs made wide curves, as if gliding by a large, roundish mass of tissue, later to resume the original direction. Very often the large cells were found either in clusters or in rings of different diameters, mostly next to RFBs. A thorough examination of these relationships throughout the entire extent of putamen and the caudate nuclei revealed their regular intrinsic structural organization. This was exemplified

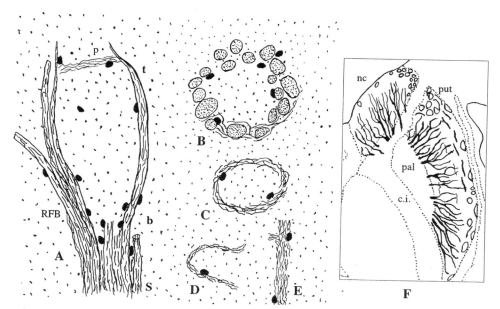

Figure 3. Scheme of the structural elements of the human striatum and their localization. Type I moduli, cut longitudinally (A) and transversely (B) to the course of RFB, that form their shells. b - base, s – stem, t - top of the moduli; p – fibrous transverse partition. (C) Type II module. (D) Curved shapes of thinnest fibers. (E) Fibrous walls. Large pointed shapes – butt-ends of RFBs; scattered dots – small cells; filled shapes – large cells. (F) Scheme of distribution of Type I moduli (area of localization of RFBs, surrounding pallidum), of Type II moduli (open shapes), and of fibrous "walls" (lines) in caudate nucleus (nc) and putamen (put); horizontal section. c.i. – c.interna, pal – pallidum.

by roundish or elongated formations of varying sizes, composed of small-celled central cores surrounded by fibrous shells. We considered these to be specific structural moduli of striatum. Based on the structure of their fibrous shells or, rather, depending on whether their shells were made of RFBs or a thin fiber layer, these formations were grouped as Type I and Type II moduli, respectively (Figure 3A, B and C). The two types of moduli were localized in different regions of striatum (Figure 3F), and in some striatal areas, the moduli were not readily apparent.

Type I moduli occupy the main volume of the inner parts of both striatal nuclei pierced by RFBs (Figure 3F). They are formed by single or a few tightly-packed thick RFBs (modular stem) that at some point give rise to several thinner dispersing bundles (modular base), which encompass a more or less salient tissue column. At some distance from the base, the bundles approach each other again to form a modular top.[*]

The shape of these voluminous moduli resembles the bud of a flower (Figure 3A). The moduli look different depending on whether the sections were cut longitudinally or transversely to the course of RFBs. If cut longitudinally, they may appear as described

---

[*] The words "modular stem, base, or top" or the expression "entrance of RFBs to striatal nuclei" were used only for convenience in describing the modular structure, without any relation to the real "direction" of RFB fibers, which are known to belong mostly to medium spiny cell axons (and thus it would be more correct to say "the exit of RFBs from the striatal nuclei").

above, even in a single or in a few consecutive sections. In sections cut transversely to the course of RFBs, they appear as rings of fiber bundle butt-ends (Figures 3B and 4A). These patterns enabled identification of Type I moduli under the microscope in longitudinal or transverse sections. Since RFBs in the human striatum surround pallidum like spokes in a wheel (Figure 3F), the moduli in these two section planes could only be seen in some sections (or even only in some parts of sections) of the whole series through the striatum in whatever plane it might be cut. In sections cut obliquely to the RFBs the moduli are more difficult to identify.

Identification of Type I moduli in sections cut longitudinally to the RFBs is easier in the caudate nucleus if the RFBs are traced from c.interna, and easier in the putamen if they are traced from the pallidum. The RFBs usually enter these nuclei as a thick single bundle or several tightly packed bundles. Some of the RFBs divide almost immediately into thinner bundles that disperse and surround roundish, more or less protuberant columns of small-celled tissue and continue running parallel to each other in the direction of the nuclear interior. In this way they form a fibrous shell around a cell column, the long axis of which is aligned with the fiber bundles. Some other thick RFBs do not divide right away after entering striatal nuclei, but instead run straight into their interiors to disperse and form moduli there. Some of these RFBs pass within the fibrous shells of the earlier moduli. On the top of the moduli, where RFBs come close to each other again, clearly transverse partitions composed of very thin fibers were observed, connecting the opposing RFBs of the modular shell. These fiber partitions do not cross the bundles that they connect. On the contrary, at these points the bundles curve slightly inside, as if having been tied to the partition (Figures 4C and 5B). Thus, these fiber partitions can not be considered thin passing-by fibers. Rather, they look like branches of RFB fibers. Other partitions observed were made of a layer of thin fiber bundle butt-ends (Figure 4B), or such butt-ends were seen in close proximity to the transverse fiber partitions (Figure 4C). The layers of butt-ends did cross the modular shell. Could they pierce through several moduli and thus integrate their activity? Large cells were seen everywhere next to RFBs and to the partitions in all parts of the moduli. They were most numerous at their bases and very often their somata literally adhered to the bundles (Figures 4A and 5). Also, small cells were always scattered among the bundles of fibrous shells, and single large cells could be found in the small-celled central cores of the moduli and in less structured areas of the striatum. The described structure of Type I modulus is presented schematically in Figure 3.

In the caudate nucleus the majority of longitudinally sectioned Type I moduli, both in the material stained for calbindin (Figure 4B) and in that stained with Luxol-Blue (Figure 4C), displayed very narrow elongated shapes resembling the bud of a lily. The fibrous shells of these moduli were seen to originate from a thick fiber stem issuing from c.interna and dividing into thinner bundles (Figure 4B), which dispersed at sharp angles and ran an almost parallel course to form the modular shell. At the point where they approached the modular "top," both types of partitions could be present: a transverse partition of thinnest fibers (Figure 4C) or a layer of butt-ends of thin fiber bundles (Figure 4B). These partitions could represent a boundary between the two adjacent moduli of similar structure (Figure 4C, adjacent moduli "a" and "b"). One could find three to four subsequent layers of lily-like moduli in between c.interna and the ventricular surface of the caudate nucleus. Some other moduli next to c.interna were much wider and more salient, resembling the bud of a tulip (Figure 4D). Similar tulip-like moduli in the dog's caudate nucleus next to c.interna appeared especially distinct after reconstruction of their shape and position from serial sections (Leontovich, 1983).

The formation and structure of the narrow "lily-like" moduli of the caudate nucleus (as well as of the Type I moduli in general) was also evident in sections cut transversely to the course of their RFBs and, hence, to their long axes (Figure 6). Several thick RFBs were seen to issue from c.interna at right angles, and to enter the caudate nucleus (modular stem) with a cluster of large cells next to them (sections 1, 2). Farther away, these bundles began

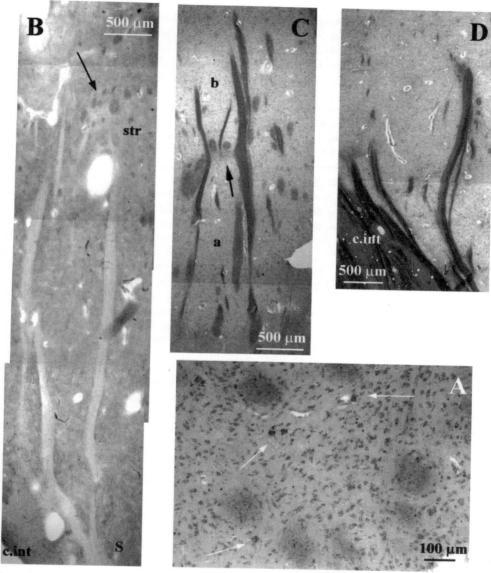

Figure 4. (A) Ring of butt-ends of RFBs, indicative of transversely cut Type I module; arrows show large cells. Caudate nucleus. Luxol-Fast–Blue. Digital camera. (B), (C), (D) Type I moduli of caudate nucleus cut longitudinally to the course of their RFBs: narrow lily–like moduli in a calbindin stained section (B), and in Luxol-Fast-Blue (C); and a tulip-like module in Luxol-Fast-Blue (D). Arrows show fiber partitions within the moduli, made from thinnest fibers, sometimes with single butt-ends of bundles next to it (in C) and only from butt-ends of thin fiber bundles (in B). c.int. – c.interna. S – stem of the moduli. str – striosome. Digital camera.

to disperse, acquiring an ovoid shape and taking different directions, again with clusters of large cells among them (sections 3, 4, modular base); these bundles eventually took a course perpendicular to the section plane and, by dividing into a greater number of bundles, formed two small rings (a, b) of butt-ends of thinner RFBs. These rings maintained almost identical shape in several subsequent sections (5–7) manifesting the modular shell. Some large cells were also seen attached to the butt-ends of the thinner bundles along their entire course (Figure 5A). At the end of the module the outlines of these rings became smooth and were replaced by a group of butt-ends of very thin bundles (section 8, modular top). Under higher magnification it was possible to discern some very thin fibers at these sites that ran in the section plane at right angles to the course of RFBs. Could this be a transverse thin fiber partition, which was so evident in the previously described (Figure 4C) lily-like modulus when sectioned longitudinally?

The images of the two sections illustrated in Figure 6 are critical for discerning Type I moduli under the microscope in any part of the striatum since they are very typical of cross-sectioned Type I moduli, regardless of their size or shape. They also allowed for identification of the specific part of the modulus sectioned. The cluster of oblique dispersing butt-ends of thick bundles with large cells attached points to the modular base (section 3), and the ring of butt-ends of bundles to the modular body (sections 4-7).

In the putamen, the shape and size of Type I moduli seemed to vary more than in the caudate nucleus. As seen in the low power Golgi image of Figure 7A, in sections cut longitudinally to RFBs, very thick bundles entered putamen from the pallidum to form "a column" of four subsequent layers of tulip-like shapes (and a smaller shape next to pallidum) that extended up to c.externa at an equal distance and close to each other, leaving free only a narrow tissue layer along c.externa. In frontal Luxol-Fast-Blue sections through the middle part of the putamen, where the thick RFBs were cut almost longitudinally, five to seven such columns of four subsequent layers of tulip-like moduli could be discerned between pallidum and c.externa, also leaving free a narrow strip of tissue next to it. Figure 7B shows such a column of four layers of almost longitudinally or slightly obliquely cut moduli under higher magnification in a Luxol-Blue slice. These moduli were divided first by a transverse fiber partition (p) and farther on by linear layers of oblique butt-ends of thin fiber bundles. The latter apparently represent the fibrous shells of the next layer of tulip-like moduli. In such a column, the first innermost layer is attached to pallidum and subsequent layers build from these comparatively short and wide moduli of varying shape and size. These moduli are bordered on their sides by thick, straight, passing-by RFBs destined for to c.externa. The outline of the latter was formed by the thinnest RFBs (Figure 7B, # 2-4). In all these moduli, large cells were always found attached to RFBs or situated in close proximity to them. Large cells were especially numerous in the modular base and next to the transverse fiber partitions (Figure 5B). In sections cut transversely to RFBs, for the interior of the putamen, and laterally by transverse thin fiber partitions (Figures 5B, 7B # 1, 8A, B). Sometimes an extensive network of these strange moduli is seen next to the pallidum (Figure 8A,B). Moduli of the outer layers of this column were tulip-like. Each consecutive module was smaller than the previous one, with the smallest being the closest to c.externa. The outline of the latter was formed by the thinnest RFBs (Figure 7B, # 2-4). In all these moduli, large cells were always found attached to RFBs or situated in close proximity to them. Large cells were especially numerous in the modular base and next to the transverse fiber partitions (Figure 5B). In sections cut transversely to RFBs, for Text lost instance, in sagittal sections through the middle area of putamen, numerous rings of butt-ends of RFBs indicative of cross-sectioned Type I moduli were observed. In more lateral sections, these rings were smaller and had thinner butt-ends. In some other sections where RFBs were cut obliquely, several clusters of dispersing butt-ends of thick RFBs were observed. These were distributed in a manner similar to that of the above described tulip-like moduli and were indicative of their bases. In some other parts of putamen, Type I moduli of different size and shape were found, although they were constructed according to the above described principle.

An unexpected feature of the myeloarchitectonics of putamen was noticed in some sections. This consisted of several layers of fibrous partitions or walls, stretching parallel to

c. externa and perpendicular to RFBs through the entire cross-section of the putamen (Figure 8B). The most lateral two layers were undoubtedly the fibrous walls commonly found next to c.externa, and the ones next to the pallidum were apparently transverse fiber partitions of moduli, but what the partitions were in the depth of putamen was unclear. Could they represent fiber partitions of the middle layers of the tulip-like moduli, or were they fibers connecting these moduli with each other, or some afferent fiber systems supplying them?

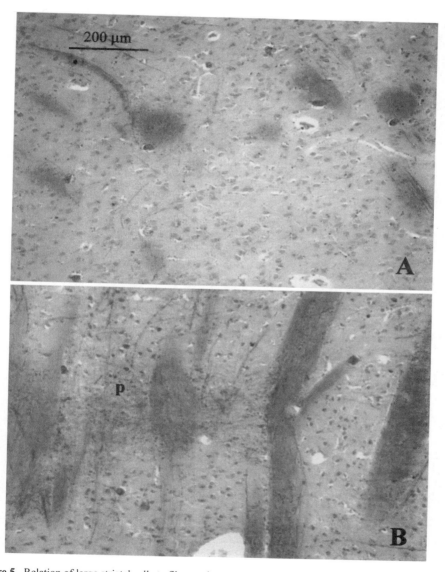

Figure 5. Relation of large striatal cells to fibrous elements: (A) Large cells, clasped to the butt-ends of RFBs in a segment of their ring in transversely cut Type I module; caudate nucleus. (B) Large cells, clasped to RFBs and to transversel fiber partition (p) in longitudinally cut module, presented in Figure 7B; putamen. Luxol-Fast- Blue. Digital camera.

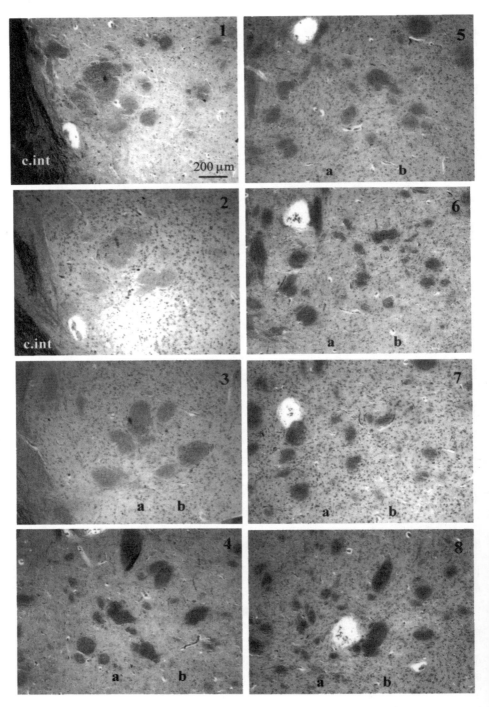

Figure 6. Serial sections of two (a, b) narrow, lily-like Type I moduli of the caudate nucleus, cut transversely to the course of their RFBs. Round or elongated shapes – butt-ends of RFB. c.int. – c.interna. Luxol-Fast-Blue. Digital camera.

Type II moduli in all kinds of sections appeared as roundish shapes with shells made of very thin, almost unmyelinated, fibers that wrapped around small-celled central cores. Single large cells were attached to these shells (Figure 3C). Such moduli were distributed in more lateral parts of striatal nuclei that were free of RFBs and of Type I moduli (Figure3F).

In the caudate nucleus they were usually smaller, and were especially numerous in its dorso-lateral part. In the putamen, they were distributed along the entire extent of c.externa as elongated or disc-like shapes. They were abundant in the anterolateral part of putamen and in its dorsolateral tip, where they displayed especially varied shapes and sizes. These moduli apparently correspond to the "dense cell islands" with their cell-free circles described in monkey striatum (Goldman-Rakic, 1982; Selemon et al., 1994), as well as to the "fiber-defined shapes" of Quinn and Graybiel (1994).

DiI application to subcortical white matter resulted in anterograde labeling of fibers in the shells of Type II moduli, both in the caudate nucleus (especially in the moduli situated between str. subcallosum and c.interna) and in the putamen (next to c.externa). These cortical fibers literally encircled Type II moduli, participating in the formation of their fibrous shells. This correlates with the data of Goldman-Rakic (1982), who traced cortical fibers into the cell-poor capsules of her "cell islands" in the monkey striatum.

Extremely thin, curved, fibro-glial contours, mostly open, embracing regions of small-celled tissue, were found in the caudate nucleus and putamen, mostly in the anterolateral part of the latter and along c.externa. Single large cells could also be found at these contours (Figure 3D). In both striatal nuclei, though mostly in putamen, we have observed numerous linear fibrous "walls," sometimes rather wide and compact, with abundant glial cells. Some of these walls were composed of very thin, very poorly myelinated fibers only, while others were composed with participation of the butt-ends of some thin fiber bundles that crossed this wall at right angles. In the caudate nucleus, these structures were seen close to c.interna. In the putamen, they were numerous and oriented parallel to c.externa in a strip of tissue free of RFBs, extending from the most oral part of this nucleus to its very caudal end and cutting the tissue into one or two layers (Figures 3E, F; 7B and 8B). The lateral region of the anterior putamen was also rich in these walls.

The distribution, specific structure, and size of both types of moduli, as well as of different fibrous "walls," were very specific and were found to be consistent in both striatal nuclei in all specimens of the human caudate nucleus and putamen studied. From case to case, it was possible to predict observation of moduli of a specific shape and size in specific parts of these nuclei. For instance, a very large ovoid Type I modulus was situated along the latero-ventral boundary of the anterior part of pallidum, with its long axis oriented sagittally. Its position corresponded to the "face area" in monkey putamen, established in hodological (Künzle, 1975) and physiological (Alexander and DeLong, 1985) studies.

Relationship of Striosomes to RFB and Structural Moduli

How the structural differentiation of mammalian striatum correlates with its striosomes is not quite clear. In the monkey, the "cell islands" of Selemon et al. (1994) and "fiber defined shapes" of Quinn and Graybiel (1994) correlated with striosomes only partially. In the cat and human putamen, correspondence of striosomes with our Type II moduli was described (Revischin et al., 2000). A thorough examination of serial sections of the human striatum stained alternatively for myelin fibers and striosomes revealed a close relationship between striosomal tissue and fibrous elements, mainly RFBs but also the fibrous walls and transverse fiber partitions. This relationship was evident even in our sections stained for calbindin, where dark brown-stained RFBs were fairly well distinguished from a lighter background and especially sharp in contrast with pale-staining striosomes.

A close relationship of striosomes to fibrous elements was obvious in all four examined brains, both in the caudate nucleus and putamen, especially in sections stained for

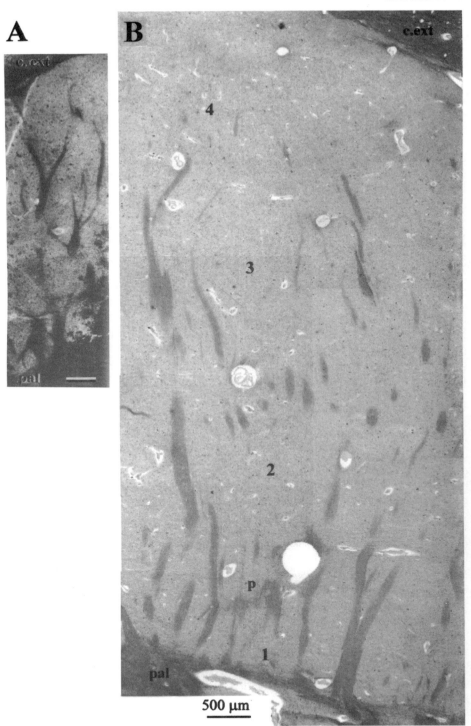

Figure 7. (A) Column of thick stems of RFBs (cut longitudinally), which subsequently divided within putamen, forming tulip-like shapes. Golgi. Photomicrograph. (B) Column of Type I moduli in the putamen (1,2,3,4) cut slightly obliquely to the course of its RFBs. p – transversel fiber partition. c.ext. – c.externa. pal - pallidum. Luxol-Fast–Blue. Digital camera.

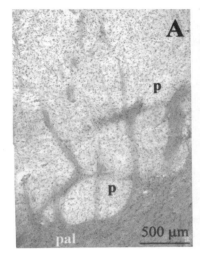

Figure 8. (A) Groups of small moduli with transversel fiber partitions in putamen (p), and next to pallidum (pal). Luxol-Fast–Blue. Digital camera. (B) Several layers of fibrous walls or partitions, stretching within the whole of putamen longitudinally to c.externa (c.ext) and complex moduli next to pallidum. Luxol-Fast–Blue. Photomicrograph.

calbindin and cut transversely to RFBs. In these sections, practically every butt-end of RFBs had some striosomal tissue attached to it. The larger bundles had a larger, and the thinner bundles a smaller, amount of such tissue associated with it. Sometimes the striosomal tissue encircled the bundles, and sometimes it appeared to be attached to the bundles (Figure 9A). A relationship of striosomes to RFBs was also manifested in serial frontal sections through the middle part of putamen, alternatively stained with Luxol-Blue and calbindin immunochemistry, where the bundles were sectioned obliquely. In the calbindin stained sections an unexpectedly strictly ordered distribution of 4-6 columns of striosomes of distinct shape was observed, which stretched from pallidum to c.externa (Figure 10A). If cut longitudinally, these columns contained four layers of striosomes. In other section planes more or less full fragments of striosomes were seen, but ordered in the same way. Each column, cut longitudinally, started from pallidum with a large triangular

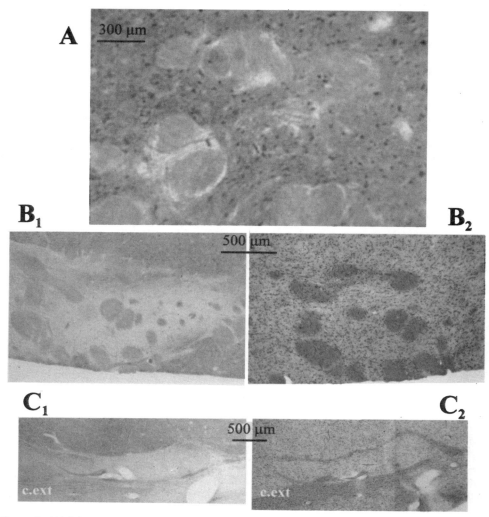

Figure 9. (A) Pale striosomal tissue surrounding the dark roundish butt-ends of RFB or attaching to them as smaller or larger pieces in transversely cut sections stained for calbindin; putamen. (B₁) Roundish striosome with a ring of butt-ends of thick RFBs at its perimeter in a transversely cut section, stained for calbindin; putamen. (B₂) Ring of butt-ends of thick RFBs, indicative of Type I module, from the same location in section adjacent to B₁ Luxol-Blue section. (C₁) Pale striosome in the putamen in a section stained for calbindin. (C₂) Type II module of the same shape in the same spot in adjacent Luxol-Blue section. c.ext. – c.externa. D. cam.

striosome attached with its longer side to pallidum and the apex pointing towards the depth of putamen. Several such striosomes were distributed at some distance from one other along the entire extent of the pallidal border. Obliquely cut, thick RFBs oriented transversely to the long bases of the triangles and to the pallidal border were very distinct inside each of these striosomes, contrasting with their pale background. These first-layer striosomal triangles gave rise to a column of three other layers of similar striosomes, with similarly oriented thick RFBs inside. Each subsequent striosome was smaller and containedthinner fiber bundles, which were still rather thick. The last very small striosome appeared as a wide stripe perpendicular to c.externa, with some thin pale threads and small pale spots next to it. It approached the zone next to c.externa but did not enter it. Thin threads of pale striosomal tissue extended from the apex of the preceding striosome and connected it with the following one. The threads were positioned along the few fiber bundles that extended between the striosomes. In the adjacent section stained with Luxol-Fast-Blue, in that very spot a column of four layers of clusters of obliquely cut RFBs was present (Figure 10B), and each subsequent cluster was also smaller and contained thinner bundles. Comparison of these two columns revealed that the position, shapes, and sizes of striosomes and RFB clusters in each of their four layers were identical (Figure 10A and B).

The described relationships indicate that striosomal tissue accompanies each RFB as a thinner or thicker strip wrapping around them or just attaching to one of their surfaces, whereas the large striosomes are situated in the area of clustering of thick RFBs. These data point also to a specific relationship between striosomes and the structural moduli, specifically Type I moduli, since the RFBs constitute their shells.

Thus, in sections stained for calbindin of putamen and the caudate nuclei with cross-sectioned RFBs, many of their buttends are distributed in a ring-like fashion, suggesting that they belong to the shell of the cross-sectioned Type I module. Since practically all bundles of these rings, even very small ones, had some striosomal tissue attached, it suggests that striosomes participate in formation of the modular shells. In the sections of putamen, several large striosomes were also observed at a great distance from one another, connected subsequently in a column-like fashion by thin strands of the tissue with two much smaller striosomes of irregular shape. Such appearance of the column suggests that it was cut in a plane, close to a transversel one, and that the smaller striosomes could be peripheral parts of larger striosomes. In this column, the first large striosome had an ovoid shape and a ring of thick butt-ends of RFBs incorporated in its periphery (Figure 9B$_1$). In the adjacent Luxol-Blue-stained section, a similar ring of butt ends of thick RFB surrounding a small-celled core was seen in the corresponding spot (Figure 9B$_2$). These rings affirm that the large striosomes are identical to the cross-sectioned Type I moduli, including both their central cores and fibrous shells. The thickness of the cut RFBs suggests that the sectioned part of the module was close to its base. These columns of striosomes seem to be similar to the ones illustrated in Figure 10A, but were cut in a more transverse plane, which enabled observation of the first layer striosome of these columns in a transversel section. The closer the section of putaminal tissue was to the longitudinal section of RFBs, the more regular the striosomal columns looked, with the more regular triangular-shaped striosomes. The correlation in putamen of the striosomes and of Type I moduli is also manifested in the correspondence by position and size of the four layers of striosomes illustrated in Figure 10A to the above mentioned four layers of Type I moduli (Figure 7B).

It is not easy to understand the strictly ordered position and shape of striosomes arranged in columns in putamen (Figure 10A). Most likely these are determined by the relationship between the plane of the section and the position of the four subsequent layers of the tulip-like moduli (with their striosomes inside), which are slightly shifted relative to one another in the horizontal plane (Figure 7A). Because of such a shift in the longitudinal section plane relative to the RFBs (as in Figure 7A), if the first layer striosome was cut in its maximal diameter, then the moduli of each subsequent layer (with its striosomes inside) will be sectioned at a greater distance from its maximal diameter. Thus, the striosomes look smaller in each subsequent layer. The final fourth layer module (and its striosome) may be cut only through its fibrous shell, which appears in the section as a simple pale strip. The triangular or almost semicircular shape of the striosomes could result from the oblique sectioning of their column-like shapes inside the moduli. Certain intervals between striosomes in the columns could correspond to sections of modular areas free of striosomes and/or to sections of intermodular intervals. The thin pale strips of striosomal tissue in these

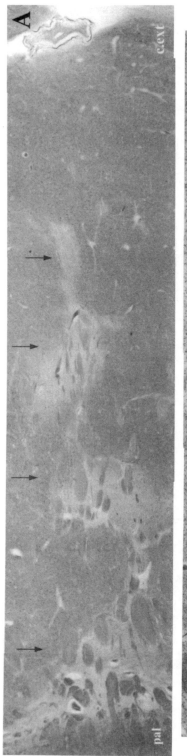

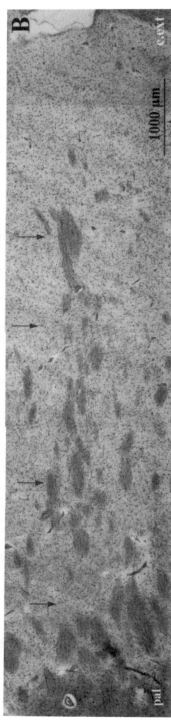

Figure 10. Column of four subsequent layers of striosomes, starting from pallidum in an obliquely cut slice, stained for calbindin (**A**) and a column of four layers of RFB clusters in the same spot of adjacent Luxol- Blue- stained slice (**B**). c.ext. – c.externa. pal. – pallidum. Arrows show the corresponding spots of both columns. Digital camera.

intervals would then correspond to modular fibrous shells and to the rest of the striosomal tissue, which fuses with one of the adjacent moduli.

In the caudate nucleus, the relationships between Type I moduli and striosomes seem to be based on the same principle as in the putamen. In sections stained for calbindin, similar columns of several striosomes with obliquely cut RFBs inside were observed, stretching from c.interna into the depth of the caudate nucleus. In some sections we could distinguish very thin pale threads, connecting consequent striosomes of the same column with one another. But striosomes in the caudate nucleus columns were rounder, much smaller, contained thinner RFBs inside, and were farther away from one another than in theputamen. In a section cut longitudinally to RFBs, it was possible to see such small roundish striosomes within a lily-like Type I module incorporating a linear layer of dark buttends of thin fiber bundles (Figure 5B). The described relationships imply that certain parts of Type I moduli (both their small-celled central cores and attached fibrous shells) are immersed within large striosomes that issue an uninterrupted net of tissue along the fiber bundles of their shell and around fibrous partitions of other parts of the moduli to reach and fuse with another large striosome within the modulus of the next layer. Thus, the modular shell is constructed not only from RFBs, but also from a net of attached striosomal tissue.

In the area where Type II moduli are concentrated, numerous striosomes of different shapes and sizes are found. In caudate nucleus, they were distributed in its dorso-lateral part and next to its ventricular surface. In the putamen, the striosomes were numerous and especially heterogeneous in its dorso-lateral tip and also along c.externa. Some of these striosomes coincided with Type II moduli, apparently occupying both their central small-celled core and fibrous shell, as was evident from comparison of adjacent calbindin and Luxol-Fast-Blue sections (Figure 9C_1 and C_2). These striosomes apparently corresponded to the fiber-defined striosomes of Quinn and Graybiel (1994). In other Type II moduli, striosomal tissue was not seen, though it may be found in case of very precise reconstructions of these moduli from serial sections.

Thus, our data suggest that certain parts of Type I moduli (and at least some of Type II moduli) correspond to striosomal tissue. Other parts of Type I moduli should correspond to matrix tissue. Because of this, the modular structural unit should possess all different histochemical properties, receptor types, and afferent and efferent connections, which were established for these two different histochemical striatal compartments in numerous studies. One may surmise specific functions of the modular unit, which can be realized by using all these possibilities. The revealed close relationships of RFBs with the striosomal tissue, as well as with the large striatal cells, including large efferent reticular cells, make it possible for them to integrate, by their long dendrites, activity of both these histochemical compartments within a single module. The revealed maximal concentration of large cells in the modular base could enable greater control of the input and output of the moduli. Our data suggest that structural moduli of the striatum could be looked upon as complex morpho-functional units, comparable to cortical columns.

ACKNOWLEDGMENTS: This work was supported by a Russian Foundation of Basic Research grant (# 98–04–49550) and an INTAS grant (# 97-1916). The author expresses deep gratitude to the late Dr. Ivan Divac, whose kind assistance helped to carry out the immunological approach. The author also thanks Dr. A. Revischin, who took part in calbindin studies, and Dr. A. Fedorov for his invaluable help in preparing the illustrations and formatting the text.

REFERENCES

Alexander, G., and DeLong, M., 1985, Microstimulation of the primate neostriatum. II. Somatotopic organization of striatal microexcitable zones and their relation to neuronal response properties, *J. Neurophysiol.* 53:1417.

Braak, H., and Braak, E., 1982, Neuronal types in the striatum of man, *Cell Tiss. Res.* 227:319.

DiFiglia, M., and Carey, J., 1986, Large neurons in the primate neostriatum examined with the combined Golgi-electron microscopic method, *J. Compar. Neurol.* 244:36.

Dimova, R., and Usunoff, K., 1989, Cortical projection of giant neostriatal neurons in the cat. Light and electron microscopic horseradish peroxidase study, *Brain Res. Bull.* 22:489.

Eder, M., Vizkelety, T., and Tombol, T., 1980, Nerve cells of the rabbit, cat, monkey and human caudate nucleus: Golgi study, *Acta. morphol. Acad. Sci. Hung.* 28:337.

Ermolenko, S., Kasabov, G., Pavlidis, T., and Lebedeva, N., 1976, Afferent projections of caudate nucleus to the monkey neocortex, *Arch. Anat., Histol., Embriol.* 71:9 (In Russian).

Goldman-Rakic, P., 1982, Cytoarchitectonic heterogeneity of the primate neostriatum: subdivision into island and matrix cellular compartments, *J. Comp. Neurol.* 205:398.

Graveland, G., Williams, R., and DiFiglia, M., 1985, A Golgi study of the human neostriatum: neurons and afferent fibers, *J. Comp. Neurol.* 234:317.

Hassler R., Chung J., Wagner, A., and Renne, V., 1977, Experimental demonstration of intrinsic synapses in cat's caudate nucleus, *Neurosci. Let.* 5:117.

Holt, D.J., Graybiel, A.M, and Saper, C.B., 1997, Neurochemical architecture of the human striatum, *J. Comp. Neurol.* 384:1.

Jayaraman, A., 1980, Anatomical evidence for cortical projections from the striatum in the cat, *Brain Res.* 195:29.

Kluver, H., and Barrera, E., 1953, A method for the combined staining of cells and fibers in the nervous system, *J. Neuropath. a. Exp. Neurol.* 12:400.

Künzle, H., 1975, Bilateral projection from precentral motor cortex to the putamen. An autoradiographic study in macaca fascicularis, *Brain Res.* 88:192.

Leontovich, T.A., 1954, Minute structure of subcortical ganglia, *J. Neuropath. Psychiatry*, 5:168 (In Russian).

Leontovich, T.A., 1969, The neurons of magnocellular neurosecretory nuclei of the dog's hypothalamus, *J. Hirnforschung*,11:499.

Leontovich, T.A., 1973, Methodik zur quantitativen Beschreibung subcorticaler Neurone, *J. Hirnforschung*, 14:59.

Leontovich, T.A., 1975 Quantitative analysis and classification of subcortical forebrain neurons, in: *Golgi Centennial Symposium,* M. Santini, ed., Raven Press, New York.

Leontovich, T.A., 1978, *Neuronal Organization of Subcortical Forebrain Formations*, Medizina, Moscow (In Russian).

Leontovich, T.A., 1983, Spatial organization of tissue elements in dog's caudate nucleus, *Neurophysiology*, 15:474(In Russian).

Leontovich, T.A., 1994, Intrinsic morphological differentiation of putamen in carnivora (in a dog) and in man, *J. Physiology*, 80:23 (In Russian).

Leontovich, T.A., and Zhukova, G.P., 1963, The specificity of the neuronal structure and topography of the reticular formation of the brain and spinal cord of carnivora, *J. Comp. Neurol.* 121:347.

Majsky, V.A., 1983, *Structural Organization and Integration of Centrifugal Neuronal Systems of the Brain and Spinal Cord*, Naukova Dumka, Kiev (In Russian).

Mukhina, J.K., Mukhin, E.I., and Leontovich, T.A., 1986, Efferent connections of striatum with the area EP of the cat's temporal cortex, *Arkhiv. Anat. Histol. Embryol.* 91:5 (In Russian).

Quinn, B., and Graybiel, A., 1994, Myeloarchitectonics of the primate caudate-putamen, in: *The Basal Ganglia IV*, G. Percheron, ed., Plenum Press, New York.

Ramon-Moliner, E., and Nauta, W., 1966, The isodendritic core of the brain stem, *J. Comp. Neurol.* 126:311.

Ramon y Cajal, S., 1911, *Histologie du Systeme Nerveux de l'Homme et des Vertebres,* Maloine, Paris.

Revischin, A.V., Radionova, E.I., and Leontovich, T.A., 2000, Relation of striosomes and structural moduli in cat and human striatum, *Dokladi Akademii Nauk.* (In Russian) In press.

Selemon, L., Gottlieb, J., and Goldman-Rakic, P., 1994, Islands and striosomes in neostriatum of the rhesus monkey - nonequivalent compartments, *Neuroscience,* 58:183.

Yelnik, J., François, C., Percheron, G., and Tandé, D., 1991, Morphological taxonomy of the neurons of the primate striatum, *J. Comp. Neurol.* 313:273.

EFFERENT CONNECTIONS OF THE HUMAN STRIATUM AND PALLIDUM: A NAUTA DEGENERATION STUDY

ANNE MOREL, MICHEL MAGNIN, AND DANIEL JEANMONOD[*]

INTRODUCTION

Knowledge of the organization and connections of the basal ganglia has greatly advanced in recent years, and, concurrently, understanding and neurosurgical treatment of movement disorders of basal ganglia origin, primarily Parkinson's disease, have significantly improved. Models of basal ganglia-thalamo-cortical circuitry have been developed to account for both normal function and motor disorders (DeLong, 1990), but these rely primarily on hodological data extrapolated from experimental animals. Direct investigations of basal ganglia output connections in the human brain are rare and often impeded by lesions encroaching upon the capsule, thus rendering interpretations of the origin of degeneration difficult (Martinez, 1961; Beck and Mignami, 1968; Rye et al., 1995). Moreover, these studies were conducted in the brains of parkinsonian patients already presenting pathological features due to the disease. The present study on the distribution of axonal degeneration resulting from lesions confined to the striatum and pallidum provides additional significant information on the pallidofugal pathways in man, especially on those from the external segment, which may have particular relevance in the physiopathology of movement disorders.

MATERIAL AND METHODS

Anterograde degeneration was investigated in the autopsy brain of a 14 year old subject who had suffered accidental bilateral striato-pallidal lesions and died 19 months later from generalized atheto-dystonia. The brain was autopsied less than 24 hrs after death, and blocks through the basal ganglia and thalamus were provided by the Neuropathology Department. The blocks were fixed several weeks in 10% formalin, then cryoprotected and cut frozen in 40μm frontal sections. Adjacent series were stained for Nissl, myelin, and processed for anterograde degeneration with a modified Nauta selective silver impregnation

---

[*] ANNE MOREL, MICHEL MAGNIN, AND DANIEL JEANMONOD • Department of Functional Neurosurgery, Neurosurgery Clinic, University Hospital Zurich, Zurich CH-8091, Switzerland.

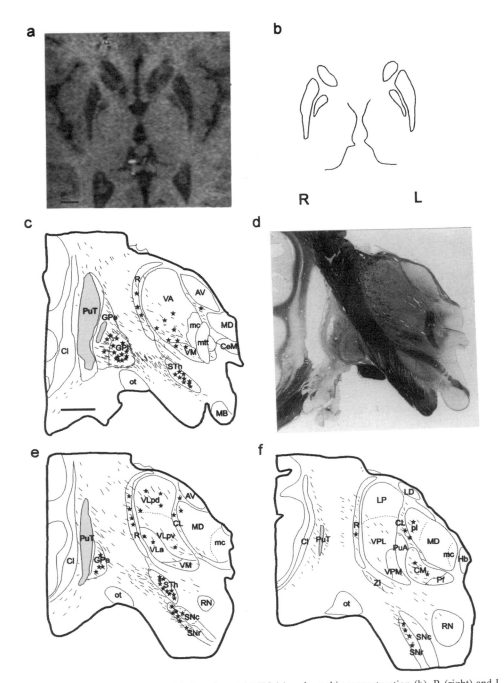

Figure 1. Extent of striato-pallidal lesions in axial MRI (a) and graphic reconstruction (b). R (right) and L (left) indicate hemisphere sides. The relative density of degenerated fibers (thin lines) and terminals (stars) is represented on frontal sections of the basal ganglia and thalamus, from rostral (c) to caudal (f). Intervals are 4 mm between sections c and e, and 6 mm between e and f. Myelin stained section adjacent to c is shown in d. Nomenclature according to Morel et al. (1997). Scale bars: 10 mm in a, and 5 mm in c.

method (Miklossy et al., 1991). Distribution of degenerated fibers and terminals was analyzed under high power microscope and traced with a computerized plotting system (Neurolucida, MicroBrightField). The plots were then superimposed on adjacent Nissl or myelin stained sections for basal ganglia compartmentalization and thalamic parcellation.

RESULTS

Lesion Localization

The extent of lesions is illustrated in axial MRI and graphic reconstruction in Figures 1a and b, respectively, and outlined on frontal sections of the basal ganglia and thalamus (gray areas in Figures 1c, e, and f). The lesions encompass most of the putamen and caudate nucleus, except for the tail of the caudate and the most rostro-ventral part of the striatum comprising the accumbens nucleus. In the pallidum, the lesions involved the rostral and dorsal portion of the external pallidum (GPe), and a very restricted area in the anterior internal pallidum (GPi). No extension of the lesions into the internal capsule or other surrounding structures was observed.

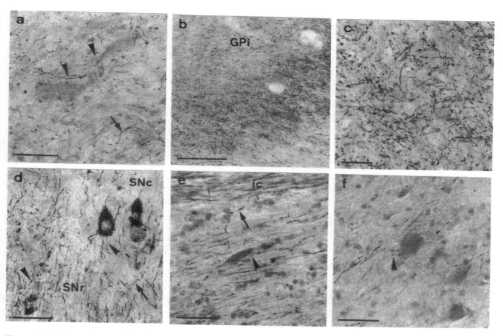

Figure 2. Photomicrographs of degenerated fibers (arrows) and terminals (arrowheads) in GPi (a) and along its ventral aspect (b), in subthalamic nucleus (c), the substantia nigra (d), the reticular nucleus (d) and ventral anterior nucleus (f). Scale bars: 50 μm in (a) and (c) to (f), 100 μm in (b).

Efferent Projections

The pattern of axonal degeneration was studied in detail in the left hemisphere, where the lesions were almost exclusively confined to the striatum and GPe with only minimal damage (~1mm diameter) to the most rostral GPi. As shown in Figure 1, bundles of degenerated fibers were traced from the lesions across GPe and GPi. Within GPi numerous fine degenerated fibers and terminals were concentrated over the dorsal and central regions of the nucleus (see also Figure 2a). Densely distributed degenerated fibers were also observed along the ventral surface of both pallidal segments, including part of the ansa lenticularis (see also Figure 2b). They joined together at the medial tip of GPi to form thick bundles of degenerating fibers that traversed the cerebral peduncle toward the dorsal aspect of the subthalamic nucleus and ventrally, toward the substantia nigra (SN). The subthalamic nucleus contained significant terminal degeneration, particularly in the central part of the nucleus (Figures 1 and 2c). Other degenerated fibers descended further caudally in the cerebral peduncle. In the substantia nigra, degenerating fibers and terminals could be traced along most part of pars reticulata (SNr). Degenerating fibers were less dense, but still clearly present, in the pars compacta (SNc) (Figures 1 and 2d).

In the thalamus, axonal degeneration was relatively profuse in several nuclei, including the ventral anterior (VA), ventral lateral (VL), ventral medial (VM), reticular (R), mediodorsal (MD), centre médian (CM), and central lateral (CL). In all nuclei, we also found clear degenerated terminals, more numerous in lateral (including R) than medial nuclei (Figures 1 and 2e-f). Many of the degenerating fibers reaching the thalamus took a course across the internal capsule in loosely arranged bundles dorsal to the levels of those directed toward subthalamic nucleus and substantia nigra, giving off terminals to the reticular nucleus. Degenerating fibers seen along the dorsal aspect of subthalamic nucleus (part of H2 field of Forel), though not densely distributed, may also have contributed to the terminal degeneration found in the thalamus, particularly in VM and VA nuclei.

DISCUSSION

The present study confirms the validity of the Nauta technique to trace connections in the human brain even after a relatively long (19 months) survival time, as reported earlier (Miklossy et al., 1991). The results demonstrate that major efferents from the human striatum and pallidum (GPe) are similar to those described in non-human primates, and, in addition, confirm a pallido-reticular pathway in the thalamus that may be of particular significance in the physiopathology of some movement disorders.

The efferents to the subthalamic nucleus and substantia nigra follow a trajectory corresponding to the "comb system," comprising the fasciculus subthalamicus and striatonigral projections (Nauta and Mehler, 1966). The very high density of terminal degeneration found in the subthalamic nucleus and concentrated over more central parts of the nucleus originates from the sensori-motor and associative rostral two thirds of the GPe (see Aldheid et al., 1990, and Parent and Hazrati, 1995, for reviews). In the substantia nigra, the terminal field comprises a large portion of the pars reticulata and extends also to pars compacta, which is in accordance with the topography of combined projections from nearly the entire striatum and part of the GPe (Lynd-Balta and Haber, 1994; see also Parent and Hazrati, 1995, for review).

One significant finding of this study is the relative profusion of pallido-thalamic projections, particularly to the reticular nucleus. A pallido-reticular pathway has been described in both humans and nonhuman primates, with an origin essentially, if not solely, in the GPe (Martinez, 1961; Hazrati and Parent, 1991; Asanuma, 1994). The transcapsular course of degenerated fibers reaching the nucleus also resembles that described by Hazrati and Parent (1991). Projections of the GPe to thalamic nuclei other than the reticulate nucleus, though reported in the literature (see, for example, Alheid et al., 1990, for review) are generally considered minor compared to those from the GPi, and less extensive than those presented here. Although a small contribution of the lesion in GPi cannot be entirely excluded, it is unlikely to represent the entire thalamic projection seen in this report, especially considering its very rostral location (Sidibé et al., 1997). A possible source of thalamic degeneration could involve basal forebrain neurons interspersed in the medullary laminae and partially affected by the lesions. However, the absence of degeneration in medial part of MD and in anterior nuclei which represent major thalamic targets of basal forebrain projections (Parent et al., 1988) speaks against this possibility.

ACKNOWLEDGMENTS: We wish to thank S. Richter and V. Streit for their skillful histological work, and R. Stillhard for photographic assistance. This work was supported by the Swiss National Science Foundation; Grant numbers 31-36330.92 and 31-47238.96.

REFERENCES

Aldheid, G.F., Heimer, L., and Switzer, R.C., 1990, Basal ganglia, in: *The Human Nervous System*, G. Paxinos, ed., Academic Press Inc., San Diego.

Asanuma, C., 1994, GABAergic and pallidal terminals in the thalamic reticular nucleus of squirrel monkeys, *Exp. Brain Res.* 101:439.

Beck, E., and Bignami, A., 1968, Some neuro-anatomical observations in cases with stereotactic lesions for the relief of parkinsonism, *Brain*, 91:589.

DeLong, M.R., 1990, Primate models of movement disorders of the basal ganglia, *Trends Neurosci.* 13:281.

Hazrati, L.-N., and Parent, A., 1991, Projection from the external pallidum to the reticular thalamic nucleus in the squirrel monkey, *Brain Res.* 550:142.

Lynd-Balta, E., and Haber, S.N., 1994, Primate striatonigral projections: A comparison of the sensorimotor-related striatum and the ventral striatum, *J. Comp. Neurol.* 345:562.

Martinez, A., 1961, Fiber connections of the globus pallidus in man, *J. Comp. Neurol.* 117:37.

Miklossy, J., Clarke, S., and Van der Loos, H., 1991, The long distance effects of brain lesions: Visualization of axonal pathways and their terminations in the human brain by the Nauta method, *J. Neuropathol. Expt. Neurol.* 50:595.

Morel, A., Magnin, M., and Jeanmonod, D., 1997, Multiarchitectonic and stereotactic atlas of the human thalamus, *J. Comp. Neurol.* 387:588.

Nauta, W.J.H., and Mehler, W.R., 1966, Projections of the lentiform nucleus in the monkey, *Brain Res.* 1:3.

Parent, A., and Hazrati, L.-N., 1995, Functional anatomy of the basal ganglia. II. The place of the subthalamic nucleus and external pallidum in basal ganglia circuitry, *Brain Res. Rev.* 20:128.

Rye, D.B., Vitek, J., Bakay, R.A.E., Kaneoke, Y., Hashimoto, T., Turner, R., Mirra, S., and DeLong, M., 1995, Termination of pallidofugal pathways in man, *Soc. Neurosci. Abstr.* 21:676.

Sidibé, M., Bevan, M.D., Bolam, J.P., and Smith, Y., 1997, Efferent connections of the internal globus pallidus in the squirrel monkey: 1. Topography and synaptic organization of the pallidothalamic projection, *J. Comp. Neurol.* 382:323.

COGNITIVE ASPECTS OF THE MOTOR
FUNCTION OF THE STRIATUM

Boris F. TOLKUNOV, Aleksandr A. ORLOV, and
Sergey V. AFANAS'EV[*]

INTRODUCTION

Changes in spike activity associated with motor function are seen in many neurons of the striatum, but these reactions are generally associated less with the movements themselves than with the conditions under which they are made. The striatum contains neurons that respond during slow movements but not during rapid repetitive movements (DeLong, 1973). Striatal neurons are also activated when movements are performed under visual control (Rolls et al., 1983; Ueda and Kimura, 1999) or in order to obtain a food reward (Nishio et al., 1991). The conditions under which motor responses of the striatal neurons occur can be very complex. For example, Fukuda et al. (1993) demonstrated responses in striatal neurons of monkey to the sight of food presented for 2-4s through a glass door. Neuron response was significantly reduced, however, if the time between the demonstration and taking of the food was shortened to 0.5 s. Also, specialized movements such as saccades are accompanied by responses of the striatal neurons mainly if they are directed toward the point in the visual field at which the animal expects to see a previously demonstrated target (Hikosaka et al., 1989). Many investigators believe that variation of the conditions under which a movement's related activity appears in the striatum indicate that the striatum is involved in the programming of movements, their initiation and sensory control, memory traces, etc. We have previously shown that each putamenal neuron can be involved in performing different fragments of a multistage behavioral task, and the more the test varies, the more "polymodal" the neuron appears to be (Tolkunov et al., 1998).

This multiplicity makes the problem of identifying the function of striatal neurons very complicated. Data have been obtained from different types of experiments in which different neurons were recorded, and the impossibility of comparing such results is one of the reasons why identifying function is so difficult. The goal of the work presented here was twofold: (1) to investigate neuronal activity by comparing the character of participation of the same neurons in different behavioral events, and (2) to determine the degree

---

[*] Boris F. Tolkunov, Aleksandr A. Orlov, and Sergey V. Afanas'ev • Sechenov Institute of Evolutionary Physiology and Biochemistry, St. Petersburg 194223, Russia.

of participation of the putamen in different stages of behavior, not only by comparing the number of involved neurons, but also by measuring the degree of reorganization of their efferent influences.

The second goal is based on the premise that knowledge of the number of activated neurons is still insufficient for estimation of the influence of a given brain structure on a behavior. If the behavior is modified, a different neuronal population can become involved, even while the number of activated neurons remains unchanged. The level of reorganization of a mosaic of neuronal activity is directly related to changes in the degree of influence these cells have on behavioral events. Therefore, reorganization of the unit activities is a more reliable characteristic of neuron involvement in the control of behavior than the number of activated cells.

METHODS

A special behavioral program was designed to allow for comparison of the activity of the same neuron during various actions of the animal (Tolkunov et al., 1997). The monkey (Macaca nemestrina) was placed in a restraining chair. The behavioral program was triggered by opening a screen that covered the levers. The monkey had to use both hands to press two levers simultaneously. This pressing turned on a small light, which appeared on the left or right after a short pause (300ms). The light served as the conditioned stimulus whose position indicated which of the two feeders (right or left) contained the food reward. In order to obtain the reward, the monkey had to complete one of two sequential movements: either (1) to keep the right-sided lever in the pressed position, transferring the left hand to a special manipulator and using the fingers to open the left feeder, or (2) to carry out the mirror-symmetrical sequence of movements towards the feeder on the right. Feeder activation was accompanied by the click of a solenoid. The monkey grasped a pellet of bread and passed it to its mouth through an opening in the primate chair. If the monkey failed to select the feeder indicated by the conditioned signal, the click was absent and the reward not provided.

Spike activity was recorded by the microelectrode system (Orlov et al., 1989; Tolkunov et al., 1997). This system allowed independent movement of seven sharpened platinum-iridium microelectrodes insulated with quartz glass. The zone in which neuron activity was recorded in the left putamen was at coordinates A = 20-21, L = 7-8, and at the microelectrode insertion depth of 8-13 mm from the coordinate origin. Potentials were digitized and recorded in a computer. Continuous recordings were made at nine successive stages of the behavioral program, which were distinguished using event markers on acto-grams: 1 - the response to the start signal, 2 - pressing of the two levers with the two hands; 3 - the 300 ms pause before releasing the conditioned stimulus; 4 - the response to the conditioned stimulus; 5 - the premotor period of making the decision about choosing the required hand; 6 - positioning of the chosen hand on the manipulator; 7 - grasping of the manipulator by fingers of the chosen hand in order to turn on the feeder; 8 - response to the feeder click (or its absence); 9 - taking of the reward.

Each group of neurons was recorded during 40-50 performances of the behavioral program. Four sets of data were arranged for statistical analysis to correspond to each type of behavioral program performance: two sets of data for cases in which errors occurred, and two sets for when the animal made no mistakes. Data sets were analyzed by constructing relative time histograms, which allowed neuron activity to be analyzed in the continuum corresponding to the continuum of the animal's acting (Tolkunov, 1997), and by subjecting them to discriminant analysis, as adapted for investigations of neuronal activity (Gochin et al., 1994; Tolkunov et al., 1999).

A standard program for discriminant analysis (Statistica for Windows 4.5) was used for each individually analyzed stage of the behavioral program to calculate an overall parameter for collective neuron activity. For individual performances, this parameter was a point in some multidimensional (determined by the number of analyzed neurons) space. The resulting points were classified according to the four possible forms of the animal's behavior. The position of the "center of gravity" (or centroid) of each region occupied by each class of points was established, and the significance with which the centroids of different regions failed to correspond was determined. The locations of these regions were plotted in coordinates defined by the first two discriminant functions. This analysis was repeated for each fragment of the neuron activity.

RESULTS

Each recorded group of neurons was tested repeatedly during 40-50 runs of the program. In total, 247 histograms of distribution of activity of 63 neurons along the nine stages of the program were obtained. Changes in the firing rate at a given stage of performance of the program were determined with a significance value $P < 0.05$. A neuron was considered related to a given stage of the program if authentic change of activity was observed in one of the histograms (i.e. in one of the variants of the experiment) at that stage.

Most of the recorded neurons responded to more than one stage of the program. In addition, the same neuron could respond in a different manner to the same action made by the animal under different conditions. The number of activated neurons at different stages of the program varied, but these distinctions were not significant. Several cases of depression of the neuronal activity at different stages of the program were more differentiated. Inhibition of activity was practically absent in reactions to the start signal but sharply increased at the stage of the movement of the chosen hand (Figure 1A,B).

Some of the neurons differentiated the task, i.e. changed the firing rate during either the left- or the right-hand tasks. A few such differentiating neurons were detected before the conditioned stimulus was delivered, but at these early stages they manifested irregularly. It is remarkable that the number of differentiating neurons remained the same at the stage of response to the conditioned stimulus (4), but increased significantly at the stage of a choice of a working hand (5). However, the number of differentiating neurons fell to earlier levels when the monkey moved the right and left hands. Significant increases in the number of differentiating neurons occurred again when the click of the solenoid indicated that the problem was solved correctly.

Discriminant analysis was used for 58 pair-wise comparisons from 19 testing cycles. The resulting points were grouped in the four sets of data according to the four possible forms of the animal's behavior: correct responses to the "left" signal, correct responses to the "right" signal, erroneous responses to the "left" signal, and erroneous responses to the "right" signal. The responses of each group of cells at each stage of the behavior program were characterized by particular locations of the areas occupied by these four blocks. Examples illustrating the collective responses of one group of neurons at the stages of the behavioral program are shown in Figure 2.

During responses to the start signal (Figure 2, 1), points characterizing mosaics of neuron group responses in performances of different types of tasks formed strongly overlapping areas. Thus, at the stage when the conditioned stimulus had not yet been presented and the defined sequence of movements had not been selected, the activity mosaics for different performances were similar, as expected. Similar overlap of the areas is visible at the stage of movement of both hands, during the pause, and even at the appearance of the conditioned stimulus (Figure 2, 2-4).

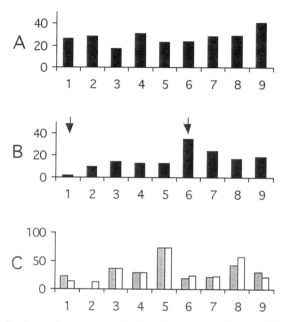

Figure 1. Number of putamenal neurons involved in the control of sequential stages of the behavioral program. A - Activated neurons. B - Inhibited neurons. C - Neurons involved in the control of either the left-hand (gray bars) or right-hand (white bars) task. The ordinate: A and B - number of neurons; C - number of differential neurons, expressed as a percentage of the number of cells responding at this stage. Abscissa - stages of the program: 1 - the start signal; 2 - the movements of two hands; 3 - the pause before releasing of the conditioned stimulus; 4 - releasing of the conditioned stimulus; 5 - the movement of one chosen hand; 6 - finger movements; 7 - the click of the feeder's solenoid; 9 - receiving of the reward. Black arrows indicate statistically significant (p < 0.05) differences.

During the premotor period (Figure 2, 5), regions corresponding to correct solutions of the task did not overlap, and the differences in their centroid positions were significant (p<0.001), while the regions for erroneous performances of the "right" and the "left" task overlapped. This result indicates that in trials with erroneous decisions, the ensemble activity of neurons was different from that seen in trials with correct solutions, although at this stage the animal did not know that the error would be made. The "right" erroneous region overlapped mostly with the correct "left" region, and the erroneous "left" region with the correct "right" region, i.e. each error region overlapped more strongly with the region for correct decisions corresponding to the same hand movement. The same patterns were seen when the selected hand was moved to the manipulator (Figure 2, 6). During this period, however, these features were significantly less pronounced and were statistically insignificant in most cases. Only two experiments stood out; the differences in response mosaics appeared only in this period and no differences were seen in the preceding period. The pattern changed qualitatively at the stage immediately following completion of the

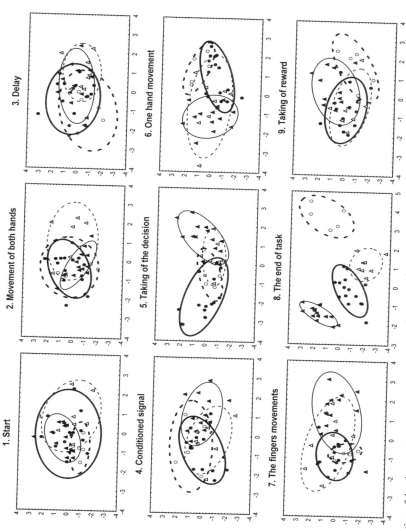

Figure 2. Distribution of the characteristics of the collective responses of one group of neurons at the different stages of the behavior program. 1-9 - responses of the same group of neurons. Black triangles correspond to correct responses to the "right" signal, white triangles erroneous responses to the "left" signal, responses to the "left" signal and white circles to erroneous responses to the "right" signal. Continuous lines show correct trials and dotted lines show erroneous trials. The axes show the first and second discrimnant functions respectively.

required sequence of movements by the animal (Figure 2, 8). All behavioral situations could now be distinguished. In most experiments, the nature of individual response mosaics could be used to identify the situation in which they were recorded, with a probability of 85-100%. The duration of this period was 400 ms. At this time, the animal's hand had not yet reached the feeder and the animal had not yet ascertained the presence or absence of the reward. Hence, these responses were associated with the presence or absence of the feeder click. In the subsequent periods (Figure 2, 9), the differences in the response mosaics disappeared.

The generalized data on all investigated groups of neurons are presented in Figure 3. The diagram shows that the significant differences of the mosaics appear when the monkey makes a correct decision about the choice of "working" hand and again when the monkey solves the problem. When differences of the mosaics at the stage of decision-making are not significant, the monkey makes an error. This means that the monkey works correctly when the putamenal neurons can "well distinguish" the left- and right-side task.

Mosaics become different again at the end of the program. The different mosaics appear at this stage in spite of the absence of any special signal that could indicate what kind of error has been made. Hence, the putamenal neurons can associate the fact of error with the conditioned stimulus that was received before.

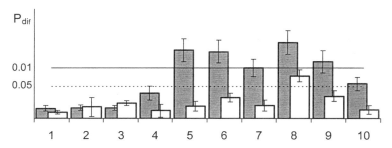

Figure 3. The differences between the neuronal mosaics during the right- and left-hand tasks in the correct (gray bars) and the erroneous (white bars) trials. The ordinate - level of significance of the differences. For further details see Legend to Figure 1.

DISCUSSION

The results show that the putamenal neurons can be involved in different kinds of behavior and that the form of this involvement can be different. The individual activities of the striatal units reflect the character of the animal's actions by the quantity of reactions and by different interrelations between the excitatory and inhibitory reactions. The number of activated neurons at different stages of behavior is more stable than the number of inhibited neurons. This probably reflects relatively constant glutamatergic inflow from the cortex and the functionally selective activation of the intrinsic GABAergic connections.

The putamen, which receives extensive inputs from the motor areas of the cortex, is traditionally regarded as a structure related to supporting movements (Cools et al., 1984; Marsden, 1987; Crutcher and Alexander, 1990). As a rule, in the activity of the caudate nucleus and putamen units related to a movement, investigators distinguish activation of the neurons during the premotor period prior to the beginning of performance of the movement. Our data also show that a number of neurons are activated in the premotor period. Most of these neurons differentiate the left- and the right-hand tasks. However, a sharp reduction of the number of such neurons during movement performance indicates that the unit activity is more related to the preparation of a movement than to its execution. It is likely that the activity of such neurons is related to some other processes that determine the direction of further actions but not their organization as such. Such premotor activity is considered a part of programming or initiation of movement (Vrijmoed de Vries and Cools, 1985; Boussaoud and Kermadi, 1997; Lebedev and Nelson, 1999; Matsumoto et al., 1999).

Another period of distinction of the right and left-hand task by the putamenal neurons was observed after the end of the execution of behavioral program, when the animal received the signal about the correctness or inaccuracy of the problem-solving. Both of these periods were identified by both kinds of analyses used: determination of the quantity of involved neurons and discriminant analysis. The task reactive neurons at a given stage of the behavior program were determined with the help of the relative time histogram by the average of their activity in all trials. Discriminant analysis was carried out according to the unit activity in each trial separately. The results of the two kinds of data analysis are conterminous, but the discriminant analysis demonstrates some additional details.

Collective reactions reflect different actions of the animal in different forms of mosaics of the responsive neurons. Mosaics of the activated neurons, as well as the quantity of differentiating neurons, shows that different experimental situations correspond to different forms of activation of neurons in the putamen. The mosaics of activity of the putamenal neurons discriminate the experimental alternative. They are significantly different during decision-making and at the end of the task, when the monkey finds out about its correctness or inaccuracy. The differences between the mosaics are especially clear when the behavior program is over. At this time even the mosaics of the neuronal activity recorded in erroneous trials are different.

The demonstrated distinction of the mosaics can be related to the distinction between the cognitive processes that guide the actions of the animal at the left and right tasks. This distinction can also be related to the performance of the right-hand and left-hand tasks, i.e. to motor functions. The data obtained in experiments described here provide evidence for the former assumption.

It is rather easy to explain the differences between the mosaics in correct trials. In one case, the monkey hears the click of the right feeder and takes a food pellet by the right hand, and in the other case, it hears the click of the left feeder and takes a food pellet by the left hand. However, such situations arise only after the task has been performed correctly. In erroneous trials the feeders do not work, there are no clicks, and lateralized movements are also absent. Even so, the greatest distinctions of mosaics of the unit activity are observed at the end of performance of the task, when the mosaics of the neuronal activity recorded in erroneous trials are different.

The differential mosaics after performance of the behavioral program in erroneous trials can arise only as a result of comparisons of the signal about the absence of a reward with the memory trace about the meaning of the task and the hand that carried it out. The results of such comparisons can arrive at the neural network of the putamen from the cortex or other structures of the brain. A memory trace about what kind of task was presented can also be saved in the putamen itself. In all cases, the configuration of mosaics of the unit activity in the putamen reflects all four variants of the problem solving (correct right, correct left, erroneous right, erroneous left). It is well known that the striatum is involved in the processes of learning and memory (Graybiel, 1998; Matsumoto et al., 1999). The data obtained in this study show that this participation can be based on the formation in the putamen (and probably in others areas of the striatum) of a neuronal model of the created situation, which can be used for correction of subsequent actions on the basis of the results received in previous experiences.

Thus, the study of the spike activity of several units in parallel in the monkey putamen shows that significant changes of unit activities are observed during all stages of the behavior of the monkey. The maximal distinctions between the individual activity of neurons and interrelations between the neurons in the group of recorded cells in parallel are observed prior to the beginning of movement, during decision-making about the choice of the working hand but not at the time of movements by different hands. A successful solution of the problem is dependent upon the level of discrimination of the task by putamenal neurons. Errors appear if this level of discrimination is not sufficient.

CONCLUSIONS

The results of this study suggest that the activities of the neurons in the striatum are more related to cognitive processes determining the purpose of the general direction of actions than they are to the specific organization of those actions. These properties of the putamenal units correspond more closely to the concept that the striatum is a neuronal network where specialized corticofugal signals from different and dispersed cortical areas converge and interact. Created in the network, neuronal mosaics appear as a generalized reflection of the dynamics of cortical activities, and they may play a role as an information base for many kinds of animal's actions. Two peaks of distinction of the experimental task - in the pre-movement period and after the end of the program - specify an opportunity for involvement of the putamen in the comparison of the decision made and the result achieved. This can explain the role of the striatum in the mechanisms of learning and memory.

ACKNOWLEDGMENTS: This work was supported by grants from the Russian Foundation for Basic Research 00-04-48477.

REFERENCES

Boussaoud, D., and Kermadi, I., 1997, The primate striatum: neuronal activity in relation to spatial attention versus motor preparation, *Eur. J. Neurosci.* 9:2152.

Cools, A.R., Jaspers, R., Schwarz, M., Sontag, K.H., Vrijmoed-de Vries, M., and Wilcock, N., 1984, Basal ganglia and switching motor programs, in: *The Basal Ganglia,* J. S. McKenzie, R. E. Kemm, and L. N. Wilcock, eds., London.

Crutcher, M. D., and Alexander, G.E., 1990, Movement-related neuronal activity selectively coding either direction or muscle pattern in three motor areas of the monkey, *J. Neurophysiol,* 64:151.

DeLong, M.R., 1973, Putamen activity of single units during slow and rapid movements, *Science,* 179:1240.

Fukuda, M., Ono, T., Hishijo, H., and Tabuchi, E., 1993, Neuronal responses in monkey anterior putamen during operant bar-press behavior, *Brain Res. Bull.* 32:227.

Gochin, P.M., Colombo, M., Dorfman, G.A., Gerstein, G.L., and Gross, C.G., 1994, Neural ensemble coding in inferior temporal cortex, *J. Neurophysiol.* 71:2325.

Graybiel, A.M., 1998, The basal ganglia and chunking of action repertoires, *Neurobiol. Learn. Mem.* 70:119.

Hikosaka, A.O., Sakamoto, M., and Ushui, S., 1989, Functional properties of monkey caudate neurons. I. Activities related to saccadic eye movements, *J. Neurophysiol.* 61:780.

Lebedev, M.A., and Nelson, R.J., 1999, Rhythmically firing neostriatal neurons in monkey: activity patterns during reaction-time hand movements, *J. Neurophysiol.* 82:1832.

Matsumoto, N., Hanakawa, T., Maki, S., Graybiel, A.M., and Kimura, M., 1999, Nigrostriatal dopamine system in learning to perform sequential motor tasks in a predictive manner, *J. Neurophysiol.* 82:978.

Marsden, C.D., 1987, What do the basal ganglia tell premotor cortical areas? in: *Motor Areas of the*

Cerebral Cortex, Chichester, New York.

Nishio, H., Hattory, S., Makamoto, K., and Ohno, T., 1991, Basal ganglia activity during operand feeding behavior in the monkey: relation to sensory integration and motor execution. *Brain Res. Bull.* 27:463.

Orlov, A.A., Shefer, V.I., and Mochenkov, B.P., 1989, A miniature multichannel micromanipulator for the independently moving microelectrodes in bundles, *Fiziol. Zh. SSSR.* 75:1275, (In Russian).

Rolls, E.T., Thorp, S.I., and Madisson, S.P., 1983, Responses of striatal neurons in the behaving monkey. 1. Head of the caudate nucleus, *Behav. Brain Res.* 7:179.

Tolkunov, B.F., 1997, Question of time in studies of the neuronal correlates of behavior, *Neurosci. Behav. Physiol.* 28:447.

Tolkunov, B.F., Orlov, A.A., and Afanas'ev , S.V., 1997, Neuron activity in the monkey striatum during parallel performance of actions, *Neurosci. Behav. Pysiol.* 27:297.

Tolkunov, B.F., Orlov, A.A., Afanas'ev , S.V., and Selezneva, E.V., 1998, Involvement of striatum (putamen) neurons in motor and nonmotor behavior fragments in monkeys, *Neurosci. Behav. Physiol.* 28:224.

Tolkunov, B.F., Orlov, A.A., and Afanas'ev , S.V., 1999, Studies of the functional characteristics of central neurons of the brain in a behavioral experiment, *Neurosci. Behav. Physiol.* 29:645.

Ueda, Y., and Kimura, M., 1999, Characteristic of primate putamen neuron activity during performance of sequence motor tasks with and without visual guidance, *Soc. Neurosci. Abstr.* 25:1928.

Vrijmoed-de Vries, M.C., and Cools, A.R., 1985, Further evidence for the role of the caudate nucleus in programming motor and nonmotor behaviour in Java monkeys, *Exp. Neurol.* 87:58.

NEUROANATOMICAL ORGANIZATION AND CONNECTIONS OF THE MOTOR THALAMUS IN PRIMATES

IGOR A. ILINSKY AND KRISTY KULTAS-ILINSKY[*]

INTRODUCTION

Motor thalamus is a widely applied term in neuroscience research as well as in clinical practice. However, in contrast to other thalamic regions such as visual, auditory, limbic, somatosensory, where afferent inputs fit neatly within the cytoarchitectonic boundaries of respective thalamic subdivisions, delineations of the motor thalamus are more ambiguous irrespective of the criteria on which they are based.

Strictly defined, the motor thalamus encompasses thalamic nuclei receiving subcortical afferents from the basal ganglia and cerebellum. Under this definition, it is comprised of quite a few thalamic nuclei, including some that are traditionally ascribed to other systems, such as the mediodorsal nucleus (MD) or some intralaminar nuclei, for example. In many instances, however, the term motor thalamus is restricted to the ventral thalamic region, which in mammals encompasses cerebellar, pallidal, and nigral afferent territories. In the primate thalamus these three projection zones are entirely segregated (see reviews by Asanuma et al., 1983; Ilinsky and Kultas-Ilinsky, 1987; Ilinsky et al., 1993b; Percheron et al., 1996; Macchi and Jones, 1997), whereas in nonprimate species basal ganglia and cerebellar afferents overlap at least partially (Ilinsky, et al., 1982; Ilinsky and Kultas-Ilinsky, 1984; Kultas-Ilinsky and Ilinsky, 1986).

There are several classifications of the primate thalamus which utilize different nomenclatures and delineations of motor thalamic subdivisions (see Table 1). The most widely used among them are Hassler's nomenclature (1959) for the human thalamus and Olszewski's nomenclature (1952) for the monkey thalamus. Recently, many investigators have applied data that became available with the development of modern hodological and histochemical techniques in an attempt to revise the boundaries of motor thalamic subdivisions (Percheron, 1977, Asanuma et al., 1983; Ilinsky and Kultas-Ilinsky, 1987; Hirai and Jones, 1989; Hirai et al., 1989; Percheron et al., 1996; Macchi et al., 1997). In our neuroanatomical studies we found that the cerebellar projection zone confines to the distinct cytoarchitectonic entity in the ventral thalamic region, which surrounds the basal ganglia afferent projection zone from posterior, dorsal, and in some parts from medial aspects

[*] IGOR A ILINSKY AND KRISTY KULTAS-ILINSKY • Department of Anatomy and Cell Biology, University of Iowa College of Medicine, Iowa City, Iowa 52242.

(Ilinsky and Kultas-Ilinsky, 1987; Ilinsky et al., 1993b). Based on the cytoarchitectonic distinction and the extent of projections, we designated the cerebellar afferent territory in the monkey as the ventral lateral nucleus (VL) as opposed to the basal ganglia territory, which was designated as the ventral anterior nucleus (VA). In the primate VA, nigral and pallidal afferents also occupy distinct cytoarchitectonic subdivisions, magnocellular (VAmc) and parvi- and densi-cellular (VApc and VAdc), respectively (Kim et al., 1976; Carpenter et al., 1976; Ilinsky et al., 1985; Ilinsky and Kultas-Ilinsky, 1987).

Table 1. Nomenclatures of motor thalamic nuclei and their relationships to major subcortical afferents.

Present study	E.G. Jones ('85)	R. Hassler ('59)	J. Olszewski ('52)	C. & O. Vogt ('41)	A.E. Walker ('38)	Subcortical afferents
VAmc	VAmc	Lpo.mc	VAmc	--	--	SNr
VApc VAdc	VApc VLa	Lpo Voa	VApc VLo	-- Voe	VA VL	GPm, medial part GPm, lateral part
VL	VLp	Vop Vim	Area X VPLo VLps VLc	Vim	VL	Deep cerebellar, nuclei

    Topographically, the pallidal afferent territory is anterior and lateral to the nigral afferent territory. In our view such a delineation strategy helps to avoid the confusion commonly encountered in the literature with respect to the existence of overlap between the basal ganglia and cerebellar afferent territories in the thalamus. In nomenclatures where these afferent territories are combined under one umbrella, for example, VLa and VLp in Jones' (1985) and Voa and Vop in Hassler's classifications, an overlap is often implied, causing interpretational inaccuracies, especially on the part of those not well adept in thalamic connections and classifications. The boundaries of the primate motor thalamus as delineated in our neuroanatomical tracing experiments have been described by Ilinsky and Kultas-Ilinsky (1987), and Ilinsky et al. (1993b) and applied in our studies on the organization of neuronal circuits in the VA and VL. At present our laboratory is in the process of preparing a three-dimensional, interactive computer-based atlas of the rhesus monkey thalamus in which the boundaries within the motor thalamus have been further refined. A significant advantage of 3D reconstructions is the ability to demonstrate spatial relationships between different afferent projection zones, which is especially important in the motor thalamus where the adjacent territories interdigitate profusely (Ilinsky et al., 1993b). Such interdigitation makes reliance on coordinates alone perilous and has led investigators in several neurophysiological studies to abandon nomenclatures and resort instead to generic descriptions, such as pallidal afferent-receiving zone and cerebellar afferent-receiving zone, both of which are identified by stimulation of afferent structures (Anderson and Turner, 1991; Vitek et al., 1994, 1996).

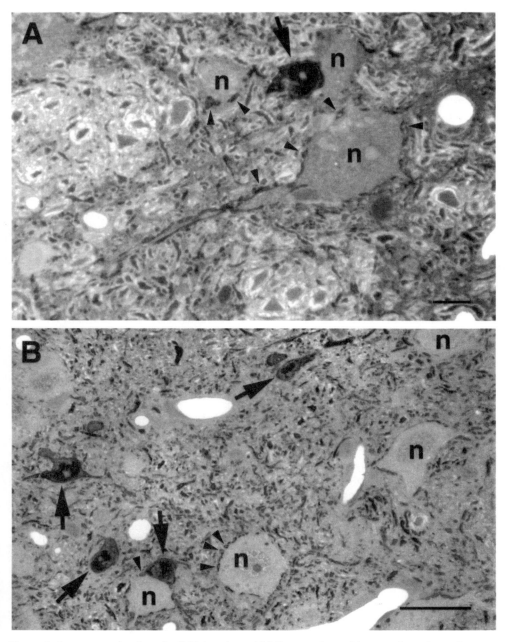

Figure 1. Immunocytochemistry for GABA on 1 µm thick Epon sections of the ventral anterior nucleus pars densicellularis (VAdc) - A and ventral lateral nucleus (VL) - B. Note the difference in soma sizes between the GABA-positive LCN (small, dark-stained neurons) and immunonegative (n) projection neurons (PN). Immuno-positive elongated profiles with light cores aligned on somata and dendrites (arrowheads) are GABAergic axon terminals mostly from the medial globus pallidus in VAdc and from the thalamic reticular nucleus in the VL. Solid, stained processes are LCN dendrites. Dark-stained structures with light rims are GABAergic myelinated axons. Bar in A = 10 µm, bar in B =50 µm. The tissue was lightly counterstained with thionin.

THALAMOCORTICAL PROJECTION NEURONS AND LOCAL CIRCUIT NEURONS

Like all other thalamic nuclei in primates, the VL and VA are comprised of two types of neurons: thalamocortical projection neurons (PN) and local circuit neurons (LCN). PN are identified by retrograde labeling from the cortex with HRP and LCN by positive immunocytochemistry for glutamic acid decarboxylase or gamma-aminobutyric acid (GABA) (Figure 1). The dendritic branching patterns of the two cell types are radically different (Figure 2), and soma sizes do not overlap. The number of LCN in the VL is greater than in the VA with an LCN/PN ratio of 1:3 in the VL and 1:10 in the VA (Ilinsky and Kultas-Ilinsky, 1990; Kultas-Ilinsky and Ilinsky 1991; Kultas-Ilinsky et al., 1997). A specific feature of thalamic LCN is that they contain synaptic vesicles in their dendrites and therefore can be not only postsynaptic to afferents but also presynaptic either to projection neurons' dendrites and soma or to other LCN. We have demonstrated the initial parts of LCN axons in motor thalamic nuclei of the monkey and cat at the electron microscopic level (Ilinsky and Kultas-Ilinsky, 1990; Kultas-Ilinsky and Ilinsky, 1988, 1991), but the precise synaptic sites of the terminals of these axons are still unknown. This is an important issue in view of some earlier suggestions (Lieberman, 1973; Yen et al., 1985; Ohara et al., 1987) that thalamic LCN do not have axons and operate by bidirectional conduction of their dendrites similar to retinal interneurons. Our studies show that this is definitely not the case in the motor thalamic nuclei. The initial axon segments of PN are potential synaptic sites for LCN axons. Axo-axonic synapses, formed by terminals displaying features of inhibitory synapses, were found on the initial axon segments of PN in the VA, VL and the centromedian nucleus of the monkey thalamus (Ilinsky and Kultas-Ilinsky, 1990; Kultas-Ilinsky and Ilinsky, 1991; Ballercia et al., 1996; Kultas-Ilinsky et al., 1997). In view of the paramount functional importance of the input to these sites, its origin needs to be identified to fully understand the thalamic circuitry.

ORGANIZATION OF THE BASAL GANGLIA INPUTS TO THE VA

Substantia nigra pars reticularis (SNr) and the medial part of globus pallidus (GPm) convey information from the basal ganglia to the cortex via three cytoarchitecturally distinct VA subdivisions and the mediodorsal nucleus. Pallidal information also reaches the centromedian thalamic nucleus and directly the brain stem. (Nauta and Mehler, 1966; Kuo and Carpenter, 1973; Ilinsky et al., 1985; Ilinsky and Kultas-Ilinsky, 1987).

Although certain topography in the distribution of pallidothalamic afferents has been noted in the above studies, our recent experiments with small biotinylated dextran amine (BDA) injections in the GPm showed that the labeled terminal fields resulting from these small injections distributed widely throughout the VAdc and VApc. Moreover, single GPm fibers were found to give rise to relatively large sized terminal fields at different VA locations (Ilinsky et al., 1993a; Kultas-Ilinsky et al., 1997). These data correlate well with the findings obtained in other laboratories (Arecchi-Bouchhioua et al., 1996, Sidibe et al., 1997). Small BDA injections in the SNr also resulted in a wide distribution of labeled terminals in the VAmc, although the terminal fields themselves appeared smaller than those formed by pallidal afferents (Tai et al., 1996). It has been well established that both nigral and pallidal outputs are GABAergic and inhibitory to thalamocortical projection neurons (DiChiara et al., 1979; Penney and Young, 1981; Deniau et al., 1978; Uno et al., 1978; Ueki, 1983). Double-labeling experiments using HRP injections in the GPm with postembedding GABA immunocytochemistry of the VA tissue further confirmed the GABAergic nature of pallidal terminals (Ilinsky et al., 1997). Similar to most inhibitory synapses throughout the brain, pallidal and nigral boutons contain pleomorphic synaptic vesicles and display symmetric synaptic contacts (Figure 3B). The size of pallidal and nigral boutons varies greatly from medium to large with some reaching up to 12μm in length, and each terminal forms more than one synaptic contact (Kultas-Ilinsky and Ilinsky, 1990, Kultas-Ilinsky et al., 1997).

EM studies with tritiated aminoacid or HRP injections into SNr and GPm demonstrated that in primates both nigral and pallidal inputs target somata and proximal dendrites of thalamocortical projection neurons (Kultas-Ilinsky and Ilinsky, 1990; Ilinsky and Kultas-Ilinsky, 1990; Ilinsky et al., 1997; Kultas-Ilinsky et al., 1997). Both pallidal and nigral terminals also form synapses on LCN but in smaller proportions than on PN. The anatomical arrangement of pallidothalamic afferents allows a single pallidal fiber to directly inhibit one group of thalamic neurons while disinhibiting another via its connections with LCN (Ilinsky et al., 1997). Moreover, pallidal terminals form so called triads, where the pallidal bouton makes one synapse on a projection neuron dendrite and another on a LCN dendrite with the latter synapsing on the same PN dendrite (Ilinsky et al., 1997). The difference between these triads and the classic thalamic triads (see Jones, 1985, for review) is that all three synapses are inhibitory. Unlike pallidal terminals, triads with nigrothalamic terminals were not found in the VAmc (Kultas-Ilinsky and Ilinsky, 1990).

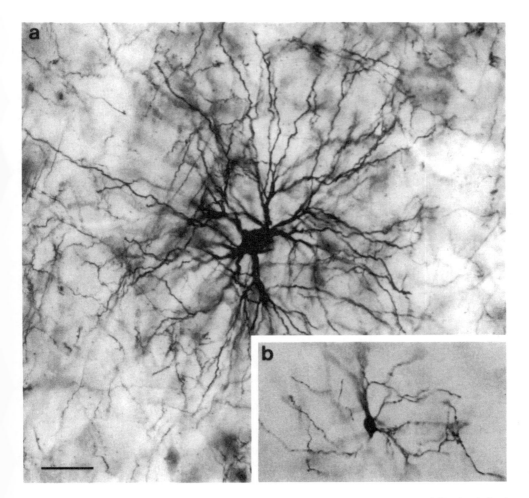

Figure 2. Examples of two types of Golgi impregnated neurons: (a) - projection neuron and (b) - local circuit neuron in the monkey VL. Bar=50 mm (from Clement Fox collection at the Wayne State University).

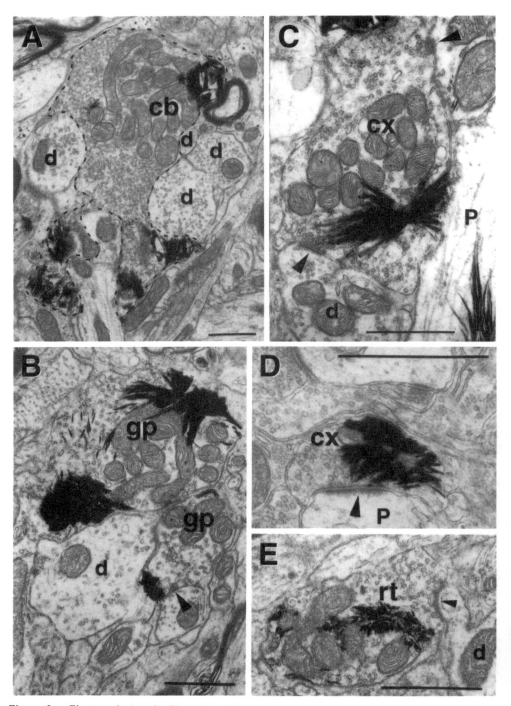

Figure 3. Electronmicrographs illustrating HRP-labeled terminals of afferents to the thalamus from cerebellum (cb) – A; medial globus pallidus (gp) – B; cortex (cx) – C, D; and reticular nucleus (rt) – E. Arrowheads point to synaptic contacts, d -vesicle containing LCN dendrites, P - projection neuron dendrites, Bars = 1μm

ORGANIZATION OF CEREBELLAR INPUT TO THE VL

Distribution of cerebellothalamic afferents in the monkey has been intensively studied over the years with a variety of neuroanatomical techniques (see Ilinsky and Kultas-Ilinsky, 1987; Percheron et al., 1996 for references). More recent studies with BDA have allowed researchers to evaluate the size and distribution of terminal fields of individual axons (Mason et al., 2000). In principle, the overall distribution of cerebellothalamic afferents to the VL is similar to that of pallidothalamic afferents to the VA in that a small BDA injection in the dentate nucleus of the rhesus monkey cerebellum results in labeling of a large number of individual terminal fields widely distributed over VL. Furthermore, a single dentate fiber can generate up to ten terminal fields, some of which are at a distance of several millimeters from one another. Three-dimensional reconstruction of a terminal field of a cerebellothalamic fiber in the monkey revealed that it had the shape of a disc elongated in the sagittal plane and flattened in the coronal plane (Mason et al., 2000). Compared to the terminal fields of pallidal fibers the terminal fields of dentate axons are more focal. The total number of synaptic boutons in all thalamic terminal fields of a single cerebellar fiber in the monkey can reach up to 300, which is remarkable compared to only 10-12 boutons formed by cerebellar fibers in the rat thalamus (Auman and Horne, 1996)

Similar to the basal ganglia terminals in the VA, cerebellothalamic terminals in the monkey VL are represented by large boutons, which are distributed to the most proximal locations of thalamocortical projection neurons and also synapse on LCN (Harding and Powell, 1977; Kultas-Ilinsky and Ilinsky, 1991; Mason et al., 1996). The principal morphological difference between basal ganglia and cerebellar terminals is in the type of synaptic contacts and synaptic vesicles. Cerebellar boutons contain preferentially round vesicles and display asymmetric synaptic contacts, i.e., synaptic active zones with distinct postsynaptic densities. This correlates well with the excitatory nature of the cerebellar afferents demonstrated in electrophysiological experiments (Uno et al., 1970; Yamamoto et al., 1983). In the EM studies cerebellar terminals are known as LR boutons, i.e., large boutons with round vesicles. EM analysis of serial sections through cerebellothalamic boutons showed that a single cerebellar terminal can establish up to 30 synapses on a single PN dendrite and one synapse on each of several LCN dendrites that in turn synapse on each other and the PN dendrite (Kultas-Ilinsky and Ilinsky, 1991; Mason et al., 1996). Three-dimensional reconstructions from these sections revealed interesting spatial relationships between various synaptic contacts: numerous active zones of a single cerebellar bouton occupy a roundish spot on the postsynaptic PN dendrite. This core is surrounded by dendro-dendritic synapses formed by several LCN dendrites on this PN dendrite, whereas at the perimeter of this area are situated several symmetric type synapses formed by a population of boutons that display ultrastructural features of reticulothalamic terminals (Mason et al., 1996). Such spatial arrangement suggests that the spread of postsynaptic current generated by a cerebellar terminal is controlled by several inhibitory systems, one provided by LCN dendrites and another by axon terminals of yet unidentified origin but most likely from the reticular thalamic nucleus.

ORGANIZATION OF THE CORTICAL INPUT TO THE VA AND VL

Regarding the topography of cortical projections to the motor thalamus in primates, the classical long-standing view on association of the VA with the premotor cortex and VL with the primary motor cortex has been confirmed with modern neuroanatomical methodology (Schell and Strick, 1984; Ilinsky et al., 1985; Inase and Tanji, 1995; Matelli and Luppino, 1996; Matelli et al., 1989). In particular, the pallidal territory of the motor thalamus (VAdc and VApc) is the main recipient of projections from the supplementary motor cortex and other parts of Brodman's area 6. The VL receives its major cortical afferent input from the primary motor cortex, whereas the nigral afferent territory of the thalamus (VAmc) is preferentially connected with area 8 (prefrontal eye field) and other prefrontal regions. Our recent studies with BDA tracing of corticothalamic fibers from the premotor and primary motor cortices revealed many additional details related to the topography and fiber organization of the terminal fields (Maldonado et al., 1999; Loukianova et al., 1999). After small BDA injections in different locations of the primary motor or premotor cortices, anterograde labeling was found throughout a rather large extent of both the VL and VA. Moreover, in contrast to the basal ganglia and cerebellar terminal fields, which faithfully confine to their respective territorial domains, corticothalamic afferents from area 6 and area 4 cross cytoarchitectonic boundaries of thalamic nuclei to terminate not only in the VA and VL but also in several other thalamic subdivisions, such as MD, intralaminar nuclei, lateroposterior nucleus, and medial pulvinar. This confirms earlier observations by Kievit and Kuypers (1977) and provides more details of the organization of terminal fields because of better resolution of the BDA technique. These observations also render further support for the inclusion of at least the mediodorsal and intralaminar nuclei under the umbrella of motor thalamus, perhaps in the capacity of motor association nuclei due to the multimodal cortical and subcortical inputs these subdivisions receive.

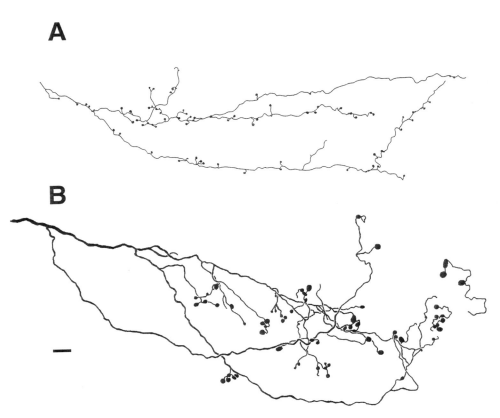

Figure 4. Terminal parts of two basic types of BDA-labeled corticothalamic axons to the motor thalamus. Small caliber fiber with small boutons and diffuse terminal fields – (A), and large caliber fiber with large boutons and focal terminal field – (B). Bar =10μm.

It is well known that in all mammals corticothalamic afferents are represented by diffuse nets of thin axons with small terminals that in EM studies have been designated as SR type boutons, i.e. small boutons with round vesicles and asymmetric synaptic contacts. These cortical boutons are distributed to the most distal dendrites of thalamocortical projection neurons and have been considered to function as the corticothalamic feedback system (see reviews by Jones, 1985; Kultas-Ilinsky and Ilinsky, 1986). Our recent studies demonstrated that the primary motor cortex (Brodman's area 4) actually sent two types of fibers to both VA and VL nuclei (Loukianova et al., 1999, Maldonado et al., 1999). One type was represented by classic corticothalamic fibers that were thin and formed diffuse terminal fields with small boutons (Figure 4A) that corresponded to SR type terminals (Figure 3D). The second type was represented by large diameter axons that formed focal terminal fields of large boutons (Figure 4B). Under EM the latter displayed typical features of LR type boutons (Figure 3C), making them quite similar to the cerebellar afferent terminals (Figure 3A). In both VA and VL, the LR boutons of cortical origin were located at more proximal locations on thalamocortical projection neurons than SR type boutons. Both types of cortical afferent terminals also formed synapses on LCN dendrites, but the LR type boutons participated in complex synaptic arrangements with LCN dendrites more often than SR type terminals.

Large-sized corticothalamic afferent terminals in primates have been described earlier, mainly in association thalamic nuclei such as pulvinar and MD (Ogren and Hendrickson, 1979; Schwartz et al., 1991). The data presented above indicate that this is not an exclusive feature of thalamic nuclei interconnected with higher order association cortices but that similar type fibers are derived also from motor cortical areas and distributed not only to the VA and VL but also to MD and intralaminar nuclei. Interestingly, large and small caliber corticothalamic fibers originating from the same injection site distribute in different proportions to different thalamic nuclei. For example, area 6 projections to the VA and VL are mostly composed of small diameter fibers, whereas the same injections give rise mostly to large caliber fibers destined to the MD, centromedian, and some other nuclei. Likewise, the posterior parietal cortex input to the VL was also represented by small caliber fibers, whereas the same injections resulted in labeling of large caliber fibers and terminals in the adjacent lateral posterior nucleus (Taktakishvili et al., 2000).

Corticothalamic input is excitatory (Deschenes and Hu, 1990), and like other afferents, it exerts dual influence on thalamocortical projection neurons via connections with LCN (Frigyesi and Schwartz, 1972). The cortical input provided by the large caliber fibers may have a very strong impact on the activity of thalamocortical neurons, because they are fast conducting, form focal terminal fields, and establish synapses closer to soma than small caliber fibers. Moreover, in each given nucleus they will influence not only the group of PN receiving direct input from them, but also adjacent groups of neurons via their synapses on LCN. The functional properties of this component of the cortical input have not yet been explored in primates or any other species, for that matter.

ORGANIZATION OF THE RETICULAR NUCLEUS INPUT TO THE VA AND VL

Among all thalamic nuclei, nucleus reticularis thalami (NRT) is quite unique in that it does not send its axons to the cortex but instead innervates all other thalamic nuclei (Scheibel and Scheibel, 1966; Jones, 1975). This feature, combined with the GABAergic nature of NRT neurons (Houser et al., 1980), makes them similar to LCN to some extent, except that there is no evidence that LCN processes extend beyond the limits of each given thalamic nucleus.

Our light microscopic studies with small iontophoretic BDA injections in the anterior pole of the monkey NRT revealed specific features of axons originating from this segment of the nucleus (Kultas-Ilinsky et al., 1995; Tai et al., 1995; Ilinsky et al., 1999). The NRT fibers branch repeatedly and are characterized by varicosities distributed along their length at irregular intervals. EM analysis confirmed that these axonal varicosities are indeed synaptic boutons *en passant,* which established long symmetric type synaptic contacts and were immunopositive for GABA. In most cases, one such bouton formed synapses on several postsynaptic structures. In the monkey, both LCN and PN receive NRT input in approximately equal proportions in all nuclei studied (Kultas-Ilinsky et al., 1995; Tai et al., 1995; Ilinsky et al., 1999). NRT terminals preferentially contact distal dendrites of projection neurons in most of the nuclei of the motor thalamus. However, in the VL and MD, NRT terminals were also found at proximal PN locations.

Terminal plexuses of individual NRT fibers are rather extensive, encompassing large territories within each nucleus of the motor thalamus and thus providing input to a large number of neurons (Ilinsky et al., 1999). It is important to note that one NRT segment in the anterior pole sends axons to several thalamic nuclei including VA, VL, and MD. Nonetheless, individual axons form plexuses in either one of these three nuclei without extending into adjacent regions except for the paracentral nucleus. Thus, as of now there is no evidence showing innervation of two or more motor thalamic nuclei by a single NRT axon. Nonetheless, since adjacent neurons in the anterior pole contact each other by their axon collaterals (Williamson et al., 1994), there may be intrathalamic interaction between the two nuclei via these connections.

SPINAL CORD AND OTHER POTENTIAL SUBCORTICAL INPUTS TO THE MOTOR THALAMUS

Aside from the spinal cord projections to the somatosensory and intralaminar thalamic nuclei, there is also an indication that spinal cord afferents are sparsely distributed to the motor thalamus (Craig and Burton, 1985; Albe-Fessard et al., 1985; Berkely, 1980). In our unpublished studies in the cat with WGA-HRP injections in the spinal cord, we were able to trace individual spinothalamic fibers to the VL, where they provided diffuse low density innervation to the entire nucleus. These boutons terminated as LR type boutons on thalamocortical projection neurons similar to the spinothalamic afferents in the somatosensory thalamus as described by Ralston et al. (1994). The physiological evidence for spinothalamic input to the motor thalamus is discussed in detail in the chapter by R. Mackel in this volume.

Scattered reports on other inputs to the motor thalamus have appeared in the literature. Afferents from the vestibular nuclear complex to the motor thalamic nuclei have been suggested by Hassler (1948), but they were not confirmed in later neuroanatomical tracing study by Tarlov (1969). The latter study suggested that the vestibular input reaching the motor thalamus may be mediated by brain stem thalamic connections. Diffuse, low density cholinergic innervation of the thalamus, including motor thalamic nuclei, from a variety of brain stem cell groups and basal forebrain has been demonstrated by Heckers et al. (1992). More studies are needed on these afferent inputs to the motor thalamus to understand their role in shaping the activity of thalamocortical projection neurons. An issue related to this connectivity is the input from the lateral globus pallidus to the reticular thalamic nucleus and other thalamic nuclei, as described by Parent and Hazrati (1995) (see also the chapter by Morel in this volume). In our tracing experiments we have not obtained any evidence for this connection. At this point one can not exclude the possibility that the labeling observed in these studies might have been from the basal forebrain neurons that closely impinge on the globus pallidus.

Finally, a significant input to the VAmc, i.e., nigral afferent territory, from the superior colliculus was reported by Harting et al. (1980). To study this projection in detail we performed a series of experiments with injections of various tracers in the rhesus monkey superior colliculus. We did confirm a projection from the superior colliculus to the MD.

However, the anterograde labeling in the VAmc turned out to be due to the thalamic collaterals of nigrocollicular axons. The BDA picked up from the injection sites in the superior colliculus by branches of nigral efferents was retrogradely transported to the axon branching site and from their anterogradely to the VAmc (Tai et al., 1996).

SUMMARY ON ORGANIZATION OF AFFERENT INPUTS TO THE MOTOR THALAMUS

There seems to be a common blueprint of anatomical organization of the motor thalamic nuclei and their afferent inputs in primates in that practically all afferents terminate on projection and local circuit neurons. This means that each afferent system delivers dual input to PN, one direct and another indirect, mediated by GABAergic LCN.

The main differences between the motor thalamic nuclei are accounted for by differences between the specific neurotransmitters associated with the major subcortical afferent systems, as well as by the varying degrees of complexity of the LCN circuitry. Some afferents are excitatory (i.e., cortical and cerebellar), and some are inhibitory (nigral, pallidal and reticular). The indirect inputs from these systems, mediated by LCN, may have a variety of effects on PN depending upon the complexity of interposed LCN connections specific for each afferent system and nucleus. The LCN connections are the most complex in the VL, not to mention MD, and the least complex in the VAmc.

Topography of afferent terminal fields of single pallidal and cerebellar axons in respective thalamic territories is such that it allows them to influence different groups of thalamic neurons and hence movements of different joints. This is in clear agreement with functional studies (Anderson and Turner, 1991; Rispal-Padel, 1993; Vitek et al., 1996).

There appears to be more integration of cortical information taking place in each motor thalamic nucleus than previously thought, since several functionally different cortical regions reach VA and VL. The dual motor cortex input to VA and VL neurons, provided by large and small fibers respectively, calls for re-evaluation of the functional significance of this connection and thalamic processing in general.

Thus, despite the common organizational principle underlying the structure of the nuclei of the motor thalamus, the mechanisms of information processing in them should be different due to differences in the intrinsic circuitry and cortical inputs. These differences should definitely manifest, not only in the normal output from these nuclei, but also in pathological states, i.e., various movement disorders.

SOME FUNCTIONAL AND CLINICAL CONSIDERATIONS

In fact, a number of indirect clinical observations suggest the existence of two functionally distinct areas in the human motor thalamus. It has been noted during stereotactic surgery in parkinsonian patients that the effect of low frequency electrical stimulation on the existing tremor depends on the topographical location of stimulating electrodes within the motor thalamus (Narabayashi, 1968; Ilinsky, 1970). The latter studies revealed that tremor amplitude showed a sharp increase as a result of electrical stimulation applied within the VL (Vop+Vim area in Hassler's nomenclature), whereas a significant decrease of the tremor amplitude or tremor arrest was noted in the same patient using identical stimulation parameters when the electrode was positioned in the VA (Lpo+Voa area in Hassler's nomenclature). The mechanism of this phenomenon is still open for speculation.

Based on our current knowledge of the anatomical organization of neuronal circuits in the VL and VA, one can assume impairment or alteration in physiological properties of basal ganglia inputs to motor thalamus in parkinsonian patients. Long-term structural and/or functional changes in the circuitry can trigger synaptic reorganization. We predict that the local neuronal circuits will be the first to react to pathologic alterations of afferent inputs. The resulting abnormal thalamic circuits may contribute to tremor manifestation in parkinsonian patients and for that matter to other clinical manifestations in a wide spectrum of movement disorders. Reorganization of connections in somatosensory thalamus of dystonia patients has been demonstrated in clinical studies by Lenz et al. (1999), (see also chapter by Lenz et al. in this volume). Similar reorganization could be occurring in the motor thalamic nuclei. Moreover, one can expect that reorganization may proceed differently in the VA and VL due to differences in normal neuronal organization. This may be one of the possible mechanisms of differential effects of thalamic stimulation upon parkinsonian tremor in the basal ganglia and cerebellar afferent territories of the motor thalamus.

REFERENCES

Albe-Fessard, D., Berkley, K.J., Kruger, L., Ralston, H.J. III, and Willis, W.D. Jr., 1985, Diencephalic mechanisms of pain sensation, *Brain Res. Rev.* 9:217.

Anderson, M.E., and Turner, R.S., 1991, Activity of neurons in cerebellar-receiving and pallidal-receiving areas of the thalamus of the behaving monkey, *J. Neurophysiol.* 66:879.

Arecchi-Bouchhioua, P., Yelnik, J., François, C., Percheron, G., and Tandé, D., 1996, 3-D tracing of biocytin labeled pallidothalamic axons in the monkey, *Neuro. Report.* 7:981.

Asanuma, C., Thach, W.T., and Jones, E.G., 1983, Anatomical evidence for segregated focal groupings of efferent cells and their terminal ramifications in the cerebellothalamic pathway of the monkey, *Brain Res. Rev.* 5:267.

Auman, T.D., and Horne, M.K., 1996, Ramification and termination of single axons in cerebellothalamic pathway of the rat, *J. Comp. Neurol.* 376:420.

Ballercia, G., Kultas-Ilinsky, K., Bentivoglio, M., and Ilinsky, I.A., 1966, Neuronal and synaptic organization of the centromedian nucleus of the monkey thalamus: A quantitative ultrastructural study with tract tracing and immunohistochemical observations, *J. Neurocytol.* 25:267.

Berkley, K.J., 1980, Spatial relationships between the terminations of somatic sensory and motor pathways in the rostral brain stem of cats and monkeys. I. Ascending somatic sensory inputs to lateral diencephalon, *J. Comp. Neurol.* 193:283.

Carpenter, M.B., Nakano, K., Kim, R., 1976, Nigrothalamic projections in the monkey demonstrated by autoradiographic techniques, *J. Comp. Neurol.* 165:401.

Craig, A.D., and Burton, H., 1985, The distribution and topographical organization in the thalamus of anterogradely-transported horseradish peroxidase after spinal injections in cat and raccoon, *Exp. Brain Res.* 58:227.

Deniau, J.M., Lackner, D., Feger, J., 1978, Effect of substantia nigra stimulation on identified neurons in the VL-VA thalamic complex: comparison between intact and chronically decorticated cats, *Brain Res.* 145:27.

Deschenes, M., and Hu, B., 1990, Electrophysiology and pharmacology of the corticothalamic input to the lateral thalamic nuclei: an intracellular study in the cat, *Eur. J. Neurosci.* 2:140.

DiChiara, G., Porceddu, M.L., Morelli, M., Mulas, M.L., Gessa, G.L., 1979, Evidence for a GABAergic projection from the substantia nigra to the ventromedial thalamus and to the superior colliculus of the rat, *Brain Res.* 176:273.

Frigyesi, T.L., and Schwartz, R., 1972, Cortical control of thalamic sensorimotor relay activities in the cat and the squirrel monkey, in: *Corticothalamic Projections and Sensorymotor Activities*, T.L. Frigyesi, E. Rinvik, and M.D. Yahr, eds., Raven, New York.

Harding, B.N., and Powell, T.P.S., 1977, An electron microscopic study of the centre-median and ventrolateral nuclei of the thalamus in the monkey, *Philos. Trans. R. Soc. Lond. (Biol)* 279:359.

Harting, B.N., Huerta, A.J., Franfurter, N.L., Strominger, N.L., and Royce, G.J., 1980, Ascending pathways from the monkey superior colliculus: An autoradiographic analysis, *J. Comp. Neurol.* 192:853.

Hassler, R., 1948, Forels Haubenfascizikel als vestibulare Empfindungsbahn mit Bemerkungen uber einige andere secundare Bahnen des Vestibularis und Trigeminus, *Arch. Psychiatr.* 194:23.

Hassler, R., 1959, Anatomy of the thalamus, in: *Introduction to Stereotactic Operations with an Atlas of the Human Brain.*, G. Schaltenbrand and P. Bailey, eds., Thieme, Stuttgart.

Heckers, S., Geula, C., and Mesulam, M., 1992, Cholinergic innervation of the human thalamus: Dual origin and differential nuclear distribution, *J. Comp. Neurol.* 325:68.

Hirai, T., Ohye, C., Nagaseki, Y. et al., 1989, Cytometric analysis of the thalamic ventralis intermedius nucleus in humans, *J. Neurophysiol.* 61:478.

Hirai, T., and Jones, E.G., 1989, A new parcelation of the human thalamus on the basis of histochemical staining, *Brain Res. Rev.* 14:1.

Houser, C.R., Vaughn, J.E., Barber, R.P., Roberts, E., 1980, GABA neurons are the major cell type of the nucleus reticularis thalami, *Brain Res.* 200:341.

Ilinsky, I.A., 1970, Identification of the ventrolateral nucleus of the thalamus by electrical stimulation in parkinsonian patients during stereotactic surgery; Ph.D. thesis, Burdenko Institute of Neurosurgery, Academy of Medical Sciences, Moscow.

Ilinsky, I.A., and Kultas-Ilinsky K., 1984, An autoradiographic study of topographical relationships between pallidal and cerebellar projections to the cat thalamus, *Exp. Brain Res.* 45:95.

Ilinsky, I.A., and Kultas-Ilinsky, K., 1987, Sagittal cytoarchitectonic maps of the *Macaca mulatta* thalamus with a revised nomenclature of the motor-related nuclei validated by observations on their connectivity, *J. Comp. Neurol.* 262:331.

Ilinsky, I.A., and Kultas-Ilinsky, K., 1990, Fine structure of the magnocellular subdivision of the ventral anterior thalamic nucleus (VAmc) of *Macaca mulatta*: I. Cell types and synaptology, *J. Comp. Neurol.* 294:455.

Ilinsky, I.A., Ambardekar, A.V., and Kultas-Ilinsky, K., 1999, Organization of projections from the anterior pole of the nucleus reticularis thalami (NRT) to subdivisions of the motor thalamus: Light and electron microscopic studies in the Rhesus monkey, *J. Comp. Neurol.* 409:369.

Ilinsky, I.A., Jouandet, M.L., Goldman-Rakic, P.S., 1985, Organization of the nigrothalamocortical system in the rhesus monkey, *J. Comp. Neurol.* 236:315.

Ilinsky, I.A., Kultas-Ilinsky, K., and Smith, K.R., 1982, Organization of the basal ganglia inputs to the thalamus. A light and electron microscopic study in the cat, *Appl. Neurophysiol.* 45:230.

Ilinsky, I., Tourtellotte, W.G., and Kultas-Ilinsky, K., 1993a, Anatomical distinctions between two basal ganglia afferent territories in the primate motor thalamus, *Stereotact. Funct. Neurosurg.* 60:62.

Ilinsky, I.A., Toga, A.W., and Kultas-Ilinsky, K., 1993b, Anatomical organization of internal neuronal circuits in the motor thalamus, in: *Thalamic Networks for Relay and Modulation,* D. Minciacchi, M. Molinari, G. Macchi and E.G. Jones, eds., Pergamon Press, New York.

Ilinsky, I.A., Yi, H., and Kultas-Ilinsky, K., 1997, The mode of termination of pallidal afferents to the thalamus: a light and electron microscopic study with anterograde tracers and immunocytochemistry in *Macaca mulatta*, *J. Comp. Neurol.* 384:603.

Inase, M, and Tanji, J., 1995, Thalamic distribution of projection neurons to the primary motor cortex relative to afferent terminal fields from the globus pallidus in the macaque monkey, *J. Comp. Neurol.* 353:415.

Jones, E.G., 1975, Some aspects of the organization of the thalamic reticular complex, *J. Comp. Neurol.* 162:385.

Jones, E.G., 1985, *The Thalamus*, Plenum Press, New York.

Kievit, J., and Kuypers, H.G.J.M., 1977, Organization of the thalamocortical connections to the frontal lobe in the Rhesus monkey, *Exp. Brain Res.* 29:299.

Kim, R., Nakano, K., Jayaraman, A., Carpenter, M.B., 1976, Projections of the globus pallidus and adjacent structures: an autoradiographic study in the monkey, *J. Comp. Neurol.* 169:263.

Kultas-Ilinsky, K., and Ilinsky, I.A., 1990, Fine structure of the magnocellular subdivision of the ventral anterior thalamic nucleus (VAmc) of *Macaca mulatta*: II. Organization of nigrothalamic afferents as revealed with EM autoradiography, *J. Comp. Neurol.* 294:479.

Kultas-Ilinsky, K. Reising, L., Yi, H., and Ilinsky, I.A., 1997, Pallidal afferent territory of the *Macaca mulatta* thalamus: neuronal and synaptic organization of the VAdc, *J. Comp. Neurol.* 386:573.

Kultas-Ilinsky, K., and Ilinsky, I.A., 1986, Neuronal and synaptic organization of the motor nuclei of mammalian thalamus, in: *Current Topics in Research on Synapses*, D.G. Jones, ed., Wiley-Liss, New York.

Kultas-Ilinsky, K., and Ilinsky, I.A., 1988, GABAergic systems of the feline motor thalamus: neurons, synapses and receptors, in: *Cellular Thalamic Mechanisms*, M. Bentivoglio and R. Spreafico, eds., Elsevier, Amsterdam.

Kultas-Ilinsky, K., and Ilinsky, I.A., 1991, Fine structure of the ventral lateral nucleus (VL) of the *Macaca mulatta* thalamus: Cell types and synaptology, *J. Comp. Neurol.* 314:319.

Kultas-Ilinsky, K., Yi, H., and Ilinsky, I.A, 1995, Nucleus reticularis thalami input to the anterior thalamic nuclei in the monkey: light and electron microscopy study, *Neurosci. Lett.* 186:25.

Kuo, J.S., and Carpenter, M.B., 1973, Organization of pallidothalamic projections in the Rhesus monkey, *J. Comp. Neurol.* 151:201.

Lenz, F.A., Jaeger, C.J., Seike, M.S., Lin, Y.C., Reich, S.G., DeLong, M.R., Vitek, J.L., 1999, Thalamic single neuron activity in patients with dystonia: Dystonia-related activity and somatic sensory reorganization, *J. Neurophysiol.* 82:2372.

Lieberman, A.R., 1973, Neurons with presynaptic perikarya and presynaptic dendrites in the rat lateral geniculate nucleus, *Brain Res.* 59:35.

Loukianova, E., Ilinsky, I.A., and Kultas-Ilinsky, K., 1999, Four types of afferent terminals from the primary motor cortex in the ventral lateral thalamic nucleus (VL): BDA tracing study in *Macaca Mulatta, Soc. Neurosci. Abstr.* 25:1407.

Macchi, G., and Jones, E.G., 1997, Toward an agreement on terminology of nuclear and subnuclear divisions of the motor thalamus, *J. Neurosurg.* 86:670.

Maldonado, S., Loukianova, E., Kultas-Ilinsky, K., and Ilinsky, I.A., 1999, Connections of the ventral anterior thalamic nucleus (VAdc) with motor and premotor cortices: light and electron microscopic study in *Macaca Mulatta, Soc. Neurosci. Abstr.* 25:1407.

Mason, A., Ilinsky, I.A., Beck, S., and Kultas-Ilinsky, K., 1996, Reevaluation of synaptic relationships of cerebellar terminals in the ventral lateral nucleus of the rhesus monkey thalamus based on serial section analysis and three-dimensional reconstruction, *Exp. Brain Res.* 109:219.

Mason, A., Ilinsky, I.A., Maldonado, S., and Kultas-Ilinsky, K., 2000, Thalamic terminal fields of individual axons from the ventral part of the dentate nucleus of the cerebellum in *Macaca mulatta, J. Comp. Neurol.* 421:412.

Matelli, M., and Luppino, G., 1996, Thalamic input to mesial and superior area 6 in the macaque monkey, *J. Comp. Neurol.* 372:59.

Matelli, M., Luppino, G., Fogassi, L., and Rizzolatti, G., 1989, Thalamic input to inferior area 6 and area 4 in the macaque monkey, *J. Comp. Neurol.* 280:468.

Narabayashi, H., 1968, Functional differentiation in and around the ventrolateral nucleus of the thalamus based on experience in human stereoencephalotomy, *Johns Hopkins Med. J.* 122:295.

Nauta, H.J.W., and Mehler, W.R., 1966, Projections of the lentiform nucleus in the monkey, *Brain Res.* 1:3.

Ogren, M.P., and Hendrickson, A.E., 1979, The morphology and distribution of striate cortex terminals in the inferior and lateral subdivisions of the Macaca monkey pulvinar, *J. Comp. Neurol.* 188:179.

Ohara, P.T., Ralston, H.J. III, Ralston, D.D., 1987, The morphology of neurons and synapses in the somatosensory thalamus of the cat and monkey, in: *Thalamus and Pain*, J.-M.,Besson, G. Guilbaud, and M. Peschanski, eds., Elsevier, Amsterdam.

Olszewski, J., 1952, *The Thalamus of Macaca Mulatta. An Atlas for Use with Stereotactic Instrument.* Karger, Basel.

Parent, A., and Hazrati, L.-N., 1995, Functional anatomy of the basal ganglia. II. The place of the subthalamic nucleus and external pallidum in basal ganglia circuitry, *Brain Res. Rev.* 20:128.

Penney, J.B., Young, A.B., 1981, GABA as the pallidothalamic neurotransmitter: implications for basal ganglia function, *Brain Res.* 207:195.

Percheron, G., 1977, The thalamic territory of cerebellar afferents and the lateral region of the thalamus of the macaque in stereotaxic ventricular coordinates, *J. Hirnforsch.* 18:376.

Percheron, G., François, C., Talbi, B., Yelnik, J., and Fénelon, G., 1996, The primate motor thalamus, *Brain Res. Rev.* 22:93.

Ralston, H.J. III, and Ralston, D.D., 1994, Medial lemniscal and spinal projections to the Macaque thalamus: An electron microscopic study of differing GABAergic circuitry serving thalamic somatosensory mechanisms, *J. Neurosci.* 14:2485.

Rispal-Padel, L., 1993, Contribution of cerebellar efferents to the organization of motor synergy, *Rev. Neurol.* 149:716.

Scheibel, M.E., and Scheibel, A.B., 1966, The organization of the nucleus reticularis thalami: a Golgi study, *Brain Res.* 1:43.

Schell, G.R., Strick, P.L., 1984, The origin of thalamic inputs to the arcuate premotor and supplementary motor areas, *J. Neurosci.* 4:539.

Schwartz, M.L., Dekker, J., and Goldman-Rakic, P.S., 1991, Dual mode of corticothalamic synaptic termination of the mediodorsal nucleus of the rhesus monkey, *J. Comp. Neurol.* 309:289.

Sidibe, M., Bevan, M.D., Bolam, J.P., and Smith, Y. 1997, Efferent connections of the internal globus pallidus in the squirrel monkey: I. Topography and synaptic organization of the pallidothalamic projection, *J. Comp. Neurol.* 382:323.

Tai, Y., Yi, H., Ilinsky, I.A., and Kultas-Ilinsky, K., 1995, Nucleus reticularis thalami connections with the mediodorsal thalamic nucleus: a light and electron microscopy study, *Brain Res. Bull.* 38:475.

Tai, Y., Kultas-Ilinsky, K., and Ilinsky I.A., 1996, Superior colliculus projections to mediodorsal (MD) and ventral anterior (VAmc) thalamic nuclei in the monkey, *Soc. Neurosci. Abstr.* 22:2031.

Taktakishvili, O., Sivan-Loukianova, E., Ilinsky, I.A., and Kultas-Ilinsky, K., 2000, Posterior parietal cortex connections with the ventrolateral and lateroposterior thalamic nuclei of the rhesus monkey, in: *Abstracts of the International Workshop on Basal Ganglia and Thalamus in Health and Movement Disorders*, Moscow, p. 52.

Tarlov, E., 1969, The rostral projections of the primate vestibular nuclei: An experimental study in macaque, baboon and chimpanzee, *J. Comp. Neurol.* 135:27.

Ueki, A., 1983, The mode of nigrothalamic transmission investigated with intracellular recording in the cat, *Exp. Brain Res.* 49:116.

Uno, M., and Yoshida, M., 1970, The mode of cerebellothalamic relay transmission investigated with intracellular recording from cells of the ventrolateral nucleus of the cat's thalamus, *Exp. Brain Res.* 10:121.

Uno, M., Ozawa, N., and Yoshida, M., 1978, The mode of pallidothalamic transmission investigated with intracellular recording from the cat thalamus, *Exp. Brain Res.* 33:493.

Vitek, J.L., Ashe, J., DeLong, M., and Alexander, G.E., 1994, Physiologic properties and somatotopic organization of the primate motor thalamus, *J. Neurophysiol.* 71:1498.

Vitek, J.L, Ashe, J., DeLong, M., and Kaneoke, Y., 1996, Microstimulation of primate motor thalamus: somatotopic organization and differential distribution of evoked motor responses among subnuclei, *J. Neurophysiol.* 75:2486.

Williamson, A.M., Ohara, P.T., Ralston, D.D., Milroy, A.M., and Ralston, H.J., 1994, Analysis of gamma-aminobutyric acidergic synaptic contacts in the thalamic reticular nucleus of the monkey, *J. Comp. Neurol.* 349:182.

Yamamoto, T., Hassler, R., Huber, C., Wagner, A., and Sasaki, K., 1983, Electrophysiologic studies on the pallido- and cerebello-thalamic projections in squirrel monkeys (*Saimiri sciureus*), *Exp. Brain Res.* 51:77.

Yen, C. T., Conley, M., and Jones E.G., 1985, Morphological and functional types of neurons in cat ventral posterior thalamic nucleus, *J. Neurosci.* 5:1316.

PALLIDAL AND CORTICAL DETERMINANTS
OF THALAMIC ACTIVITY

M. E. ANDERSON, MARK RUFFO, JOHN A. BUFORD, AND
MASAHIKO INASE[*]

INTRODUCTION

Output from the basal ganglia reaches the cerebral cortex via the thalamus. In the primate, GABAergic axons from the internal pallidal segment (GPi) and the substantia nigra, pars reticulata (SNr) provide an inhibitory input to the ventral anterior (VA), oral, and some of the caudal portions of the ventrolateral (VLo and VLc), the ventromedial (VM), and to some extent, the dorsomedial (MD) thalamic nuclei (Sakai et al., 1996; Kultas-Ilinsky et al., 1997; Ilinsky et al., 1993; DeVito and Anderson, 1982). Changes in this inhibition are presumed to be major contributors to the symptoms of Parkinson's disease (Hutchison et al., 1994; Wichmann et al., 1999), and several therapeutic interventions, such as pallidotomy or deep brain stimulation, are directly targeted at modification of pallidal output.

The activity of thalamocortical neurons will be the product of all of their synaptic inputs plus their intrinsic membrane properties. The basal ganglia are not the only sources of input to their target thalamic nuclei. As is the case with other nuclei of the thalamus, a major source of glutamatergic axon terminals is the cerebral cortex (Fonnum et al., 1981; McCormick and von Krosigk, 1992; Vidnyanszky et al., 1996). Inhibitory inputs also originate from the thalamic reticular nucleus and local circuit inhibitory neurons of the thalamus (Kultas-Ilinsky et al., 1995; Kultas-Ilinsky et al., 1997). Other potential sources of input are not well documented in the primate, although prior studies in carnivores present possibilities (Anderson and DeVito, 1987; Nakano et al., 1985; Jimenez-Castellanos and Reinoso-Suarez, 1985).

We have sought to determine whether there are other potential sources of synaptic input that might converge with basal ganglia input to the thalamus in primates. We also have examined the changes in thalamic cell activity - and in arm position and movement -

[*] MARJORIE E. ANDERSON • Departments of Rehabilitation Medicine and Physiology and Biophysics, and Program in Neurobiology and Behavior, University of Washington, Seattle, WA 98195. MARK RUFFO • Program in Neurobiology and Behavior, University of Washington, Seattle, WA 98195. JOHN A. BUFORD • Division of Physical Therapy, The Ohio State University, Columbus, Ohio 43210. MASAHIKO INASE • Department of Physiology, Kinki University School of Medicine, Osaka-Sayama, Osaka 589-8511, Japan.

when pallidal activity is manipulated by injection of the inhibitory agent muscimol or by microstimulation within the globus pallidus with short or long duration stimulus trains.

ANATOMICAL IDENTIFICATION OF POTENTIAL CONVERGING INPUTS TO BASAL GANGLIA-RECEIVING (BG-R) THALAMUS

To identify neurons whose axons could potentially provide input that converged with that from the basal ganglia, we have injected wheat germ agglutinated horseradish peroxidase (WGA-HRP) into the rostral thalamus. Injections were centered either at sites chosen stereotaxically or at positions just rostral to the microexcitable zones of the thalamus that we and others have shown to receive afferents primarily from the cerebellar nuclei (Buford et al., 1996). The centers of two such injections in one animal are shown in Figure 1. Both injections were centered quite dorsally, in VA and VLc. That on the right extended more ventrally. (A cannula track lesion distorts the tissue below the right injection.)

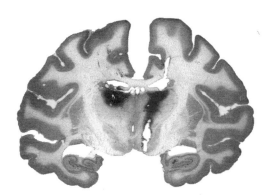

Figure 1. WGA-HRP injection sites into the anterior thalamus.

Neurons in GPi were labeled on both sides, as shown in Figure 2A. Curiously, the labeled neurons were distributed in a ring-like distribution on the outer borders of each GPi. These dorsal injections resulted in little labeling of neurons in SNr, especially on the left.

Neither of these injections was exclusively in BG-R regions of the thalamus, however. As shown in Figure 2B, neurons in the deep cerebellar nuclei also were labeled retrogradely. This would be expected based on the adjacent and interdigitating location of pallidal and cerebellar-receiving areas of thalamus, as shown by other techniques (DeVito and Anderson, 1982; Sakai et al., 1996; Sidibe et al., 1997).

What other neurons were labeled retrogradely by injections into these thalamic sites? Quantitatively, the cerebral cortex provided the greatest number of neurons with input to the thalamic injection sites. As illustrated in Figure 3A, neurons labeled by the injections shown in Figure 1 were located in the pre-SMA, the dorsal bank of the cingulate cortex, and the dorsal prefrontal cortex. Some brainstem sites also had limited numbers of labeled neurons, potential sources of afferents to BG-R thalamus. As illustrated in Figure 3B, these included the superior colliculus, the pedunculopontine region, and a limited area of the pretectum (not illustrated). Neurons in the hypothalamus were also labeled retrogradely, probably because the injection impinged on the anterior thalamic nuclei (Somogyi, 1978). All of these sites had also contained retrogradely labeled neurons after small thalamic injections of HRP in an earlier study in the cat (Anderson and DeVito, 1987), although those in the pretectum only appeared in cases in which cells in the cerebellar nuclei, as well as the basal ganglia, were labeled. Thus, the cerebral cortex, usually the cortical area to which thalamocortical connections from the same areas of thalamus are made, provides the major source of information converging with basal ganglia output at the thalamus. At present we know almost nothing about the signals carried by these corticothalamic neurons. We do know that they originate from both layers V and VI (Jones, 1985), and Pare et al. (1996) have shown that at least some of those from layer V appear to send branched axons

A
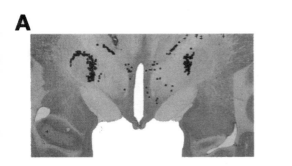

B
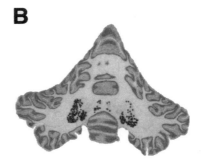

Figure 2. The locations of neurons labeled retrogradely after the WGA-HRP injections in the thalamus shown in Figure 1. (A) Crescent-shaped arrangement of labeled neurons in GPi, shown at this level on the left. Labeled neurons are also present in the reticular thalamic nucleus and scattered in the hypothalamus. (B) Labeled neurons in the deep cerebellar nuclei.

to the striatum and the thalamus. Swadlow and his colleagues, who have identified corticothalamic axons antidromically in awake rabbits, have shown that those from layer VI cortex are very slowly conducting (Swadlow, 1994) and show little modulation in their activity during locomotion.

A
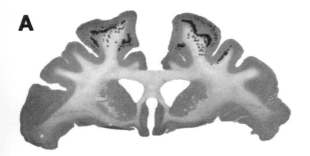

B
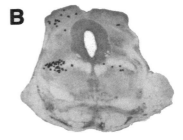

Figure 3. The locations of neurons labeled retrogradely in the cerebral cortex and brainstem after the injections of WGA-HRP shown in Figure 1. (A) Retrogradely labeled neurons in the prefrontal and dorsal cingulate cortex. Large dots: Retrogradely labeled neurons. Small dots: anterogradely labeled terminal fields. Crosses: labeled fibers. (B) Retrogradely labeled neurons in superior colliculus and pedunculopontine region.

CONSEQUENCES OF INHIBITION OF PALLIDAL DISCHARGE

Several investigators have used local injection of the GABA$_A$ agonist muscimol to determine the consequences of reversible inactivation of a portion of the pallidal output (Inase et al., 1996; Mink and Thach, 1991; Wenger et al., 1999). We have used this technique in monkeys trained to make reaching movements, and we have simultaneously examined the kinematic characteristics of hand position and the activity of thalamic neurons in the BG-R region (Inase et al., 1996).

Two behavioral changes followed muscimol injection within 5-45 minutes, with the latency depending on the amount and site of the injection. First, as shown in Figure 4, the hand drifted from the center start position until the center light changed color. When this happened, the animal returned his hand to the center zone, whereupon it began to drift again. This drift, which most commonly was in the flexor direction, was an active process, as shown by the accompanying activity of the biceps muscle, also illustrated in Figure 4. The biceps activity ceased when the center light changed color, and the hand was returned to the "on" position.

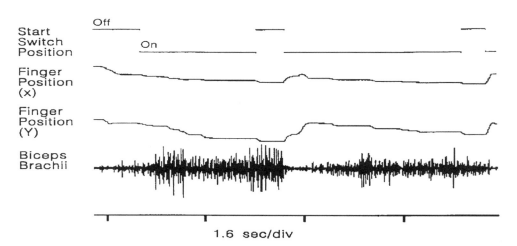

Figure 4. Active drift of hand position after injection of muscimol into GPi. (From Inase et al., 1996.)

When the animal held the home position long enough to activate a trigger tone, movements to peripheral targets were slowed. This was most consistently manifested as a decrease in peak velocity for movements in multiple target directions. Reaction time changed less consistently and, in contrast to reports of eye movements studied when muscimol was injected into the substantia nigra (Hikosaka and Wurtz, 1985), changes in peak velocity occurred during hand movements made to visible, as well as remembered, target locations.

Neurons in the pallidal-receiving areas of the thalamus showed changes in tonic activity after the injection of muscimol (Inase et al., 1996). An example of the usual increase in tonic activity is shown in Figure 5. Of fifteen cells whose activity was recorded before and after injection of muscimol, eight had a significant change in tonic discharge rate after the injection while the hand was in the home zone, and for seven of the eight, this was an increase in rate. These included all cells studied during injections that produced clear drift within 5-10 minutes (although two cells had short latency increases in tonic rate without clear postural drift).

In none of the cells were task-related phasic changes in discharge abolished after injection of muscimol. As shown in Figure 5, however, the phasic activity could be superimposed on an increased tonic rate, giving a decrease in the signal-to-noise ratio.

Pre-Muscimol   **Post-Muscimol**

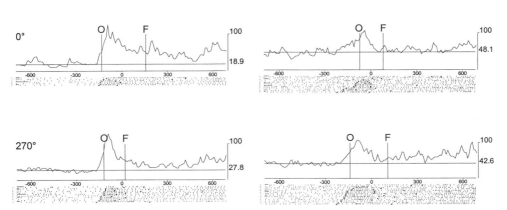

Figure 5. Activity of a thalamic neuron before and after injection of muscimol into GPi. Left column: Before injection. Right column: After injection. Upper panels: Trials in which movements of the right arm were made to the right (target at 0 degrees). Lower panels: Trials in which right arm movements were made toward the animal (target at 270 degrees). All data are aligned on the initiation of movement (time 0), and rasters are reordered from shortest (top) to longest (bottom) movement times. Average discharge rate during the hold time before the trigger tone is indicated by the horizontal line through the average instantaneous firing rate record at the top of each panel. O: Average time of movement initiation. F: Average time of movement termination. (From Inase et al., 1996.)

The increase in tonic discharge rate of thalamic neurons would be predicted when the usually highly active GPi neurons are inhibited, and we hypothesized that this led to a destabilization that resulted in drift of the arm. The slowed arm movements, which also have been reported by others following injections of kainic acid or muscimol into GPi (Horak and Anderson, 1984a; Mink and Thach, 1991; Inase et al., 1996; Wenger et al, 1999), may have been a primary consequence of turning off pallidal inhibition at the thalamus (and other targets), but they also may have been a compensatory mechanism used to stabilize the arm and reduce the postural drift described above.

Slowed arm movements following pallidal inactivation contrast with the speeded saccadic eye movements reported after injection of muscimol into SNr, especially to remembered target locations (Hikosaka and Wurtz, 1985). It is important to recognize that stabilization against varying loads is a requirement for the arms, and co-contraction of antagonistic muscles or "braking" contraction of the antagonists is available for these functions. In contrast, saccadic eye movements are made against predictable loads with no antagonist burst to terminate the eye on the target. Thus, the circuitry "released" when inhibitory control by the basal ganglia is removed must be different for eye and arm movements.

CONSEQUENCES OF HIGH-FREQUENCY STIMULATION IN GPi

Stimulation in GPi is another way to disrupt the normal signals that emanate from the basal ganglia during behavior. Horak and Anderson (1984b) used this technique to probe the effect of high-frequency stimulation on reaching behavior. We found that the behavioral consequence of microstimulation depended on both the location and the timing of the stimulus train applied to GPi.

At most stimulus locations in GPi and GPe, long duration stimulus trains applied during the reaction and movement times (monopolar, 50-100 µA, 300 Hz for 500 ms) re-

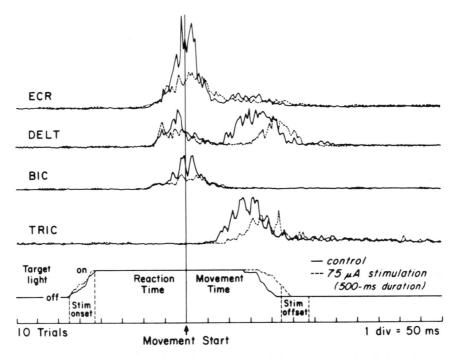

Figure 6. Stimulus-induced slowing of reaching movements. ECR, extensor carpi radialis; DELT, anterior deltoid; BIC, biceps brachii; TRIC, triceps brachii. (From Horak and Anderson, 1984b.)

sulted in prolonged movement times. This was accompanied by a decrease in the peak amplitude of EMG bursts in muscles with early activity and a delay in the activity of muscles normally activated later during the reach (see Figure 6). The magnitude of the increase in movement time depended on the stimulus intensity (Figure 7), and there was a critical time period, approximately 50 to 150 ms prior to movement initiation, during which the stimulus train had to be applied to elicit slowed movements (Anderson and Horak, 1985).

During stimulation at some stimulus locations, however, particularly those in ventroposterolateral GPi, movements became faster instead of slower than controls (see Figure 8). Our speculation at the time of this early study was that the differential effects of high-frequency stimulation on movement times were a consequence of direct activation of projection neurons in GPi from some stimulus positions and synaptic inhibition of the projection neurons from other stimulus positions. The numerous inhibitory axons that originate from the striatum or GPe and course through these two structures could be activated from stimulation at many sites en route to their synaptic destinations on GPi projection neurons (Horak and Anderson, 1984b).

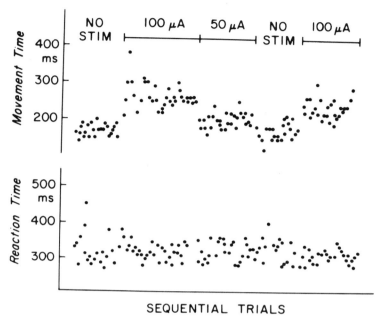

Figure 7. Prolongation of movement times is dependent on the intensity of stimulation in the globus pallidus. (From Horak and Anderson, 1984b.)

More recently we have applied longer duration high-frequency stimulus trains between two electrodes positioned in GPi and have examined the effects of this stimulation on the activity of target thalamic neurons in VA and VLo. In our preliminary studies, in which stimuli have been applied during the hold period prior to visual and auditory signals used to elicit targeted reaching movements, changes in the neuronal activity of target thalamic neurons also have been mixed. Examples are shown in Figures 10 and 11.

The activity of thalamic neurons can change in complex ways in response to stimulation in GPi. Stimulation with 1-2 pulses in GPi produced an initial depression, with or without a subsequent facilitation, in most thalamic neurons studied (see Figure 9) (Buford et al., 1996; Inase et al., 1996). Suppression followed by facilitation was the response in 31 of the 51 thalamic neurons examined in the latter study. Another 11 showed suppression only, 5 showed only facilitation, and 4 had more complex oscillatory responses. Suppression occurred at a latency of 6 ms or less in 2/3 of the cases, whereas facilitation, either alone or following suppression, began at longer latencies (4-22 ms).

Suppression of thalamic activity during high-frequency stimulation in GPi is shown in Figure 10. Stimulation at 240 Hz (100 μA) suppressed the large amplitude spikes present before and after the 100 pulse stimulus train (5 trials superimposed). In this particular record, small amplitude, short duration spikes, consistent with those usually recorded from fibers, are consistently present between shock artifacts. (See the expanded time base in B.) During short stimulus trains (1-3 pulses), these small spikes were consistently evoked at a latency of 12 ms (not illustrated). It is tempting to speculate that these spikes are recorded from the inhibitory axons of GPi neurons excited by the stimuli and responsible for the suppression of thalamic neuronal activity illustrated.

In other cases, stimulation in GPi activated thalamic neurons. One example is shown in Figure 11. In this case, a short duration stimulus train (5 pulses, 120 Hz, 100 μA) evoked a burst of spikes that lasted for about the same duration as the 5-pulse train. In some cases (not illustrated), the activity of thalamic neurons is increased during longer duration stimulus trains.

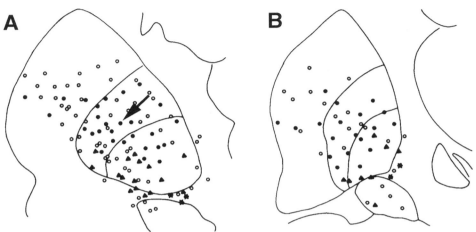

Figure 8. Stimulus positions in GPi, GPe, and the putamen at which stimulation changed movement times during a reaching task. Filled circles: Increased movement time. Filled triangles: Decreased movement time. Open circles: No significant effect. (From Horak and Anderson, 1984b.)

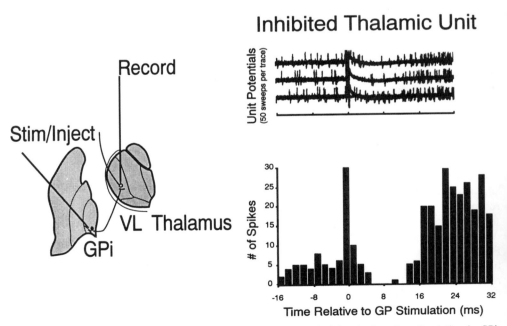

Figure 9. Suppression and facilitation of thalamic discharge evoked by single pulse stimulation in GPi. (From Inase et al., 1996.)

DISCUSSION

Because basal ganglia output reaches the cerebral cortex via the thalamus, the activity of basal ganglia-receiving thalamic neurons must be a major determinant of both normal and abnormal influences of the basal ganglia on movement and cognitive function. The GABAergic inhibitory output from the basal ganglia converges at the thalamus primarily

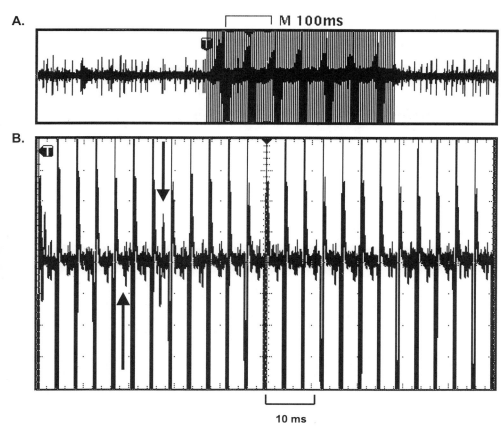

Figure 10. Suppression of thalamic activity during high-frequency stimulation in GPi. (A) Slow sweep (see 100 ms scale). (B) Expanded time base for the time period indicated by the time calibration bracket in (A). Upward arrow: small fiber-like spikes. Downward arrow: Large cellular type spike. Five trials superimposed.

with input from the cerebral cortex, which others have shown is primarily excitatory. Determination of the information carried by these corticothalamic fibers is an important goal for future investigations.

Other potential sources of synaptic input include the pretectum, the superior colliculus, and the pedunculopontine region. Our earlier study in the cat, in which small injections of HRP were made at different thalamic sites and the ratio of neurons labeled retrogradely in the basal ganglia and the cerebellar nuclei was determined, indicated that retrograde labeling in the pretectum increased as cerebellar labeling increased (Anderson and DeVito, 1987). Thus, the pretectum may primarily provide input to cerebellar-receiving, rather than basal ganglia-receiving thalamic neurons.

In the cat, retrograde labeling in the superior colliculus was especially marked when SNr neurons were labeled retrogradely, although these cases also had significant label in the deep cerebellar nuclei (Anderson and DeVito, 1987). Since the superior colliculus and the substantia nigra both have important roles in the control of saccadic eye movements, it is possible that there is convergent input from these two sources at the thalamus.

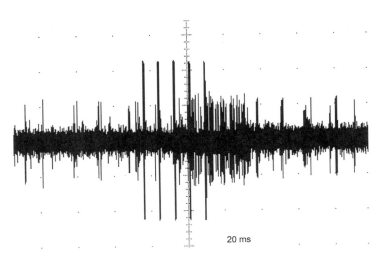

Figure 11. Activation of thalamic neurons by 5-pulse stimulation in GPi.

The cholinergic neurons in the pedunculopontine region have been well documented to project to the thalamus, including the ventrolateral thalamus, and there is some evidence for non-cholinergic projections there as well (Hallanger et al., 1987; Lee et al., 1988; Steriade et al., 1988). This input may provide a modulatory influence on the BG-R thalamus, as well as on other thalamic nuclei.

Examination of the effects of inhibition of normal GPi activity is particularly interesting in light of the resurgence of pallidotomy for symptomatic treatment of Parkinson's disease. The theoretical basis for pallidotomy is that it eliminates excessive or abnormally patterned inhibitory output from GPi that is found in Parkinson's disease or in the MPTP-induced model of parkinsonism in the non-human primate (Filion and Tremblay, 1991; Wichmann et al., 1999). In those parkinsonian patients with the best outcomes, bradykinesia, rigidity, and tremor are all reduced, especially in the limbs (Baron et al., 2000). At first glance, it is difficult to reconcile this improvement in motor function after pallidotomy in parkinsonian patients with the positional instability and slowed movements found when GPi is inhibited or ablated in otherwise normal monkeys (Inase et al., 1996; Horak and Anderson, 1984a; Hore and Vilis, 1980; Mink and Thach, 1991; Wenger et al., 1999). Perhaps the final explanation will be that the long-term abnormal pallidal output in parkinsonism is part of a very different functional network than that present in the normal brain, and disruption of the appropriate part of this abnormal network at any point, GPi, the subthalamic nucleus, or the thalamus, may allow the remaining network to regain a more normal balance between suppression of inappropriate motor patterns (or stabilization) and activation of others (controlled destabilization). In contrast, inhibition of GPi output that had contributed to a normally functioning network results in an imbalance between the desired stabilization and controlled destabilization. As a consequence, imbalanced co-contraction can produce both postural drift and slowed movements.

Even more paradoxical is the apparently similar effect of lesioning or stimulating the globus pallidus, both on the symptoms of Parkinson's disease (Siegfried, 1994; Davis et al., 1997) and on movements made by monkeys (Horak and Anderson, 1984a). Fundamental to understanding what high-frequency stimulation in GPi really does is the determination of its effect on target neurons in structures such as the thalamus. Our results to date using microstimulation at 100 μA intensities have shown that the consequences of high-frequency stimulation on thalamic discharge are mixed. Even long stimulus trains (100 pulses) commonly result in a reduction in the discharge of thalamic neurons, which would be consistent with stimulus-induced activation of inhibitory pallido-thalamic axons. In other cases, stimulation in GPi clearly resulted in activation of thalamic neurons, which would be unexpected if the consequence of stimulation was a reduction in the activity of GPi output neurons.

Others have reported that, in the MPTP-treated monkey, high intensity (350 μA) macrostimulation in GPi of monkeys reduces the firing frequency of GPi neurons located up to several mm from the stimulating electrodes (Boraud et al., 1996). The most parsimonious explanation for this depression of activity seems to be that the stimulus activates inhibitory axons presynaptic to the GPi cells whose activity was recorded. It remains to be determined whether higher intensity macrostimulation in GPi would more consistently produce activation of target thalamic neurons. On the other hand, activation of GPi cells and resultant inhibition of the thalamus would be a plausible mechanism for stimulation-induced relief of dystonia (Kumar et al., 1999).

The complex interaction of thalamic membrane properties and synaptic input from the basal ganglia and other sources (including inhibitory reticular thalamic and local circuit interneurons) remains to be determined. Alteration of this dynamic interaction, by disease, long term synaptic plasticity, or therapeutic intervention, must underlie the degree to which symptoms are manifest in basal ganglia disorders.

ACKNOWLEDGMENTS: The authors' work was supported in part by grants NS-15017, NS-38228, and RR-00166 from the National Institutes of Health.

REFERENCES

Anderson, M. E., and DeVito, J. L., 1987, An analysis of potentially converging inputs to the rostral ventral thalamic nuclei of the cat, *Exp. Brain Res.* 68:260.

Anderson, M. E., and Horak, F. B., 1985, Influence of the globus pallidus on arm movements in monkeys. III. Timing of movement-related information, *J. Neurophysiol.* 54:433.

Baron, M. S., Vitek, J. L., Bakay, R. A., Green, J., McDonald, W. M., Cole, S. A., and DeLong, M. R., 2000, Treatment of advanced Parkinson's disease by unilateral posterior GPi pallidotomy: 4-year results of a pilot study, *Mov. Disord.* 15:230.

Boraud, T., Bezard, E., Bioulac, B., and Gross, C., 1996, High frequency stimulation of the internal globus pallidus (GPi) simultaneously improves parkinsonian symptoms and reduces the firing frequency of GPi neurons in the MPTP-treated monkey, *Neurosci. Lett.* 215:17.

Buford, J. A., Inase, M., and Anderson, M. E., 1996, Contrasting locations of pallidal-receiving neurons and microexcitable zones in primate thalamus, *J. Neurophysiol.* 75:1105.

Davis, K. D., Taub, E., Houle, S., Lang, A. E., Dostrovsky, J. O., Tasker, R. R., and Lozano, A. M., 1997, Globus pallidus stimulation activates the cortical motor system during alleviation of parkinsonian symptoms [see comments], *Nat. Med.* 3:671.

DeVito, J. L., and Anderson, M. E., 1982, An autoradiographic study of efferent connections of the globus pallidus in *Macaca mulatta*, *Exp. Brain Res.* 46:107.

Filion, M., and Tremblay, L., 1991, Abnormal spontaneous activity of globus pallidus neurons in monkeys with MPTP-induced parkinsonism, *Brain Res.* 547:142.

Fonnum, F., Storm-Mathisen, J., and Divac, I., 1981, Biochemical evidence for glutamate as neurotransmitter in corticostriatal and corticothalamic fibres in rat brain, *Neuroscience*, 6:863.

Hallanger, A. E., Levey, A. I., Lee, H. J., Rye, D. B., and Wainer, B. H., 1987, The origins of cholinergic and other subcortical afferents to the thalamus in the rat, *J. Comp. Neurol.* 262:105.

Hikosaka, O., and Wurtz, R. H., 1985, Modification of saccadic eye movements by GABA-related substances. II. Effects of muscimol in monkey substantia nigra pars reticulata, *J. Neurophysiol.* 53:292.

Horak, F. B., and Anderson, M. E., 1984a, Influence of globus pallidus on arm movements in monkeys. I. Effects of kainic acid-induced lesions, *J. Neurophysiol.* 52:290.

Horak, F. B., and Anderson, M. E., 1984b, Influence of globus pallidus on arm movements in monkeys. II. Effects of stimulation, *J. Neurophysiol.* 52:305.

Hore, J., and Vilis, T., 1980, Arm movement performance during reversible basal ganglia lesions in the monkey, *Exp, Brain Res.* 39:217.

Hutchison, W. D., Lozano, A. M., Davis, K. D., Saint-Cyr, J. A., Lang, A. E., and Dostrovsky, J. O., 1994, Differential neuronal activity in segments of globus pallidus in Parkinson's disease patients, *Neuroreport*, 5:1533.

Ilinsky, I. A., Tourtellotte, W. G., and Kultas-Ilinsky, K., 1993, Anatomical distinctions between the two basal ganglia afferent territories in the primate motor thalamus, *Stereotact. Funct. Neurosurg.* 60:62.

Inase, M., Buford, J. A., and Anderson, M. E., 1996, Changes in the control of arm position, movement, and thalamic discharge during local inactivation in the globus pallidus of the monkey, *J. Neurophysiol.* 75:1087.

Jimenez-Castellanos, J., Jr., and Reinoso-Suarez, F., 1985, Topographical organization of the afferent connections of the principal ventromedial thalamic nucleus in the cat, *J. Comp. Neurol.* 236:297.

Jones, E. G., 1985, *The Thalamus*, Plenusm Press, New York.

Kultas-Ilinsky, K., Reising, L., Yi, H., and Ilinsky, I. A., 1997, Pallidal afferent territory of the *Macaca mulatta* thalamus: neuronal and synaptic organization of the VAdc, *J. Comp. Neurol.* 386:573.

Kultas-Ilinsky, K., Yi, H., and Ilinsky, I. A., 1995, Nucleus reticularis thalami input to the anterior thalamic nuclei in the monkey: a light and electron microscopic study, *Neurosci. Lett.* 186:25.

Kumar, R., Dagher, A., Hutchison, W. D., Lang, A. E., and Lozano, A. M., 1999, Globus pallidus deep brain stimulation for generalized dystonia: clinical and PET investigation, *Neurology*, 53:871.

Lee, H. J., Rye, D. B., Hallanger, A. E., Levey, A. I., and Wainer, B. H., 1988, Cholinergic vs. noncholinergic efferents from the mesopontine tegmentum to the extrapyramidal motor system nuclei, *J. Comp. Neurol.* 275:469.

McCormick, D. A., and von Krosigk, M., 1992, Corticothalamic activation modulates thalamic firing through glutamate "metabotropic" receptors, *Proc. Natl. Acad. Sci. (U S A)* 89:2774.

Mink, J. W., and Thach, W. T., 1991, Basal ganglia motor control. III. Pallidal ablation: normal reaction time, muscle cocontraction, and slow movement, *J. Neurophysiol.* 65:330.

Nakano, K., Kohno, M., Hasegawa, Y., and Tokushige, A., 1985, Cortical and brain stem afferents to the ventral thalamic nuclei of the cat demonstrated by retrograde axonal transport of horseradish peroxidase, *J. Comp. Neurol.* 231:102.

Pare, D., and Smith, Y., 1996, Thalamic collaterals of corticostriatal axons: their termination field and synaptic targets in cats, *J. Comp. Neurol.* 372:551.

Sakai, S. T., Inase, M., and Tanji, J., 1996, Comparison of cerebellothalamic and pallidothalamic projections in the monkey (*Macaca fuscata*): a double anterograde labeling study, *J. Comp. Neurol.* 368:215.

Sidibe, M., Bevan, M. D., Bolam, J. P., and Smith, Y., 1997, Efferent connections of the internal globus pallidus in the squirrel monkey: I. Topography and synaptic organization of the pallidothalamic projection, *J. Comp. Neurol.* 382:323.

Siegfried, J., and Lippitz, B., 1994, Bilateral chronic electrostimulation of ventroposterolateral pallidum: A new therapeutic approach for alleviating all parkinsonian symptoms, *Neurosurgery*, 35:1126.

Somogyi, G., Hajdu, F., Tombol, T., and Madarasz, M., 1978, *Acta. Anat.* 102:68.

Steriade, M., Pare, D., Parent, A., and Smith, Y., 1988, Projections of cholinergic and non-cholinergic neurons of the brainstem core to relay and associational thalamic nuclei in the cat and macaque monkey, *Neuroscience*, 25:47.

Swadlow, H. A., 1994, Efferent neurons and suspected interneurons in motor cortex of the awake rabbit: axonal properties, sensory receptive fields, and subthreshold synaptic inputs, *J. Neurophysiol.* 71:437.

Vidnyanszky, Z., Gorcs, T. J., Negyessy, L., Borostyankio, Z., Knopfel, T., and Hamori, J., 1996, Immunocytochemical visualization of the mGluR1a metabotropic glutamate receptor at synapses of corticothalamic terminals originating from area 17 of the rat, *Eur. J. Neurosci.* 8:1061.

Wenger, K. K., Musch, K. L., and Mink, J. W., 1999, Impaired reaching and grasping after focal inactivation of globus pallidus pars interna in the monkey, *J. Neurophysiol.* 82:2049.

Wichmann, T., Bergman, H., Starr, P. A., Subramanian, T., Watts, R. L., and DeLong, M. R., 1999, Comparison of MPTP-induced changes in spontaneous neuronal discharge in the internal pallidal segment and in the substantia nigra pars reticulata in primates, *Exp. Brain Res.* 125:397.

PHYSIOLOGICAL EVIDENCE FOR SPINAL CORD INPUT TO THE MOTOR THALAMUS[*]

ROBERT MACKEL[*]

INTRODUCTION

The motor thalamus can be divided into a pallidal-receiving and a cerebellar-receiving part (Ilinsky and Kultas-Ilinsky, 1987). According to this definition, the nuclei of the pallidal-receiving part can be grouped together as VA (ventral anterior), those of the cerebellar-receiving part as VL (ventral lateral). The cerebellar-receiving VL is traditionally viewed as a relay center through which the cerebellum sends information to motor cortical areas (Massion and Sasaki, 1979). In reality, it is more than a simple relay because it contains interneurons (Ilinsky et al., 1993) and receives projections from various sources, such as the cerebral cortex, intrathalamic nuclei, midbrain, brain stem, and spinal cord (c.f. Jones, 1985; Nakano et al., 1985; Anderson and DeVito, 1987, Ilinsky et al., 1999). Information from a number of sources (e.g. cerebral cortex, Uno et al., 1970; thalamic reticular nucleus, Ando et al., 1995; spinal cord, Mackel and Noda, 1988) converges with cerebellar information in VL neurons and can, therefore, modulate the information that the cerebellum sends to the motor cortex.

In 1979, Asanuma and collaborators reported that the receptive fields of neurons in the motor cortex of cats and monkeys do not change significantly when the sensory cortex is removed. They suggested that a major portion of peripheral sensory inputs reach the motor cortex directly from the thalamus and not through the sensory cortex (Asanuma et al., 1979a). They also identified neurons in cat VL[1] and corresponding monkey VPLo that projected to the motor cortex and discharged in response to peripheral stimulation (Asanuma et al., 1979b; Asanuma et al., 1980). These thalamocortical projections were considered an alternative to the projections from the sensory to motor cortex, and their functional significance was seen in sensory feedback control of cortically induced movements. This alternate route may explain, in part, why removal of the sensory cortex

---

[*] This paper is dedicated to Professor Hiroshi Asanuma.

[*] ROBERT MACKEL • The Rockefeller University, New York, NY 10021.

[1] VL refers to the cat, VPLo to the monkey, Vim to human (for different terminologies, see Jones, 1985).

produces few motor deficits, why sensory-evoked potentials can still be recorded from the motor cortex when the sensory cortex is removed, and why receptive fields of neurons in the motor cortex do not significantly change when the sensory cortex is removed (cf. Asanuma and Mackel, 1989).

The idea that sensory information reaches the motor cortex directly via thalamocortical projections was challenged on anatomical and physiological bases. The anatomical controversy arose from difficulty in demonstrating that the main spinal sensory paths (i.e. the dorsal column-medial lemniscal system and the spinothalamic tract) project to VL relay neurons. In particular, some authors could not demonstrate projections from the dorsal column nuclei to VPLo (Tracey et al., 1980) or VL (Hirai and Jones, 1988). Other authors (Berkley, 1983; Tamai et al., 1984) found projections from the dorsal column nuclei only to a small border region between VL and somatosensory nucleus VPL, not to the main body of VL. Although spinothalamic projections were known to terminate in VL (Berkley, 1983; Craig and Burton, 1985), no overlap with neurons projecting to the motor cortex was found (Greenan and Strick, 1986).

The physiological controversy derived from inconsistencies in the ability to drive VL (or VPLo) neurons with natural stimuli. A number of studies conducted in alert or anesthetized animals found no or little responsiveness of VL and VPLo neurons to natural stimuli (Strick, 1976; Schmied et al., 1979; McPherson et al., 1980; Anner-Baratti et al., 1986), or responsiveness only in a limited region (Asanuma et al., 1979b; Lemon and van der Burg, 1979; Ohye et al., 1989). These studies were conducted with extracellular recording, and the neurons were located anatomically. Because of the difficulties of delineating the anatomical borders of VL (or VPLo, e.g. Jones, 1985), it was not clear to what extent the recording sites were located in VL. Furthermore, it was not known whether the studied neurons were VL relay neurons, i.e. projected to the motor cortex and received input from the cerebellum.

To determine whether or not VL relay neurons in fact do receive and convey sensory information, it was evidently necessary to both (1) optimize the conditions for detecting spinal input to VL neurons and (2) identify the VL neurons as relay neurons. The physiological studies described below set out to meet these requirements. Neurons were fully-identified as VL relay neurons by their monosynaptic excitation from the cerebellum and antidromic activation from the motor cortex. Conditions for detecting spinal input to VL relay neurons were optimized by using intracellular recording to detect subthreshold responses, and by using electrical stimulation of sensory tracts in the spinal cord to activate many afferents simultaneously (Mackel and Noda, 1988, 1989; Mackel et al., 1991, 1992; Mackel and Miyashita, 1992, 1993). As detailed below, considerable evidence for spinal input to VL relay neurons was found.

These findings have been confirmed in contemporaneous and subsequent anatomical and physiological studies. Anatomical work has demonstrated that there is overlap between spinothalamic terminals and thalamocortical projection neurons in VL (Hirai and Jones, 1988). Physiological studies using extracellular recording (in alert monkeys and human patients, as well as in anesthetized cats) have shown that cerebellar-receiving neurons (Anderson and Turner, 1991) and neurons located anatomically in cerebellar-receiving regions of the motor thalamus (Yen et al., 1991; Butler et al., 1992; Vitek et al., 1994; Lenz et al., 1999) can be driven with natural stimuli. It is now generally accepted that sensory information reaches VL neurons.

The following presents the evidence from intracellular studies in the anesthetized cat that spinal input reaches VL relay neurons. Routes that convey spinal input to VL are also described.

SPINAL INPUT TO VL NEURONS

Inputs from the main spinal sensory paths, the dorsal columns and the spinothalamic tract, were studied. The dorsal columns (DC) were stimulated in isolation. Since the spinothalamic tract ascends in the ventral quadrant (VQ) together with other afferents (e.g. spinomesencephalic, spinoreticular, and spinocerebellar), they are also stimulated when VQ is stimulated.

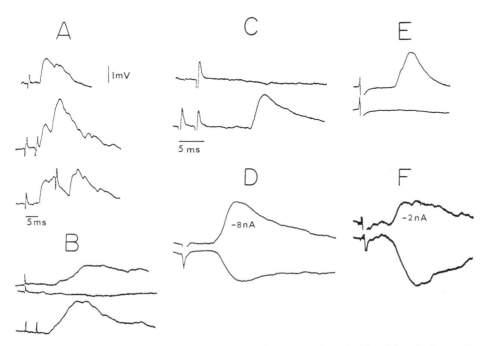

Figure 1. Postsynaptic responses elicited by stimulation of ventral quadrant (A-D) and dorsal columns (E and F) at mid-cervical level. A: early and late EPSPs to single (top trace) and double shocks (middle and bottom traces). Notice absence of temporal facilitation of early response to double volleys. In the middle trace, the early response to the 2nd shock adds to the late response of the 1st shock. B and C: occurrence of temporal facilitation to double volleys. B: response with double shocks shows a shortening of latency and an enhancement of the early component of the response. Trace in the middle is the extracellular field. C: a postsynaptic response is elicited with double-shock but not single shock (top trace) stimulation. D: postsynaptic inhibition. IPSP is reversed by injecting hyperpolarizing current through the electrode. E and F: postsynaptic excitation (E) and inhibition (F) to stimulation of the dorsal columns. IPSP in F is also shown reversed. Extracellular field below the response in E. Calibrations for A and B in A and for C-F in C. Voltage calibrations (A), same for all records. From Mackel et al., 1992, with permission.

It was found that 85% of VL neurons responded with post-synaptic potentials to stimulation of ascending paths. Input from the VQ was more common than input from the DC (present in 85% versus 58% of cells tested). All neurons that responded to stimulation of the DC also responded to stimulation of the VQ, indicating that there is considerable convergence among ascending inputs. Excitatory postsynaptic responses were more common (88%) than inhibitory postsynaptic responses (12%) (Mackel and Noda, 1988; Mackel et al., 1992). Examples of postsynaptic responses are illustrated in Figure 1. Short-latency responses were often mixed with longer latency (1A), and responses often displayed several components (1B and E).

Many responses were long latency and therefore likely to be polysynaptic. To examine this possibility, double shock stimulation was routinely used to test for temporal facilitation in the spinal pathways to VL. Presence of temporal facilitation of a postsynaptic response (shortening of latency and size increased beyond pure summation) is an indication of polysynaptic transmission (Hongo and Jankowska, 1967). Absence of temporal facilitation (fixed latency and pure summation of response) is one indication of monosynaptic transmission. Examples of double shock stimulation are also illustrated in Figure 1. The facilitation is manifest as shortened latency and/or enhancement of the response (1B and C). In contrast, the responses to double shocks are additive and latencies of responses stable in Figure 1A.

The VQ (including the spinothalamic tract)

Latencies of the postsynaptic responses to stimulation of the VQ at midcervical level ranged between 2.9 and 18 msec (median 6.2 msec). Approximately 30% of the neurons responded with short latency excitation (2.9-7msec) and did not display temporal facilitation in response to double shock stimulation. An example is illustrated in Figure 1A. These neurons were additionally tested with an extrapolation procedure which allows estimates of synaptic delays in the target neurons (Mackel et al., 1992). Synaptic delays in the order of 0.5 msec, a typical value for monosynaptic transmission, were measured in those neurons. It can be concluded that monosynaptic input from the spinal cord reaches VL relay neurons via spinothalamic afferents.

Most of the input from the VQ displayed temporal facilititation in response to double shock stimulation (Figure 1B and C). Some of this input was probably mediated via spinomesencephalic and spinoreticular projections (possibly also indirect spinothalamic projections) and relayed in subcortical structures en route to VL (see below).

Rise times of monosynaptic spinothalamic EPSPs were significantly slower than rise times of cerebellothalamic and corticothalamic monosynaptic EPSPs (Figure 2). This would indicate that the synapses of the spinothalamic fibers are located on more distal dendrites (Tsukahara et al., 1975) than the synapses of cerebellothalamic and corticothalamic projections. Preliminary ultrastructural data support this conclusion (Ilinsky et al., 1991). Synaptic contacts from single spinothalamic fibers have been found on the distal dendrites of VL projection neurons (as well as on local circuit neurons). This arrangement seems to be the same as in the thalamic somatosensory nucleus VPL, where spinothalamic terminals are found on distal dendrites of projection neurons (Ralston and Ralston, 1994).

The dorsal column-medial lemniscal system

Stimulation of the isolated dorsal columns was also very effective in eliciting postsynaptic responses in VL neurons. The responses were mostly long latency (median 10.5 msec, range 3-19 msec). Examples of responses to dorsal column stimulation are shown in Figure 1E and F. In a series of experiments (Mackel and Miyashita, 1992), the dorsal column nuclei were directly stimulated in order to bypass the afferents' synapses in the brain stem and to determine whether the dorsal column nuclei project directly to VL neurons (see Introduction).

It was found that 75% of the neurons responded with excitation and/or inhibition to stimulation of the dorsal column nuclei, with excitation much more common (90% of the responses). The latencies of the postsynaptic responses ranged from 2.0 to 20 msec (median 10.0 msec), similar to the latency range measured to stimulation of the DC at midcervical level. The latency of only two responses (2.0 and 3.0 msec) out of 80 responses measured fell within a range considered to be monosynaptic (1.5-3.0 msec, see Mackel and Miyashita, 1992). The latencies of the remaining responses (> 3 msec) were too long for monosynaptic transmission, and the responses typically displayed temporal facilitation to

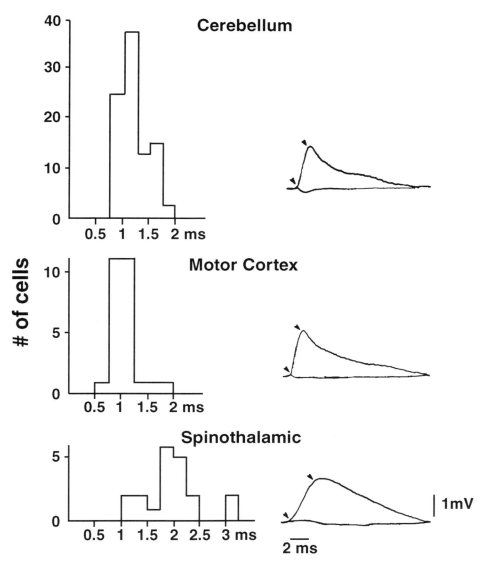

Figure 2. Measurement of rise times of monosynaptic cerebellothalamic, corticothalamic, and spinothalamic EPSPs. Extracellular fields underneath the responses illustrated on the right. Rise times were measured from the point where the EPSPs leave the extracellular fields to peak response, as indicated by arrows. Each record is an average of 3 sweeps. Calibrations the same for all records. On the left, histograms of rise time measurements. Notice that the rise times of spinothalamic EPSPs are twice as slow as those of cerebello- and corticothalamic EPSPs.

double shock stimulation. The recording sites covered VL extensively, including the border region with VPL. The transition from VL to VPL was abrupt. As soon as the electrode tip penetrated VPL, monosynaptic EPSPs from the cerebellum were replaced by monosynaptic EPSPs from the dorsal column nuclei.

Thus, the findings show little evidence for direct input from the dorsal column nuclei to VL relay neurons, in agreement with anatomical work (Hirai and Jones, 1988; Aumann et al., 1996). Dorsal column-medial lemniscal input reaches the neurons, but it does so indirectly, disynaptically, and/or polysynaptically.

SUBCORTICAL RELAYS BETWEEN THE SPINAL CORD AND VL

Most of the spinal input to VL neurons was long latency and indirect, indicating that it was relayed en route to VL. Although there are many ways for long latency input to reach VL, the cerebral cortex and cerebellum do not appear to be significantly involved in the transfer of information. In a limited number of experiments, Mackel and Noda (1988) observed that the spinal input to VL neurons is not obviously altered when the sensorimotor cortex and/or cerebellum are removed. These observations would support Asanuma's observations that the sensory receptive fields of motor cortical neurons are not significantly altered when the sensory cortex is removed, because the sensory route through VL is intact (Asanuma et al., 1979).

Anatomical work guided the present studies to the midbrain and brain stem. The pretectal region and underlying reticular formation of the midbrain project rostrally to VL (and other thalamic nuclei, Nakano et al., 1985; Anderson and DeVito, 1987) and receive input from the dorsal column nuclei (Berkley and Mash, 1978; Yoshida et al., 1992) and the VQ (Bjorkeland and Boivie, 1984; Yezierski, 1988). Nucleus Z of the brain stem (Brodal and Pompeiano, 1957) also projects to VL (and VPL, Grant et al., 1973; Mackel and Miyashita, 1992) and receives input from the ipsilateral hindlimb (primarily muscle, but also tendons, joints, and skin) via collaterals of the dorsal spinocerebellar tract (Landgren and Silfvenius, 1971; Low et al., 1986).

The Pretectal Region

Lesions in the pretectal region and underlying reticular formation were found to be effective in reducing the occurrence of postsynaptic responses in VL neurons to electrical stimulation of the DC and VQ (Mackel and Noda, 1989), suggesting that spinal input to VL passes through this region. In a series of experiments, an array of stimulating electrodes was positioned into VL, and neurons in the midbrain were antidromically identified as projecting to VL, using extracellular recording. Input from the spinal cord to these neurons was tested by stimulation of the DC, dorsal column nuclei, and VQ. Collision techniques were used to confirm that the antidromic and orthodromic responses were obtained from the same neurons (Mackel et al., 1991).

The locations of the recording sites are illustrated in Figure 3. It can be seen that the neurons were located primarily in the posterior pretectal nucleus and underlying reticular formation. Caudally, recording sites extended to the transitional area between the pretectum and superior colliculus.

Many midbrain neurons (75%) projecting to VL discharged with short and/or long latency to stimulation of the DC or dorsal column nuclei. Short latency responses (<3 msec) to stimulation of the dorsal column nuclei were seen in roughly half of the sample. Approximately 25% of these neurons discharged additionally to stimulation of the VQ (about half with short latency, i.e. < 5 msec), reemphasizing the substantial convergence of different inputs en route to VL. The short latency responses were probably conveyed directly via the anatomically described projections from the dorsal column nuclei and from the spinal cord (spinomesencephalic, spinoreticular) to the midbrain (see above). Figure 4 illustrates examples of discharges in midbrain neurons in response to stimulation of the spinal cord and dorsal column nuclei.

A few neurons were tested for, and showed responses to, pressure on limb or trunk muscles. Cutaneous input was not assessed (dorsal columns were cut in these preparations).

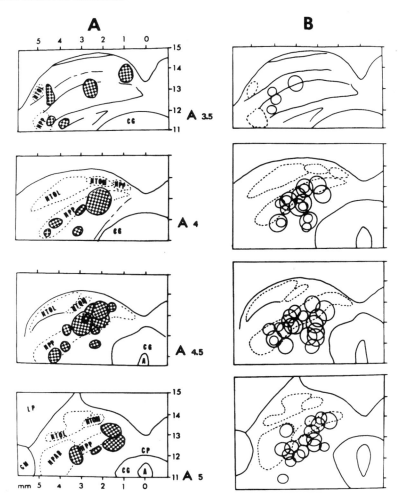

Figure 3. Marked (A) and reconstructed (B) recording sites in the pretectal region with stereotaxic coordinates indicated. NPP, N. pretectalis posterior; NPAR, N. pretectalis anterior; NTOL, N. tracti optici, pars lateralis; NTOM, N. tracti optici, pars medialis; CP, commisura posterior; GM, corpus geniculatus medialis; LP, N. lateralis thalami, pars posterior; CG, central gray; A, aqueduct. From Mackel et al., 1991, with permission.

Nucleus Z

As in the study of the midbrain, an array of stimulating electrodes was positioned into VL and recordings were obtained from neurons that were antidromically identified as projecting to VL. Spinal input to these neurons was investigated by stimulation of the dorsolateral funiculus (DLF), where the dorsal spinocerebellar tract (DSCT) ascends, and by natural stimulation of the hindlimb (dorsal columns intact). Collision techniques were used to ascertain that the recordings were obtained from the same neurons (Mackel and Miyashita, 1993).

The locations of the recording sites in nucleus Z are illustrated in Figure 5. Although nucleus Z is anatomically a small nucleus, it can be easily located physiologically because its neurons readily respond to natural stimulation of the ipsilateral hindlimb (Landgren and

Sylfvenius 1971). Responses to natural stimulation of the forelimb indicate that the recording electrode is located caudally in the cuneate nucleus. No responses to natural stimulation of the limbs indicates that the recording electrode is located rostrally in the vestibular nuclei.

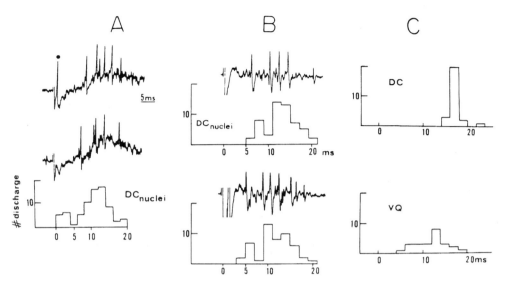

Figure 4. Poststimulus time histograms from 15 consecutive sweeps in response to single (A and C) and double volleys (B) to ascending pathways. A shows early (dot over action potential) and late responses of input to a neuron in the pretectal region following stimulation of the dorsal column nuclei. Early responses are not always elicited (lower trace and smaller earlier elevation in the histogram). Later discharges occur in each sweep. B, responses of a different neuron to single (upper insets) and double volleys (lower insets). C, responses of a third neuron to stimulation of afferents in the dorsal columns (DC) and in the ventral quadrant (VQ). From Mackel et al., 1991, with permission.

All neurons in nucleus Z responded with short latency discharges (1.1. to 4.4. sec) to electrical stimulation of the DLF, confirming earlier reports (Landgren and Sylvenius, 1971). Almost all neurons (81/84) had receptive fields in deeper tissues (muscle and joints), occasionally (3 neurons) in the skin, findings consistent with other studies (Landgren and Silfvenius, 1971; Johansson and Silvenius, 1977). An example is illustrated in Figure 6. The data suggest that the route through nucleus Z conveys primarily proprioceptive information to VL.

SUMMARY AND CONCLUSIONS

The data show that many VL neurons receive input from the spinal cord. There is direct input, carried by the spinothalamic tract, but most is indirect and ascends along afferents in the ventral quadrant, the dorsal columns, and collaterals of the dorsal spinocerebellar tract.

The indirect input from the VQ could have been mediated by other afferents ascending in the VQ (e.g. spinomesencephalic, spinoreticular, spinocerebellar, other spinothalamic) as well as by more complicated routes involving propriospinal and local interneuronal systems with their relay structures. Because the dorsal columns were cut and stimulated in isolation at C_3, descending and segmental interactions can largely be excluded. Most interneuronal

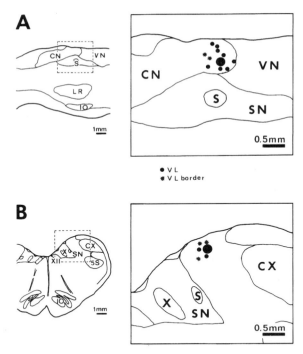

Figure 5. Schematic illustration of a number of reconstructed recording sites from two typical experiments. A, sagittal section through nucleus Z. B, frontal section. Recording sites were reconstructed with respect to a dye injection mark (large dots). Also indicated are the thalamic projection sites of the neurons. CN, cuneate N., VN, vestibular nuclei, CX, external cuneate N., SN, N. solitary tract, S, solitary tract, X, dorsal motor N. of X. From Mackel et al., 1993, with permission.

relays were probably located supraspinally. Some of the input from the VQ and the dorsal columns passed through the pretectal region and underlying reticular formation. Neurons in nucleus Z were also found to convey input from the spinal cord, in all likelihood via collaterals of the DSCT. Part of the input that is relayed in the pretectal region and all of the input that is relayed in nucleus Z appears to be of proprioceptive origin (muscles, tendons and joints), the type of information that is generated during active movement. Interestingly, recordings from VL and VPLo neurons have shown that they also respond preferentially to stimulation of deeper structures of the limb and trunk (Anderson and Turner, 1991; Yen et al., 1991; Vitek et al., 1994), suggesting that some of that information might have been conveyed through the relay structures presently identified.

The present findings and other recent data (see Introduction) indicate that there is sensory input in the path from VL to motor cortex. Under normal conditions, sensory input to VL may serve as a feedback modulation of cerebellar commands to the motor cortex. It is also a route for sensory information to reach the motor cortex. As suggested in the introduction, the sensory input that reaches the motor cortex through VL can partly explain why there are few motor deficits when sensory cortex is removed. The motor cortex also receives inputs from other cortical sources (e.g. SII, premotor, supplementary motor, association, commissural), which may also play a role in substituting sensory cortical input to the motor cortex. These projections have been mostly studied anatomically and their functions are still largely unknown.

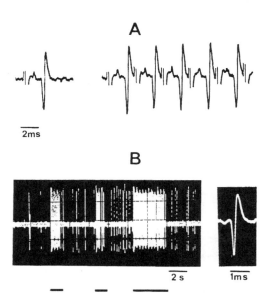

Figure 6. Antidromic and orthodromic discharges of a nucleus Z neuron projecting to VL. Orthodromic discharges (B) were elicited to palpation of a thigh muscle in the ipsilateral hindlimb (indicated by bars underneath the oscillographic records). Antidromic discharges (A) display short, constant latency and ability to follow high frequency (333 Hz) stimulation to electrical stimulation in VL. Notice that shapes of antidromic and orthodromic (fast time base in the lower right) action potentials were similar. From Mackel et al., 1993, with permission.

Sensory input to VL may also play a role in recovery from cerebellar motor deficits. A number of studies have shown that lesions of sensory structures (e.g. dorsal roots, Gilman et al., 1976; dorsal columns, Carpenter and Correll, 1961; sensory cortex, Mackel, 1987) reverse the recovery process from cerebellar lesions. It was suggested (Mackel, 1987) that, among other possibilities, sensory drive of cerebellar-deprived VL neurons helps restore motor cortical excitability reduced after cerebellar injury (c.f. Dow and Moruzzi, 1958). Withdrawal of this drive could interfere with the recovery process.

It becomes increasingly clear that sensory input to motor thalamus does not only play a role in the normal control of movement but also in cortical, cerebellar, and other (e.g. Lenz et al., 1999) types of pathologies.

ACKNOWLEDGMENTS: The author would like to thank Dr. Emily Brink for editorial help and many helpful comments.

REFERENCES

Anderson, M.E., and DeVito, J.L., 1987, An analysis of potentially converging inputs to the rostral ventral thalamic nuclei of the cat, *Exp. Brain Res.* 68:260.

Anderson, M.E., and Turner, R.S., 1991, Activity of neurons in cerebellar-receiving and pallidal-receiving areas of the thalamus of the behaving monkey, *J. Neurophysiol.* 66:879.

Ando, N., Izawa, Y., and Shinoda, Y., 1995, Relative contributions of thalamic reticular nucleus neurons and intrinsic interneuons to inhibition of thalamic neurons projecting to the motor cortex, *J. Neurophysiol.* 73:2470.

Anner-Baratti, R., Allum, J.H.J., and Hepp-Reymond, M.-C., 1986, Neural correlates of isometric force in the "motor" thalamus, *Exp. Brain Res.* 63:567.

Asanuma, H., Larsen, K.D., and Yumiya, H., 1979a, Direct sensory pathway to the motor cortex: A basis of cortical reflexes, in: *Integration in the Nervous System*, H. Asanuma and V.J. Wilson, eds., Igaku shoin, Tokyo, New York.

Asanuma, H., Larsen, K.D., and Yumiya, H., 1979b, Receptive fields of thalamic neurons projecting to the motor cortex in the cat, *Brain Res.* 172:217.

Asanuma, H., Larsen, K.D., and Yumiya, H., 1980, Peripheral input pathways to the monkey motor cortex, *Exp. Brain Res.* 38:349.

Asanuma, H., and Mackel, R., 1989, Direct and indirect sensory input pathways to the motor cortex; its structure and function in relation to learning of motor skills, *Jap. J. Physiol.* 39:1.

Aumann, R.D., Rawson, J.A., Pichitpornchai, C., and Horne, M.K., 1996, Projections from the cerebellar interposed and dorsal column nuclei to the thalamus in the rat: a double anterograde labelling study, *J. Comp. Neurol.* 368:608.

Berkley, K.J., and Mash, D.C., 1978, Somatic sensory projections to the pretectum of the cat, *Brain Res.* 158:445.

Berkley, K.J., 1983, Spatial relationships between the terminations of somatic sensory and motor pathways in the rostral brain stem of cats and monkeys. II. Cerebellar projections compared with those of the ascending somatic sensory pathways in lateral diencephalon, *J. Comp. Neurol.* 220:229.

Bjorkeland, M., and Boivie. J., 1984, The termination of spinomesencephalic fibers in cat. An experimental anatomical study, *Anat. Embryol.* 170:265.

Brodal, A., and Pompeiano, O., The vestibular nuclei in the cat, *J. Anat.* 91:438.

Butler, E.G., Horne, M.K., and Rawson, J.A., 1992, Sensory characteristics of monkey thalamic and motor cortex neurons, *J. Physiol.* 445:1.

Carpenter, M.B., and Correll, J.W., 1961, Spinal pathways mediating cerebellar dyskenesia in rhesus monkey, *J. Neurophysiol.* 24:535.

Craig, A.D., and Burton, H., 1985, The distribution and topographical organization in the thalamus of anterogradely-transported horseradish peroxidase after spinal injections in cat and raccoon, *Exp. Brain Res.* 58:227.

Dow, R.S., and Moruzzi, G., 1958, *The Physiology and Pathology of the Cerebellum*, University of Minnesota Press, St. Paul.

Greenan, T.J., and Strick, P.L., 1986, Do thalamic regions which project to rostral primate motor cortex receive spinothalamic input?, *Brain Res.* 362:384.

Gilman, S., Carr, D., and Hollenberg, J., 1976, Kinematic effects of deafferentation and cerebellar ablation, *Brain* 99:311.

Grant, G., Boivie, J., and Silfenius, H., 1973, Course and terminations of fibers from the nucleus Z of the medulla oblongata. An experimental light microscopical study in the cat, *Brain Res.* 55:55.

Hirai, T., and Jones, E.G., 1988, Segregation of lemniscal and motor cortex outputs in cat ventral thalamic nuclei: application of a novel technique, *Exp. Brain Res.* 71:329.

Hongo, T., and Jankowska, E., 1967, Effects from the sensorimotor cortex on the spinal cord in cats with transsected pyramids, *Exp. Brain Res.* 3:117.

Ilinsky, I.A., Mackel, R., Miyashita, E., and Kultas-Ilinsky, K., 1991, Ultrastructural features of spinal cord afferents to the intralaminar area of the cat, *Soc. Neurosci. Abstr.* 477.6, 1204.

Ilinsky, I., and Kultas-Ilinsky, K., 1987, Sagittal cytoarchitectonic maps of the *Macacca Mulatta* thalamus with a revised nomenclature of the motor-related nuclei validated by observations on their connectivity, *J. Comp. Neurol.* 262:331.

Ilinsky, I.A., Toga, A.W., and Kultas-Ilinsky, K., 1993, Anatomical organization of internal neuronal circuits in the motor thalamus, in: *Thalamic Networks for Relay and Modulation*, M. Molinari, G. Macchi, and E.G. Kones, eds., Pergamon, Oxford.

Ilinsky, I.A., Ambardekar, A.V., and Kultas-Ilinsky, K., 1999, Organization of projections from the anterior pole of the nucleus reticularis thalami (NRT) to subdivisions of the motor thalamus: Light and electron microscopic studies in the rhesus monkey. *J. Comp. Neurol.* 409:369.

Johansson, H., and Silfvenius, H., 1977, Input from ipsilateral proprio- and exteroceptive hind limb afferents to nucleus Z of the cat medulla oblongata, *J. Physiol.* 265:371

Jones, E.G., 1985, *The Thalamus*, Plenum Press, New York.

Landgren, S., and Silfvenius, H., 1971, Nucleus Z, the medullary relay in the projection path to the cerebral cortex of group I muscle afferents from the cat's hind limb, *J. Physiol.* 218:551.

Low, J.S.T., Mantle-St.John., L.A., and Tracey, D.J., 1986, Nucleus Z in the rat: spinal afferents from collaterals of spinocerebellar tract neurons, *J. Comp. Neurol.* 243:510.

Lemon, R.N., and Vander Burg, J., 1979, Short-latency peripheral input to thalamic neurons projecting to the motor cortex in the monkey, *Exp. Brain Res.* 36:445.

Lenz, F.A., Jaeger, C.J., Seike, M.S., Lin, Y.C., Reich, S.G., DeLong, M.R., and Vitek, J.L., 1999, Thalamic single neuron activity in patients with dystonia: Dystonia-related activity and somatic sensory reorganization, *J. Neurophysiol.* 82:2372.

Mackel, R., 1987, The role of the monkey sensory cortex in the recovery from cerebellar injury, *Exp. Brain Res.* 66:638.

Mackel, R., and Noda, T., 1988, Sensory input to cerebellocerebral relay neurons in the cat thalamus, *Brain Res.* 440:348.

Mackel, R., and Noda, T., 1989, The pretectum as a site for relaying dorsal column input to thalamic VL neurons, *Brain Res.* 476: 135.

Mackel, R., Iriki, A., Jorum, E., and Asanuma, H., 1991, Neurons of the pretectal area convey spinal input to the motor thalamus of the cat, *Exp. Brain Res.* 84:12.

Mackel, R., Iriki, A., and Brink, E.E., 1992, Spinal input to thalamic VL neurons: Evidence for direct spinothalamic effects, *J. Neurophysiol.* 67:132.

Mackel, R., and Miyashita, E., 1992, Dorsal column input to thalamic VL neurons: an intracellular study in the cat, *Exp. Brain Res.* 88:551.

Mackel, R., and Miyashita, E., 1993, Nucleus Z: A somatosensory relay to motor thalamus, *J. Neurophysiol.* 69:1607.

MacPherson, J.M., Rasmusson, D.D., and Murphy, J.T., 1980, Activities of neurons in "motor" thalamus during control of forelimb movement in the primate, *J. Neurophysiol.* 44:11.

Massion,J., and Sasaki,K., 1979, *Cerebrocerebellar Interactions*, Elsevier, Amsterdam.

Nakano, K., Kohno, M., Hasegawa, Y., and Tokushige, A., 1985, Cortical and brain stem afferents to the ventral thalamic nuclei of the cat demonstrated by retrograde axonal transport of horseradish peroxidase. *J. Comp. Neurol.* 231:102.

Ohye, C., Shibazaki, T., Hirai, T., Wada, H., Hirato, M., and Kawashima, Y., 1989, Further physiological observations on the ventralis intermedius neurons in the human thalamus, *J. Neurophysiol.* 61:488.

Ralston, H.J., and Ralston, D.D., 1994, Medial lemniscal and spinal projections to the macaque thalamus: an electron microscopic study of differeing GABAergic circuitry serving thalamic somatosensory mechanisms, *J. Neurosci.* 14:2485.

Schmied, A., Benita, M., Conde, H., and Dormont, J.F., 1979, Activity of ventrolateral thalamic neurons in relation to a simple reaction time task in the cat. *Exp. Brain Res.* 36:285.

Strick, P., 1976, Activity of ventrolateral thalamic neurons during arm movement, *J. Neurophysiol.* 39:1032.

Tamai, Y., Waters, R.S., and Asanuma, H., 1984, Caudal cuneate nucleus projection to the direct thalamic relay to motor cortex in cat: an electrophysiological and anatomical study, *Brain Res.* 323:360

Tracey, D.J., Asanuma, C., Jones, E.G., and Porter, R., 1980, Thalamic relay to motor cortex: Afferent pathways from brain stem, cerebellum, and spinal cord in monkeys, *J. Neurophysiol.* 44:532.

Tsukahara, N., Hultborn, H., Murakami, F., and Fujito, Y., 1975, Electrophysiological study of formation of new synapses and collateral sprouting in red nucleus neurons after partial denervation, *J. Neurophysiol.* 38:1359.

Uno, M., Yoshida, M., and Hirota, I., 1970, The mode of cerebello-thalamic relay transmission investigated with intracellular recording from cells of the ventrolateral nucleus of the cat's thalamus, *Exp. Brain Res.* 10:121.

Vitek, J.L., Ashe, J., DeLong, M.R., Alexander, G.E., 1994, Physiologic properties and somatotopic organization of the primate motor thalamus, *J. Neurophysiol.* 71:1498.

Yen, C.-T., Honda, C.N., and Jones, E.G., 1991, Electrophysiological study of spinothalamic inputs to ventrolateral and adjacent thalamic nuclei of the cat, *J. Neurophysiol.* 66:1033.

Yoshida, A., Sessle, B.J., Dostrovsky, J.O., and Chiang, C.Y., 1992, Trigeminal and dorsal column nuclei projections to the anterior pretectal nucleus in the rat, *Brain Res.* 590:81.

Yezierski, R.P., Spinomesencephalic tract: Projections from the lumbosacral spinal cord of the rat, cat, and monkey, *J. Comp. Neurol.* 267:131.

PART III. NEUROTRANSMITTERS, RECEPTORS AND THEIR ROLE IN MOTOR BEHAVIOR

CHEMICAL ANATOMY AND SYNAPTIC CONNECTIVITY OF THE GLOBUS PALLIDUS AND SUBTHALAMIC NUCLEUS

YOLAND SMITH, ALI CHARARA, JESSE E. HANSON, GEORGE W. HUBERT, AND MASAAKI KUWAJIMA[*]

INTRODUCTION

Since the pioneering work of Whittier and Mettler (1949) showing that small electrolytic lesion of the subthalamic nucleus (STN) induces violent involuntary movements of the contralateral limbs in non-human primates, the exact mechanisms by which the STN plays such a powerful role in the control of motor behaviors have been the subject of intensive research. In the 1980's, the introduction of sensitive immunocytochemical and tract-tracing techniques, combined with single-unit recording in normal and parkinsonian monkeys, led to major breakthroughs in our knowledge of the critical role of this nucleus in the functional circuitry of the basal ganglia (Albin et al., 1989; Bergman et al., 1990; Wichmann and DeLong, 1996). The STN, which is the only excitatory glutamatergic structure of the basal ganglia (Smith and Parent, 1988), is now considered to be a major source of excitatory drive to basal ganglia output nuclei (Kitai and Kita, 1987). Observations in 1-methyl-4-phenyl-1,2,3,6-tetrahydropyridine (MPTP)-treated monkeys showing that the STN activity is increased after dopamine depletion led to the development of novel surgical therapies aimed at silencing STN outflow to basal ganglia output nuclei (Bergman et al., 1990; Aziz et al., 1991). Such therapies are currently used worldwide as efficient treatment for Parkinson's disease in humans (Starr et al., 1998). The recent demonstration that the interconnections between the globus pallidus (GP) and the STN act as a pacemaker of neuronal oscillations observed in various basal ganglia structures after dopamine depletion is further evidence that the STN plays a critical role in mediating basal ganglia functions in both normal and pathological conditions (Plenz and Kitai, 1999).

Over the past few years, we and others studied in detail the connectivity, synaptology, neurotransmitter content, and receptor localization of the STN and GP in primates and non-primates. The present review summarizes some of these findings and discusses the possibility that drugs which target various subtypes of metabotropic glutamate and gamma-

[*] YOLAND SMITH, ALI CHARARA, JESSE E. HANSON, GEORGE W. HUBERT, AND MASAAKI KUWAJIMA • Division of Neuroscience, Yerkes Regional Primate Center, and Department of Neurology, Emory University, Atlanta, GA 30322.

aminobutyric acid (GABA) receptors might be used as novel therapeutic agents in Parkinson's disease.

Due to the limited space of this paper, we will not discuss all aspects of basal ganglia anatomy. Readers are referred to recent extensive reviews for specific information not covered in this manuscript (Joel and Weiner, 1994, 1997; Parent and Hazrati, 1995; Chesselet and Delfs, 1996; Gerfen and Wilson, 1996; Levy et al., 1997; Smith et al., 1998a, 1998b; Wilson, 1998)

CHEMICAL PHENOTYPE OF NEURONAL POPULATIONS IN STN AND GP

The STN is largely composed of a homogeneous population of glutamatergic projection neurons (Smith and Parent, 1988) which, in some cases, contain calcium binding proteins. In the rat, parvalbumin (PV)-positive neurons are confined to the lateral two thirds of the nucleus which, on the other hand, is devoid of calbindin D 28 K (CABP) and calretinin (CR) immunoreactivity (Hontanilla et al., 1997). In monkeys, intensely stained PV-containing neurons lying in a rich immunostained neuropil are found throughout the whole extent of the STN, whereas CABP immunoreactivity is much lighter and confined to a small group of neurons in the dorsolateral part of the nucleus (Parent et al., 1996). On the other hand, whereas the STN is devoid of CR-containing neurons in rodents, a small population of CR-positive neurons is confined to the medial tip of the STN in monkeys (Parent et al., 1996).

In the rat GP, approximately two-thirds of GABAergic neurons express PV immunoreactivity (Kita, 1994b). In general, PV-positive neurons, which have larger somata than PV-negative cells, are located laterally in an area devoid of CABP-containing fibers from the striatum (Rajakumar et al., 1994). Such a complementarity between the distribution of PV and CABP immunoreactivity was also seen in the entopeduncular nucleus (EPN) where the ventral and medial sectors are enriched in CABP-positive neuropil but devoid of PV-immunoreactive cells, whereas CABP-positive neurons abound in the rostral part of the nucleus (Hontanilla et al., 1997). The GP neuropil is also rich in PV-positive axon terminals which, for the most part, display ultrastructural features of pallidal boutons and form symmetric synapses with the proximal part of PV-positive and PV-negative neurons (Kita, 1994b). On the other hand, CR-positive neurons are expressed in subsets of large and small neurons distributed according to a decreasing rostrocaudal gradient throughout the pallidal complex in monkeys (Parent et al., 1996). A subpopulation of pallidal neurons enriched in GABA and parvalbumin also expresses strong immunoreactivity for a neurotensin-related hexapeptide named LANT-6 (Reiner and Carraway, 1987). The exact function of the various calcium binding proteins and LANT-6 in pallidal and STN neurons remains to be established.

ARE THERE INTERNEURONS AND INTRINSIC AXON COLLATERALS IN GP AND STN?

Although the firing of pallidal and STN neurons is uncorrelated in normal monkeys, oscillatory activity and synchronized firing appears in animals rendered parkinsonian (Bergman et al., 1998). A potential substrate to mediate such cross-talk between groups of GP and STN neurons are interneurons or intrinsic axon collaterals of projection neurons. Due to the incomplete impregnation of axonal projections, early Golgi studies led to controversial data regarding the existence of interneurons in the STN (Rafols and Fox, 1976; Iwahori, 1978; Yelnik and Percheron, 1979). Whereas Rafols and Fox originally proposed that the monkey STN contained a population of small interneurons, subsequent studies in cats, monkeys, and humans concluded that the STN was a relatively homogeneous structure largely composed of projection neurons (Yelnik and Percheron, 1979). These early findings were later supported by intracellular labeling experiments showing that the axons of all labeled STN neurons could be traced beyond the boundaries of the nucleus in rats (Kita et al., 1983). Interestingly, more than half of these projection neurons had intranuclear axon collaterals which extended outside the dendritic domains of the parent neurons, suggesting that they may serve as a feed-forward circuit in the STN (Kita et al., 1983).

The situation is essentially the same in the GP where Golgi and intracellular- or juxtacellular-labeling studies revealed that most GP neurons give off intranuclear collaterals which terminate on the cell bodies and proximal dendrites of neighboring neurons (Kita and Kitai, 1994; Bevan et al., 1998; Sato et al., 2000). In addition, a population of small interneurons with relatively short dendrites and sparse synaptic innervation was found in the pallidum of rat, monkey, and guinea-pig (Fox et al., 1974; Iwahori and Mizuno, 1981; DiFiglia et al., 1982; Shink and Smith, 1995; Nambu and Llinas, 1997). Therefore, it appears that both STN and GP have intranuclear connections, which might play an important integrative role in the processing of information by pools of projection neurons.

SYNAPTIC INTERCONNECTIONS BETWEEN STN AND GP NEURONS

The main inputs to both pallidal segments arise from medium-sized GABAergic neurons in the striatum. Although both striato-external pallidum (GPe) and striato-internal pallidum (GPi) neurons use GABA as a neurotransmitter, these two populations of neurons can be differentiated by their peptide content and relative level of expression of dopamine receptor subtypes (Gerfen et al., 1990). Striatofugal neurons are the source of more than 80% of the total number of synaptic terminals in contact with individual pallidal neurons in monkeys (Shink and Smith, 1995). Two other major sources of synaptic afferents to the pallidum are from glutamatergic neurons in the STN or local connections between the two pallidal segments (Figure 1). The STN glutamatergic afferents, which account for about 10% of the total synaptic innervation of GPe and GPi, are homogeneously distributed on the soma and dendrites of pallidal neurons. In contrast, GABAergic terminals from GPe terminate almost exclusively on the soma and proximal dendrites of GPi neurons, a strategic location to modulate basal ganglia outflow (Smith et al., 1994; Shink and Smith, 1995). A similar pattern of innervation by GP terminals was found in the rat substantia nigra pars reticulata (SNr) (Smith and Bolam, 1990, 1991). On the other hand, GP terminals are homogeneously distributed on the somata, proximal dendrites, and distal dendrites of STN neurons. In the rat, 31% of GP terminals form synapses with perikarya, 39% with large dendrites, and 30% with small dendrites (Smith et al., 1990) (Figure 1).

Elucidation of the neuronal network underlying the reciprocal connections between the GPe and the STN, and the relationships between these structures and the GPi/EPN, is of critical importance in our understanding of how the indirect pathways influence neurons of the GPi/EPN and, hence, the output of the basal ganglia. Data from our laboratory demonstrated highly ordered and specific relationships between neurons of the GPe, STN, and GPi in squirrel monkeys (Shink et al., 1996). Furthermore, the organizational principle of this neuronal network holds true for different functional regions of the GPe, STN, and GPi. Thus, populations of neurons within sensorimotor, cognitive, and limbic territories in the GPe are reciprocally connected with populations of neurons in the same functional territories of the STN, and the neurons in each of these regions then, in turn, innervate the same functional territory of the GPi. An additional interesting and important feature of this network is that neurons in the STN possess axon collaterals that innervate both segments of the pallidal complex (Shink et al., 1996; Smith et al., 1998a). The interconnections between the GPe, the STN, and output neurons of the basal ganglia in the GPi are, thus, much more specific than previously suspected. The basic circuit is such that interconnected groups of neurons in the GPe and STN innervate, via axon collaterals, the same population of neurons in the internal pallidum, i.e. output neurons of the basal ganglia. The topography of the tight interconnections respects the functional organization of the GPe, the STN, and the GPi. However, it is important to keep in mind that all STN neurons do not necessarily project to both pallidal segments. It is thus likely that there are additional organizational principles that do not respect the functional topography of the indirect network and underlie a system for integration of functionally diverse information (Joel and Weiner, 1997; Smith et al., 1998a).

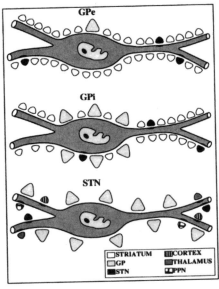

Figure 1. Schematic representation of the distribution and relative densities of GABAergic and glutamatergic inputs to striato-external pallidum, striato-internal pallidum, and subthalamic nucleus neurons.

Other afferents to the pallidum include glutamatergic inputs from the intralaminar thalamic nuclei, serotonergic inputs from the dorsal raphe, glutamatergic and/or cholinergic inputs from the pedunculopontine tegmental nucleus, and dopaminergic inputs from the substantia nigra pars compacta (Pasik et al., 1984; Lee et al., 1988; Smith et al., 1989; Mouroux et al., 1997). In turn, besides the STN, GPi/EPN, and SNr, the GPe/GP neurons project to the striatum, the reticular thalamic nucleus and, the SNc. In rats, the pallidostriatal projection is quite specific and terminates preferentially on striatal GABAergic interneurons (Bevan et al., 1998). In monkeys, GPe neurons are divided into four distinct types based on their projection sites: (1) those that project to STN, GPi, and SNr, (2) those that project to GPi and STN, (3) those that project to SNr and STN, and (4) those that project to the striatum (Sato et al., 2000). Interestingly, GP neurons are much more collateralized and belong to a single population which projects to all targets in rats (Bevan et al., 1998). It is worth noting that the existence of the pallidoreticular projection, which was demonstrated by retrograde and anterograde labeling methods (Hazrati and Parent, 1991; Asanuma, 1994), could not be observed in either rats or monkeys with unicellular labeling techniques (Bevan et al., 1998; Sato et al., 2000).

THE CORTICAL AND THALAMIC INPUTS TO STN: OTHER ENTRANCES TO THE BASAL GANGLIA CIRCUITRY

As with the striatum, the STN also receives excitatory glutamatergic projections from the cerebral cortex. In primates, the cortico-subthalamic projection is exclusively ipsilateral and arises mainly from the primary motor cortex (area 4), with a minor contribution of prefrontal and premotor cortices (Künzle, 1975; Hartman-von Monakow et al., 1978; Afsharpour, 1985; Nambu et al., 1996). Somatosensory and visual cortical areas do not project to the STN but innervate the striatum (Künzle, 1977; McGeorge and Faull, 1989; Saint-Cyr et al., 1990; Flaherty and Graybiel, 1991). In both rat and monkey, the cortico-subthalamic projection is topographically organized so that afferents from the primary motor cortex (M1) are confined to the dorsolateral part of the STN (Hartmann-von Monakow et al., 1978; Nambu et al., 1996), whereas the premotor (areas 8, 9 and 6), supplementary motor, and pre-supplementary motor areas, as well as adjacent frontal cortical regions, innervate preferentially the medial third of the nucleus (Hartmann-von Monakow et al., 1978; Künzle, 1978; Nambu et al., 1996; Inase et al., 1999). On the other hand, inputs from prefrontal-limbic cortices are confined to the medialmost tip of STN (Hartmann-von Monakow et al., 1978; Afsharpour, 1985; Berendse and Groenewegen, 1989; Maurice et al., 1998). By virtue of its cortical inputs, the dorsolateral sector of the STN is involved in the control of motor behaviors, whereas the ventromedial sector processes oculomotor, associative, and limbic information (Delong et al., 1985; Matsumara et al., 1992; Wichmann et al., 1994). Like cortical afferents to the striatum, the cortico-subthalamic projection from M1 is somatotopically organized; the face area projects laterally, the arm area centrally, and the leg area medially (Künzle, 1975; Hartman-von Monakow et al., 1978; Nambu et al., 1996). Interestingly, the arrangement of somatotopical representations from the supplementray motor area (SMA) to the medial STN is reversed against the ordering from M1 to the lateral STN in macaque monkeys (Nambu et al., 1996). Therefore, the cerebral cortex imposes a specific functional segregation not only upon the striatum, but also at the level of the STN (Alexander and Crutcher, 1990; Wichmann et al., 1994). However, it is worth noting that STN neurons have long dendrites that may cross boundaries of functional territories imposed by cortical projections in rats (Bevan et al., 1997). This anatomical arrangement opens up the possibility for some functionally segregated information at the level of the cerebral cortex to converge on individual STN neurons in rodents.

Another source of excitatory inputs to the STN arises from the centromedian parafascicular nuclear complex (CM/PF). No other thalamic nuclei are known to innervate the STN. The thalamo-subthalamic projection arborizes ipsilaterally in discrete portions of the STN. This input respects the functional organization of the STN, i.e. sensorimotor neurons in CM terminate preferentially in the dorsolateral part of the nucleus, whereas limbic- and associative-related neurons in the parafascicular nucleus (PF) project almost exclusively to the medial STN (Sadikot et al, 1992; Féger et al., 1995, 1997). In rats, the thalamo-subthalamic projection is excitatory and tonically drives the activity of STN neurons (Mouroux et al., 1995). Although some PF neurons that project to the striatum send axon collaterals to the STN (Deschênes et al., 1996), the thalamo-subthalamic and thalamo-striatal projections largely arise from segregated populations of PF neurons in rats (Féger et al., 1994).

Even if cortical and thalamic inputs are relatively sparse and terminate on the distal dendrites and spines of STN neurons (Moriizumi et al., 1987; Bevan et al., 1995) (Figure 1), electrophysiologic experiments show that activation of these inputs results in very strong short latency monosynaptic excitatory postsynaptic potentials (EPSP) in STN neurons (Kita, 1994a; Feger et al., 1997). Furthermore, it is worth noting that the information flowing through the STN reaches basal ganglia output structures much faster than information transmitted along the striatofugal pathways (Kita, 1994a). These anatomical and electrophysiological data, therefore, suggest that the STN is another main entrance of information to the basal ganglia circuitry.

Additional synaptic inputs to the STN arise from glutamatergic and cholinergic neurons in the pedunculopontine nucleus (PPN), serotonergic neurons in the dorsal raphe, and dopaminergic neurons in the SNc (Mori et al., 1985; Bevan and Bolam, 1995; Smith and Kieval, in press). On the other hand, it is noteworthy that, apart from GPe, GPI, and SNr, STN neurons project to the SNc, PPN, and striatum (Parent and Smith, 1987b; Smith and Grace, 1992).

THE DOPAMINERGIC NIGROPALLIDAL AND NIGROSUBTHALAMIC PROJECTION

Although the striatum is, by far, the major target of midbrain dopaminergic neurons, the GP, ventral pallidum, and STN also receive significant dopaminergic inputs from SNc and VTA. Various anatomical and immunocytochemical data support the existence of dopaminergic nigropallidal and nigrosubthalamic projections: (1) both pallidal segments, but particularly the GPi and the STN are enriched in small varicose processes immunoreactive for tyrosine hydroxylase or dopamine which form *en passant* type symmetric synapses with pallidal and STN neurons (Parent and Smith, 1987a, Smith and Kieval, in press), (2) retrogradely labeled cells are found in the SNc/VTA complex after tracer injections in GPe, GPi, or STN (Smith et al., 1989; Hassani et al., 1997), and (3) anterograde tracing of nigrofugal neurons leads to labeled axon terminals in both pallidal segments and the STN (Hanley and Bolam, 1997; Gauthier and Parent, 1999). Retrograde and anterograde labeling studies revealed that the nigropallidal and nigrostriatal projections largely arise from separate neuronal populations in monkeys (Smith and Parent, 1989; Gauthier and Parent, 1999). The nigropallidal projection, which is relatively spared in animal models of Parkinson's disease (Parent et al., 1990), seems to play a major role in functional recovery induced by intranigral glial-derived neurotrophic factor (GDNF) administration in parkinsonian monkeys (Gash et al., 1996). In line with these anatomical observations, various functional data indicate that dopamine modulates the activity of pallidal neurons and regulates their expression of c-fos (Pan and Walters, 1988; Ruskin and Marshall, 1997). Furthermore, bilateral infusions of D1 or D2 antagonists in GP induce akinesia (Hauber and Lutz, 1999). At the level of the STN, most dopamine-mediated effects on firing rates and c-fos expression involve D1 and D2 dopamine receptors (Ruskin and Marshall, 1995; Kreiss et al.,1996; Hassani and Féger, 1999), but the exact localization of the dopamine receptors involved in these functional effects is still a matter of debate (Weiner et al., 1991; Flores et al., 1999; Smith and Kieval, in press).

IONOTROPIC *vs* METABOTROPIC GLUTAMATE RECEPTORS IN GP AND STN

As discussed above, GABA and glutamate are the two main transmitters that mediate activity along the direct and indirect pathways of the basal ganglia (Albin et al., 1989; Bergman et al., 1990). The effects of GABA and glutamate on pallidal neurons depend on the subtype and subunit composition of the ionotropic receptors expressed by the postsynaptic neurons and their spatial relationships to glutamate and GABA release sites. In this respect, pallidal neurons express various N-methyl-D-aspartate (NMDA) and α-amino-3-hydroxy-5-methyl-4-isoxazolepropionate (AMPA) receptor subunits concentrated in the core of asymmetric glutamatergic synapses in the rat GP and EPN (Bernard and Bolam, 1998; Clarke and Bolam, 1998) (Figure 2). Pallidal neurons are also enriched in group I (mGluR1 and mGluR5) metabotropic glutamate receptors' mRNAs (Testa et al.,

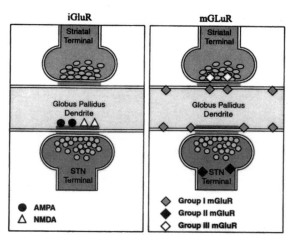

Figure 2. Subsynaptic localization of ionotropic and metabotropic glutamate receptor immunoreactivity in the globus pallidus.

1994). Surprisingly, electron microscopic analysis of group I mGluRs localization revealed that both mGluR1 and mGluR5 immunoreactivity is found postsynaptically in the core of striatopallidal GABA-ergic synapses and at the edges of subthalamopallidal glutamatergic synapses in monkeys (Hanson and Smith, 1999) (Figures 2, 3A,B). These data raise questions about the functions and mechanisms of activation of group I mGluRs at GABAergic synapses in the pallidal complex (see Hanson and Smith, 1999; and Smith et al., 2000, for details). Pallidal neurons express low levels of group II mGluRs which, on the other hand, abound in glial cells (Ohishi et al., 1993a,b; Testa et al., 1994, 1998; Petralia et al., 1996). Finally, the group III mGluRs (mGluR4a and mGluR7a,b) are mostly expressed pre-synaptically in striatopallidal GABAergic terminals where they may act as heteroreceptors to modulate GABA release from striatal terminals (Kinoshita et al., 1998; Bradley et al., 1999; Kosinki et al., 1999) (Figure 2). It is worth noting that mGluR7a, but not mGluR4a, is also expressed pre-synaptically in striatonigral terminals. The selective localization of mGluR4a in the GP makes this receptor an attractive target to reduce the release of GABA from overactive striatopallidal terminals in Parkinson's disease (PD) (see below). The expression of postsynaptic mGluR7a immunoreactivity in dendrites of GP and SNr neurons was also found in rats (Kosinki et al., 1999).

STN neurons express high levels of immunoreactivity for various NMDA and AMPA glutamatergic receptor subunits. Ultrastructural analysis revealed that both types of ionotropic glutamate receptors are expressed preferentially in the postsynaptic membrane of putative glutamatergic synapses, though AMPA receptor subunit immunoreactivity was also found at symmetric GABAergic synapses in rats (Clarke and Bolam, 1998). Although the synaptic localization of mGluRs has not been studied in great detail in the STN, preliminary data indicate that group I mGluRs are found post-synaptically at the edges of asymmetric glutamatergic synapses or in the main body of symmetric GABAergic synapses in rats and monkeys (Smith et al., 2000) (Figure 5A,B). A particular feature that characterizes STN neurons in rats is their strong expression of group II (mGluR2) mGluR mRNAs relative to other populations of basal ganglia neurons (Testa et al., 1994). Consistent with this, group II mGluR agonists reduce STN-mediated excitatory postsynaptic potentials (EPSPs) in rat SNr neurons (Bradley et al., 2000). These effects are likely to be mediated by activation of presynaptic mGluR2 receptors on STN axons and terminals in the SNr. Very low levels of group III mGluR mRNAs and immunoreactivity were found in STN neurons (Testa et al., 1994; Ohishi et al., 1995).

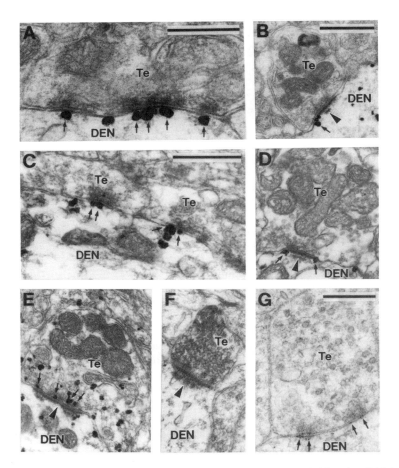

Figure 3. Examples of Group I mGluRs (A-B), GABA$_B$R1 (C-E), GABA$_B$R2 (F) and α_1GABA$_A$ receptor subunit (G) immunoreactivity in the monkey globus pallidus. (A) mGluR5 immunoreactivity in the main body of a symmetric synapse established by a striatal-like GABAergic terminal (Te). (B) mGluR5 immunoreactivity at the edge of an axo-dendritic asymmetric synapse formed by a putative glutamatergic STN terminal (Te). (C) GABA$_B$R1 immunoreactivity at symmetric axo-dendritic synapses. (D) GABA-R1 immunoreactivity at the edge of an asymmetric axo-dendritic synapse. (E) Pre-synaptic GABA$_B$R1 immunoreactivity in a STN-like terminal. (F) Pre-synaptic GABA$_B$R2 immunoreactivity in a putative glutamatergic terminal forming an asymmetric axo-dendritic synapse (arrowhead). (G) Post-synaptic α_1GABA$_A$ receptor subunit immunoreactivity at a striatopallidal axo-dendritic synapse. Arrows indicate immunogold labeling, arrowheads point out asymmetric synapses. Scale bars: A: 0.3 µm; B-F: 0.5 µm; G: 0.3 µm.

IONOTROPIC *VS* METABOTROPIC GABA RECEPTORS IN GP AND STN

As expected, pallidal neurons also express moderate to high levels of GABA$_A$ and GABA$_B$ receptors (Zhang et al., 1991; Wisden et al., 1992; Fritschy and Möhler, 1995; Waldvogel et al., 1998). The GABA$_A$ receptor subunits α1, β2/3, and γ2 are mostly found postsynaptically in the core of symmetric striatopallidal GABAergic synapses (Somogyi et al., 1996; Charara and Smith, 1998; Waldvogel et al., 1998; Fujiyama et al., 1998) (Figures 3G, 4), whereas the GABA$_B$R1 receptor subunits are present on the post-synaptic membrane of striatopallidal synapses, perisynaptic to asymmetric synapses and presynaptic in subthalamic glutamatergic terminals (Charara et al., 2000a; Smith et al., in press) (Figures 3C-E, 4). Recent immunocytochemical data revealed that GABA$_B$R2 subunits follow a similar pattern of distribution (Charara et al., 2000b) (Figure 3F), which suggests that both GABA$_B$ receptor subunits may form functional heterodimers at pre- and postsynaptic sites to modulate the GABAergic and glutamatergic transmission in the pallidal complex.

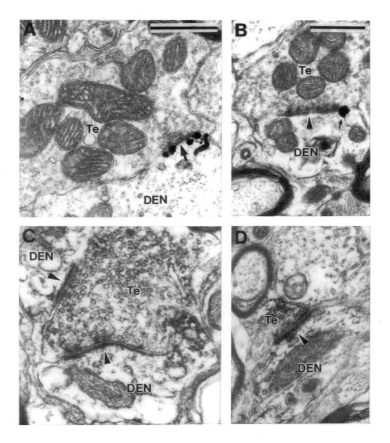

Figure 4. Subsynaptic localization of GABA$_A$ and GABA$_B$R1 receptors' immunoreactivity in the globus pallidus.

As is the case for the pallidum, STN neurons display strong immunoreactivity for various GABA$_A$ receptor subunits (Zhang et al., 1991; Wisden et al., 1992; Fritschy and Möhler, 1995; Charara and Smith, 1998), which is consistent with electrophysiological studies showing that the pallidal inhibition of STN neurons is largely mediated by GABA$_A$ receptor activation. In addition, STN neurons display moderate immunoreactivity for GABA$_B$ receptor subunits (Charara et al., 2000a,b) (Figure 5C,D). At the electron microscope level, GABA$_B$ receptors are expressed post-synaptically on dendrites of STN neurons and pre-synaptically in putative glutamatergic axon terminals (Charara et al.,

2000a,b) (Figure 5C,D). Together, these data indicate that both $GABA_A$ and $GABA_B$ receptors are likely to mediate postsynaptic inhibition from GPe in STN neurons. In addition, $GABA_B$ receptors may also control the activity of STN neurons by presynaptic inhibition of neurotransmitter release from extrinsic and/or intrinsic glutamatergic terminals.

METABOTROPIC GLUTAMATE AND $GABA_B$ RECEPTORS: NOVEL THERAPEUTIC TARGETS FOR PARKINSON'S DISEASE

An imbalance of activity between the direct and indirect striatofugal pathways in favor of the indirect pathway is thought to underlie most symptoms of PD (DeLong, 1990). The increased activity of the glutamatergic subthalamopallidal and, possibly, corticostriatal projections in animal models of PD led various groups to test the potential therapeutic benefits of ionotropic glutamate antagonists in PD (see Starr, 1995 for a review). Systemic administration of NMDA and non-NMDA antagonists with a subthreshold dose of L-DOPA or D2 dopamine receptor agonist has proven to ameliorate symptoms in primate models of PD (Starr, 1995; Blandini et al., 1996). Data reported in this review strongly suggest that interactions with metabotropic glutamate and $GABA_B$ receptors may also have beneficial effects in PD. Drugs interacting with these receptors are expected to influence the induction and progression of the symptoms of the disease without hampering the efficiency of fast glutamatergic and GABAergic synaptic transmission, thereby reducing unwanted side effects commonly seen with drugs that target ionotropic receptors (Starr, 1995).

At first glance, activation of the group II mGluR, mGluR4, seems to be an ideal strategy to alleviate symptoms of PD. It is well established that activation of presynaptic group III mGluRs reduces neuroransmitter release in the hippocampus (Conn and Pin, 1997). If such is also the case in GP, activation of these receptors in parkinsonism should reduce the activity of the overactive indirect pathway by reducing GABA release at striatopallidal synapses, thereby inhibiting subthalamopallidal neurons which, in turn, relieve basal ganglia output neurons in GPi and SNr from their tonic STN excitatory drive. The final outcome of such therapy should be an increased activity of thalamocortical neurons and facilitation of motor behaviors. Another mGluR subtype of interest for PD therapy is mGluR2, which was found to be expressed on subthalamonigral terminals in rats (Bradley et al., 2000). Furthermore, activation of these receptors in brain slices reduces glutamatergic transmission at subthalamonigral synapses (Bradley et al., 2000). In line with the hypothesis that activation of these receptors might alleviate parkinsonian symptoms, systemic administration of group II agonist reverses haloperidol- induced catalepsy in rats (Bradley et al., 2000). Finally, the group I mGluRs located perisynaptically at STN synapses in GPe and GPi should also be considered as a potential target in PD because the perisynaptic mGluR1a and mGluR5 are likely to be activated by excessive amounts of glutamate released during hyperactivity of subthalamopallidal synapses in parkinsonism. Group I mGluR activation might, then, lead to increased activity of basal ganglia output neurons through various mechanisms including potentiation of ionotropic glutamatergic transmission, reduction of K^+ conductances, and so forth (see Conn and Pin, 1997 for details). Group I mGluRs antagonists should, therefore, reduce the over-excitatory drive generated by the STN in pallidal neurons. So far, only a few specific agents for group II (LY354740 and LY379268) and group I (MPEP) mGluRs were found to produce central pharmacological actions when administered systemically in animals (Helton et al., 1998; Moghaddam and Adams, 1998; Bordi and Ugolini, 1999; Schoepp et al., 1999). However, the potential therapeutic benefits of such agents will likely drive the development of additional compounds that could be administered systemically for novel medical purposes.

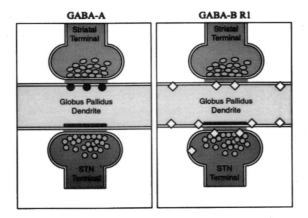

Figure 5. Group I mGluRs (A-B) and GABAB receptor (C-D) immunoreactivity in the monkey STN. (A) mGluR1a at a symmetric axo-dendritic synapse (arrow) established by a putative GABAergic terminal from GPe (Te). (B) mGluR1a immunoreactivity at the edge of an asymmetric axo-dendritic synapse (arrowhead). (C-D) GABAB-R1 (C) and GABAB-R2 (D) immunoreactivity in putative glutamatergic terminals forming asymmetric axo-dendritic synapses (arrowheads). Scale bars: A-D: 0.5 μm.

The expression of GABA$_B$ receptors in subthalamopallidal and subthalamonigral terminals (Charara et al., 2000a,b) suggest that activation of these pre-synaptic heteroreceptors might attenuate the overflow of glutamate released by STN neurons in PD. In support of this hypothesis, application of baclofen was found to decrease the efflux of glutamate in the rat globus pallidus *in vivo* (Singh, 1990) and reduce the evoked synaptic currents mediated by glutamate in the rat SNr *in vitro* (Shen and Johnson, 1997). The current use of GABA$_B$ agonists in therapeutics is mostly restricted to baclofen in the treatment of spasticity (Porter, 1997). In fact, the beneficial antispastic effect of baclofen is believed to derive from the suppression of excitatory neurotransmitter release to motor neurons in the spinal cord (Fox et al., 1978; Davies, 1981; Bonnano et al., 1998). Future behavioral studies of baclofen in animal models should help determine the potential therapeutic efficacy of this drug for Parkinson's disease and clarify the functions of GABA$_B$ in modulating glutamatergic neurotransmission in the basal ganglia circuitry. Interestingly, baclofen was found to reduce haloperidol-induced dyskinesias without causing gross motor depression in squirrel monkeys (Neale et al., 1984).

ACKNOWLEDGMENTS: The authors acknowledge Jean-François Paré and Jeremy Kieval for technical assistance, Frank Kiernan for photography, and Peggy Plant for clerical work. The authors also thank Dr. Allan I. Levey and Mr. Craig Helman for their help with the purification of the GABAB-R1 antiserum. Thanks are also due to Drs. R. Shigemoto and A. Kulik for the generous gift of the GABA$_B$R2 antiserum. This work was supported by the NIH base grant to the Yerkes Regional Primate Research Center (RR00165) and NINDS grants R01 NS 37423-03 and P50 NS38399-01.

REFERENCES

Afsharpour, S., 1985, Topographical projections of the cerebral cortex to the subthalamic nucleus, *J. Comp. Neurol.* 236:14.

Albin, R.L., Young, A.B., and Penney, J.B. Jr., 1989, The functional anatomy of basal ganglia disorders, *Trends Neurosci.* 12:366.

Alexander, G.E., and Crutcher, M.E., 1990, Functional architecture of basal ganglia circuits: neural substrates of parallel processing, *Trends Neurosci.* 13:266.

Asanuma, C., 1994, GABAergic and pallidal terminals in the thalamic reticular nucleus of squirrel monkeys, *Exp. Brain Res.* 101:439.

Aziz, T.Z., Peggs, D., Sambrook, M.A., and Crossman, A.R., 1991, Lesion of the subthalamic nucleus for the alleviation of 1-methyl-4-phenyl-1,2,3,6-tetrahydropyridine (MPTP)-induced parkinsonism in the primate, *Movement Dis.* 6:388.

Berendse, H.W., and Groenewegen, H.J., 1989, The connections of the medial part of the subthalamic nucleus in the rat: evidence for a parallel organization, in: *The Basal Ganglia III*, G. Bernardi, M.B. Carpenter, G. Di Chiara, M. Morelli, and P. Stanzione, eds., Plenum Press, New York.

Bergman, H., Feingold, A., Nini, A., Raz, A., Slovin, H., Abeles, M., and Vaadia, E., 1998, Physiological aspects of information processing in the basal ganglia of normal and parkinsonian primates, *Trends Neurosci.* 21:32.

Bergman, H., Wichmann, T., and DeLong, M.R., 1990, Reversal of experimental parkinsonism by lesions of the subthalamic nucleus, *Science,* 249:1436.

Bernard, V., and Bolam, J.P., 1998, Subcellular and subsynaptic distribution of NR1 subunit of the NMDA receptor in the neostriatum and globus pallidus of the rat: co-localization at synapses with the GluR2/3 subunit of the AMPA receptor, *Eur. J. Neurosci.* 10:3721.

Bevan, M.D., and Bolam, J.P., 1995, Cholinergic, GABAergic, and glutamate-enriched inputs from the mesopontine tegmentum to the subthalamic nucleus of the rat, *J. Neurosci.* 15: 7105.

Bevan, M.D., Booth, P.A.C., Eaton, S.A., and Bolam, J.P.,1998, Selective innervation of neostriatal interneurons by a subclass of neurons in the globus pallidus of the rat, *J. Neurosci.* 18:9438.

Bevan, M.D., Clarke, N.P., and Bolam, J.P., 1997, Synaptic integration of functionally diverse pallidal information in the entopeduncular nucleus and subthalamic nucleus in the rat, *J. Neurosci.* 17:308.

Bevan, M.D., Francis, C.M., and Bolam, J.P., 1995, The glutamate-enriched cortical and thalamic input to neurons in the subthalamic nucleus of the rat: convergence with GABA-positive terminals, *J. Comp. Neurol.* 361:491.

Blandini, F., Porter, H.P., and Greenamyre, J.T., 1996, Glutamate and Parkinson's disease, *Molec. Neurobiol.* 12:73.

Bonanno, G., Fassio, A., Sala, R., Schmid, G., and Raiteri, M., 1998, GABA$_B$ receptors as potential targets for drugs able to prevent excessive excitatory amino acid transmission in the spinal cord, *Eur. J. Pharmacol.* 362:143.

Bordi, F., and Ugolini, A., 1999, Group I metabotropic glutamate receptors: Implications for brain diseases, *Prog. Neurobiol.* 59:55.

Bradley, S.R., Marino, M.J., Wittmann, M., Rouse, S.T., Awad, H., Levey, A.I., and Conn, P.J., 2000, Activation of group II metabotropic glutamate receptors inhibits synaptic excitation of the substantia nigra pars reticulata, *J. Neurosci.* 20:3085.

Bradley, S.R., Standaert, D.G., Rhodes, K.J., Rees, H.J., Testa, C.M., Levey, A.I., and Standaert, D.G., 1999, Immunohistochemical localization of subtype 4a metabotropic glutamate receptors in the rat and mouse basal ganglia, *J. Comp. Neurol.* 407:33.

Charara, A, Heilman, C., and Levey, A.I., 2000a, Pre- and Post-synaptic GABA$_B$ receptors in the basal ganglia in monkeys, *Neuroscience,* 95:127.

Charara, A., Kulik, A., Shigemoto, R., and Smith, Y., 2000b, Cellular and subcellular localization of GABA$_B$R2 receptors in the basal ganglia in monkeys, *Soc. Neurosci. Abstr.* 26: 6071.

Charara, A., and Smith, Y., 1998, Subsynaptic distribution of GABA$_A$ receptor subunits in the globus pallidus and subthalamic nucleus in monkeys, *Soc. Neurosci. Abstr.* 24:650.

Chesselet, M.-F., and Delfs, J.M., 1996, Basal ganglia and movement disorders: an update, *Trends Neurosci.* 19:417.

Clarke, N.P., and Bolam, J.P., 1998, Distribution of glutamate receptor subunits at neurochemically characterized synapses in the entopeduncular nucleus and subthalamic nucleus of the rat, *J. Comp. Neurol.* 397:403.

Conn, P.J., and Pin, J.P., 1997, Pharmacology and functions of metabotropic glutamate receptors, *Ann. Rev. Pharmacol. Toxicol.* 37:205.

Davies, J., 1981, Selective depression of synaptic excitation in cat spinal neurones by baclofen: an iontophoretic study, *Br. J. Pharmacol.* 72:373.

DeLong, M.R., 1990, Primate models of movement disorders of basal ganglia origin, *Trends Neurosci.* 13:281.

DeLong, M.R., Crutcher, M.D., and Georgopoulos, A.P., 1985, Primate globus pallidus and subthalamic nucleus functional organization, *J. Neurophysiol.* 53:530.

Deschênes, M., Bourassa, J., Doan, V.D., and Parent, A., 1996, A single-cell study of the axonal projections arising from the posterior intralaminar thalamic nuclei in the rat, *Eur. J. Neurosci.* 8:329.

DiFiglia, M., Pasik, P., and Pasik, T., 1982, A Golgi and ultrastructural study of the monkey globus pallidus, *J. Comp. Neurol.* 212:53.

Féger, J., Bevan, M., and Crossman, A.R., 1994, The projections from the parafascicular thalamic nucleus to the subthalamic nucleus and the striatum arise from separate neuronal populations: a comparison with the corticostriatal and corticosubthalamic efferents in a retrograde fluorescent double-labeling study, *Neuroscience,* 60:125.

Féger, J., Hassani, O.-K., and Mouroux, M., 1995, The relationships between subthalamic nucleus, globus pallidus and thalamic parafascicular nucleus, in: *The Basal Ganglia V,* C. Ohye, M. Kimura, and J.S. McKenzie, eds., Plenum Press, London.

Féger, J., Hassani, O.-K., and Mouroux, M., 1997, The subthalamic nucleus and its connections. New electrophysiological and pharmacological data, in: *The Basal Ganglia and New Surgical Approaches for Parkinson's Disease,* J.A. Obeso, M.R. DeLong, C. Ohye, and C.D. Marsden, eds., Lippincott-Raven Publishers, Philadelphia.

Flaherty, A.W., and Graybiel, A.M., 1991, Corticostriatal transformations in the primate somatosensory system. Projections from physiologically mapped body-part representations, *J. Neurophysiol.* 66:1249.

Flores, G., Liang, J.J., Sierra, A., Martinez-Fong, D., Quirion, R., Aceves, J., and Srivastava, L.K., 1999, Expression of dopamine receptors in the subthalamic nucleus of the rat: characterization using reverse transcriptase-polymerase chain reaction and autoradiography, *Neuroscience,* 91:549.

Fox, C.A., Andrade, A.N., LuQui, I.J., and Rafols, J.A., 1974, The primate globus pallidus: a Golgi and electron microscopic study, *J. Hirnforschung,* 15:75.

Fox, S., Krnjevic, K., Morris, M.E., Puil, E., and Werman, P., 1978, Action of baclofen on mammalian synaptic transmission, *Neuroscience,* 3:445.

Fritschy, J.-M., and Möhler, 1995, GABA$_A$ receptor heterogeneity in the adult rat brain: differential regional and cellular distribution of seven major subunits, *J. Comp. Neurol.* 359:154.

Gash, D.M., Zhang, Z., Ovadia, A., Cass, W.A., Yi, A., Simmerman, L., Russell, D., Martin, D., Lapchak, P.A., Collins, F., Hoffer, B.J., and Gerhardt, G.A., 1996, Functional recovery in parkinsonian monkeys treated with GDNF, *Nature,* 380: 252.

Gauthier, J., and Parent, A. (1999) The axonal arborization of single nigrostriatal neurons in rats, *Brain Res.* 834:228.

Gerfen, C.R., Engber, T.M., Mahan, L.C., Susel, Z., Chase, T.N., Monsma, F.J., and Sibley, D.R., 1990, D1 and D2 dopamine receptor-regulated gene expression of striatonigral and striatopallidal neurons, *Science,* 250:1429.

Gerfen, C.R., and Wilson, C.J., 1996, The basal ganglia, in: *Handbook of Chemical Neuroanatomy, Integrated Systems of the CNS, Part III,* A. Björklund, T. Hökfelt, and L. Swanson, eds., Elsevier, Amsterdam.

Hanley, J.J., and Bolam, J.P., 1997, Synaptology of the nigrostriatal projection in relation to the compartmental organization of the neostriatum in the rat, *Neuroscience,* 81:353.

Hanson, J.E., and Smith, Y., 1999, Group I metabotropic glutamate receptors at GABAergic synapses in monkeys, *J. Neurosci.* 19:6488.

Hartmann-von Monakow, K., Akert, K., and Künzle, H., 1978, Projections of the precentral motor cortex and other cortical areas of the frontal lobe to the subthalamic nucleus in the monkey, *Exp. Brain Res.* 33:395.

Hassani, O.-K., and Féger, J., 1999, Effects of intrasubthalamic injection of dopamine receptor agonists on subthalamic neurons in normal and 6-hydroxydopamine-lesioned rats: an electrophysiological and c-fos study, *Neuroscience,* 92: 533.

Hassani, O-K., François, C., Yelnik, J., and Féger, J., 1997, Evidence for a dopaminergic innervation of the subthalamic nucleus in the rat, *Brain Res.* 749:88.

Hauber, W. and Lutz, S., 1999, Dopamine D1 or D2 receptor blockade in the globus pallidus produces akinesia in the rat, *Behav. Brain Res.* 106:143.

Hazrati, L.-N., and Parent, A., 1991, Projection from the external pallidum to the reticular thalamic nucleus in the squirrel monkey, *Brain Res.* 550:142.

Helton, D.R., Tizzano, J.P., Monn, J.A., Schoepp, D.D., and Kallman, M.J., 1998, Anxiolytic and side-effect profile of LY354740: A potent, highly selective, orally active agonist for group II metabotropic glutamate receptors, *J. Pharmacol. Exp. Ther.* 284:651.

Hontanilla, B., Parent, A., and Giménez-Amaya, J., 1997, Parvalbumin and calbindin D-28k in the entopeduncular nucleus, subthalamic nucleus, and substantia nigra of the rat as revealed by double-immunohistochemical methods, *Synapse,* 25:359.

Inase, M., Tokuno, H., Nambu, A., Akazawa, T., and Takada, M., 1999, Corticostriatal and corticosubthalamic input zones form the presupplementary motor area in the macaque monkey: comparison with the input zones from the supplementary motor area, *Brain Res.* 833:191.

Iwahori, N., 1978, A Golgi study on the subthalamic nucleus of the cat, *J. Comp. Neurol.* 182:383.

Iwahori, N., and Mizuno, N., 1981, A Golgi study on the globus pallidus of the mouse, *J. Comp. Neurol.* 197:29.

Joel, D., and Weiner, I., 1994, The organization of the basal ganglia-thalamocortical circuits: open interconnected rather than closed segregated, *Neuroscience,* 63:363.

Joel, D., and Weiner, I., 1997, The connections of the primate subthalamic nucleus: indirect pathways and the open-interconnected scheme of basal ganglia-thalamocortical circuitry, *Brain Res. Rev.* 23:62.

Kinoshita, A., Shigemoto, R., Ohishi, H., van der Putten, H., and Mizuno, N., 1998, Imunohistochemical localization of metabotropic glutamate receptors, mGluR7a and mGluR7b, in the central nervous system of the adult rat and mouse: a light and electron microscopic study, *J. Comp. Neurol.* 393:332.

Kita, H., 1994a , Physiology of two disynaptic pathways from the sensorimotor cortex to the basal ganglia output nuclei, in: *The Basal Ganglia IV, New Ideas and Data on Structure and Function*, G. Percheron, J.S. McKenzie, and J. Féger, Plenum Press, New York.

Kita, H., 1994b , Parvalbumin-immunopositive neurons in ratglobus pallidus: a light and electron microscopic study, *Brain Res.* 657:31.

Kita, H., Chang, H.T., and Kitai, S.T., 1983, The morphology of intracellularly labeled rat subthalamic neurons: a light microscopic analysis, *J. Comp. Neurol.* 215:245.

Kita, H., and Kitai, S.T., 1994, The morphology of globus pallidus projection neurons in the rat: an intracellular staining study, *Brain Res.* 636:308.

Kitai, S.T., and Kita, H., 1987, Anatomy and physiology of the subthalamic nucleus: a driving force of the basal ganglia, in: *The Basal Ganglia II, Structure and Function: Current Concepts*, M.B. Carpenter, and A. Jayaraman, eds., Plenum Press, New York.

Kosinki, C.M., Bradley, S.R., Conn, P.J., Levey, A.I., Landwermeyer, G.B., Penney, J.B., Young, A.B., and Standaert, D.G., 1999, Localization of metabotropic glutamate receptor 7 mRNA and mGluR7a protein in the rat basal ganglia, *J. Comp. Neurol.* 415:266.

Kreiss, D.S., Anderson, L.A., and Walters, J.R. 1996, Apomorphine and dopamine D1 receptor agonists increase the firing rates of subthalamic nucleus neurons, *Neuroscience*, 72: 863.

Künzle, H., 1975, Bilateral projections from precentral motor cortex to the putamen and other parts of the basal ganglia. An autoradiographic study in *Macaca fascicularis, Brain Res.* 88:195.

Künzle, H., 1977, Projections from the primary somatosensory cortex to basal ganglia and thalamus in the monkey, *Brain Res.* 30:481.

Künzle, H., 1978, An autoradiographic analysis of the efferent connections from premotor and adjacent prefrontal regions (areas 6 and 9) in *Macaca fascicularis, Brain Behav. Evol.* 15:185.

Lee, H.J., Rye, D.B., Hallanger, A.E., Levey, A.I., and Wainer, B.H., 1988, Cholinergic vs. noncholinergic efferents from the mesopontine tegmentum to the extrapyramidal motor system nuclei, *J. Comp. Neurol.* 275:469.

Levy, R., Hazrati, L.-N., Herrero, M.-T., Vila, M., Hassani, O.K., Mouroux, M., Ruberg, M., Asensi, H., Agid, Y., Féger, J., Obeso, J.A., Parent, A., and Hirsch, E.C., 1997, Re-evaluation of the functional anatomy of the basal ganglia in normal and parkinsonian states, *Neuroscience*, 76:335.

Matsumura, M., Kojima, J., Gardiner, T.W., and Hikosaka, O., 1992, Visual and oculomotor functions of the monkey subthalamic nucleus, *J. Neurophysiol.* 67:1615.

Maurice, N., Deniau, J.-M., Glowinski, J., and Thierry, A.-M., 1998, Relationships between the prefrontal cortex and the basal ganglia in the rat: physiology of the corticosubthalamic circuits, *J. Neurosci.* 18:9539.

McGeorge, A.J., and Faull, R.L., 1989, The organization of the projection from the cerebral cortex to the striatum in the rat, *Neuroscience*, 29:503.

Moghaddam, B., and Adams, B.W., 1998, Reversal of phencyclidine effects by group II metabotropic glutamate receptor agonist in rats, *Science*, 281: 1349.

Mori, S., Takino, T., Yamada, H., and Sano, Y., 1985, Immunohistochemical demonstration of serotonin nerve fibers in the subthalamic nucleus of the rat, cat and monkey, *Neurosci. Lett.* 62:305.

Moriizumi, T., Nakamura, Y., Kitao, Y., and Kudo, M., 1987, Ultrastructural analyses of afferent terminals in the subthalamic nucleus of the cat with a combined degeneration and horseradish peroxidase tracing method, *J. Comp. Neurol.* 265: 159.

Mouroux, M., Hassani, O.-K., and Féger, J., 1997, Electrophysiological and Fos immunohistochemical evidence for the excitatory nature of the parafascicular projection to the globus pallidus, *Neuroscience*, 81:387.

Nambu, A., and Llinas, R., 1997, Morphology of globus pallidus neurons: its correlation with electrophysiology in guinea pig brain slices, *J. Comp. Neurol.* 277:85.

Nambu, A., Takada, M., Inase, M., and Tokuno, H., 1996, Dual somatotopical representations in the primate subthalamic nucleus: evidence for ordered but reversed body-map transformations from the primary motor cortex and the supplementary motor area, *J. Neurosci.* 16:2671.

Neale, R., Gerhardt, S., and Liebman, J.M., 1984, Effects of dopamine agonists, catecholamine depletors, and cholinergic and GABAergic drugs on acute dyskinesias in squirrel monkeys, *Psychopharmacology*, 82:20.

Ohishi, H., Akazawa, C., Shigemoto, R., Nakanishi, S., and Mizuno, N., 1995, Distribution of the mRNAs for L-2-amino-4-phosphonobutyrate-sensitive metabotropic glutamate receptors, mGluR4 and mGluR7, in the rat brain, *J. Comp. Neurol.* 360:555.

Ohishi, H., Shigemoto, R., Nakanishi, S., and Mizuno, N., 1993a, Distribution of the mRNA for a metabotropic glutamate receptor (mGluR3) in the rat brain: an *in situ* hybridization study, *J. Comp. Neurol.* 335:252.

Ohishi, H., Shigemoto, R., Nakanishi, S., and Mizuno, N., 1993b, Distribution of the messenger RNA for a metabotropic glutamate receptor, mGluR2, in the central nervous system of the rat, *Neuroscience*, 53:1009.

Pan, H.S., and Walters, J.R., 1988, Unilateral lesion of the nigrostriatal pathway decreases the firing rate and alters the firing pattern of globus pallidus neurons in the rat, *Synapse*, 2:650.

Parent, A., Fortin, M., Côté, P.-Y., and Cicchetti, F., 1996, Calcium-binding proteins in primate basal ganglia, *Neurosci. Res.* 25:309.

Parent, A., Hazrati, L.-N., 1995, Functional anatomy of the basal ganglia. I. The cortico-basal ganglia-thalamo-cortical loop, *Brain Res. Rev.* 20:91.

Parent, A. et al., 1990, The dopaminergic nigropallidal projection in primates: Distinct cellular origin and relative sparing in MPTP-treated monkeys, *Adv. Neurol.* 53:111.

Parent, A., and Smith Y., 1987a, Differential dopaminergic innervation of the two pallidal segments in the squirrel monkey (*Saimiri sciureus*), *Brain Res.* 426:397.

Parent, A., and Smith, Y., 1987b, Organization of efferent projections of the subthalamic nucleus in the squirrel monkey as revealed by retrograde labeling methods, *Brain Res.* 436:296.

Pasik, P., Pasik, T., Pecci-Saavedra, J., Holstein, G.R., and Yahr, M.D., 1984, Serotonin in pallidal neuronal circuits: an immunohistochemical study in monkeys, in: *Advances in Neurology, Vol. 40*, R.G. Hassler, and J.F. Christ, eds., Raven Press, New York.

Petralia, R.S., Wang, Y.-X., Niedzielski, A.S., and Wenthold, R.J., 1996, The metabotropic glutamate receptors, mGluR2 and mGluR3, show unique postsynaptic, presynaptic and glial localizations, *Neuroscience,* 71:949.

Plenz, D., and Kitai, S.T., 1999, A basal ganglia pacemaker formed by the subthalamic nucleus and external globus pallidus, *Nature,* 400:677.

Porter, B., 1997, A review of intrathecal baclofen in the management of spasticity, *Br. J. Nurs.* 6:253.

Rafols, J.A., and Fox, C.A., 1976, The neurons in the primate subthalamic nucleus: a Golgi and electron microscopic study, *J. Comp. Neurol.* 168:75.

Rajakumar, N., Rushlow, W., Naus, C.C.G., Elisevich, K., and Flumerfelt, B.A., 1994, Neurochemical compartmentalization of the globus pallidus in the rat: an immunocytochemical study of calcium-binding proteins, *J. Comp. Neurol.* 346:337.

Reiner, A., and Carraway, R.E., 1987, Immunohistochemical and biochemical studies on Lys[8]-Asn[9]-neurotensin[8-13] (LANT6)-related peptides in the basal ganglia of pigeons, turtles, and hamsters, *J. Comp. Neurol.* 257:453.

Ruskin, D.N., and Marshall, J.F., 1995, D1 dopamine receptors influence Fos immunoreactivity in the globus pallidus and subthalamic nucleus of intact and nigrostriatal-lesioned rats, *Brain Res.* 703:156.

Ruskin, D.N., and Marshall, J.F., 1997, Differing influences of dopamine agonists and antagonists of fos expression in identified populations of globus pallidus neurons, *Neuroscience,* 81:79.

Sadikot, A.F., Parent, A., and Francois, C., 1992, Efferent connections of the centromedian and parafascicular thalamic nuclei in the squirrel monkey: a PHA-L study of subcortical projections, *J. Comp. Neurol.* 315:137.

Saint-Cyr, J.A., Ungerleider, L.G., and Desimone, R., 1990, Organization of visual cortical inputs to the striatum and subsequent outputs to the pallido-nigral complex in the monkey, *J. Comp. Neurol.* 298:129.

Sato, F., Lavallée, P., Lévesque, M., and Parent, A., 2000, Single-axon tracing study of neurons of the external segment of the globus pallidus in primate, *J. Comp. Neurol.* 417:17.

Schoepp, D.D., Jane, D.E., and Monn, J.A., 1999, Pharmacological agents acting at subtypes of metabotropic glutamate receptors, *Neuropharmacology,* 38:1431.

Shen, K.Z., and Johnson, S.W., 1997, Presynaptic GABA$_B$ and adenosine A$_1$ receptors regulate synaptic transmission to rat substantia nigra reticulata neurones, *J. Physiol.* 505:153.

Shink, E., Bevan, M.D., Bolam, J.P., and Smith, Y., 1996, The subthalamic nucleus and the external pallidum: two tightly interconnected structures that control the output of the basal ganglia in the monkey, *Neuroscience,* 73:335.

Shink, E., and Smith, Y., 1995, Differential synaptic innervation of neurons in the internal and external segments of the globus pallidus by the GABA- and glutamate-containing terminals in the squirrel monkey, *J. Comp. Neurol.* 358:119.

Singh, R., 1990, GABA$_B$ receptors modulate glutamate release in the rat caudate and globus pallidus, *Soc. Neurosci. Abstr.* 16:1041.

Smith, Y., Bevan, M.D., Shink, E., and Bolam, J.P., 1998a, Microcircuitry of the direct and indirect pathways of the basal ganglia, *Neuroscience,* 86:353.

Smith, Y., and Bolam, J.P., 1990, The output neurons and the dopaminergic neurones of the substantia nigra receive a GABA-containing input from the globus pallidus in the rat, *J. Comp. Neurol.* 296:47.

Smith, J., and Bolam, J.P., 1991, Convergence of synaptic inputs from the striatum and the globus pallidus onto identified nigrocollicular cells in the rat: a double anterograde labelling study, *Neuroscience,* 44:45.

Smith, J., Bolam, J.P., and von Krosigk, M., 1990, Topographical and synaptic organization of the GABA-containing pallidosubthalamic projection in the rat, *Eur. J. Neurosci.* 2:500.

Smith, Y., Charara, A., Hanson, J.E., Paquet, M., and Levey, A.I., in press, GABA$_B$ and group I metabotropic glutamate receptors in the striatopallidal complex in primates, *J. Anat.* 196:555.

Smith, I.D., and Grace, A.A., 1992, Role of the subthalamic nucleus in the regulation of nigral dopamine neuron activity, *Synapse,* 12:287.

Smith, Y., and Kieval, J.Z., in press, Anatomy of striatal and extrastriatal dopamine systems in the basal ganglia, *Trends Neurosci.*

Smith, Y., Lavoie, B., Dumas, J., and Parent, A., 1989, Evidence for a distinct nigropallidal dopaminergic projection in the squirrel monkey, *Brain Res.* 482:381.

Smith, Y., and Parent, A., 1988, Neurons of the subthalamic nucleus in primates display glutamate but not GABA immunoreactivity, *Brain Res.* 453:353.

Smith, Y., Shink, E., and Sidibé, M., 1998b, Neuronal circuitry and synaptic connectivity of the basal ganglia, *Neurosurg. Clin. N. Am.* 9:203.

Smith, Y., Wichmann, T., and DeLong, M.R., 1994, Synaptic innervation of neurones in the internal pallidal segment by the subthalamic nucleus and the external pallidum in monkeys, *J. Comp. Neurol.* 343:297.

Somogyi, P., Fritschy, J.-M., Benke, D., Roberts, J.D.B., and Sieghart, W., 1996, The $\gamma 2$ subunit of the GABA$_A$ receptor is concentrated in synaptic junctions containing the $\alpha 1$ and $\beta 2/3$ subunits in hippocampus, cerebellum and globus pallidus, *Neuropharmacology*, 35: 1425.

Starr, M.S., 1995, Glutamate/dopamine D1/D2 balance in the basal ganglia and its relevance to Parkinson's disease, *Synapse*, 19:264.

Starr, P.A., Vitek, J.L., and Bakay, R.A.E., 1998, Ablative surgery and deep- brain stimulation for Parkinson's disease, *Neurosurgery*, 43:989.

Testa, C.M., Friberg, I.K., Weiss, S.W., and Standaert, D.G., 1998, Immunohistochemical localization of metabotropic glutamate receptors mGluR1a and mGluR2/3 in the rat basal ganglia, *J. Comp. Neurol.* 390:5.

Testa, C.M., Standaert, D.G., Young, A.B., and Penney, J.B., 1994, Metabotropic glutamate receptor mRNA expression in the basal ganglia of the rat, *J. Neurosci.* 14:3005.

Waldvogel, H.J., Fritschy, J.-M., Mohler, H., and Faull, R.L.M., 1998, GABA$_A$ receptors in the primate basal ganglia: An autoradiographic and a light and electron microscopic immunohistochemical study of the $\alpha 1$ and $\beta 2,3$ subunits in the baboon brain, *J. Comp. Neurol.* 397: 297.

Weiner, D.M. et al., 1991, D1 and D2 dopamine receptor mRNA in rat brain, *Proc. Natl. Acad. Sci.* 88:1859.

Whittier, J.R., and Mettler, F.A., 1949, Studies on the subthalamus of the rhesus monkey. II. Hyperkinesia and other physiologic effects of subthalamic lesions with special reference to the subthalamic nucleus of Luys, *J. Comp. Neurol.* 90:319.

Wichmann, T., Bergman, H., and DeLong, M.R., 1994, The primate subthalamic nucleus. I. Functional properties in intact animals, *J. Neurophysiol.* 72:494.

Wichmann, T., and DeLong, M.R., 1996, Functional and pathophysiological models of the basal ganglia, *Curr. Opin. Neurobiol.* 6:751.

Wilson, C.J., 1998, Basal ganglia, in: *The Synaptic Organization of the Brain,* Oxford University Press, New York.

Wisden, W., Laurie, D.J., Monyer, H., and Seeburg, P.H., 1992, The distribution of 13 GABA$_A$ receptor subunit mRNAs in the rat brain, I. Telencephalon, diencephalon, mesencephalon, *J. Neurosci.* 12:1040.

Yelnik, J., and Percheron, G., 1979, Subthalamic neurons in primates: a quantitative and comparative analysis, *Neuroscience*, 4:1717.

Zhang, J.-H., et al., 1991, Region-specific expression of the mRNAs encoding subunits (β_1, β_2 and β_3) of GABA$_A$ receptor in the rat brain, *J. Comp. Neurol.* 303:637.

EFFECTS OF DOPAMINE RECEPTOR STIMULATION ON BASAL GANGLIA ACTIVITY

JUDITH R. WALTERS, DEBRA A. BERGSTROM, LANCE R. MOLNAR, LAUREN E. FREEMAN, AND DAVID N. RUSKIN[*]

INTRODUCTION

The importance of dopamine in the regulation of basal ganglia function and the control of movement has been appreciated for several decades. In the mid 1960's, histochemical studies established the presence of a dopaminergic nigrostriatal pathway, and neuropathological investigations demonstrated a correlation between dopamine cell loss and Parkinson's disease (PD) (Anden et al., 1964; Dahlström and Fuxe, 1964; Hornykiewicz, 1966; Poirier and Sourkes, 1966). In the intervening years, a large body of research has provided support for dopamine's critical role in a range of cognitive and motor functions. In particular, evidence that changes in dopamine receptor stimulation can affect the balance between hypo- and hyperkinetic states has led to considerable interest in how dopamine receptor stimulation ultimately modulates basal ganglia output to permit effective motor control.

One strategy for gaining insight into how dopamine-mediated transmission regulates activity in the basal ganglia nuclei has involved investigation of effects of systemically administered dopamine agonists on neuronal activity in these areas. The beneficial effect of dopamine replacement therapy in PD suggests that tonic, agonist-induced stimulation of dopamine receptors can provide much of the same regulation of postsynaptic systems as normal dopaminergic transmission. Thus, studies of the effects of systemically administered dopamine agonists on firing rate, pattern, and synchronization of activity in basal ganglia nuclei have the potential to shed light on the role of dopamine-mediated transmission in basal ganglia function. Early studies of the *in vivo* effects of dopamine agonists on basal ganglia activity nuclei postsynaptic to the nigra dopamine projection were conducted in monkeys with electrolytic lesions of the nigrostriatal pathway and in intact rats. Subsequently, similar studies were carried out in 1-methyl-4-phenyl-1,2,3,6-tetrahydropyridine (MPTP)-treated monkeys and rats with 6-hydroxydopamine (6OHDA)-induced unilateral lesions of the nigrostriatal dopamine system - animal models of PD.

[*] JUDITH R. WALTERS, DEBRA A. BERGSTROM, LANCE R. MOLNAR, LAUREN E. FREEMAN, AND DAVID N. RUSKIN • Experimental Therapeutics Branch, National Institute of Neurological Disorders and Stroke, National Institutes of Health, Building 10 Room 5C103, 10 Center Drive, MSC 1406, Bethesda, MD 20892-1406.

Most recently, similar recordings have been performed in PD patients in conjunction with electrode placement for pallidotomy or deep brain stimulation. The availability of data from PD patients, monkeys, and rodents makes it possible to compare results across species and gain insight into the physiology of PD and the relevance of the animal models of PD. This chapter will review these studies. Since the primate data available are from awake subjects, and since a number of studies indicate that systemic anesthesia reduces the effects of dopamine and sensory stimulation on basal ganglia activity (Bergstrom et al., 1984; Waszczak et al., 1984; Kreiss et al., 1996; Ruskin et al., 1999a, 2000; West, 1998), discussion will focus on data from subjects which are not systemically anesthetized. In addition, evidence from rodents and primates regarding the effects of dopamine receptor stimulation on aspects of neuronal activity other than firing rate, such as firing pattern and degree of synchronization, will be considered.

EFFECTS OF DOPAMINE RECEPTOR STIMULATION ON BASAL GANGLIA NEURONAL FIRING RATES

Because of the relatively dense dopaminergic innervation of the striatum and the high concentration of D1 and D2 dopamine receptor subtypes in this structure, the classical view is that dopamine exerts its main effect on basal ganglia output through modulation of striatal activity. Striatal efferents, through what is frequently referred to as the "direct" pathway, exert a direct effect on basal ganglia output to the thalamus and lower brain stem via projections to the internal globus pallidus (GPi), or entopenduncular nucleus (EPN) as it is called in the rodent, and the substantia nigra pars reticulata (SNpr). Striatal efferents also affect the GPi and SNpr through projections to the external globus pallidus (GPe), referred to as the "indirect" pathway. The GPe projects to the GPi and SNpr, and to the subthalamic nucleus (STN) which, in turn, also projects to the GPi and SNpr.

The main hypothesis which has been explored in single unit recording studies of dopamine agonist effects on basal ganglia activity, is the idea that dopamine receptor stimulation facilitates locomotor activity by reducing the level of inhibitory output from the basal ganglia to the thalamus through both the "indirect" and "direct" pathways (Albin et al., 1989; DeLong, 1990). Evidence supporting this "dual circuit" model emerged from investigations of the effects of dopamine cell lesion on GABA receptors at the sites of termination of the striatonigral and striatopallidal neurons (Pan et al., 1985), and from behavioral studies involving lesions of specific basal ganglia nuclei (Scheel-Krüger, 1986). Also critical were studies that mapped the distribution of the two main dopamine receptor subtypes in the striatum with respect to these striatal efferents (Gerfen et al., 1990; Le Moine et al., 1991). Additional biochemical and immunohistochemical data (Gerfen et al., 1990; Robertson et al., 1992) led to the view that dopamine acts primarily to inhibit activity in the so-called "indirect" pathway through stimulation of the D2 receptors expressed by striatopallidal neurons, and to enhance activity in the "direct" pathway through the D1 receptors expressed by striatal neurons projecting to the SNpr and the GPi /EPN. The general ideas derived from these data were: (1) D1 and D2 receptors act separately and oppositely on the "direct" and "indirect" pathways, respectively; (2) the STN acts primarily as a relay between the external globus pallidus (GPe) and the SNpr and GPi/EPN; (3) dopamine receptor stimulation increases activity in the GPe and reduces activity in the STN, GPi/EPN and SNpr; and (4) inhibition of the SNpr and GPi/EPN neuronal activity leads to behavioral activation by disinhibiting thalamic systems regulating cortical activity. This basic view of how dopamine affects basal ganglia output has served as a primary frame of reference for generating hypotheses and interpreting results, providing testable explanations for how dopamine - acting primarily in the striatum - might induce changes in the thalamus and motor cortex, which could critically modulate our ability to move. Thus, studies examining effects of systemic administration of dopamine agonists on neuronal activity in basal ganglia nuclei have typically explored the various predictions of this model of basal ganglia function with respect to these nuclei.

Effects of Dopamine Agonists on Striatal Firing Rates

Studies examining the effects of dopamine agonists or uptake blockers on firing rate in the striatum in awake rats (Skirboll et al., 1979; Warenycia and McKenzie, 1984; Kelland and Walters, 1992; Rosa-Kenig et al., 1993; West et al., 1997; Chen et al., 1999; Kish et al., 1999) support the view that dopamine receptor stimulation exerts a range of effects on the activity of striatal neurons. They also highlight the fact that anatomical and neurophysiological considerations make it difficult to accurately assess the distribution of responses within any of the anatomically or histochemically defined striatal subpopulations. The medium-sized striatonigral and striatopallidal neurons are not readily distinguished from one another in *in vivo* recording studies and typically fire quite slowly, making it likely there will be a sampling bias favoring the faster firing and larger neurons which are thought to be the cholinergic interneurons (Wilson et al., 1990; Aosaki et al., 1994; Kawaguchi et al., 1995; Raz et al., 1996). In fact, one study using firing rate and autocorrelation as a basis for distinguishing between interneurons (referred to as "tonically active neurons" or TANs) and medium spiny neurons reports finding that approximately 30% of the cells recorded can be characterized as interneurons, a number far greater than expected on the basis of anatomical studies (Kish et al., 1999). In another study in which apomorphine was administered systemically to awake behaving rats with chronic indwelling electrodes in the striatum, changes in rate were induced (> 15% baseline) in approximately 75% of the neurons. Both increases and decreases in rate were observed, and both were reversed in the majority of the cases by subsequent administration of haloperidol, a D2 family dopamine receptor antagonist (Molnar and Walters, unpublished observations). The D2 antagonist's ability to reverse the increases in rate induced by apomorphine in medium spiny-like neurons (neurons showing slow irregular firing patterns) in this study contradicts the classical view, described above, that increases in the firing rates of medium spiny neurons are mediated solely by the D1 receptor subtype. These results are consistent with earlier studies of amphetamine effects on striatal activity in behaving animals showing both D1 and D2 antagonist-mediated reversals of amphetamine-induced rate increases in movement-related striatal neurons (Rosa-Kenig et al., 1993), and with recent immunocytochemical studies demonstrating D1/D2 cooperativity in activation of striatonigral neurons (Ruskin and Marshall, 1994; Wang and McGinty, 1996).

In 6-OHDA-lesioned rats, chronic recording techniques have shown that the loss of dopamine is associated with increased tonic activity of striatal neurons and enhanced responses to apomorphine, although there is no current consensus with respect to the effect of this lesion on responses of striatal subpopulations to dopamine agonists (Chen et al., 1999; Kish et al., 1999).

These results highlight the complexity of the neurophysiological responses of striatal neurons to dopamine receptor stimulation, and are, in this sense, consistent with *in vitro* studies, which have demonstrated that D1 and D2 dopamine receptor subtypes, through second messenger-mediated processes, exert modulatory effects on a variety of processes in striatal neurons, including responses to stimulation of glutamate and GABA receptors, voltage-gated calcium channels, and gap junction formation (Hernandez-Lopez et al., 1997; Yan et al., 1997; Cepeda and Levine, 1998; Greengard et al., 1999; Flores-Hernandez et al., 2000; Snyder et al., 2000).

Effects of Dopamine Agonists on GPe Firing Rates

The basic "dual circuit" model of basal ganglia function, as discussed above, predicts that dopamine agonist administration should induce a tonic increase in activity of GPe neurons. In neurophysiological studies in awake, immobilized rats, responses recorded from the predominant GPe neuronal subtype are consistent with this prediction. These neurons [characterized as Type II neurons on the basis of their extracellularly recorded waveform shape according to Kelland et al. (1995)] consistently exhibit marked increases in firing rate in response to D1/D2 dopamine agonist administration in intact rats (Bergstrom and Walters, 1981; Bergstrom et al., 1982; Carlson et al., 1987, 1990; Walters et al., 1987; Kelland et al., 1995; Ruskin et al., 1998) (Figure 1). Similar rate increases have been observed in freely moving rats following dopamine agonist administration (Everett et al., 1984; Hooper et al., 1997). Another GPe neuronal class, Type I, responds to systemically administered apomorphine with a mean decrease in rate (Kelland et al., 1995).

In rats with unilateral 6-OHDA lesions, pallidal neurons show significant increases in burstiness and decreases in rate, changes which are evident one week after lesion (Pan and Walters, 1988). In rats studied 6 - 10 weeks after lesion, about ~60% of the type II neurons show significant increases in rate after apomorphine administration; 28% of the neurons show significant decreases in rate (Figure 1).

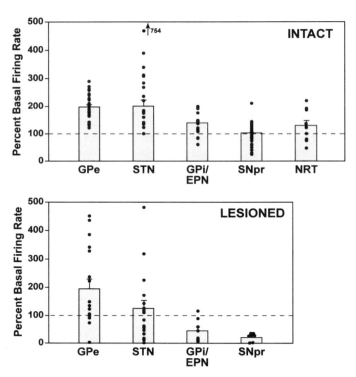

Figure 1. Effects of D1/D2 receptor stimulation on the firing rates of neurons in the basal ganglia and thalamus. Apomorphine (0.32 mg/kg) or the combination of a D1 and a D2 receptor agonist (SKF 38393 1.3 mg/kg and quinpirole 0.26 mg/kg) was injected intravenously to either neurologically intact rats (top panel) or rats with 6-OHDA-induced lesions of the nigrostriatal dopaminergic projection (bottom panel), while single unit activity was recorded extracellularly in locally anesthetized, immobilized rats. Dopamine agonist-induced firing rate changes in GPe, STN, GPi/EPN, and the SNpr of the intact animals were qualitatively different from those observed in lesioned animals. Apomorphine-induced firing rate changes are also depicted for NRT (nucleus reticularis thalami) in the intact preparation (top panel). Data are expressed as percent basal firing rate; dots indicate the responses of individual neurons and bar graphs indicate the mean ± SEM response of the group. Dashed lines indicate the 100% basal firing rate.

Firing patterns of GPe neurons recorded after dopamine cell lesion have also been reported to be altered in awake monkeys (Miller and DeLong, 1987; Filion and Tremblay, 1991; Boraud et al., 1998), with Filion and Tremblay (1991) reporting an increase in the mean proportion of intervals longer than 100 ms and Boraud et al. (1998) reporting only a slight increase in burstiness, as measured by the technique of Kaneoke and Vitek (1996). In addition, apomorphine administration induced an increase in firing rates of GPe neurons in monkeys with electrolytical or MPTP-induced dopamine cell lesions and in PD patients. Filion (1979) reported that 7 out of 10 rapidly firing GPe neurons showed increases in rate after apomorphine administration in a study in monkeys with electrolytic ventromedial midbrain lesion. Subsequently, in a larger study in MPTP-treated monkeys, 34 out of 39 GPe neurons displayed increases in rate after apomorphine administration (Filion et al., 1991). Two studies have examined the effects of levodopa on GPe neurons in these monkeys: Boraud and coworkers (1998, 2000) reported a reduction in the proportion of neurons firing with a bursting pattern but no significant effect on GPe firing rates with doses which did not induce dyskinesias, while Papa et al. (2000) reported increased GPe firing rates after administration of doses of levodopa which did induce dyskinesias. Increases in GPe firing rates have also been observed after apomorphine administration in recordings from patients undergoing unilateral pallidotomy for PD (Hutchison et al., 1997; Lenz et al., 1998; Lozano et al., 2000). A second GPe neuronal type characterized by slow firing rates and burstiness is inhibited by apomorphine and has elevated firing rates after MPTP-treatment (Filion and Tremblay, 1991; Filion et al., 1991). This cell type may correspond to the GPe type I neurons recorded in rodents.

These studies show substantial concordance between results observed in the primate and rodent studies with respect to effects of dopamine receptor stimulation and GPe firing rate and pattern. In addition, to the extent that a substantial number of GPe neurons in both rodent and primate show rate increases after systemic administration of drugs like apomorphine, these studies support the prediction of the "dual circuit" model that dopamine receptor stimulation will increase tonic firing rate in the GPe. However, the model does not account for the population of GPe neurons whose firing rates are inhibited by dopamine. Also, additional studies in rodents have raised questions about the relative roles of the dopamine receptor subtypes mediating these effects. The hypothesis that the effects of dopamine on the "indirect" pathway is controlled independently by D2 receptors acting to reduce the activity of inhibitory striatopallidal neurons is inconsistent with observations that drugs, such as quinpirole and RU 24926, which selectively stimulate D2 receptors, induce only modest rate effects on GPe neuronal activity in awake, locally anesthetized, gallamine-immobilized rats (Carlson et al., 1987; Walters et al., 1987). Data from the rodent studies indicate that combined stimulation of D1 and D2 receptor subtypes induces a more robust increase in activity in the GPe than does stimulation of the individual receptors. Also relevant to these considerations are emerging data identifying several anatomically and immunohistochemically defined neuronal subpopulations in the GPe (Ruskin and Marshall, 1997; Kita, 1994; Kita et al., 1999; Bolam et al., 2000). The recent appreciation that an estimated 1/3 of GPe neurons project back to the striatum has drawn attention to the complex circuitry of this area and the reality that information flow in the basal ganglia is not solely striatofugal (Kita et al., 1999; Bolam et al., 2000; Sato et al., 2000).

Effects of Dopamine Agonists on STN Firing Rates

The STN has been the focus of considerable attention in current research on PD, since, as reviewed elsewhere in this volume, deep brain stimulation in this nucleus has emerged as an effective treatment in advanced PD. Basic models of basal ganglia function, as described above, predict that increases in dopamine receptor stimulation will reduce activity in the STN, as a result of increased activity in the inhibitory GPe projection to this region. Consistent with these predictions, it has been demonstrated in both primate (Miller and

DeLong, 1987; Bergman et al., 1994) and rodent models of PD (Hassani et al., 1996; Kreiss et al., 1997) that the loss of dopamine is associated with an increase in tonic firing rates in the STN. This later finding has been quite reassuring, in fact, with respect to the relevance of the rodent model of PD to the primate model.

In view of these considerations, it has been somewhat surprising that in awake immobilized intact rats systemic administration of apomorphine and selective D1 agonists induces marked increases in STN activity (Kreiss et al., 1996; 1997; Olds et al., 1999; Allers et al., 2000b) (Figure 1). D2 agonists either have no effect or induce modest increases in STN rate. These results are inconsistent with predictions that D2 receptors, expressed by striatopallidal neurons and therefore positioned to influence activity in the striatopallidal and pallido-subthalamic projections, are more likely than D1 receptors to affect STN activity. They call attention to projections through which D1 receptors may affect STN activity, such as the cortical input to the STN, and the efferents from the striatonigral neurons projecting to the GPe. Evidence for direct dopaminergic projections to the STN has also been described, although they appear relatively sparse (Meibach and Katzman, 1979; Campbell et al., 1985; Allers et al., 1997; Hassani et al., 1997; Augood et al., 1999; Cossette et al., 1999).

In rats with unilateral 6-OHDA-induced lesions of the nigrostriatal dopaminergic pathway, somewhat different effects are induced by dopamine agonist administration. While systemic administration of D1 agonists still brings about substantial increases in STN firing rates in these lesioned animals, levodopa and apomorphine have a significantly greater tendency to inhibit firing activity in the STN after 6-OHDA lesion (Kreiss et al., 1997; Allers et al., 2000b) (Figure 1), observations more consistent with the initial predictions regarding effects of dopamine receptor stimulation on STN activity. Effects of nigrostriatal lesion on firing pattern in the STN are somewhat controversial: in anesthetized preparations, a number of authors report low burstiness in the intact preparation and increased burstiness after 6-OHDA lesions (Hollerman and Grace, 1992; Hassani et al., 1996; Vila et al., 2000). In contrast, studies in awake rodents and primates report substantial burstiness in STN firing rates in the intact animals; following lesion, the amount of burstiness has been reported to be slightly increased (Bergman et al., 1994) or decreased (Kreiss et al., 1997). Since anesthesia appears to be a potential confound regarding burstiness, more work in awake preparations is needed.

As with studies of effects of dopamine agonists on activity in the GPe, results quite similar to those observed in the rodent model of PD have been observed in studies in MPTP-treated monkeys and in PD patients. Papa et al. (2000) have reported that levodopa, at doses which result in an "on" state, induces rate decreases in 28% of the STN neurons recorded, and no change in 72%. Doses inducing dyskinesia in these monkeys decreased rates in 57% of the neurons and induced an averaged 53% decrease in STN firing rates, compared to pre-drug baseline. Apomorphine administration reduced the activity of 2 out of 8 neurons recorded in the STN of PD patients (Lozano et al., 2000).

Effects of Dopamine Agonists on GPi/EPN and SNpr Firing Rates

The idea that agonist-induced increases in dopamine receptor stimulation are associated with decreases in the activity of the inhibitory GPi/EPN and SNpr basal ganglia output nuclei has served for many years as a working hypothesis for explaining how drugs, acting on dopamine receptors in the striatum, might - through disinhibition of thalamocortical neurons - produce a state of behavioral hyperactivity. In intact rats, however, single unit recording studies have not supported such a scenario. Firing rate changes in the basal ganglia output nuclei are variable after systemic administration of dopamine agonists such as apomorphine, with either little net change in the SNpr (Waszczak et al., 1984; Ruskin et al., 1999a) or a net increase in the GPi/EPN (Ruskin et al., 1999a,c) (Figure 1). These results are consistent with observations, described above, showing that systemically administered D1/D2 agonists induce increases in both excitatory and inhibitory inputs to the basal ganglia output nuclei in intact rats: D1/D2 receptor stimulating agonists induce net increases in STN activity - which would result in increased excitatory input to the output nuclei - in addition to increases in the activity of many GPe and striatal efferents, resulting in increased inhibitory input to the output nuclei. Firing rates are also not consistently affected in the nucleus reticularis thalami (NRT) by apomorphine administration (Figure 1).

A different picture emerges, however, with administration of dopamine agonists to rats with unilateral nigrostriatal lesions. In these animals, firing rates in the GPi/EPN and SNpr ipsilateral to the lesion are consistently depressed by systemically administered D1/D2 agonists. Simultaneous activation of D1 and D2 receptors produces robust decreases in SNpr neuronal activity; this effect involves a synergistic D1/D2 interaction (Weick and Walters, 1987). These single unit recording data illustrate that the loss of striatal dopamine innervation induces both quantitative and qualitative changes in the response of basal ganglia neurons to the acute application of dopamine receptor stimulating agents. The net effects of dopamine agonists on firing rates in the SNpr and GPi/EPN in lesioned rats can be explained by denervation-enhanced dopamine receptor mediated excitation of inhibitory sriatonigral/striatoentopeduncular neurons (Robertson et al., 1992; Weick and Walters, 1988; Huang and Walters, 1994, 1996), along with the altered effects of dopamine agonists on excitatory STN output. They are also consistent with predictions derived from basic models of the basal ganglia, and with effects of apomorphine and levodopa administration in primates. Several studies have now documented that levodopa and apomorphine administration exerted similar inhibitory effects on single unit activity in the GPi of monkeys with electrolytic or MPTP-induced dopamine cell lesions (Filion, 1979; Filion et al., 1991; Boraud et al., 1998, 2000; Papa et al., 1999, 2000) and in PD patients (Hutchinson et al., 1997; Merello et al., 1999; Stefani et al., 1999; Lozano et al., 2000). This effect was considerably more notable in MPTP-lesioned monkeys following doses of apomorphine which induced dyskinesia (Papa et al., 1999), although Verhagen Metman et al. (2000) observed that the occurrence of dyskinesia is not necessarily associated with reduction in GPi firing rates beyond those observed with emergence of an "on" state. Other studies have emphasized the role of changes in firing pattern and "center surround" phenomena with respect to the GPi's role in generating dyskinesias in PD patients (Matsumura et al., 1995; Obeso et al., 2000, see also chapter by Filion in this volume).

Single unit recording studies in intact and 6-OHDA lesioned rats have not lent clear support to the view that drugs that increase dopamine receptor stimulation produce increased behavioral activity in rodents by reducing the net activity of the basal ganglia output nuclei. Even in rats with 6-OHDA lesions where the predicted decreases in net activity in the GPi/EPN and SNpr nuclei can be induced by D1 or D1/D2 agonists, low doses of D1 or D2 agonists can produce motor effects (i.e., rotation) without net decreases in SNpr and GPi/EPN firing rates (Ruskin et al., 1999b). Similarly, the rate of onset of decreases in activity in SNpr neurons following subcutaneous administration of apomorphine does not correlate with the rate of onset of rotation in rats with unilateral 6-OHDA lesions (Ruskin et al., 1999b), as shown in Figure 2. Thus, while these studies certainly support the idea that dopamine receptor stimulation can modulate overall rates of activity in basal ganglia nuclei, the hypothesis that an overall decrease in the activity of the GABAergic basal ganglia output neurons mediates the increase in behavioral activity induced by these drugs is not consistent with the *in vivo* data. In a similarly unexpected manner, reduction of basal ganglia output via pallidotomy or deep brain stimulation not only reduces bradykinesia in PD patients, it also reduces the dyskinesias induced by levodopa (for review, see Obeso et al., 2000).

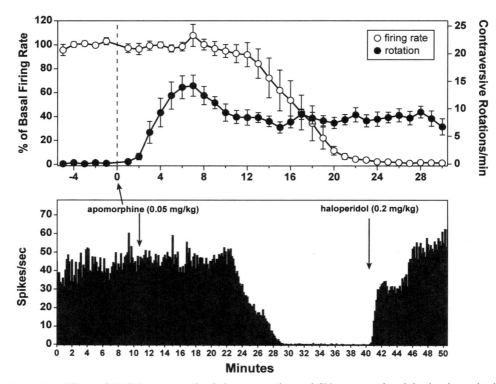

Figure 2. Effects of D1/D2 receptor stimulation on rotation and SNpr neuronal activity in nigrostriatal lesioned rats. Apomorphine (0.05 mg/kg, sc) induced significant contraversive rotation from 3 minutes post-injection onwards while SNpr firing rate was not significantly changed until 16 minutes post-injection. Data averaged from the rotation experiments and unit recordings of SNpr neurons are plotted in the top panel. Apomorphine was administered at the time indicated by the arrows. Below, a firing rate histogram illustrates the response to apomorphine of a single SNpr neuron recorded ipsilateral to the nigrostriatal lesion; a gradual yet robust rate decrease began 12 minutes post-injection. Inhibition of firing activity was antagonized by haloperidol (0.2 mg/kg, iv). Rate data are shown in 1 second bins.

With respect to baseline firing characteristics following nigrostriatal lesions, significant firing pattern changes have been noted in the GPi and SNpr of MPTP-treated monkeys with either no significant changes in GPi and SNpr firing rates (Bergman et al., 1994; Wichmann et al., 1999b) or with GPi baseline rate increases (Miller and DeLong, 1987; Filion et al., 1991). Baseline firing rates in the SNpr and EPN of rats with unilateral 6-OHDA lesions are also not significantly altered 6 - 10 weeks after lesion (Walters et al., 1997; Ruskin and Walters, unpublished observations), but recent examination of firing pattern in the EPN suggests further parallels with primate data (Ruskin, Bergstrom and Walters, submitted).

EFFECTS OF DOPAMINE RECEPTOR STIMULATION ON BASAL GANGLIA NEURONAL FIRING PATTERNS

The studies described above highlight the need to consider effects of tonic dopamine receptor stimulation on aspects of neuronal activity in addition to rate, such as firing pattern and synchronization of spiking activity within and between functionally related circuits, in order to achieve a more thorough understanding of how dopamine receptor stimulation relates to the proper functioning of the basal ganglia. Several studies have suggested a role for dopamine as a modulator of the network properties of the basal ganglia, affecting the extent to which activity in different components may be synchronized on different time scales (Bergman et al., 1998; Allers et al., 1999, 2000b; Onn and Grace, 1999; Ruskin et al., 1999a,c,d, 2000). Grace and coworkers (Onn and Grace, 1999) have shown that modulation of dopaminergic transmission with agonists or nigrostriatal lesion affects the activity of gap junctions between striatal neurons. Bergman and coworkers (Raz et al., 1996; Bergman et al., 1998) have demonstrated changes in the degree of synchronous activity in the globus pallidus and striatum after nigrostriatal lesion, and other studies have examined burst firing properties of STN neurons after 6-OHDA lesion (Hollerman and Grace, 1992; Bergman et al., 1994; Hassani et al., 1996; Kreiss et al., 1997; Allers et al., 2000b; Vila et al., 2000).

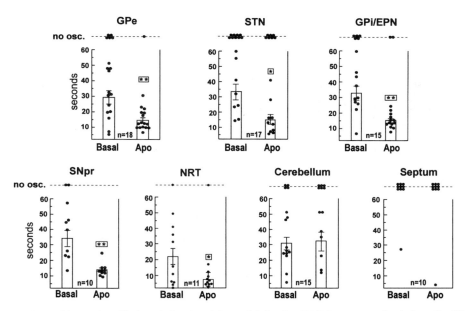

Figure 3. Multisecond oscillations in basal activity and following D1/D2 receptor stimulation. Significant periodic oscillations in basal activity were observed in 50 - 90% of spike trains from basal ganglia and thalamic nuclei. Oscillatory periods in baseline averaged ~ 30 seconds. Apomorphine (0.32 mg/kg, iv) induced significant decreases in the period of spike trains from the GPe, STN, GPi/EPN, SNpr and NRT. Increased incidence of oscillations was also seen after apomorphine. Many spike trains from cerebellar neurons displayed basal oscillations, but these were not consistently altered by apomorphine. Septal spike trains displayed little multisecond oscillatory activity. Multisecond oscillations were not observed in baseline or following apomorphine administration in spike trains from animals anesthetized with urethane, ketamine/xylazine, or chloral hydrate (data not shown, see Ruskin et al., 1999a, 2000 and Allers et al., 2000b). Spike trains with no significant oscillations are indicated above each graph (no osc.). Data are expressed as the oscillatory period in seconds; dots indicate the periods of individual neurons, and the bar graphs indicate the mean ± SEM. *p<0.05, ** p<0.01, *** p<0.001 vs. basal.

Additional evidence suggesting dopamine may affect the synchronization of neuronal activity in the basal ganglia comes from recent single unit recording studies in several nuclei that have shown effects of systemically administered dopamine agonists on periodic variations in firing rate at time scales of seconds. Recordings from awake, locally anesthetized, immobilized rats have demonstrated that many SNpr, GPe, EPN, and STN neurons exhibit slow periodic changes in rate within the range of 2 to 60 s (Figure 3). During baseline recording intervals, oscillations are significantly periodic in 50 - 75% of the neurons recorded (means: 25 - 35 s) (Ruskin et al., 1999a,c,d, 2000; Allers et al., 2000b) (Figure 3,4). Although the regularity and amplitude of these periodic fluctuations in firing rate vary greatly from unit to unit, the fluctuations can be quite large. These multisecond oscillations are not present in recordings from rats under general anesthesia with chloral hydrate, ketamine/xylazine, or urethane (Ruskin et al., 1999a, 2000; Allers et al., 2000b). Interestingly, multisecond oscillations are also observed in recordings from the NRT and cerebellum but were rarely observed in the septum (Figure 3) (Allers et al., 2000a; Walters et al., 2000b,c).

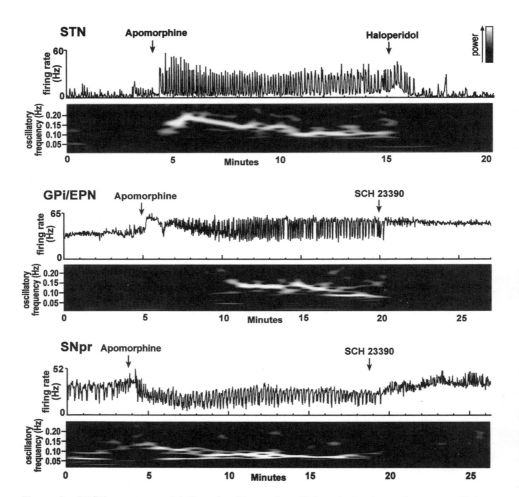

Figure 4. D1/D2 receptor modulation of multisecond oscillations in basal ganglia neurons. Firing rate histograms from SNpr, GPi/EPN, and STN neurons illustrate firing rate and pattern changes induced by apomorphine administration (0.32 mg/kg, iv). Wavelet-based scalograms (time-vs.-frequency representation of oscillatory power, with greater power depicted by the redder hues) illustrate the increase in oscillatory power and frequency (varying between 0.05 and 0.20 Hz in these examples) following apomorphine. Apomorphine's effects were effectively reversed by either the D1 antagonist SCH 23390 (0.5 mg/kg, iv) or the D2 antagonist haloperidol (0.2 mg/kg, iv). Of these 3 units, only the SNpr neuron had notable multisecond oscillations prior to apomorphine administration. Similar results were observed in GPe and NRT neurons (data not shown).

These multisecond oscillations in basal ganglia neuronal firing rates are strikingly modulated by increased dopamine receptor stimulation (Figure 3,4). Systemic dopamine agonist administration induces a dramatic increase in the frequency and, in most cases, the regularity of these firing rate oscillations, and can induce oscillatory activity in neurons that do not have significant baseline oscillations (Ruskin et al., 1999a,c,d, 2000; Allers et al., 2000b). As shown in Figure 4, strong, regular multisecond oscillations appear after apomorphine injection, superimposed on drug-induced changes in mean firing rate. Apomorphine's effects on the frequency and regularity of the oscillations are reversed by either D1 or D2 antagonists (Ruskin et al., 1999a,d; Allers et al., 2000b). Further investigation with selective D1 and D2 agonists (e.g., SKF 81297 and quinpirole) and antagonists confirmed that both dopamine receptor subtypes contribute to the effects of dopamine agonists on these oscillations (Ruskin et al., 1999c,d) and suggests a synergistic interaction between these receptor subtypes. The reduced range of oscillatory periods found after dopamine agonist treatment suggests the possibility that basal ganglia neurons become more synchronized at these time scales. In support of this view, highly correlated multisecond oscillations were observed in about 60% of the pairs of simultaneously recorded basal ganglia neurons, some in opposite hemispheres, following dopamine agonist administration (Allers et al., 1999).

The multisecond oscillations observed in baseline firing rates of basal ganglia and thalamic nuclei are markedly slower than most oscillatory activity typically studied in neuronal activity from basal ganglia, thalamus, or cortex. However, there are indications in the literature that oscillations in central activity with similar time periods can be seen in other species and under other conditions. Early reports by Aladjalova (1964), Norton and Jewett (1965), and Kropotov and Gretchin (1975) described multisecond oscillations in pO_2 and DC-potentials in deep brain electrode recordings and EEG recordings. Notably, some of these DC-potential recordings were performed in PD patients in whom electrodes were implanted for diagnosis and therapy (Kropotov and Gretchin, 1975). Similar low-frequency oscillations have been detected in humans undergoing functional magnetic resonance imaging (Biswal et al., 1997). Slow multisecond oscillations in spike trains recorded from SNpr neurons in the awake monkey have been reported by Wichmann and colleagues (1999a). The suggestion that these oscillations play a role in brain mechanisms relating to attention is provided by the finding that EEG recordings in monkeys show oscillations in alpha rhythm with similar periodicities correlating with periods of increased vigilance and activity (Ehlers and Foote, 1984). Preliminary observations from this laboratory demonstrate that multisecond oscillations in GP, STN, and NRT firing rates are clearly related to similarly slow variations in the power of EEG theta (4-7 Hz) rhythm (Allers et al., 2000a; Walters et al., 2000a). In sum, the existing data demonstrate that variations in activity at time scales of many seconds occur in several brain regions of waking humans and animals, and that these oscillations in cerebral cortex EEG and basal ganglia/thalamic spiking activity appear to be functionally related.

CONCLUSIONS

The investigations reviewed here demonstrate impressive correlations between neurophysiological data from the rodent model of PD and data from MPTP-treated monkeys and PD patients and, hence, support the relevance of the findings emerging from rodent studies with respect to basal ganglia function in humans and non-human primates. Overall, the studies in primates and rodents strongly support the idea that tonic changes in dopamine receptor stimulation can modulate neuronal firing rates in basal ganglia nuclei. However, the hypothesis that a net decrease in the firing rates of the GABAergic basal ganglia output neurons mediates the increase in behavioral activity induced by these drugs is not consistent with the *in vivo* data from rodents or with results from pallidotomy and deep brain stimulation in PD patients. These observations have drawn attention to the effects of dopamine receptor stimulation on aspects of neuronal activity such as burstiness, synchronization, and oscillatory activity, suggesting a more complex integrated basal ganglia network than envisioned in early models. Multisecond oscillations may act to organize information in the time domain in distributed circuits, facilitating the propagation and synchronization of faster oscillatory activity or, alternatively, limiting information throughput in a particular circuit. The observation that these oscillations are markedly affected by dopamine agonists suggests that coordinated changes in neuronal activity at time scales longer than commonly investigated play a role in the cognitive and motor processes affected by dopamine.

REFERENCES

Aladjalova, N.A., 1964, Slow electrical processes in the brain, *Prog. Brain Res.* 7:1.

Albin, R.L., Young, A.B., and Penney, J.B., 1989, The functional anatomy of basal ganglia disorders, *Trends Neurosci.* 12:366.

Allers, K.A., Juncos, J.L., and Rye, D.B., 1997, Anatomical investigation of the dopaminergic innervation of the rat subthalamic nucleus, *Soc. Neurosci. Abstr.* 23:197.

Allers, K.A., Ruskin, D.N., Bergstrom, D.A., Molnar, L.R., and Walters, J.R., 1999, Correlations of multisecond oscillations in firing rate in pairs of basal ganglia neurons, *Soc. Neurosci. Abstr.* 25:1929.

Allers, K.A., Ghazi, L.J., Freeman, L.E., Ruskin, D.N., Bergstrom, D.A., Tierney, P.L., and Walters, J.R., 2000a, Multisecond oscillations in firing rates of rat subthalamic and thalamic reticular nucleus neurons are correlated with bursts of 4-7 Hz cortical activity, *Soc. Neurosci. Abstr.* 26:690.

Allers, K.A., Kreiss, D.S., and Walters, J.R., 2000b, Multisecond oscillations in the subthalamic nucleus: effects of apomorphine and dopamine cell lesion, *Synapse,* 38:38.

Anden, N.-E, Carlsson, A., Dahlström, A., Fuxe, K., Hillarp, N.A., and Larsson, K., 1964, Demonstration and mapping out of the nigro-neostriatal dopamine neurons, *Life Sci.* 3:523.

Aosaki, T., Tsubokawa, H., Ishida, A., Watanabe, K., Graybiel, A.M., and Kimura, M., 1994, Responses of tonically active neurons in the primate's striatum undergo systematic changes during behavioral sensorimotor conditioning, *J. Neurosci.* 14:3969.

Augood, S.J., Hollingsworth, Z.R., Standaert, D.G., Emson, P.C., and Penney, J.B., 2000, Localization of dopaminergic markers in the human subthalamic nucleus, *J. Comp. Neurol.* 421:247.

Bergman, H., Wichmann, T., Karmon, B., and DeLong, M.R., 1994, The primate subthalamic nucleus. II. Neuronal activity in the MPTP model of parkinsonism, *J. Neurophysiol.* 72:507.

Bergman, H., Feingold, A., Nini, A., Raz, A., Slovin, H., Abeles, M., and Vaadia, E., 1998, Physiological aspects of information processing in the basal ganglia of normal and parkinsonian primates, *Trends Neurosci.* 21:32.

Bergstrom, D.A., and Walters, J.R., 1981, Neuronal responses of the globus pallidus to systemic administration of d-amphetamine: investigation of the involvement of dopamine, norepinephrine, and serotonin, *J. Neurosci.* 1:292.

Bergstrom, D.A., Bromley, S.D., and Walters, J.R., 1982, Apomorphine increases the activity of rat globus pallidus neurons, *Brain Res.* 238:266.

Bergstrom, D.A., Bromley, S.D., and Walters, J.R., 1984, Dopamine agonists increase pallidal unit activity: attenuation by agonist pretreatment and anesthesia, *Eur. J. Pharmacol.* 100:3.

Biswal, B., Hudetz, A.G., Yetkin, F.Z., Haughton, V.M., and Hyde, J.S., 1997, Hypercapnia reversibly suppresses low-frequency fluctuations in the human motor cortex during rest using echo-planar MRI, *J. Cereb. Blood Flow Metab.* 17:301.

Bolam, J.P., Hanley, J.J., Booth, P.A.C., and Bevan, M.D., 2000, Synaptic organization of the basal ganglia, *J. Anat.* 196:527.

Boraud, T., Bezard, E., Guehl, D., Bioulac, B., and Gross, C.E., 1998, Effects of L-DOPA on neuronal activity of the globus pallidus externalis (GPe) and globus pallidus internalis (GPi) in the MPTP-treated monkey, *Brain Res.* 787:157.

Boraud, T, Bezard, E., Stutzmann, J.M., Bioulac, B., and Gross, C.E., 2000, Effects of riluzole on the electrophysiological activity of pallidal neurons in the 1-methyl-4-phenyl-1,2,3,6-tetrahdropyridine-treated monkey, *Neurosci. Lett.* 281 (2-3):75.

Campbell, G.A., Eckardt, M.J., and Weight, F.F., 1985, Dopaminergic mechanisms in subthalamic nucleus of rat: analysis using horseradish peroxidase and microiontophoresis, *Brain Res.* 333:261.

Carlson, J.H., Bergstrom, D.A., and Walters, J.R., 1987, Stimulation of both D1 and D2 dopamine receptors appears necessary for full expression of postsynaptic effects of dopamine agonists: a neurophysiological study, *Brain Res.* 400:205.

Carlson, J.H., Bergstrom, D.A., Demo, S.D., and Walters, J.R., 1990, Nigrostriatal lesion alters neurophysiological responses to selective and nonselective D-1 and D-2 dopamine agonists in rat globus pallidus, *Synapse,* 5:83.

Cepeda, C., and Levine, M.S., 1998, Dopamine and N-methyl-D-aspartate receptor interactions in the neostriatum, *Dev. Neurosci.* 20:1.

Chen, M.T., Hoffer, B.J., Morales, M., Borlongan, C.V., Hoffman, A.F., and Janak, P.H., 1999, Extracellular single-unit recording from the striatum of freely-moving unilateral 6-OHDA lesioned rats, *Soc. Neurosci. Abstr.* 25:331.

Cossette, M., Levesque, M., and Parent, A., 1999, Extrastriatal dopaminergic innervation of human basal ganglia, *Neurosci. Res.* 34:51.

Dahlström, A., and Fuxe, K., 1964, Evidence for the existence of monoamine neurons in the central nervous system. I Demonstration of monoamines in the cell bodies of brain stem neurons, *Acta Physiol. Scand.* 62:1.

DeLong, M.R., 1990, Primate models of movement disorders of basal ganglia origin, *Trends Neurosci.* 13:281.

Ehlers, C. L., and Foote, S.L., 1984, Ultradian periodicities in EEG and behavior in the squirrel monkey (*Saimiri sciureus*), *Am. J. Primatol.* 7:381.

Everett, P.W., Kemm, R.E., and McKenzie J.S., 1984, Neural activity in basal ganglia output nuclei and induced hypermotility, in: *The Basal Ganglia Structure and Function*, G.M. McKenzie, R.E. Kemm, and L.N. Wilcock, eds., Plenum Press, New York.

Filion, M., 1979, Effects of interruption of the nigrostriatal pathway and of dopaminergic agents on the spontaneous activity of globus pallidus neurons in the awake monkey, *Brain Res.* 179:425.

Filion, M., and Tremblay, L., 1991, Abnormal spontaneous activity of globus pallidus neurons in monkeys with MPTP-induced parkinsonism, *Brain Res.* 547:142.

Filion, M., Tremblay, L., and Bedard, P.J., 1991, Effects of dopamine agonists on the spontaneous activity of globus pallidus neurons in monkeys with MPTP-induced parkinsonism, *Brain Res.* 547:152.

Flores-Hernandez, J., Hernandez, S., Snyder, G.L., Yan, Z., Fienberg, A.A., Moss, S.J., Greengard, P., and Surmeier, D.J., 2000, D-1 dopamine receptor activation reduces GABA(A) receptor currents in neostriatal neurons through a PKA/DARPP-32/PP1 signaling cascade, *J. Neurophysiol.* 83:2996.

Gerfen, C.R., Engber, T.M., Mahan, L.C., Susel, Z., Chase, T.N., Monsma, F.J., Jr., and Sibley, D.R., 1990, D_1 and D_2 dopamine receptor-regulated gene expression of striatonigral and striatopallidal neurons, *Science,* 250:1429.

Greengard, P., Allen, P.B., and Nairn, A.C., 1999, Beyond the dopamine receptor: the DARPP-32/Protein phosphatase-1 cascade, *Neuron,* 23:435.

Hassani, O.K., Mouroux, M., and Féger, J., 1996, Increased subthalamic neuronal activity after nigral dopaminergic lesion independent of disinhibition via the globus pallidus, *Neuroscience,* 72:105.

Hassani, O.K., François, C., Yelnik, J., and Féger, J., 1997, Evidence for a dopaminergic innervation of the subthalamic nucleus in the rat, *Brain Res.* 749:88.

Hernandez-Lopez, S., Bargas, J., Surmeier, D.J., Reyes, A., and Galarraga, E., 1997, D_1 receptor activation enhances evoked discharge in neostriatal medium spiny neurons by modulating an L-type Ca^{2+} conductance, *J. Neurosci.* 17:3334.

Hollerman, J.R., and Grace, A.A., 1992, Subthalamic nucleus cell firing in the 6-OHDA-treated rat: basal activity and response to haloperidol, *Brain Res.* 590:291.

Hooper, K.C., Banks, D.A., Stordahl, L.J., White, I.M., and Rebec, G.V., 1997, Quinpirole inhibits striatal and excites pallidal neurons in freely moving rats, *Neurosci. Lett.* 237:69.

Hornykiewicz, O., 1966, Dopamine (3-hydroxytyramine) and brain function, *Pharmacol. Rev.* 18:925.

Huang, K.-X., and Walters, J.R., 1994, Electrophysiological effects of SKF 38393 in rats with reserpine treatment and 6-hydroxydopamine-induced nigrostriatal lesions reveal two types of plasticity in D1 dopamine receptor modulation of basal ganglia output, *J. Pharmacol. Exp. Ther.* 271:1434.

Huang, K.-X., and Walters, J.R., 1996, Dopaminergic regulation of AP-1 transcription factor DNA binding activity in rat striatum, *Neuroscience,* 75:757.

Hutchison, W.D., Levy, R., Dostrovsky, J.O., Lozano, A.M., and Lang, A.E., 1997, Effects of apomorphine on globus pallidus neurons in parkinsonian patients, *Ann. Neurol.* 42:767.

Kaneoke, Y., and Vitek, J.L., 1996, Burst and oscillation as disparate neuronal properties, *J. Neurosci. Meth.* 68:211.

Kawaguchi, Y., Wilson, C.J., and Augood, S.J., 1995, Striatal interneurones: chemical, physiological and morphological characterization, *Trends Neurosci.* 18:527.

Kelland, M.D., and Walters, J.R., 1992, Apomorphine-induced changes in striatal and pallidal neuronal activity are modified by NMDA and muscarinic receptor blockade, *Life Sci.* 50:L179.

Kelland, M.D., Soltis, R.P., Anderson, L.A., Bergstrom, D.A., and Walters, J.R., 1995, In vivo characterization of two cell types in the rat globus pallidus which have opposite responses to dopamine receptor stimulation: comparison of electrophysiological properties and responses to apomorphine, dizocilpine, and ketamine anesthesia, *Synapse,* 20:338.

Kish, L.J., Palmer, M.R., and Gerhardt, G.A., 1999, Multiple single-unit recordings in the striatum of freely moving animals: effects of apomorphine and D-amphetamine in normal and unilateral 6-hydroxydopamine-lesioned rats, *Brain Res.* 833:58.

Kita, H., 1994, Parvalbumin-immunopositive neurons in rat globus pallidus: a light and electron microscopic study, *Brain Res.* 657:31.

Kita, H., Tokuno, H., and Nambu, A., 1999, Monkey globus pallidus external segment neurons projecting to the neostriatum, *Neuroreport,* 10:1467.

Kreiss, D.S., Anderson, L.A., and Walters, J.R., 1996, Apomorphine and dopamine D_1 receptor agonists increase the firing rates of subthalamic nucleus neurons, *Neuroscience,* 72:863.

Kreiss, D.S., Mastropietro, C.W., Rawji, S.S., and Walters, J.R., 1997, The response of subthalamic nucleus neurons to dopamine receptor stimulation in a rodent model of Parkinson's disease, *J. Neurosci.* 17:6807.

Kropotov, J.D., and Gretchin, V.B., 1975, Correlations between slow oscillations of electrical potential and pO2 in the human brain, *Sov. J. Physiol.* 61:331.

Le Moine, C., Normand, E., and Bloch, B., 1991, Phenotypical characterization of the rat striatal neurons expressing the D_1 dopamine receptor gene, *Proc. Natl. Acad. Sci. USA* 88:4205.

Lenz, F.A., Suarez, J.I., Verhagen Metman, L., Reich, S.G., Karp, B.I., Hallett, M., Rowland, L.H., and Dougherty, P.M., 1998, Pallidal activity during dystonia: somatosensory reorganisation and changes with severity, *J. Neurol. Neurosurg. Psychiat.* 65:767.

Lozano, A.M., Lang, A.E., Levy, R., Hutchison, W., and Dostrovsky, J., 2000, Neuronal recordings in Parkinson's disease patients with dyskinesias induced by apomorphine, *Ann. Neurol.* 47:S141.

Matsumura, M., Tremblay, L., Richard, H., and Filion, M., 1995, Activity of pallidal neurons in the monkey during dyskinesia induced by injection of bicuculline in the external pallidum, *Neuroscience,* 65:59.

Meibach, M.R.C., and Katzman, R., 1979, Catecholaminergic innervation of the subthalamic nucleus: evidence for a rostral continuation of the A9 (substantia nigra) dopaminergic group, *Brain Res.* 173:364.

Merello, M., Balej, J., and Delfino, M., 1999, Apomorphine induces changes in GPi spontaneous outflow in patients with Parkinson's disease, *Mov. Disorders,* 14:45.

Miller, W.C., and DeLong, M.R., 1987, Altered tonic activity of neurons in the globus pallidus and subthalamic nucleus in the primate MPTP model of parkinsonism, in: *The Basal Ganglia*, M.B. Carpenter and A. Jayarman, eds., Plenum Press, New York.

Norton, S., and Jewett, R.E., 1965, Frequencies of slow potential oscillations in the cortex of cats, *Electroencephalogr. Clin. Neurophysiol.* 19:377.

Obeso, J.A., Rodriguez-Oroz, M.D., Rodriguez, M., DeLong, M.R., and Olanow, C.W., 2000, Pathophysiology of levodopa-induced dyskinesias in Parkinson's disease: problems with the current model, *Ann. Neurol. (Suppl. 1)* 47:S22.

Olds, M.E., Jacques, D.B., and Kopyov, O., 1999, Subthalamic responses to amphetamine and apomorphine in the behaving rat with a unilateral 6-OHDA lesion in the substantia nigra, *Synapse,* 34:228.

Onn, S.-P., and Grace, A.A., 1999, Alterations in electrophysiological activity and dye coupling of striatal spiny and aspiny neurons in dopamine-denervated rat striatum recorded in vivo, *Synapse,* 33:1.

Pan, H.S., Penney, J.B., and Young, A.B., 1985, γ-Aminobutyric acid and benzodiazepine receptor changes induced by unilateral 6-hydroxydopamine lesions of the medial forebrain bundle, *J. Neurochem.* 45:1396.

Pan, H.S., and Walters, J.R., 1988, Unilateral lesion of the nigrostriatal pathway decreases the firing rate and alters the firing pattern of globus pallidus neurons in the rat, *Synapse,* 2:650.

Papa, S.M., Desimone, R., Fiorani, M., and Oldfield, E.H., 1999, Internal globus pallidus discharge is nearly suppressed during levodopa-induced dyskinesias, *Ann. Neurol.* 46:732.

Papa, S.M., Mewes, K., DeLong, M.R., and Baron, M.S., 2000, Neuronal activity correlates of levodopa-induced dyskinesias in the basal ganglia, *Neurology,* 54 (Suppl. 3):A456.

Poirier, L.J., and Sourkes, T.L., 1966, Influence of the substantia nigra on the catecholamine content of the striatum, *Brain,* 88:181.

Raz, A., Feingold, A., Zelanskaya, V., Vaadia, E., and Bergman, H., 1996, Neuronal synchronization of tonically active neurons in the striatum of normal and Parkinsonian primates, *J. Neurophysiol.* 76:2083.

Robertson, G.S., Vincent, S.R., and Fibiger, H.C., 1992, D_1 and D_2 dopamine receptors differentially regulate *c-fos* expression in striatonigral and striatopallidal neurons, *Neuroscience,* 49:285.

Rosa-Kenig, A., Puotz, J.K., and Rebec, G.V., 1993, The involvement of D_1 and D_2 dopamine receptors in amphetamine- induced changes in striatal unit activity in behaving rats, *Brain Res.* 619:347.

Ruskin, D.N., and Marshall, J.F., 1994, Amphetamine- and cocaine-induced Fos in the rat striatum depends on D_2 dopamine-receptor activation, *Synapse,* 18:233.

Ruskin, D.N., and Marshall, J.F., 1997, Differing influences of dopamine agonists and antagonists on Fos expression in identified populations of globus pallidus neurons, *Neuroscience,* 81:79.

Ruskin, D.N., Rawji, S.S., and Walters, J.R., 1998, Effects of full D_1 dopamine receptor agonists on firing rates in the globus pallidus and substantia nigra pars compacta *in vivo*: tests for D_1 receptor selectivity and comparisons to the partial agonist SKF 38393, *J. Pharmacol. Exp. Ther.* 286:272.

Ruskin, D.N., Bergstrom, D.A., Kaneoke, Y., Patel, B.N., Twery, M.J., and Walters, J.R., 1999a, Multisecond oscillations in firing rate in the basal ganglia: robust modulation by dopamine receptor and anesthesia, *J.*

Neurophysiol. 81:2046.

Ruskin, D.N., Bergstrom, D.A., Twery, M.J., and Walters, J.R., 1999b, Dopamine agonist-mediated rotation in rats with unilateral nigrostriatal lesions is not dependent on net inhibitions of rate in basal ganglia output nuclei, *Neuroscience,* 91:935.

Ruskin, D.N., Bergstrom, D.A., and Walters, J.R., 1999c, Firing rates and multisecond oscillations in the rodent entopeduncular nucleus: effects of dopamine agonists and nigrostriatal lesion, *Soc. Neurosci. Abstr.* 25:1929.

Ruskin, D.N., Bergstrom, D.A., and Walters, J.R., 1999d, Multisecond oscillations in firing rate in the globus pallidus: synergistic modulation by D1 and D2 dopamine receptors, *J. Pharmacol. Exp. Ther.* 290:1493.

Ruskin, D.N., Bergstrom, D.A., Shenker, A., Freeman, L.E., Baek, D., and Walters, J.R., 2000, Drugs used in the treatment of ADHD affect postsynaptic firing rate and oscillations without preferential dopamine autoreceptor action, *Biol. Psychiat.* in press.

Sato, F., Lavallee, P., Levesque, M., and Parent, A., 2000, Single-axon tracing study of neurons of the external segment of the globus pallidus in primate, *J. Comp. Neurol.* 417:17.

Scheel-Krüger, J., 1986, Dopamine-GABA interactions: evidence that GABA transmits, modulates and mediates dopaminergic functions in the basal ganglia and the limbic system, *Acta Neurol. Scand.* Suppl. 73:1.

Skirboll, L.R., Grace, A., and Bunney, B.S., 1979, Dopamine auto- and postsynaptic receptors: electrophysiological evidence for differential sensitivity to dopamine agonists, *Science,* 206:80.

Snyder, G.L., Allen, P.B., Fienberg, A.A., Valle, C.G., Huganir, R.L., Nairn, A.C., and Greengard, P., 2000, Regulation of phosphorylation of the GluR1 AMPA receptor in the neostriatum by dopamine and psychostimulants in vivo, *J. Neurosci.* 20:4480.

Stefani, A., Stanzione, P., Bassi, A., Mazzone, P., Vangelista, T., and Bernardi, G., 1997, Effects of increasing doses of apomorphine during stereotaxic neurosurgery in Parkinson's disease: clinical score and internal globus pallidus activity, *J. Neural Transm.* 104:895.

Verhagen Metman, L., Lee, J.I., Chen, P., Dougherty, P.M., and Lenz, F.A., 2000, Apomorphine-induced dyskinesia and single-cell discharges in globus pallidus internus of parkinsonian subjects, *Neurology,* 54 (Suppl.3):A456.

Vila, M., Périer, C., Féger, J., Yelnik, J., Faucheux, B., Ruberg, M., Raisman-Vozare, R., Agid, Y., and Hirsch, E.C., 2000, Evolution of changes in neuronal activity in the subthalamic nucleus of rats with unilateral lesion of the substantia nigra assessed by metabolic and electrophysiological measurements, *Eur. J. Neurosci.* 12:337.

Walters, J.R., Bergstrom, D.A., Carlson, J.H., Chase, T.N., and Braun, A.R., 1987, D_1 dopamine receptor activation required for postsynaptic expression of D_2 agonist effects, *Science,* 236:719.

Walters, J.R., Bergstrom, D.A., Ruskin, D.N., Allers, K.A., Rawji, S.S., and Twery, M.J., 1997, Relative properties of spike trains in the substantia nigra pars reticulata and subthalamic nucleus in an animal model of Parkinson's disease, *Soc. Neurosci. Abstr.* 23:192.

Walters, J.R., Allers, K.A., Bergstrom, D.A., Freeman, L.E., Baek, D., Tierney, P.L., and Ruskin, D.N, 2000a, Multisecond periodicities in globus pallidus firing rate correlate with bursts of EEG theta rhythm in baseline and after dopamine agonist treatment, *Soc. Neurosci. Abstr.* 26:962.

Walters, J.R., Ruskin, D.N., Allers, K.A., and Bergstrom, D.A., 2000b, Pre- and postsynaptic aspects of dopamine-mediated transmission, *Trends Neurosci.* 23:S41.

Walters, J.R., Ruskin, D.N., Allers, K.A., and Bergstrom, D.A., 2000c, Effects of dopamine receptor stimulation on single unit activity in the basal ganglia, in: *The Basal Ganglia VI,* M.R. DeLong, A.M. Graybiel, and S.T. Kitai, eds., Kluwer Press, New York.

Wang, J.Q., and McGinty, J.F., 1996, D_1 and D_2 receptor regulation of preproenkephalin and preprodynorphin mRNA in rat striatum following acute injection of amphetamine or methamphetamine, *Synapse,* 22:114.

Warenycia, M.W., and McKenzie, G.M., 1984, Immobilization of rats modifies the response of striatal neurons to dexamphetamine, *Pharmacol. Biochem. Behav.* 21:53.

Waszczak, B.W., Lee, E.K., Ferraro, T., Hare, T.A., and Walters, J.R., 1984, Single unit responses of substantia nigra pars reticulata neurons to apomorphine: effects of striatal lesions and anesthesia, *Brain Res.* 306:307.

Weick, B.G., and Walters, J.R., 1987, Effects of D_1 and D_2 dopamine receptor stimulation on the activity of substantia nigra pars reticulata neurons in 6-hydroxydopamine lesioned rats: D_1/D_2 coactivation induces potentiated responses, *Brain Res.* 405:234.

Weick, B.G., and Walters, J.R., 1988, The D-1 selective agonist SKF 38393 can activate striatal neurons in 6-hydroxydopamine lesioned rats, *Soc. Neurosci. Abstr.* 14:1077.

West, M.O., Peoples, L.L., Michael, A.J., Chapin, J.K., and Woodward, D.J., 1997, Low-dose amphetamine elevates movement-related firing of rat striatal neurons, *Brain Res.* 745:331.

West, M.O., 1998, Anesthetics eliminate somatosensory-evoked discharges of neurons in the somatotopically organized sensorimotor striatum of the rat, *J. Neurosci.* 18:9055.

Wichmann, T., Bergman, H., Kliem, M.A., Soares, J., and DeLong, M.R., 1999a, Low-frequency oscillatory discharge in the primate substantia nigra pars reticulata in the normal and parkinsonian state, *Soc. Neurosci. Abstr.* 25:1928.

Wichmann, T., Bergman, H., Starr, P.A., Subramanian, T., Watts, R.L., and DeLong, M.R., 1999b,
Comparison of MPTP-induced changes in spontaneous neuronal discharge in the internal pallidal
segment and in the substantia nigra pars reticulata in primates, *Exp. Brain. Res.* 125:397.
Wilson, C.J., Chang, H.T., and Kitai, S.T., 1990, Firing patterns and synaptic potentials of identified giant
aspiny interneurons in the rat neostriatum, *J. Neurosci.* 10:508.
Yan, Z., Song, W.J., and Surmeier, D.J., 1997, D-2 dopamine receptors reduce N-type Ca^{2+} currents in rat
neostriatal cholinergic interneurons through a membrane- delimited, protein-kinase-C-insensitive
pathway, *J. Neurophysiol.* 77:1003.

PHYSIOLOGICAL ROLES OF MULTIPLE METABOTROPIC GLUTAMATE RECEPTOR SUBTYPES IN THE RAT BASAL GANGLIA

P. JEFFREY CONN, HAZAR AWAD, STEFANIA R. BRADLEY, MICHAEL J. MARINO, SUSAN T. ROUSE, AND MARION WITTMANN[*]

INTRODUCTION

Parkinson's disease (PD) is a common neurodegenerative disorder afflicting over 1% of adults over age 65. The clinical syndrome that occurs in Parkinson's patients is characterized by a disabling motor impairment that includes tremor, rigidity, and bradykinesia. A large number of basic and clinical studies reveal that the primary pathophysiological change giving rise to the symptoms of PD is a loss of substantia nigra dopaminergic neurons that are involved in modulating function of the striatum and other basal ganglia structures. Based on this, the treatment of PD has traditionally utilized strategies for replacing the lost dopamine and thereby restoring the critical dopaminergic modulation of basal ganglia function. Levodopa (L-DOPA), the immediate precursor of dopamine, was the first highly effective treatment for PD and remains the most effective drug for treating the motor manifestations of PD (see Poewe and Granata, 1997, for extensive review). However, while effective early in treatment for a majority of patients, L-DOPA and other dopamine replacement therapies have a number of serious shortcomings. Within 5 years of beginning treatment, most patients begin to experience motor fluctuations, and the efficacy of L-DOPA becomes unpredictable. In addition, patients begin to develop serious side effects that often limit therapy. Unfortunately, the therapeutic window of L-DOPA narrows as the disease progresses, making it difficult to treat patients in whom the disease is advanced. Because of these problems with current strategies for the treatment of PD, a great deal of effort has been focused on developing a detailed understanding of the circuitry and function of the basal ganglia in hopes of developing novel therapeutic approaches for restoring normal basal ganglia function in patients suffering from this disorder.

---

[*] P. Jeffrey Conn, Hazar Awad, Stefania R. Bradley, Michael J. Marino, Susan T. Rouse, and Marion Wittmann • Emory University School of Medicine, Department of Pharmacology, Atlanta, GA 30322.

FUNCTIONAL CIRCUITRY OF BASAL GANGLIA INCLUDES DIRECT AND INDIRECT PATHWAYS FROM THE STRIATUM TO THE BASAL GANGLIA OUTPUT NUCLEI

An understanding of the anatomy, physiology, and neurochemistry of basal ganglia circuits is key to developing new pharmacological approaches for treatment of PD and other movement disorders (Bergman et al, 1990; Kaneoke and Vitek, 1995; Ciliax et al., 1997; Wichmann and DeLong, 1997). The basal ganglia are a richly interconnected set of subcortical nuclei that provide a feedback loop to wide areas of the cortex as well as descending influences to brainstem motor regions. The input nucleus of the basal ganglia is the striatum (caudate, putamen, and nucleus accumbens), which receives dense innervation from the cortex and subcortical structures. The primary output nuclei of the basal ganglia in primates are the internal globus pallidus (GPi) and the related substantia nigra pars reticulata (SNr). In non primates, the GPi is referred to as the entopeduncular nucleus (EPN). Recent models of the flow of cortical information through the basal ganglia suggest that striatal neurons project to GPi and SNr through two major routes, a direct pathway and an indirect pathway (Albin et al., 1989; Bergman et al., 1990; DeLong, 1990; Ciliax et al., 1997). Striatal neurons giving rise to both of these pathways provide inhibitory GABAergic projections. The direct pathway originates from a subpopulation of GABAergic neurons that project directly to GPi/SNr and thereby inhibit activity in these output nuclei. The indirect pathway originates from a different population of GABAergic striatal neurons that project to the external globus pallidus (GPe) in primates. In turn, GPe (referred to as globus pallidus or GP in non primates) sends a GABAergic projection to the subthalamic nucleus (STN), which provides an important excitatory glutamatergic drive to GPi/SNr. Thus, the output nuclei, GPi/SNr, receive a balance of opposing inhibitory and excitatory signals from the direct and indirect pathways respectively. Figure 1 schematically illustrates the major components of the direct and indirect pathways of basal ganglia circuitry in normal conditions.

A major pathophysiological change giving rise to the symptoms of PD is an increase in excitatory output from the STN. A fine tuning of motor control requires an intricate balance of activity in the direct and indirect pathways. In normal animals, this balance is maintained, in part, by modulation of striatal neurons by dopaminergic projections from the substantia nigra pars compacta (SNc). Release of dopamine from terminals of the nigrostriatal projection facilitates transmission through the direct pathway (via activation of D1 dopamine receptors) and inhibits transmission over the indirect pathway (via activation of D2 receptors) (Gerfen et al., 1990). Recent studies suggest that loss of dopaminergic input to the striatum leads to overall increased activation of the indirect pathway and decreased activity in the direct pathway (Albin et al., 1989; DeLong, 1990). These changes lead to increased inhibition of thalamocortical neurons. Figure 1 schematically illustrates the activity changes in basal ganglia-thalamocortical circuitry that are thought to occur in PD.

Within this model, overactivity of glutamatergic projections from the STN to basal ganglia output nuclei is thought to play a major role in the pathophysiology of PD (Bergman et al., 1990; DeLong, 1990). Consistent with this, lesions or high frequency stimulation of the STN normalize metabolic activity in the output nuclei and ameliorate parkinsonian motor signs (Bergman et al., 1990; Aziz et al., 1991; Limousin et al., 1995; 1997; Wichmann and DeLong, 1997). Based on this, any pharmacological manipulation that reduces the net excitatory drive through the STN would be expected to have a therapeutic effect in the treatment of PD.

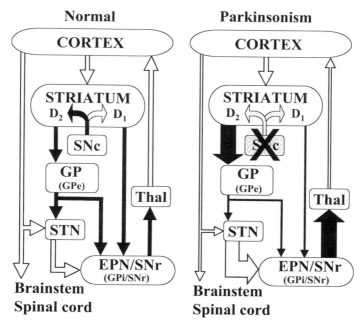

Figure 1. A model of how the Parkinson's-related loss of dopamine neurons in the SNc impacts information flow through the basal ganglia. Note the increase of glutamatergic transmission at the STN-SNr synapse. Inhibitory connections are depicted by black arrows, excitatory transmission by white arrows.

REDUCTION OF TRANSMISSION AT EXCITATORY STN SYNAPSES MAY BE EFFECTIVE IN TREATMENT OF PD

As outlined above, an increase in transmission through the indirect pathway is believed to be responsible for the motor symptoms of PD. This model has led to a major effort to develop new therapeutic strategies aimed at decreasing activity of STN neurons or decreasing transmission at excitatory STN synapses in GPi/SNr (see Starr, 1995; Blandini et al., 1996 for reviews). Fast excitatory synaptic responses at STN synapses are mediated by activation of members of a well characterized family of glutamate receptors referred to as the ionotropic glutamate receptors (iGluRs) (for reviews, see Monyer and Seeburg, 1993; Seeburg, 1993). The iGluRs are ligand-gated cation channels and have been divided into the N-methyl-D-aspartate (NMDA), kainate, and [RS]-α-amino-3-hydroxy-5-methyl-isoxazolepropionic acid (AMPA) receptor subtypes. Immunocytochemical and physiological studies suggest that both NMDA and AMPA receptors are present at STN-GPi/SNr synapses, where AMPA receptors are likely to play the most important role in generating fast excitatory postsynaptic potentials (EPSPs) (See Ciliax et al., 1997, for review). Thus far, the major focus of efforts to reduce transmission at STN synapses has been on the use of iGluR antagonists (Starr, 1995; Blandini et al., 1996). Interestingly, systemic administration of iGluR antagonists has therapeutic effects in parkinsonian monkeys (Greenamyre et al., 1994; Ciliax et al., 1997). It is not surprising, therefore, that several drugs currently in clinical use for treatment of PD (amantadine, memantine, and budipine) have varying degrees of activity as antagonists of the NMDA receptor (Bormann, 1989; Klockgether et al., 1993; Kornhuber et al., 1994; Danielczyk, 1995; Vitti et al., 1996; Ciliax et al., 1997). Furthermore, stereotactic injection of glutamate receptor antagonists into the GPi of parkinsonian monkeys has profound antiparkinsonian effects (Brotchie et al., 1991), suggesting that the therapeutic effects of systemically administered iGluR antagonists are at least partially due to blockade of receptors in this target of glutamatergic STN projections.

Despite these encouraging results, the long-term therapeutic efficacy of iGluR antagonists may ultimately be limited, because iGluRs mediate the excitatory effects at the vast majority of glutamatergic synapses in the CNS. Administration of sufficient doses of iGluR antagonists to induce a substantial reduction in transmission at STN synapses will also reduce excitatory transmission throughout the brain, which can result in serious side effects (Starr, 1995). Thus, while studies with iGluR antagonists are important, it will also be critical to develop other strategies for reducing STN activity or transmission at STN-GPi/SNr synapses while maintaining an acceptable level of transmission at excitatory synapses in other brain regions. Interestingly, recent studies suggest that a novel family of glutamate receptors, termed metabotropic glutamate receptors (mGluRs) are richly distributed in the basal ganglia, where they may play important roles in regulating transmission through the indirect pathway. These new data suggest that ligands of mGluRs could serve as novel therapeutic targets for treatment of PD.

METABOTROPIC GLATAMATE RECEPTORS MAY PROVIDE NOVEL TARGETS FOR DEVELOPMENT OF NEW THERAPEUTIC AGENTS FOR PD TREATMENT

Until recently, it was thought that all of the actions of glutamate were mediated by activation of iGluRs. However, in recent years it has become clear that glutamate also activates receptors coupled to effector systems through GTP-binding proteins. These receptors, referred to as metabotropic glutamate receptors (mGluRs), are widely distributed throughout the central nervous system, where they play important roles in regulating cell excitability and synaptic transmission (see Conn and Pin, 1997, for extensive review). One of the primary functions of mGluRs is to serve as presynaptic receptors involved in reducing transmission at glutamatergic synapses. Activation of presynaptic mGluRs has been shown to reduce transmission at glutamatergic synapses in a wide variety of brain regions including the amygdala, olfactory cortex, neocortex, spinal cord, striatum, cerebellum, nucleus of the solitary tract, and olfactory bulb, and in each major subsector of the hippocampus. The mGluRs also serve as heteroreceptors involved in reducing GABA release at inhibitory synapses. Finally, postsynaptically localized mGluRs often play an important role in regulating neuronal excitability and in regulating currents through iGluR channels. If mGluRs play these roles in the basal ganglia, and particularly at STN synapses, members of this receptor family may provide an exciting new target for drugs that could be useful in ameliorating the symptoms of PD.

MULTIPLE METABOTROPIC GLUTAMATE RECEPTOR SUBTYPES HAVE BEEN IDENTIFIED IN MAMMALIAN BRAIN AND MAY PLAY A ROLE IN MODULATING BASAL GANGLIA FUNCTION

To date, eight mGluR subtypes (designated mGluR1-mGluR8) have been cloned from mammalian brain. These mGluRs are classified into three major groups based on sequence homologies, coupling to second messenger systems, and selectivities for various agonists (see Conn and Pin, 1997, for review). Group I mGluRs (mGluR1, and 5) couple to Gq and phosphoinositide hydrolysis, while groups II (mGluR2, and 3) and III (mGluR4, 6, 7, and 8) couple to Gi/Go and related effector systems such as inhibition of adenylate cyclase. Multiple splice variants have been identified for many of the mGluR subtypes. These include at least four splice variants of mGluR1 (mGluR1a, b, c, and d), and two each of mGluR4 (mGluR4a and mGluR4b), mGluR5 (mGluR5a and b), and mGluR7 (mGluR7a and b). The receptors in each group have a pharmacological profile that is distinct from that of receptors in the other groups and highly selective mGluR agonists are now available that interact with members of one group of mGluRs at concentrations that are inactive at the other two groups. Group I mGluRs are selectively activated by 3,5-dihydroxyphenylglycine (DHPG; Ito et al., 1992; Schoepp et al., 1994; Gereau and Conn, 1995); group II mGluRs by (2S,1'R,2'R,3'R)-2-(2,3-dicarboxycyclopropyl) glycine (DCG-IV; Hayashi et al., 1993; Gereau and Conn, 1995) and group III mGluRs by (L)-2-amino-4-phosphonobutyric acid (L-AP4) (see Pin and Duvoisin, 1995; Conn and Pin, 1997, for reviews). These and other recently developed drugs provide valuable tools that can be used for dissecting the physiological roles of different mGluR subtypes.

Previous studies suggest that presynaptic group II and group III mGluRs play important roles in regulating excitatory (Lovinger, 1991; Lovinger et al., 1993; Lovinger and McCool, 1995; Pisani et al., 1997a) and inhibitory (Calabresi et al., 1992; 1993; Stefani et al., 1994) transmission in the striatum and that postsynaptic group I mGluRs regulate striatal cell excitability and NMDA receptor currents (Calabresi et al., 1992; Colwell and Levine, 1994; Pisani et al., 1997b). Moreover, intrastriatal injection of mGluR agonists induces rotational behavior (Sacaan et al., 1991; 1992; Kaatz and Albin, 1995), and this effect is abolished by lesions of the STN (Kaatz and Albin, 1995). Furthermore, multiple mGluR subtypes are involved in regulating synaptic transmission and cell excitability in the dopaminergic neurons of the substantia nigra pars compacta (SNc) (Fiorillo and Williams, 1998; Wigmore and Lacey, 1998). Despite these advances in our understanding of the roles of mGluRs in the striatum and SNc, relatively little attention has been focused on examining the roles of mGluRs in regulating transmission through the indirect pathway. However, in the past two years, we and others have made important advances in developing an understanding of the roles of mGluRs in regulating transmission through this important component of the basal ganglia motor circuit.

GROUP I METABOTROPIC GLUTAMATE RECEPTORS PLAY AN IMPORTANT ROLE IN TRANSMISSION AT THE STN-SNr SYNAPSE

A strategy that could be useful for reducing the net excitatory input from STN to SNr and EPN would be the use of ligands at postsynaptic mGluRs associated with STN synapses. Interestingly, both of the group I mGluRs, mGluR1 and mGluR5, are expressed in SNr and EPN neurons (Testa et al., 1994). In other brain regions, group I mGluRs are predominantly postsynaptic in their localization (Martin et al., 1992; Baude et al., 1993; Romano et al., 1995; Lujan et al., 1996). Consistent with this, Testa et al. (1998) recently reported that mGluR1a immunoreactivity is abundant along the surface of MAP2-immunoreactive dendritic processes in the SNr. In addition, we have recently shown that mGluR1a and mGluR5 antibodies label the SNr, and analysis at the EM level reveals that the immunoreactivity is localized postsynaptically in dendrites that form asymmetric synapses with glutamatergic terminals (Marino et al., 1999).

We have recently investigated the physiological roles mGluRs play in the modulation of the SNr. These studies have shown that activation of group I mGluRs produces a robust direct depolarization of SNr GABAergic neurons (Marino et al., 1999). This effect is blocked by the mGluR1-selective antagonist CPCCOEt, but not by the mGluR5-selective MPEP. Therefore, this effect appears to be mediated solely by mGluR1. In addition, stimulation of glutamatergic afferents in the SNr at frequencies consistent with the normal firing rate of STN neurons induces an mGluR-mediated slow EPSP which is completely blocked by CPCCOEt. Taken together, these studies indicate that the group I mGluR, mGluR1, plays a role in modulation of SNr excitability. Interestingly, we recently found that activation of group I mGluRs also reduces inhibitory transmission in the SNr by a presynaptic mechanism (Wittmann et al., 1999). This disinhibition, coupled with the direct excitatory effect of mGluR1 activation, could play an important role in the powerful excitation of the SNr exerted by the relatively sparse glutamatergic input to this nucleus from the STN.

Another site at which group I mGluRs might be involved in regulation of the STN-SNr synapse is in the STN. Both mGluR1 and 5 mRNA are present in the STN (Testa et al., 1994) and both the group I mGluR subtypes are postsynaptically localized to dendrites of STN neurons at both symmetric and asymmetric synapses (Awad et al., 2000). Activation of group I mGluRs induces a robust depolarization of STN neurons that is blocked by MPEP, but not by CPCCOEt (Figure 2A-B). Therefore, in contrast to the findings in SNr, the group I-mediated depolarization of STN neurons is solely mediated by mGluR5. In addition to directly depolarizing the STN neurons, group I mGluR activation also facilitates burst firing in the STN (Beurrier et al., 1999; Awad et al., 2000) (Figure 2C). Interestingly, a switch from single spike activity to a burst-firing mode is one of the characteristics of parkinsonian states (Hollerman and Grace, 1992; Bergman et al., 1994; Hassani et al., 1996). These data raise the possibility that mGluR5 may play a role in generation of burst firing in STN neurons.

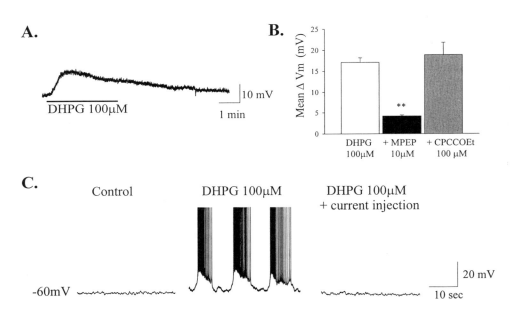

Figure 2. Activation of mGluR5 excites neurons in the STN. The group I mGluR agonist DPHG induces a depolarization (A) of STN neuron membrane potential which is selectively blocked by the mGluR5 antagonist MPEP (B). DHPG also increases the incidence of burst-firing in STN neurons (C).

GROUP II METABOTROPIC GLUTAMATE RECEPTORS MAY SERVE AS PRESYNAPTIC RECEPTORS AT GLUTAMATERGIC SYNAPSES BETWEEN STN NEURONS AND NEURONS IN THE BASAL GANGLIA OUTPUT NUCLEI

As discussed above, one of the primary functions of mGluRs is to serve as presynaptic receptors involved in reducing transmission at glutamatergic synapses. If mGluRs serve this role at terminals of STN projections to SNr and EPN (or GPi), this could provide an exciting new target for therapeutic agents useful for treatment of PD. Interestingly, *in situ* hybridization studies suggest that mGluR2 is abundantly expressed in STN neurons (Testa et al., 1994). In other brain regions, mGluR2 protein is often targeted to presynaptic sites and serves to reduce glutamate release (Hayashi et al., 1993; Ohishi et al., 1994; Neki et al., 1996; Yokoi et al., 1996; Shigemoto et al., 1997). Thus, mGluR2 is a likely candidate for an mGluR that could serve as a presynaptic receptor at STN synapses. Consistent with this, recent immunocytochemistry studies with antibodies that react with both mGluR2 and mGluR3 revealed punctate neuropil staining in rat SNr and GP (Petralia et al., 1996; Testa et al., 1998), both of which receive

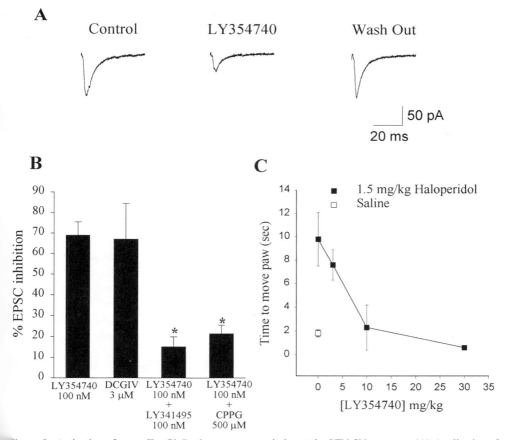

Figure 3. Activation of group II mGluRs decreases transmission at the STN-SNr synapse. (A) Application of LY354740 produces a reversible decrease in the amplitude of EPSCs evoked by stimulation of the STN. (B) Mean data demonstrating that the pharmacology of this effect is consistent with mediation by group II mGluRs. (C) Systemic application of LY345740 reduces haloperidol-induced catalepsy in a dose-dependent manner.

glutamatergic projections from STN. Furthermore, electron microscopic studies from our laboratory indicate that mGluR2/3 immunoreactivity is localized in small unmyelinated axons and presynaptic terminals that form asymmetric synapses in the SNr (Bradley et al., 2000).

Since the group II mGluRs often act presynaptically to reduce glutamate release, we tested for the ability of the highly selective group II agonist LY354740 to inhibit glutamatergic transmission at the STN-SNr synapse. Consistent with the anatomical localization of the group II receptors in the SNr, activation of these receptors produces an inhibition of transmission at the STN-SNr synapse (Bradley et al., 2000) (Figure 3A,B). This effect is selective for excitatory transmission, and is mediated by a presynaptic mechanism. This specific decrease in excitatory transmission at the STN-SNr synapse indicates that the group II mGluRs may provide an excellent target for drugs which could reduce the excessive transmission through the indirect pathway and thereby ameliorate the symptoms of PD. We tested this hypothesis in a rat model of parkinsonism using haloperidol-induced catalepsy (Ossowska et al., 1990; Schmidt et al., 1997). Systemic application of the group II agonist LY354740 produced a dose-dependent decrease in haloperidol-induced catalepsy (Figure 3C). Interestingly, a recent study has shown that intranigral injection of DCG-IV, another selective group II agonist, causes a reduction in reserpine-induced akinesia (Dawson et al., 2000). Taken together, these results indicate that reduction in transmission at the STN-SNr synapse by activation of group II mGluRs has the potential of providing relief to PD patients.

GROUP III METABOTROPIC GLUTAMATE RECEPTORS MAY REGULATE TRANSMISSION AT THE STRIATO-PALLIDAL SYNAPSE

An additional approach that could be used to pharmacologically reduce activity through the indirect pathway and reduce STN activity would be to regulate transmission at inhibitory GABAergic projections from the striatum to the GP. A common function of mGluRs that has been observed in other brain regions is a role as presynaptic heteroreceptors on inhibitory nerve terminals involved in inhibition of GABA release (Desai and Conn, 1991; Calabresi et al., 1992; Desai et al., 1994; Stefani et al., 1994; Pin and Conn, 1997). Interestingly, we have performed a series of studies that suggest that certain group III mGluRs are presynaptically localized on terminals of GABAergic striato-pallidal synapses (Kosinski et al., 1999). For instance, both mGluR4 and mGluR7 (but not mGluR6 or mGluR8) mRNA are abundantly expressed in the striatum (Nakajima et al., 1993; Testa et al., 1994; Duvoisin et al., 1995; Saugstad et al., 1997; Kosinski et al., 1999) and double-labeling *in situ* hybridization revealed that mGluR7 mRNA is in projection neurons (Kosinski et al., 1999). Furthermore, we have found that mGluR7a and mGluR4a proteins are abundant in GP neurons. Finally, immuno-EM revealed the presence of both mGluR4a and mGluR7a immunoreactivity in striatal-like terminals that form symmetric synapses in GP (Bradley et al., 1999; Kosinski et al., 1999). Taken together, these data provide strong evidence that mGluR7a and mGluR4a are presynaptically localized on striato-pallidal terminals. This would suggest that agonists of group III mGluRs would ultimately reduce STN activity and could therefore be useful in treatment of PD.

CONCLUSIONS

In summary, mGluRs are expressed throughout the basal ganglia and play important roles in the regulation of basal ganglia function. Activation of group I mGluRs has excitatory effect in several basal ganglia nuclei, including STN and SNr. Group II mGluRs play a role in regulating transmission at the STN-SNr synapse. The reduction of excitatory transmission at the STN-SNr synapse by group II mGluRs activation raises the exciting possibility that a group II mGluR agonist could be beneficial for reducing the over excitation of the output nuclei that occurs in PD. In addition, group III mGluR activation could serve presynaptically to reduce transmission at the striato-pallidal synapse, thereby serving to reduce STN activity. All of these findings provide strong evidence that mGluRs could serve as novel therapeutic targets for the treatment of PD.

REFERENCES

Albin, R.L., Young, A.B., Penney, J.B., 1989, The functional anatomy of basal ganglia disorders, *Trends Neurosci.* 12:366.

Awad, H., Hubert, G.W., Smith, Y., Levey, A., and Conn, P.J., 2000, Activation of metabotropic receptor 5 has direct excitatory effects and potentiates NMDA receptor currents in neurons of the subthalamic nucleus. *J. Neurosci. in press*

Aziz, T.Z., Peggs, D., Sambrook, M.A., Crossman, A.R., 1991, Lesion of the subthalamic nucleus for the alleviation of 1-methyl-4-phenyl-1,2,3,6-tetrahydropyridine (MPTP)-induced parkinsonism in the primate. *Mov. Disord.* 6:288.

Baude, A., Nusser, Z., Roberts, J.D.B., Mulvihill, E., McIlhinney, R.A.J., Somogyi, P., 1993, The metabotropic glutamate receptor (mGluR1α) is concentrated at perisynaptic membrane of neuronal subpopulations as detected by immunogold reaction, *Neuron.* 11:771.

Bergman, H., Wichmann, T., DeLong, M.R., 1990, Reversal of experimental parkinsonism by lesions of the subthalamic nucleus, *Science,* 249:1436.

Bergman, H., Wichmann, T., Karmon, B., DeLong, M.R., 1994, The primate subthalamic nucleus. II. Neuronal activity in the MPTP model of parkinsonism, *J. Neurophysiol.* 72:507.

Blandini, F., Porter, R.H., Greenamyre, J.T., 1996, Glutamate and Parkinson's disease, *Mol. Neurobiol.* 12:73.

Bormann, J., 1989, Memantine is a potent blocker of N-methyl-D-aspartate (NMDA) receptor channels, *Eur. J. Pharmacol.* 166:591.

Bradley, S.R., Standaert, D.G., Rhodes, K.J., Rees, H.D., Testa, C.M., Levey, A.I., Conn, P.J., 1999, Immunohistochemical localization of subtype 4a metabotropic glutamate receptors in the rat and mouse basal ganglia. *J. Comp. Neurol.* 407:33.

Bradley, S.R., Marino, M.J., Wittmann, M., Rouse, S.T., Awad, H., Levey, A.I., Conn, P.J., 2000, Activation of group II metabotropic glutamate receptors inhibits synaptic excitation of the substantia nigra pars reticulata, *J. Neurosci.* 20:3085.

Brotchie, J.M., Mitchell, I.J., Sambrook, M.A., Crossman, A.R., 1991, Alleviation of parkinsonism by antagonism of excitatory amino acid transmission in the medial segment of the globus pallidus in rat and primate, *Mov. Disord.* 6:133.

Beurrier, C., Congar, P., Bioulac, B., Hammond, C., 1999, Subthalamic nucleus neurons switch from single-spike activity to burst- firing mode, *J. Neurosci.* 19:599.

Calabresi, P., Mercuri, N.B., Bernardi, G., 1992, Activation of quisqualate metabotropic receptors reduces glutamate and GABA-mediated synaptic potentials in the rat striatum, *Neurosci. Lett.* 139:41.

Calabresi, P., Pisani, A., Mercuri, N.B., Bernardi, G., 1993, Heterogeneity of metabotropic glutamate receptors in the striatum: Electrophysiological evidence, *Eur. J. Neurosci.* 5:1370.

Ciliax, B.J., Greenamyre, T., Levey, A.I., 1997, Functional biochemistry and molecular neuropharmacology of the basal ganglia and motor systems, in: *Movement Disorders: Neurological Principles and Practice* R.L. Watts ed., McGraw-Hill, New York.

Colwell, C.S., and Levine, M.S., 1994, Metabotropic glutamate receptors modulate *N*-methyl-D-aspartate receptor function in neostriatal neurons, *Neurosci.* 61:497.

Conn, P.J., and Pin, J.P., 1997, Pharmacology and functions of metabotropic glutamate receptors, *Annu. Rev. Pharmacol. Toxicol.* 37:205.

Danielczyk, W., 1995, Twenty-five years of amantadine therapy in Parkinson's disease, *J. Neural. Transm.* 46:399.

Dawson, L., Chadha, A., Megalou, M., Duty, S., 2000, The group II metabotropic glutamate receptor agonist, DCG-IV, alleviates akinesia following intranigral or intraventricular administration in the reserpine-treated rat, *Br. J. Pharmacol.* 129:541.

DeLong, M.R., 1990, Primate models of movement disorders of basal ganglia origin, *Trends Neurosci.* 13:18.

Desai, M.A., Conn, P.J., 1991, Excitatory effects of ACPD receptor activation in the hippocampus are mediated by direct effects on pyramidal cells and blockade of synaptic inhibition, *J. Neurophysiol.*

66:40.

Desai, M.A., McBain, C., Kauer, J.A., Conn, P.J., 1994, Metabotropic glutamate receptor induced disinhibition is mediated by reduced transmission at excitatory synapses onto interneurons and inhibitory synapses onto pyramidal cells, *Neurosci. Lett.* 181:78.

Duvoisin, R.M., Zhang, C., Ramonell, K., 1995, A novel metabotropic glutamate receptor expressed in the retina and olfactory bulb, *J. Neurosci.* 15:3075.

Fiorillo, C.D., and Williams, J.T., 1998, Glutamate mediates an inhibitory postsynaptic potential in dopamine neurons, *Nature,* 394:78.

Gereau , R.W., and Conn, P.J., 1995, Roles of specific metabotropic glutamate receptor subtypes in regulation of hippocampal CA1 pyramidal cell excitability, *J. Neurophysiol.* 74:122.

Gerfen , C.R., Engber, T.M., Mahan, L.C., Susel, Z., Chase, T.N., Monsma, F.J., Sibley, D.R., 1990, D1 and D2 dopamine receptor regulated gene expression of striatonigral and striatopallidal neurons, *Science,* 250:1429.

Greenamyre, J.T., Eller, R.V., Zhang, Z., Ovadia, A., Kurlan, R., Gash, D.M., 1994, Antiparkinsonian effects of remacemide, a glutamate antagonist, in rodent and primate models of Parkinson's disease, *Ann. Neurol.* 35:655.

Hassani, O.K., Mouroux, M., Feger, J., 1996, Increased subthalamic neuronal activity after nigral dopaminergic lesion independent of disinhibition via the globus pallidus, *Neurosci.* 72:105.

Hayashi, Y., Momiyama, A., Takahashi, T., Ohishi, H., Ogawa-Meguro, R., Shigemoto, R., Mizuno, N., Nakanishi, S., 1993, Role of a metabotropic glutamate receptor in synaptic modulation in the accessory olfactory bulb, *Nature,* 366:687.

Hollerman, J.R., and Grace, A.A., 1992, Subthalamic nucleus cell firing in the 6-OHDA-treated rat: basal activity and response to haloperidol, *Brain Res.* 590:291.

Ito, I., Kohda, A., Tanabe, S., Hirose, E., Hayashi, E., Mitsunaga, S., Sugiyama, H., 1992, 3,5-dihydroxyphenylglycine: a potent agonist of metabotropic glutamate receptors, *Neuro. Report,* 3:1013.

Kaatz, K.W., and Albin, R.L., 1995, Intrastriatal and intrasubthalamic stimulation of metabotropic glutamate receptors: A behavioral and Fos immunohistochemical study, Neurosci. 66:55.

Kaneoke, Y., and Vitek, J.L., 1995, The role of basal ganglia in movement control, *Rinsho Shinkeigaku,* 35:1518.

Klockgether, T., Jacobsen, P., Loschmann, P.A., Turski, L., 1993, The antiparkinsonian agent budipine is an N-methyl-D-aspartate antagonist, *J. Neural Transm.* 1993:101.

Kornhuber, J., Weller, M., Schoppmeyer, K., Riederer, P., 1994, Amantadine and memantine are NMDA receptor antagonists with neuroprotective properties, *J. Neural Transm.* 98 Suppl. 43:91.

Kosinski, C.M., Risso, Bradley, S., Conn, P.J., Levey, A.I., Landwehrmeyer, G.B., Penney, J.B. Jr., Young, A.B., Standaert, D.G., 1999, Localization of metabotropic glutamate receptor 7 mRNA and mGluR7a protein in the rat basal ganglia, *J. Comp. Neurol.* 415:266.

Limousin, P., Greene, J., Pollak, P., Rothwell, J., Benabid, A.L., Frackowiak, R., 1997, Changes in cerebral activity pattern due to subthalamic nucleus of internal pallidum stimulation in Parkinson's disease, *Ann. Neurol.* 42:283.

Limousin, P., Pollak, P., Benazzouz, A., Hoffman, D., Broussolle, E., Perret, J.E., Benabid, A.L., 1995, Bilateral subthalamic nucleus stimulation for severe Parkinson's desease, *Mov. Disord.* 10:672.

Lovinger, D.M., 1991, *Trans*-1-aminocyclopentante-1,3-dicarboxylic acid (*t*-ACPD) decreases synaptic excitation in rat striatal slices through a presynaptic action, *Neurosci. Lett.* 129:17.

Lovinger, D.M., Tyler, E., Fidler, S., Merritt, A., 1993, Properties of a presynaptic metabotropic glutamate receptor in rat neostriatal slices, *J. Neurophysiol.* 69:1236.

Lovinger, D.M., and McCool, B.A., 1995, Metabotropic glutamate receptor-mediated presynaptic depression at corticostriatal synapses involves mGluR2 or 3, *J. Neurophysiol.* 73:1076.

Luján, R., Nusser, Z., Roberts, J.D.B., Shigemoto, R., Somogyi, P., 1996, Perisynaptic location of metabotropic glutamate receptors mGluR1 and mGluR5 on dendrites and dendritic spines in the rat hippocampus, *Eur. J. Neurosci.* 8:1488.

Marino, M.J., Bradley, S.R., Wittmann, M., Conn, P.J., 1999, Direct excitation of GABAergic projection neurons of the rat substantia nigra pars reticulata by activation of the mGluR1 metabotropic glutamate receptor, *Soc. Neurosci. Abstr.* 25:176.

Martin, L.J., Blackstone, C.D., Huganir, R.L., Price, D.L., 1992, Cellular localization of a metabotropic glutamate receptor in rat brain, *Neuron,* 9:259.

Monyer, H., and Seeburg, P.H., 1993, Constituents involved in glutamate receptor signaling, *Hippocampus,* 3 Suppl.:125.

Nakajima, Y., Iwakabe, H., Akazawa ,C., Nawa, H., Shigemoto, R., Mizuno, N., Nakanishi, S., 1993, Molecular characterization of a novel retinal metabotropic glutamate receptor mGluR6 with a high agonist selectivity for L-2-amino-4-phosphonobutyrate, *J. Biol. Chem.* 268:11868.

Neki, A., Ohishi, H., Kaneko, T., Shigemoto, R., Nakanishi, S., Mizuno, N., 1996, Pre- and postsynaptic localization of a metabotropic glutamate receptor, mGluR2, in the rat brain: An immunohistochemical study with a monoclonal antibody, *Neurosci. Lett.* 202:197.

Ohishi, H., Ogawa-Meguro, R., Shigemoto, R., Kaneko, T., Nakanishi, S., Mizuno, N., 1994, Immunohistochemical localization of metabotropic glutamate receptors, mGluR2 and mGluR3, in rat cerebellar cortex, *Neuron,* 13:55.

Ossowska, K., Karcz, M., Wardas, J., Wolfarth, S., 1990, Striatal and nucleus accumbens D1/D2 dopamine receptors in neuroleptic catalepsy, *Eur. J. Pharmacol.* 182:327.

Petralia, R.S., Wang, Y.-X., Niedzielski, A.S., Wenthold, R.J., 1996, The metabotropic glutamate recepotrs mGluR2 and mGluR3 show unique postsynaptic, presynaptic and glial localizations, *Neurosci.* 71:949.

Pin, J.P., and Duvoisin, R., 1995, Neurotransmitter receptors I: The metabotropic gutamate receptors: structure and functions, *Neuropharmacol.* 34:1.

Pisani, A., Calabresi, P., Centonze, D., Bernardi, G., 1997a, Activation of group III metabotropic glutamate receptors depresses glutamatergic transmission at corticostriatal synapse, *Neuropharmacol.* 36:845.

Pisani, A., Calabresi, P., Centonze, D., Bernardi, G., 1997b Enhancement of NMDA responses by group I metabotropic glutamate receptor activation in striatal neurones, *Br. J. Pharmacol.* 120:1007.

Poewe, W., and Granata, R., 1997, Pharmacological treatment of Parkinson's disease, in: *Movement Disorders: Neurological Principles and Practice*, R.L. Watts ed., McGraw-Hill, New York.

Romano, C., Sesma, M.A., McDonald, C.T., O'Malley, K., Van den Pol, A.N., Olney, J.W., 1995, Distribution of metabotropic glutamate receptor mGluR5 immunoreactivity in rat brain, *J. Comp. Neurol.* 355:455.

Sacaan, A.I., Monn, J.A., Schoepp, D.D., 1991, Intrastriatal injection of a selective metabotropic excitatory amino acid receptor agonist induces contralateral turning in the rat, *J. Pharmacol. Exp. Ther.* 259:1366.

Saugstad, J.A., Kinzie, J.M., Shinohara, M.M., Segerson, T.P., Westbrook, G.L., 1997, Cloning and expression of rat metabotropic glutamate receptor 8 reveals a distinct pharmacological profile, *Mol. Pharmacol.* 51:119.

Schmidt, W.J., Schuster, G., Wacker, E., Pergande, G., 1997, Antiparkinsonian and other motor effects of flupirtine alone and in combination with dopaminergic drugs, *Eur. J. Pharmacol.* 327:1.

Schoepp, D.D., Goldsworthy, J., Johnson, B.G., Salhoff, C.R., Baker, S.R., 1994, 3,5-dihydroxyphenylglycine is a highly selective agonist for phosphoinositide-linked metabotropic glutamate receptors in the rat hippocampus, *J. Neurochem.* 63:769.

Seeburg, P.H., 1993, The *TiPS/TINS* lecture: The molecular biology of mammalian glutamate receptor channels, *Trends Pharmacol. Sci.* 14:297.

Shigemoto, R., Kinoshita, A., Wada, E., Nomura, S., Ohishi, H., Takada, M., Flor, P.J., Neki, A., Abe, T., Nakanishi, S., Mizuno, N., 1997, Differential presynaptic localization of metabotropic glutamate receptor subtypes in the rat hippocampus, *J. Neurosci.* 17:7503.

Starr, M.S., 1995, Glutamate/dopamine D1/D2 balance in the basal ganglia and its relevance to Parkinson's disease, *Synapse*, 19:264.

Stefani, A., Pisani, A., Mercuri, N.B., Bernardi, G., Calabresi, P., 1994, Activation of metabotropic glutamate receptors inhibits calcium currents and GABA-mediated synaptic potentials in striatal neurons, *J. Neurosci.* 14:6734.

Testa, C.M., Standaert, D.G., Young, A.B., Penney, J.B. Jr., 1994 Metabotropic glutamate receptor mRNA expression in the basal ganglia of the rat, *J. Neurosci.* 14:3005.

Testa, C.M., Friberg, I.K., Weiss, S.W., Standaert, D.G., 1998, Immunohistochemical localization of metabotropic glutamate receptors mGluR1a and mGluR2/3 in the rat basal ganglia, *J. Comp. Neurol.* 390:5.

Vitti, R.J., Rajput, A.H., Ahlskog, J.E., Offord, K.P., Schroeder, D.R., Ho, M.M., Prasad, M., Rajput, A., Basran, P., 1996, Amantadine treatment is an independent predictor of improved survival in Parkinson's disease, *Neurology*, 46:1551.

Wichmann, T., and DeLong, M.R., 1997, Physiology of the basal ganglia and pathophysiology of movement disorders of basal ganglia origin, in: *Movement Disorders: Neurological Principles and Practice*, R.L. Watts ed., McGraw-Hill, New York.

Wigmore, M.A., and Lacey, M.G., 1998, Metabotropic glutamate receptors depress glutamate-mediated synaptic input to rat midbrain dopamine neurones in vitro, *Br. J. Pharmacol.* 123:667.

Wittmann, M., Marino, M.J., Bradley, S.R., and Conn, P.J., 1999, GABAergic inhibition of substantia nigra pars reticulata projection neurons is modulated by metabotropic glutamate receptors, *Soc. Neurosci. Abstr.* 25:176.

Yokoi, M., Kobayashi, K., Manabe, T., Takahashi, T., Sakaguchi, I., Katsuura, G., Shigemoto, R., Ohishi, H., Nomura, S., Nakamura, K., Nakao, K., Katsuki, M., Nakanishi, S., 1996, Impairment of hippocampal mossy fiber LTD in mice lacking mGluR2, *Science*, 273:645.

MUSCARINIC RECEPTORS OF THE DORSAL STRIATUM: ROLE IN REGULATION OF MOTOR BEHAVIOR

KSENIA SHAPOVALOVA[*]

INTRODUCTION

Striatum is a large paired structure of forebrain that plays a key role in the formation and realization of voluntary motor acts. The major targets of striatal influences are the substantia nigra pars reticularis, entopeduncular nucleus, and the lateral globus pallidus, which have an inhibitory, GABAergic influence on several motor and sensory structures. A peculiarity of striatum is that it has two efferent output pathways, the direct and indirect, which differ from each other in topography, net effect on targets, involvement of dopamine and muscarinic receptors, and the peptides coexisting with them (Gerfen, 1992; De Lapp et al., 1996; Wang and Mc Ginty, 1996). The transmitter in both cases, however, is GABA (Kita and Kitai, 1988).

The striatum, and primarily its dorsal part, the neostriatum (nucleus caudatus and putamen), is among the forebrain structures with the highest level of acetylcholine and the highest activities of acetylcholinesterase and cholineacetyltransferase (Contant et al., 1996). A feature specific to the striatum is that its acetylcholine source is intrinsic. A small number of striatal interneurones have large cell bodies (up to 40 µm) and aspiny dendrites and are known as large aspiny cells (Pasik et al., 1979; Bolam et al., 1984; Gerfen, 1992). In spite of the relative rarity of cholinergic neurons, which in rats, for instance, account for 5% of striatal neuronal population, they have a pronounced effect on striatum's function (Mavridis et al., 1994).

Neostriatum produces two opposite influences on its targets: (1) via the direct output pathway it inhibits its targets, thus reducing their inhibitory effects on other structures; and (2) via the indirect pathway it enhances these inhibitory influences by first inhibiting lateral globus pallidus, which in turn provides the GABAergic input to the subthalamic nucleus (STN). STN then delivers glutamatergic excitation to the targets.

Nigrostriatal dopamine and intrastriatal acetylcholine have opposite effects on these efferent striatal pathways (Albin et al., 1989; De Boer, 1992; Gerfen, 1992; Shapovalova,

---

[*] KSENIA SHAPOVALOVA • Department of Physiology, I.P. Pavlov Institute of Physiology of the Russian Academy of Sciences, St. Petersburg 199034, Russia.

1993, 2000; Wang and McGinty, 1996) by acting via different types of dopamine and muscarinic receptors located on striatal efferent neurons. The data lead to the important concept that dopamine-cholinergic interaction in the striatum provides the basis for the balanced influence of its efferents on the subcortical targets, necessary for the adequate performance and elaboration of voluntary motor acts.

The most common striatal receptors are the M1, M2, and M4 muscarinic acetylcholinol receptors, and the D1 and D2 dopamine receptors. The M1 and D2 receptors are co-localized and present on efferent neurons of the indirect pathway (Weiner et al., 1990; Harsing and Zigmond, 1998). The M4 and D1 receptors are co-localized and are present on efferent neurons of the direct pathway (De Lapp et al., 1996; Wang and Mc Ginty, 1996; Ince et al., 1997). Activation of M1 receptors leads to release of the neurotransmitter GABA (Olianas et al., 1996; Murthy and Maclouf, 1997), whereas activation of M4 receptors leads to a decrease in the level of adenylatecyclase and, as a result, to the inhibition of expression of GABA. (De Lapp et al., 1996; Takahashi and Goh, 1997; Murakami et al., 1998). The M2 muscarinic receptors are located on striatal cholinergic interneurons. Activation of M2 receptors leads to presynaptic inhibition of acetylcholine (ACh) release (Murakami et al., 1996). Activation of D1 receptors results in release of GABA, while activation of D2 receptors inhibits it (Stoof and Kebabian, 1981; Surmeir et al., 1996).

By acting on different striatal receptors, we can change the balance of striatal outputs or compensate for an imbalance, which can occur in both normal and pathological states. We will present here data obtained on dogs measuring the behavioral effects of intrastriatal microinjections of the nonselective muscarinic receptor agonist carbachol, nonselective antagonist scopolamine, selective M1 muscarinic receptor antagonist pirenzepine, and selective M2 muscarinic receptor agonist oxotremorine.

EXPERIMENTAL DETAILS

The experiments were carried out on ten mongrel dogs using an instrumental defensive reaction flexor posture maintenance model (Petropavlovsky, 1934). The dogs were fixed on a tensoplatform. The task of the animals was to avoid the electrical current applied to the left hind limb after presentation of the conditioned stimulus (metronome, 130 beats/min, M130) by lifting the limb to a certain height (8 cm) and maintaining this position during the whole conditioned stimulus duration (10 s). The current was turned on at the fifth second of the conditioned stimulus and stayed on along with the stimulus for 5 s. The study of the participation of the striatal cholinergic system in the analysis of the informative sensory (sound) stimuli was carried out using the same model, with a number of differentiating signals (i.e. not reinforced by the current. Different frequencies of metronome, i.e., M30, M60, and M90).

Activation and blockade of the neostriatal cholinergic system were performed using guided cannulae (external diameter 0.8 mm), implanted stereotaxically under sterile conditions and ketamine anaesthesia bilaterally into the dorsolateral part of the head of the caudate nucleus. The microinjectors, calibrated and filled with an agent, were introduced into the cannulae before the experiment. The effects of bilateral microinjections of a selective M2 muscarinic receptor agonist (oxotremorine, 0.001, 0.002 μg), a nonselective muscarinic receptor agonist (carbachol, 0.05, 0.1 and 0.4 μg), nonselective muscarinic receptor blocker (scopolamine, 0.5 μg) and selective M1 muscarinic receptor antagonist (pirenzepine, 0.003

μg) were compared. The agents were introduced in 1.5 μl bidistilled water over the course of 25-35 s. Solvent of the same volume was used as the control. Experiments involving injections were carried out no more frequently than once a week and were alternated with control experiments. Injections were given after presentation of 10 defensive conditioned stimuli and 6 differentional stimuli (background, 10 min). Results were analyzed every 10 min for 40-50 min after microinjection. Controls consisted of prolonged experiments (50 min) without microinjections. The coordinates of the implantation of the chemotrodes were subsequently verified histologically.

Before and after microinjections the following recordings were performed and analyzed: tensograms from four legs, amplitude and duration of the instrumental movement, latent periods of movement initiation and solution of the experimental task (the time necessary for entry into the "safety" zone), and the number of interstimulus and phasic leg raisings normally superimposed on the tonic type of instrumental response. Presentation of stimuli and collection of data were performed using a computer program. Analysis of data and presentation of the results of the experiments were performed using original computer programs for the statistical treatment and graphical presentation of the information (summarized and superimposed data of instrumental movement mechanograms, summarized data of tensograms, and the positions of the centers of gravity).

RESULTS OF EXPERIMENTS AND DISCUSSION

The elaborated instrumental response in dogs had two components: the tonic component (maintenance of the hind limb flexion of certain amplitude and duration), which plays a crucial role in the solution of the instrumental task, and the phasic component (fast jerks of the leg) reproducing a distinct response to the pain stimulation. The phasic interstimulus leg raisings were also present at the background level.

The data emphasized the importance of striatal dopaminergic/cholinergic interaction in balancing striatum's efferent influences on subcortical targets. The defensive situation is a powerful activating factor, particularly for excited dogs. In our experiments, the defensive situation lead to a marked enhancement of the phasic movement component, an increase in interstimulus raisings, postural disturbance, and general anxiety of the animals during the experiment (Figure 1,1). This could be due to an imbalance of striatal efferent outputs, possibly resulting from enhancement of the activating effect of the direct dopamine-induced pathway on the final targets. In this case, microinjections into striatum of acetylcholine receptor agonist carbachol could compensate for this imbalance, whereas injection of the antagonist scopolamine or dopamine could enhance it.

Indeed, in dogs with a completely established instrumental reflex, microinjections of dopamine unto the caudate nucleus head had been shown to impair performance of the learned reaction by increasing interstimulus leg raisings and phasic movements interposed with the tonic-type instrumental reaction. As a result, the animal ended up being in the "dangerous" zone more often and received pain reinforcement (Shapovalova, 1993). Similar data were obtained in experiments with scopolamine microinjections (Figure 1,3).

In contrast, microinjections of carbachol into the neostriatum of the dogs resulted in a soothing effect (Figure 1,2). Unilateral, predominantly contralateral carbachol microinjections into the neostriatum produced changes only on the day of the injection, whereas bilateral microinjections produced long-lasting changes in the formed pattern of the

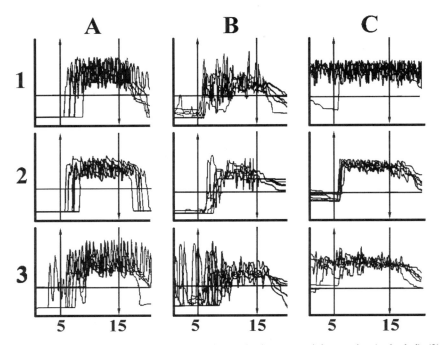

Figure 1. Effects of bilateral microinjections of nonselective muscarinic agonist (carbachol) (2) and nonselective muscarinic receptor blocker (scopolamine) (3) into neostriatum on the form of realization of instrumental reaction mechanogram related to the maintenance of flexor posture in three dogs (A,B,C). Each frame - superimposed data (n=10). Vertical arrows mark conditioned stimulus presentation time, the horizontal line marks safety zone. Abscissa - time, s.

instrumental behavior. The prolonged changes of bilateral carbachol microinjections (0.1 µg) included: an increase in a latent period of leg raising (Table 1), a decrease in the phasic component and an enhancement of the tonic component of the motor response, inhibition of the interstimulus leg raisings, streamlining of the posture, and a general soothing of the animal. Repetitive microinjections of carbachol into the neostriatum resulted in the instrumental reaction acquiring a kind of "smoothed," ramp-shaped form with increased amplitudes of postural adjustment components (Figure 2A,B).

The use of the bilateral carbachol microinjections over the course of two years lead to prolonged effects in one of the dogs, manifesting in akinesia and rigidity. During the entire experiment the dog stood with his leg slightly below the level of the "safety" zone, and usually received electrical current switched on at the fifth second of the conditioned stimulus. However, the escape reaction was complete for the remaining 5 s (Figure 3).

To eliminate the pathologic state, microinjections of a selective M1 muscarinic receptor blocker, pirenzepine (0.003 µg), were used. The pirenzepine microinjections were accompanied by an increase of the movement's phasic component, a marked rise of the movement amplitude, and the appearance of interstimulus leg raisings (Figure 3, lower panel). 30 min after the bilateral pirenzepine microinjection in this dog, 80% of the task trials were solved, whereas prior to the injection the task was never solved (0%). It is to be noted that during the pirenzepine microinjection, the dog was constantly correcting the leg position (Figure 3, lower panel). The effect of the pirenzepine injections was prolonged and observed for 10 days, with gradual decrease in effect.

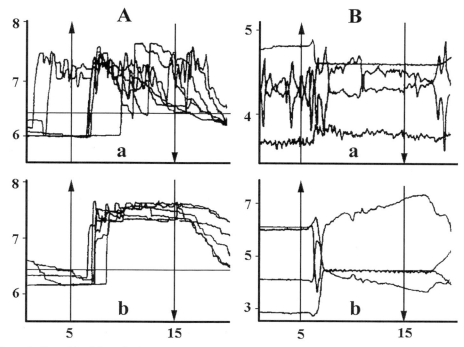

Figure 2. Repetitive bilateral microinjections of nonselective muscarinic agonist carbachol (0.1 µg) increased the tonic component of instrumental reaction mechanogram (A) and the amplitude of the postural adjustment component of tensogram (B) in two different dogs. a – background, b - 30 min after microinjection. Summarized data, n=10. Other labeling as in Figure 1.

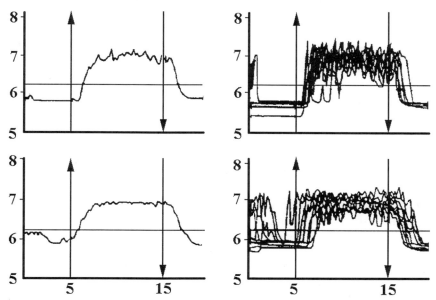

Figure 3. Bilateral microinjection of selective M1 muscarinic receptor antagonist pirenzepine (0.003 µg) into neostriatum sharply increases the phasic component and amplitude of instrumental movement mechanogram in akinetic dog. Top - before , and bottom - 30 min after microinjection. All other labeling as in Figure 1.

Bilateral pirenzepine microinjections in control dogs have also led to an increase in the phasic response component. This increase, however, was expressed as a decrease in the percentage of correct response realizations, as the animals, due to pronounced phasic raisings, entered the dangerous zone more frequently.

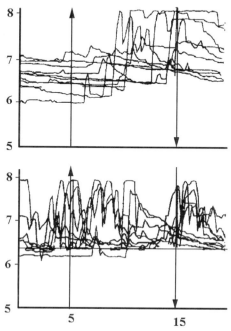

Figure 4. Effect of bilateral microinjection of oxotremorine (0.001 µg) into the neostriatum (bottom) as compared to the background (above). Summarized data (left), superposition (right). On each frame n=10. Ordinate – amplitude of mechanogram of instrumental reaction, cm; abscissa - time, s.

Bilateral microinjections of the muscarinic receptor agonist oxotremorine into neostriatum of 5 dogs were accompanied by an enhancement of the tonic movement component, and, apart from this, an increase in interstimulus leg raisings (Figure 4, lower panel). The latent period of instrumental reactions showed almost no change (Table 1), and changes of postural adjustment were observed very rarely. The tonic component changes were much less pronounced than after the carbachol microinjections. It can be suggested that carbachol's more pronounced effect is due to its action on both the M1 and M4 muscarinic receptors of neostriatum, resulting in activation of the indirect pathway via the M1 receptors and inhibition of the direct pathway via the M4 receptors. The function of oxotremorine in the neostriatum seems to be complex and is not yet fully understood. It was long considered to be a nonselective muscarinic receptor agonist. Recently, however, it was suggested that oxotremorine belongs to a group of selective M2 muscarinic receptor agonists. Its presynaptic effect on the M2 receptors has been identified as an inhibition of acetylcholine release (Murakami et al., 1996; Takahashi and Goh, 1997). According to our data, however, the oxotremorine effect seems to be more complex. On one hand, our results might reveal a possible activation of the M1 muscarinic receptors, and, as a result, enhancement of the tonic component of the instrumental response. But on the other hand, the data might reveal a possible activation of the M2 muscarinic receptors, which leads to presynaptic inhibition of acetylcholine release and, as a result, an increase in locomotor activity. This conclusion is confirmed by results of Wei et al. (1992) on expression of genes of the M1 and M2 muscarinic receptors.

Table 1. Effects of microinjections of cholinergic drugs on the time (s) of the leg maintenance in safety zone (T), and on the latent period of the leg lifting (LP) in response to defensive (M130) and differentiating (M30, M60, and M90) signals

Dogs	Signals	Background		Carbachol		Oxotremorine		Scopolamine	
		T	LP	T	LP	T	LP	T	LP
PIRAT	M130	11.0 + 0.5	0.8 + 0.5	9.1 + 0.5	0.9 + 0.4	10.9.+ 0.6	1.7 + 0.4*		
	M30	8.4 + 0.8	1.4 + 0.2	0.0 + 0.0*	10.0 + 0.0*	5.5 + 0.9*	3.6 + 2.0		
	M130	10.4 + 0.3	0.8 + 0.2	10.2 + 0.5	1.0 + 0.2	10.4 + 0.1	0.9 + 0.3		
	M30	10.1 + 0.4	0.4 + 0.1	0.6 + 0.1*	10.0 + 0.0*	7.8 + 0.6*	1.5 + 0.6		
RIJIK	M130	10.2 + 0.2	0.5 + 0.1	10.7 + 0.2	0.5 + 0.1	10.6 + 0.4	0.4 + 0.1		
	M90	4.3 + 1.8	0.9 + 0.1	1.1 + 0.1*	1.2 + 0.7	1.0 + 0.5*	4.7 + 3.8		
	M130	11.7+ 0.6	0.1 + 0.4	11.2 + 0.5	0.7 + 0.3			10.8 + 0.4	0.2+ 0.1
	M60	4.3 + 0.3	1.4 + 0.8	2.1 + 0.7*	5.2 + 2.5*			7.8 + 0.9*	0.8 + 0.4
DIMOK	M130	10.9 + 0.4	0.2 + 0.1	9.9 + 0.5	1.1 + 0.2*	9.5 + 0.4	0.5 + 0.2	10.5 + 0.2	0.0 + 0.0*
	M30	9.9 + 0.7	0.2 + 0.1	4.5 + 1.1*	2.3 + 1.4*	4.9 + 3.4	0.1 + 0.1	9.2 + 0.6	0.4 + 0.2
	M130	12.1 + 0.7	0.5 + 0.1	9.8 + 0.3	1.2 + 0.2*				
	M30	10.6 + 0.8	0.8 + 0.2	2.4 + 1.0*	5.9 + 1.9*				
CHERNI	M130	10.5 + 0.2	0.6 + 0.4	10.1 + 0.3	1.2 + 0.3*	10.2 + 0.8	0.5 + 0.4		
	M30	6.5 + 1.3	4.2 + 1.8	3.6 + 0.8*	5.7 + 4.4	2.5 + 1.6*	7.1 + 3.0		
	M130	10.9 + 0.2	0.4 + 0.4			10.3 + 0.4	1.2 + 0.3*		
	M30	6.5 + 1.1	0.2 + 1.0			3.1 + 0.9*	4.1 + 1.9*		
FOKS	M130	11.6 + 0.2	0.4 + 0.5	9.7 + 0.4*	0.5 + 0.2	10.4 + 0.2	0.6 + 0.3		
	M30	10.2 + 0.4	0.1 + 0.0	5.4 + 1.6*	0.4 + 0.1	9.0 + 0.8	0.1 + 0.2		

*   significant from background

The main effect of the bilateral carbachol and oxotremorine microinjections on movement was an enhancement of the tonic component of the instrumental response. Carbachol microinjections caused an inhibition of the phasic component of the instrumental response, an increase in amplitude of postural adjustment components (mainly of upper legs), as well as a streamlining of the posture adjustment and, judging from the position of the center of gravity, "steadier" posture as well. It was shown earlier that microinjections of dopamine and the acetylcholine receptor blocker scopolamine into neostriatum resulted in opposite effects (Shapovalova, 1993, 2000). These data indicate that the tonic and phasic components of learned movements are controlled by different neurotransmitter systems of the neostriatum, which are involved in the functioning of different striatal efferent outputs. Among the main striatal targets, the chief regulator of the hind limb muscle tone is the substantia nigra pars reticularis (Ellenbroek, 1988). Its neurons have an inhibitory (GABAergic) influence on superior colliculi, thalamic ventromedial nucleus, and pedunculo-pontine nucleus (PPN) (Garcia-Rill, 1986). The PPN receives direct (GABAergic) inputs from the globus pallidum, substantia nigra pars reticularis, and entopeduncular nucleus (Grofova and Spann, 1989) and excitatory (glutamatergic) inputs from the cerebral cortex and subthalamic nucleus (Garcia-Rill, 1988). Organization and functional significance of the two glutamatergic inputs to the PPN seem to be different. This is confirmed by the conclusion that the control of muscle tone and the control of locomotion are realized via different types of PPN glutamate receptors (Lai and Siegel, 1991). It may be suggested that the locomotor activity evoked by activation of the cortical glutamatergic input to the PPN can be inhibited, up to complete arrest, during activation of the neostriatal cholinergic system (Shapovalova, 1993, 2000). This is accompanied by inhibition of the direct (M4 receptors) and activation of the indirect (M1 receptors) outputs of the neostriatum, the latter activation being mediated by the subthalamic nucleus. This leads, apart from a restriction of locomotion (via an increase of the inhibitory GABAergic influences from the medial part of the globus pallidus and substantia nigra pars reticularis to the PPN), to an increase of the muscle tone due to direct glutamatergic influences from subthalamic nucleus to the PPN.

The microinjections of the muscarinic receptor agonist carbachol lead to significant improvement of the sound signal differentiation in the instrumental defensive reaction system (Table 1), whereas an injection of oxotremorine lead to a much smaller effect (Table 1). In contrast, administration of the cholinoreceptor blocker scopolamine produced a complete disinhibition of differentations, even in well trained animals (Shapovalova, 2000). The analysis of the time of leg maintenance in the safety zone (T) and the latent period of the leg lifting (LP) during the differentiated signal action showed that after microinjections of cholinergic agonists into neostriatum these parameters changed more frequently than during defensive signal action (Table 1). These data allowed us to conclude that the cholinergic system of neostriatum plays an important role in the process of attention (Shapovalova, 2000).

CONCLUSIONS

Based on the results presented here and literature data, we can make the following conclusion: The balance of direct and indirect striatal outputs is determined by interaction of different types of muscarinic and dopamine receptors located on neostriatal efferent neurons. The degree of involvement of these striatal mechanisms depends on the level of motivation, overall environmental stimulation, and the emotional and hormonal state of the animals. By microinjections into the neostriatum of choline and dopamine receptor agonists and blockers, we can shift this balance and thus change the motor response. Such microinjections may be of particular significance for correction of striatal pathology.

ACKNOWLEDGMENTS: This work was supported by the Russian Fund of Fundamental Investigations (grant No. 99-04-49683).

REFERENCES

Albin, R.L., Young, A.B., and Penney, J.B., 1989, The functional anatomy of basal ganglia disorders, *Trends Neurosci.* 12:366.
Bolam, J.P., Wainer, B.N., and Smith, A., 1984, Characterization of cholinergic neurons of the rat neostriatum: a combination of choline acetyltransferase immunocytochemistry, Golgi impregnation and electron microscopy, *Neurosci.* 12:711.
Contant, C., Umbriaco, D., Garcia, S., Watkins, K., and Descaries, L., 1996, Ultrastructural characterization of the acetylcholine innervation in adult rat neostriatum, *Neuroscience,* 13:937.

De Boer, P., 1992, Dopamine-acetylcholine interactions in the rat striatum, *Doctoral dissertation thesis*, Netherlands, pp. 189.

De Lapp, N.W., Eckols, K., and Shanon, H.E., 1996, Muscarinic agonist inhibition of rat striatal adenylate cyclase is enhanced by dopamine stimulation, *Life Sci.* 59:565.

Ellenbroek, B., 1988, Animal model of schizophrenia and neuroleptic drug action, *Doctoral dissertation thesis*, Netherlands, pp. 387.

Garcia-Rill, E., 1986, The basal ganglia and locomotor regions, *Brain Res. Rev.* 11:47.

Gerfen, Ch., 1992, The neostriatal mosaic organization in the basal ganglia, *Annu. Rev. Neurosci.* 15:285.

Grofova, I., and Spann, B., 1989, Involvement of the nucleus tegmenti pedunculo pontinus in descending pathways of the basal ganglia in the rat, in: *The Basal Ganglia III*, G. Bernardi, M.B. Carpenter, G. Di Chiara, M. Morelli, and P. Stanzione, eds., Plenum Press, New York, p. 79.

Harsing, L.G., and Zigmond, M.J., 1998, Postsynaptic integration of cholinergic and dopaminergic signals on medium-size GABAergic projection neurons in the neostriatum, *Brain Res. Bull.* 45:607.

Ince, E., Ciliax, E., and Levey, A., 1997, Differential expression of D1 and D2 dopamine and m4 muscarinic acetylcholine receptors protein in identified striatonigral neurons, *Synapse,* 27:357.

Kita, H., and Kitai, S.T., 1988, Glutamate decarboxylase immunoreactive neurons in rat neostriatum. Their morphological types and population, *Brain Res.* 447:346.

Lai, Y., and Siegel, J.M., 1991, Pontomedullary glutamate receptors mediating locomotion and muscle tone supression, *J. Neurosci.* 11:29.

Mavridis, M., Kayadijanian, N., and Besson, M., 1994, Cholinergic modulation of GABAergic efferent striatal neurones, in: *The Basal Ganglia IV, New Ideas and Data on Structure and Function*, G. Percheron, J.S. McKenzie, and J. Féger, Plenum Press, New York.

Murakami, Y., Matsumoto, K., Ohta, H., and Watanabe, H., 1996, Effects of oxotremorine and pilocarpine on striatal acetylcholine release as studied by brain dialysis in anesthtized rats, *Gen. Pharmacol.* 27:833.

Murthy, K.S., and Maclouf, G.M., 1997, Differential coupling muscarinic m2 and m3 receptors to adenylyl cyclase V/VI in smooth muscle. Concurrent M2-mediated inhibition via Galphai 3 and m-3 mediated stimulation via Gbetagammaq, *J. Biol. Chem.* 272:21317.

Olianas, M.C., Adem, A., Karlsson, E., and Onali, P., 1996, Rat striatal muscarinic receptors coupled to the inhibition of adenylyl cyclase activity potent block by the selective m4 ligand muscarinic toxin (MT3), *Br. J. Pharmacol.* 118:283.

Pasik, P., Pasik, T., and DiFiglia, M., 1979, The internal organization of neostriatum of mammals, in: *The Neostriatum.* I. Divac and G. Oberg, eds., Pergamon Press, Oxford.

Petropavlovsky, V.P., 1934, To the method of conditioned movement reflexes, *Sechenov Physiol. J.* 17:217 (in Russian).

Shapovalova, K.B., 1993, Possible mechanisms of participation of the neostriatum in regulation of voluntary movement, in: *Soviet Scientific Reviews,* T.M. Turpaev, ed., Harwood Academic Publishers, U.S.A.

Shapovalova, K.B., 2000, The striatal cholinergic system and instrumental behaviour, in: *Complex Brain Functions: Conceptual Advances in Russian Neuroscience,* R. Millar et al., eds., Harwood Press, London.

Stoof, J.C., and Kebabian, J.W., 1981, Opposite roles for the D1 and D2 dopamine receptors in efflux of cAMP from rat neostriatum, *Nature,* 294:366.

Surmeier, D.J., Song, W.-J., and Yan, Z., 1996, Coordinated expression of dopamine receptors in neostriatal medium spiny neurons, *Neurosci.* 16:6579.

Takahashi, T.K., and Goh, C.S., 1997, Presynaptic muscarinic cholinergic receptors in the dorsal hippocampus regulate behavioral inhibition of preweanling rats, *Brain Res.* 731:230.

Wang, L.Q., and McGinty, J.F., 1996, Muscarinic receptors regulate striatal neuropeptide gene expression in normal and amphetamine-treated rats, *Neurosci.* 75:43.

Wei, H., Roeske, W., Lai, J., et al., 1992, Pharmacological characterization of a novel muscarinic partial agonist YM796, in transfected cells expressing the m1 or m2 muscarinic receptor gene, *Life Sci.* 50:355.

Weiner, D.M., Levey, A.I., and Braun, M.R., 1990, Expression of muscarinic acetylcholine and dopamine receptor mRNAs in rat basal ganglia, *Proc. Natl. Acad. Sci. USA.* 87:7050.

PART IV. MOVEMENT AND SLEEP DISORDERS AS RELATED TO BASAL GANGLIA-THALAMIC CIRCUITS

FUNCTIONAL ORGANIZATION OF BRAINSTEM-BASAL GANGLIA INTERACTIONS AS VIEWED FROM THE PEDUNCULOPONTINE REGION

GLENDA L. KEATING AND DAVID B. RYE*

INTRODUCTION

Basal ganglia pathways to the brainstem pedunculopontine (PPN) region have been implicated in the pathophysiology of abnormal waking and nocturnal movement. This region may influence movement either via its output to basal ganglia nuclei, or by relay of basal ganglia influences to medullary and/or spinal motor systems. Many details concerning the normal anatomy and physiology of these pathways are yet to be elucidated. Degeneration of these pathways occurs in Parkinson's disease (PD) (Hirsch et al., 1987; Zweig et al., 1987; Jellinger, 1988; Gai et al., 1991), progressive supranuclear palsy (Zweig et al., 1987), and torsion dystonia (Zweig et al., 1988), yet it is not clear what significance this has to the symptomatology of these disorders.

Functional models of the basal ganglia describe a series of parallel segregated, cortical-subcortical re-entrant pathways centered upon the striatum and thalamocortical relation-ships (e.g., limbic, associative, oculomotor, and sensorimotor loops) (Alexander et al., 1986). The emphasis of these models on basal ganglia-thalamocortical circuitry largely ignores pathways between the basal ganglia and the PPN region (MEA/PPN, Figure 1). The PPN region, located at the junction of the midbrain and pons, is densely innervated by the main ouput nuclei of the basal ganglia, i.e., the internal pallidal segment (GPi) and substantia nigra reticulata (SNr), via collaterals of thalamic axons in monkeys (Harnois and Filion, 1982; Parent, 1990; Hazrati and Parent, 1991; Crutcher et al., 1994; Rye et al., 1996) and humans (Rye et al.,1995b; 1996). The region modulates thalamocortical arousal as well as REM-sleep (Jones and Cuello, 1989; Steriade and McCarley, 1990; Jones, 1991; Steriade, 1992). Stimulation and lesion studies in subprimates also implicate the PPN region in modulating locomotion and spontaneous motor activtity (Beresovskii and Bayev, 1988; Brudzynski et al., 1988; Milner and Mogenson, 1988; Bringmann and Klingberg, 1989; Mori et al., 1989; Garcia-Rill et al., 1990; Lai and Siegel, 1990), muscle tone and posture (Mogenson and Wu, 1988; Kelland and Asdourian, 1989; Lai and Siegel, 1990) and orofacial

* GLENDA L. KEATING AND DAVID B. RYE • Department of Neurology, Emory University School of Medicine, Atlanta, GA 30322.

stereotypy (Bachus and Gale, 1986; Gunne et al., 1988; Spooren et al., 1989; Inglis et al., 1994a; Allen and Winn, 1995).

Evidence that the MEA/PPN modulates normal and pathological movement also comes from primate models of PD, where enhanced utilization of 2-deoxyglucose in the terminal fields of GABAergic pallidotegmental axons suggests that excessive inhibition of the PPN region might contribute to hypokinesia/akinesia (Mitchell, et al., 1989). Single unit recordings in the PPN of the macaque monkey reveal firing rate changes preceding and coincident with voluntary arm movements in both limbs (Matsumura et al., 1997). Further study by the same investigators demonstrated that kainic acid lesions centered upon the PPN region produce a flexed posture and mild-to-moderate levels of hypokinesia in upper and lower limbs contralateral to the lesion (Kojima et al., 1997). An ill-defined multisynaptic pathway from the basal ganglia to pontine and medullary reticulospinal neurons, by way of the PPN region, modulates the acoustic startle reflex (Koch et al., 1993; Swerdlow and Geyer, 1993; Lingenhohl and Friauf, 1994). It has also been hypothesized that pathophysiological changes in this multisynaptic pathway underly the abnormal brainstem and spinal cord mediated reflexes described in PD (Penders and Delwaide, 1971; Kimura, 1973; Delwaide et al., 1991; Delwaide et al., 1993; Nakashima et al., 1993). These studies are difficult to interpret because the heterogeneity of the PPN region precludes identification of the precise neuronal populations or pathways responsible for an observed behavior. Indeed, many of the basic features of the anatomy, pharmacology, and synaptic organization of the PPN region, particularly with respect to basal ganglia output, have yet to be elucidated.

ANATOMY/PHYSIOLOGY OF BASAL GANGLIA OUPUT TO THE PPN REGION

We have previously described in detail the cytoarchitecture and chemoarchitecture of the PPN region in the albino rat (Rye et al., 1987). Most conspicuous are cholinergic neurons that extend from the caudal pole of the substantia nigra (SN) to a rostral pontine level, densely clustered lateral to the ascending limb of the superior cerebellar peduncle. Following the convention of Olzsewski and Baxter (Olszewski and Baxter, 1954; Olszewski and Baxter, 1982) in the human, other investigators include both cholinergic and non-cholinergic neurons within the PPN, and recognize two divisions: a more diffuse pars dissipata (PPN-d) located rostrally and medially, and a more cell dense pars compacta (PPN-pc) situated dorsolaterally in the caudal half of the nucleus. Defined in this manner, cholinergic neurons account for only 50% of all neurons contained within the PPN (Mesulam et al., 1983; Mesulam et al., 1984; Rye et al., 1987) and exhibit divergent projections to widespread telencephalic, diencephalic, and rhombencephalic structures (Wainer and Mesulam, 1990). These anatomical features, together with physiological data, are compatible with the view that this cholinergic subpopulation, like the brainstem dorsal raphe and locus coeruleus, is involved in modulating thalamocortical processing and thereby behavioral state (Steriade et al., 1990a; Steriade and McCarley, 1990).

In contrast, one subset of non-cholinergic neurons contained within the PPN-d displays a restricted pattern of reciprocal connectivity with the basal ganglia and, as such, was designated the midbrain extrapyramidal area (MEA) (Rye et al., 1987; Lee et al., 1988; Malin et al., 1993). Because these two cell populations are to a great degree spatially, connectionally, and neurochemically distinct, we and others (Carpenter and Jayaraman, 1990; Wirtshafter, 1994; Bevan and Bolam, 1995) prefer the MEA/PPN terminology. [The Paxinos and Watson rat atlas (Paxinos and Watson, 1986) delineates a "subpeduncular tegmental nucleus" that conforms in all respects to the MEA.] The MEA/PPN convention emphasizes that the cholinergic cell population is *not* the focus of extrapyramidal connectivity in subprimates. Our recent findings in monkeys (Crutcher et al., 1994; Rye et

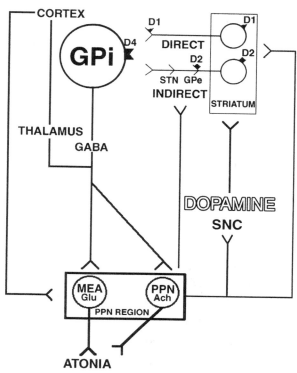

Figure 1. Schematic diagram illustrating key features of basal ganglia output from the internal pallidum (GPi) to the pedunculopontine (PPN) region (viz., MEA/PPN). The light microscopic pattern of GABA-ergic pallidotegmental axons suggests they terminate predominantly on glutamatergic (Glu) MEA neurons and to a lesser extent on cholinergic (ACh) PPN neurons. Indirect striatal and dopaminergic influences on the PPN region can be modulated by striatal-entopeduncular, striatopallidal, and entopeduncular (GPi) neurons, each of which expresses a unique dopamine receptor subtype (e.g., D1, D2 and D4, respectively). Putative projections from dopamine responsive neurons in the MEA/PPN to REM-sleep atonia centers in the medulla and/or re-innervation of the "direct" and "indirect" striatal pathways both represent means for effecting movement. Involvement of this multisynaptic pathway in the control of normal and pathological waking and nocturnal movement demands further attention.

al., 1996) and humans (Rye et al., 1995b; 1996), independently confirmed, further demonstrate this dichotomy (Smith and Shink, 1995; Shink et al., 1997).

Electron microscopic (Spann and Grofova, 1992; Grofova and Zhou, 1993; Grofova and Zhou, 1998) and electrophysiological (Noda and Oka, 1986; Kang and Kitai, 1990; Granata and Kitai, 1991) studies in rats also confirm that efferents from the primary output nuclei of the basal ganglia primarily target the MEA with a smaller contingent synapsing with the PPN. Nigrotegmental axons in the rat, for example, are most concentrated in the PPT-d, where cholinergic cells constitute a minority (~25%) of all cell types (Spann and Grofova, 1992). Within the PPT-pc, 85% of these axons make synaptic contact with non-cholinergic neural elements (Grofova and Zhou, 1993; Grofova and Zhou, 1998). A distinct subpopulation of non-cholinergic neurons receive frequent axosomatic synapses (Spann and Grofova, 1992; Steininger et al., 1997) and may correspond to the neurons contacted in a perisomatic fashion by pallidotegmental axons. Morphological and neurochemical characterizations, following intracellular recordings of midbrain tegmental neurons, also suggest that the primary output of the basal ganglia is concentrated on non-cholinergic neurons (Noda and Oka, 1986; Kang and Kitai, 1990; Granata and Kitai, 1991). Three electrophysiologically distinct neuronal populations hyperpolarized via the nigrotegmental pathway are identifiable, but only a portion of one class, constituting approximately 30% of the total recorded population, appear to be cholinergic (Kang and Kitai, 1990). Restriction of c-fos expression to non-cholinergic MEA neurons, on the intact side of unilaterally dopamine-depleted rats given an amphetamine challenge, is further evidence that the MEA represents the primary target of movement-related basal ganglia output (Wirtshafter, 1994).

In summary, the main descending output pathways of the basal ganglia (i.e., the pallidotegmental and nigrotegmental tracts) take origin from the same neurons contributing to innervation of motor thalamic nuclei (e.g. the VA/VL and centromedian (CM) thalamic nuclei) (Harnois and Filion, 1982; Parent and DeBellefeuille, 1982; Parent et al., 1988) and terminate in a convergent fashion primarily on non-cholinergic neurons in the dorsolateral mesopontine tegmentum (i.e., the MEA). Basal ganglia output is also likely to contact cholinergic PPN neurons, albeit on their distal dendrites. It is clear that basal ganglia output to these two neuronal populations will manifest behaviorally in unique ways based on differences in their efferent connections and physiological and receptor characteristics, some of which are summarized in Table 1. Despite these details, the precise projection patterns of neurons so contacted by basal ganglia output remains unknown.

The application of modern neuroanatomical techniques in several species is also beginning to differentiate further neural subpopulations in the mesopontine tegmentum. Non-cholinergic and dopaminergic neurons in the retrorubral field (RRF), for example, exhibit specific afferent (and efferent) connectivities with limbic/associative basal ganglia subcircuits (Rye, 1997). These findings further challenge the conventional wisdom that the cholinergic "PPN" is *the* monolithic brainstem locus for convergence of a panoply of limbic, visceral, and sensorimotor information. While this may yet be determined to be the case for *individual* cholinergic PPN neurons, our findings begin to paint a picture of functionally segregated circuits within the upper brainstem 'reticular core'. This concept of segregation is complemented by observations that motor behaviors resulting from stimuli applied near the "PPN" can be opposite in character depending upon the stimulus parameters or the site of stimulus application (see below). Further elucidation of the detailed synaptic organization of basal ganglia output confined to the PPN region would contribute to a better understanding of the structural and pharmacological substrates underlying many basal ganglia-mediated behaviors.

WAKING MOTOR BEHAVIORS ASSOCIATED WITH BASAL GANGLIA/PPN CIRCUITS

Electrical stimulation of the dorsal midbrain near the PPN region elicits treadmill locomotion in postmamillary-precollicular transected subprimates (i.e., the "mesencephalic locomotor region", MLR) (Garcia-Rill, 1983; Garcia-Rill et al., 1983al., 1983b). The precise anatomy of the MLR is difficult to ascertain because electrical current or pharmacologic agents are likely to spread and affect diverse neuronal populations with unique projection patterns, and because of the nonspecific nature of locomotion induction (Beresovskii and Bayev, 1988). Nevertheless, electrophysiological (Garcia-Rill, 1983; Garcia-Rill et al., 1983b) and anatomical (Garcia-Rill et al., 1983a) evidence indicate that the MLR receives inputs from the basal ganglia. Thus, locomotion is potentially modulated by basal ganglia influences that are relayed via the PPN region to pontine and medullary reticulospinal neurons or directly to the spinal cord (Milner and Mogenson, 1988; Mogenson and Wu, 1988). It is unclear how induced locomotor activity from the midbrain tegmentum specifically relates to pathways normally active in generating locomotion, to other aspects of

Table 1. Distinguishing characteristics of MEA and PPN

	MEA	PPN	References
Neurotransmitter content/ Neurochemical characteristics	Non-cholinergic; glutamate & GABA (?) parvalbumin containing	Acetylcholine; NADPH-diaphorase; CRF; Substance P	(Rye et al., 1987; Vincent and Satoh, 1984; Vincent et al., 1983a; Vincent et al., 1983b)
Synaptic contacts from GPi	numerous & perisomatic	sparse and dendritic	(Rye et al., 1987; Rye et al., 1996; Smith and Shink, 1995)
Response to GPi/SNr	IPSPs	IPSPs	(Clarke et al., 1987; Gonya-Magee and Anderson, 1983; Kang and Kitai, 1990; Scarnati et al., 1987)
Response to systemic dopamine	Immediate early gene activation	Absence of IEG activation	(Wirtshafter, 1994; Daley et al., 1997)
Efferents	Restricted; Basal ganglia, Inferior olive, PPRF Dorsal gigantocellular field, Nucleus Reticularis tegmenti pontis (of Bechterew) (NRTP)	Widespread; Thalamus, Ventral and para-giganto-cellular field (cardiac control), Ventrolateral medulla (esp. Bötzinger complex) Raphe magnus, PPRF, NRTP	(Hallanger et al., 1987; Rye et al., 1988; Rye et al., 1987; Wainer and Mesulam, 1990; Keating et al., 1999)
Physiological characteristics	Tonically active in EEG desynchronized states and many displaying low threshold Ca^{2+} spikes & bursting	Tonically active in EEG desynchronized states with few displaying low threshold Ca^{2+} spikes & bursting	(Luebke et al., 1992 ; Steriade, 1992; Steriade et al., 1990a; Steriade et al., 1990b)
Response to ACh	None or depolarizing	Hyperpolarizing	(Leonard and Llinas, 1994; Muhlethaler et al., 1990)
Muscarinic receptors	(?) m3	m2 predominates	(Rye et al., 1995a; Vilaro et al., 1994)
Response to 5-HT	None or slow hyperpolarizing	Hyperpolarizing	(Leonard and Llinas, 1994; Luebke et al., 1992; Muhlethaler et al., 1990)
5-HT 2 receptors	Absent	Present	(Morilak and Ciaranello, 1993; Morilak et al., 1993)

motor control, and specifically, to the functional role of basal ganglia/PPN pathways.

The second line of evidence implicating the PPN region in waking movement is based on observations that kainic acid lesions including, but not restricted to, the PPN region in the cat produce incomplete hindlimb extension, bradykinesia, and dyscoordination (Webster and Jones, 1988). Electrical stimulation of the PPN region decreases "postural support " in the cat (Mori et al., 1989) and increases spontaneous motor activity in the rat (Bringmann and Klingberg, 1989). Nearly opposite effects on motor behavior can be elicited either by changing the electrophysiologic or pharmacologic stimulus parameters (Lai and Siegel, 1990) or by making only a small change in the site of stimulus application (Brudzynski et al., 1988; Milner and Mogenson, 1988). Increased locomotor activity follows injections of NMDA, kainic acid, or GABA antagonists targeting the cholinergic PPN, for example, while no motor effects occur with applications of glutamate, glycine, or strychnine at the same locus (Milner and Mogenson, 1988). Moreover, locomotion is attenuated following injections of the cholinergic agonist carbachol into the PPN region *(cholinergic PPN perikarya?)*, while it is accentuated with injections immediately surrounding the *PPN (MEA?)* (Brudzynski et al., 1988; Milner and Mogenson, 1988). The evidence strongly supports the notion that distinct motor circuits can be distinguished in the PPN region on the basis of neurotransmitter receptor and spatial specificities.

The third line of research has investigated how well-defined motor behaviors mediated by dopamine and specific basal ganglia subcircuits are effected by cell-specific lesions targeted at the cholinergic PPN neuronal subpopulation (Inglis et al., 1994a, 1994b; Allen and Winn, 1995). Lesions of a multisynaptic "limbic" basal ganglia subcircuit (i.e., nucleus accumbens-ventral pallidum-PPN region) does not effect spontaneous locomotion, but rather modulates reward-related motor responses, possibly via PPN efferents that reinnervate the dopaminergic substantia nigra (Blaha and Winn, 1993; Inglis et al., 1994a). A "sensorimotor" subcircuit connecting the lateral striatum with the PPN region via a synapse in the internal pallidum appears critical in preventing dopamine-induced orofacial dyskinesias (Allen and Winn, 1995). These interpretations are complicated by the fact that lesions were centered upon cholinergic PPN neurons, which as already discussed, are not the primary recipients of basal ganglia output. Nonetheless, these studies establish: (1) that the PPN region modulates movement driven in part by dopamine's actions in the basal ganglia; and (2) that the PPN region mediates different motor behaviors dependent upon its relationship with specific dopamine sensitive basal ganglia subcircuits (e.g., limbic versus sensorimotor). The MEA can be viewed as a reflection of basal ganglia "sensorimotor" circuitry at a brainstem level, not only because it is the principal target of output from the GPi's sensorimotor territory (see above), but also because: (1) cortical innervation of the MEA derives solely from the primary motor area (Hartmann-von Monakow et al., 1979; Rye et al., in preparation); and (2) MEA pathways that reinnervate the striatum preferentially target its sensorimotor division (Malin et al., 1993).

The fourth, compelling line of evidence that the MEA/PPN is critical in modulating normal and pathological movement comes from investigations of PD. In the primate model of PD, enhanced tonic and phasic activity in the sensorimotor division of the GPi is felt to play a primary role in the development of parkinsonian motor signs by virtue of increased inhibition of thalamocortical neurons (Albin et al., 1989; Crossman, 1989; DeLong, 1990). Enhanced utilization of 2-deoxyglucose in the terminal fields of GABAergic pallido-tegmental axons also suggests that excessive inhibition of the PPN region contributes significantly to hypokinesia/akinesia (Mitchell, 1989). This hypothesis is born out by clinical experience that reversal of akinesia in PD is generally not effected by lesions of the pallidothalamic pathway or motor thalamus that leave the pallidotegmental pathway intact (Narabayashi, 1990; Iacono et al., 1994). We recently confirmed that degeneration of the pallidotegmental tract within the MEA accompanies successful relief of parkinsonian motor symptoms in a patient undergoing sensorimotor pallidotomy, consistent with the hypothesis that pallidal inhibition of the MEA contributes to the motor features of PD (Rye et al., 1996).

The entirety of the anatomical, physiological, and behavioral evidence implicate the PPN region in mediating motor activities influenced by dopamine and the basal ganglia. A comprehensive accounting of structure-function relationships is nonetheless lacking because knowlege of the anatomical/physiological relationships between the basal ganglia and the PPN region are imprecise. Our preliminary findings demonstrate immediate early gene (IEG) upregulation restricted unilaterally in a cell specific manner to MEA neurons following systemic administration of dopaminomimetic agents [A77636 (a selective D1 agonist); quinpirole (a selective D2 agonist); apomorphine, and amphetamine] to hemiparkinsonian monkeys (Daley et al., 1997). These responses presumably reflect indirect actions of dopamine at striatal D1 and D2 receptors since: (1) no dopamine terminals, receptors, or physiological responses have been described in the PPN region; and (2) destruction of nigrostriatal terminals in the contralateral hemisphere by intracarotid MPTP administration abolished the response. Inhibition of GABA-ergic basal ganglia output by direct (D1 sensitive) and indirect (D2 sensitive) striatal pathways (DeLong, 1990; Gerfen et al., 1990, 1991; Graybiel, 1990) is therefore likely to operationally activate (i.e., disinhibition) the MEA (see Figure 1). Dopaminergic innervation of the GPi (Lavoie et al., 1989) and pallidal D4 receptors (Mrzljak et al., 1996) (see Figure 1) are likely to mediate a competing disinhibition of pallidal output (Bergstrom and Walters, 1984) that would be reflected as inactivation of the MEA. It will be important to establish that IEG expression is mediated by specific sensorimotor subcircuits in the basal ganglia and does not reflect dopamine actions on alternate multisynaptic pathways to the MEA (e.g. pallido-thalamo-cortical-MEA) (see above). Such a parcellation of the PPN region will contribute to a better understanding of the structural and pharmacological bases underlying many basal ganglia-mediated behaviors and will be useful in guiding manipulations of the PPN region so that the validity of the observed effects on motor behavior is enhanced.

NOCTURNAL MOTOR CONTROL ASSOCIATED WITH BASAL GANGLIA/PPN CIRCUITS

The MEA/PPN region has also been implicated in modulating nocturnal movement (Rye and Bliwise, 1997). Glutamatergic and cholinergic neurons in the PPN region play a critical role in maintaining the atonia which accompanies rapid-eye-movement (REM) sleep (Chase and Morales, 1994). Neurons overlapping in distribution with the MEA display REM-sleep specific increases in neural discharge (Sakai, 1980), project to medullary regions essential in maintaining REM-atonia (Rye et al., 1988; Lai et al., 1993), and when lesioned release complex motor behaviors in REM-sleep [e.g., REM-sleep behavior disorder (RBD)] in animals (Sakai, 1980, 1985; Jones and Webster, 1988) and humans (Culebras and Moore, 1989; Shimizu et al., 1990; Mahowald and Schenck, 1994). Although efferents from the SNr (Nakamura et al., 1989) and GPi (von Krosigk et al., 1992) make synaptic contact with tegmental neurons, which in turn project to the ventromedial medulla, it remains unclear if MEA neurons, PPN neurons, or both participate in this multisynaptic route linking the basal ganglia with lower motor centers involved in modulating REM-atonia.

The relevance of the basal ganglia to physiological measures of nocturnal movement is becoming more clear. Neurons in the GPi (DeLong, 1969; Rye and Bliwise, 1997) and SNr (Datta et al., 1991; Steriade, 1992) discharge phasically at rates exceeding those seen in wakefulness (200Hz vs. 50Hz) in concert with phasic events of REM-sleep (e.g., eye movements and fine skeletomuscle bursts). Neural discharge in these main output nuclei of

Table 2. Relevance of basal ganglia and dopamine to nocturnal movement disorders

Nocturnal Disorder	Movements	Treatment	Aggravators	Pathophysiological Basis in BG	References
Periodic Leg Movements(PLMs)	PLMs in stage 2>>REM-sleep	L-dopa D2 agonists Opiates Benzodiazepines	pimozide (D2 antagonists) metaclopramide	Decreased affinity of striatal D2 receptors	(Montplaisir et al., 1986; Montplaisir and Godbout, 1994; Montplaisir et al., 1985; Staedt et al., 1993; Staedt et al., 1995)
REM-behavior disorder (RBD)	Phasic EMG activity in REM-sleep; PLMs	L-dopa D2 agonists	Neuroleptics Benzodiazepines	?Pre-parkinsonian	(Mahowald and Schenck, 1994; Schenck et al, 1986; Schenck et al., 1996; Schenck et al., 1987a; Schenck et al., 1987b; Schenck et al., 1987c; Schenck et al., 1992)
Narcolepsy	PLMs & RBD	L-dopa D2-like agonists	Neuroleptics	Increased D1 and D2 binding sites in GPi	(Aldrich et al., 1992; Baker et al., 1986; Schenck and Mahowald,1992)

the basal ganglia is otherwise slightly below waking discharge rates during quiet periods of REM-sleep in which atonia predominates. We hypothesize that much of this GABA-ergic basal ganglia output targets glutamatergic MEA neurons whose activation of ventromedial medullary zones promotes REM-atonia. Heightened phasic discharge of the GPi that occurs transiently or that is persistent in pathological states such as PD (Vitek et al., 1993), therefore, would be expected to excessively inhibit the MEA, thereby allowing for the expression of movement that overcomes REM-atonia (e.g., RBD). This hypothesis is borne out by clinical experience that excessive nocturnal movement in PD (Rye and Bliwise, 1997) can be reversed by removing excessive inhibition of the MEA by pallidotomy. Dopamine would also be expected to effect such a disinhibition of the MEA (see above), which is consistent with the clinical experience that pathological nocturnal movements (e.g., periodic leg movments of sleep, RBD and narcolepsy) are: (1) responsive to dopaminomimetics; (2) aggravated by dopamine antagonists; and (3) characterized by diminished dopaminergic "tone" in the basal ganglia (see Table 2). Although many aspects of pathological movement, including nocturnal movement, have been hypothesized to manifest through multisynaptic neural circuits that afford dopamine indirect access to the PPN region, a specific motor function has not yet been attributed to PPN neurons targeted by basal ganglia output pathways.

HEALTH RELATED SIGNIFICANCE OF BASAL GANGLIA/PPN CIRCUITS

Distinguishing the brainstem synaptic circuits through which dopamine and the basal ganglia modulate movement will significantly enhance understanding of the pathophysiology of hyper- and hypokinetic movement disorders (e.g., PD, PSP, dystonia and chorea). Existing functional models of basal ganglia circuitry have encouraged the development of more rational treatment strategies for disorders of movement. Incorporation of brainstem circuitry into these models would facilitate more accurate assessment and treatment for many aspects of pathological movement. Validation that the basal ganglia-brainstem neural circuits are critical in maintaining atonia in sleep will advance a plausible pathophysiological mechanism to account for the clinical observation that nocturnal movements interrupting sleep in many human diseases respond favorably to dopaminomimetics (see Table 2).

Thus, more rational and effective treatment strategies should result for pathological nocturnal movement.

ACKNOWLEDGMENTS: Supported by NIH grants NS36697, NS-35345 and an American Academy of Sleep Medicine post-doctoral fellowship to Dr. Keating.

REFERENCES

Albin, R., Young, A., and Penney, J., 1989, The functional anatomy of basal ganglia disorders, *Trends. Neurosci.* 12:366.

Aldrich, M., Hollingsworth, Z., and Penney, J., 1992, Dopamine-receptor autoradiography of human narcoleptic brain, *Neurology,* 42:410.

Alexander, G. E., DeLong, M.R., and Strick, P.L., 1986, Parallel organization of functionally segregated circuits linking basal ganglia and cortex, *Annu. Rev. Neurosci.* 9:357.

Allen, L., and Winn, P., 1995, Excitotoxic lesions of the pedunculopontine tegmental nucleus disinhibit orofacial behaviours stimulated by microinjections of d-amphetamine into rat ventrolateral caudate-putamen, *Exp. Brain Res.* 104:262.

Bachus, S., and Gale, K., 1986, Muscimol microinfused into the nigrotegmental target area blocks selected components of behavior elicited by amphetamine or cocaine, *Arch. Pharmacol.* 333:143.

Baker, T., Guilleminault, C., Nino-Murcia, G., and Dement, W., 1986, Comparative polysomnographic study of narcolepsy and idiopathic central nervous system hypersomnia, *Sleep,* 9:232.

Beresovskii, V. K., and Bayev, K.V., 1988, New locomotor regions of the brainstem revealed by means of electrical stimulation, *Neuroscience,* 3:863.

Bergstrom, D., and Walters, J., 1984, Dopamine attenuates the effects of GABA on single unit activity in the globus pallidus, *Brain Res.* 310:23.

Bevan, M., and Bolam, J., 1995, Cholinergic, GABAergic, and glutamate-enriched inputs from the mesopontine tegmentum to the subthalamic nucleus in the rat, *J. Neurosci.* 15:7105.

Blaha, C., and Winn, P., 1993, Modulation of dopamine efflux in the striatum following cholinergic stimulation of the substantia nigra in intact and pedunculopontine tegmental nucleus-lesioned rats. *J. Neurosci.* 13:1035.

Bringmann, A., and Klingberg, F., 1989, Electrical stimulation of the basal forebrain and the nucleus cuneiformis differently modulate behavioral activation of freely moving rat, *Biomed. Biochim. Acta.* 48:781.

Brudzynski, S., Wu, M., and Mogenson, G., 1988, Modulation of locomotor activity induced by injections of carbachol into the tegmental pedunculopontine nucleus and adjacent areas in the rat, *Brain Res.* 451:119.

Carpenter, M. B., and Jayaraman, A., 1990, Subthalamic nucleus of the monkey: Connections and immunocytochemical features of afferents, *J. Hirnforsch.* 31:653.

Chase, M., and Morales, F., 1994, The control of motoneurons during sleep, in: *Principles and Practice of Sleep Medicine*, M. Kryger, T. Roth, and W. Dement, eds., WB Saunders Company, Philadelphia.

Clarke, P., Hommer, D., Pert, A., and Skirboll, L., 1987, Innervation of substantia nigra neurons by cholinergic afferents from pedunculopontine nucleus in the rat: Neuroanatomical and electrophysiological evidence, *J. Neurosci.* 23:1011.

Crossman, A., 1989, Neural mechanisms in disorders of movement, *Comp. Biochem. Physiol.* 93A:141.

Crutcher, M., Turner, R., Perez, J., and Rye, D., 1994, Relationship of the primate pedunculopontine nucleus (PPN) to tegmental connections with the internal pallidum (GPi), *Soc. Neurosci. Abstr.* 20:334.

Culebras, A., and Moore, J., 1989, Magnetic resonance findings in REM sleep behavior disorder, *Neurology,* 39:1519.

Daley, J., Perez, J., Bakay, R., and Rye, D., 1997, Dopamine responsive mesopontine tegmental circuits in non-human primates, *Soc. Neurosci. Abstr.* 23:192.

Datta, S., Dossi, R.C., Pare, D., Oakson, G., and Steriade, M., 1991, Substantia nigra reticulata neurons during sleep - waking states: Relation with ponto-geniculo-occipital waves, *Brain Res.* 566:344.

DeLong, M., 1969, Activity of pallidal neurons in the monkey during movement and sleep, *The Physiologist (Abstr), 12:*207.

DeLong, M. R., 1990, Primate models of movement disorders of basal ganglia origin, *Trends Neurosci.* 13:281.

Delwaide, P., Pepin, J., and Noordhout, A.M., 1991, Short-latency autogenic inhibition in patients with parkinsonian ridigity, *Ann. Neurol.* 30:83.

Delwaide, P., Pepin, J., and Noordhout, A.M., 1993, The audiospinal reaction in parkinsonian patients reflects functional changes in reticular nuclei, *Ann. Neurol.* 33:63.

Gai, W., Halliday, G., Blumbergs, P., Geffen, L., and Blessing, W., 1991, Substance P-containing neurons in the mesopontine tegmentum are severely affected in Parkinson's Disease, *Brain,* 114:2253.

Garcia-Rill, E., 1983, Connections of the mesencephalic locomotor region (MLR). III. Intracellular recordings, *Brain Res. Bull.* 10:73.

Garcia-Rill, E., Kinjo, N., Atsuta, Y., Ishikawa, Y., Webber, M., and Skinner, R., 1990, Posterior midbrain induced locomotion, Brain Res. Bull. 24:499.

Garcia-Rill, E., Skinner, R., Gilmore, S., and Owings, R., 1983a, Connections of the mesencephalic locomotor region (MLR). II. Afferents and efferents, *Brain Res. Bull.* 10:63.

Garcia-Rill, E., Skinner, R.D., Jackson, M.B., and Smith, M.M., 1983b, Connections of the mesencephalic locomotor region (MLR) I. substantia nigra afferents, *Brain Res. Bull.* 10:57.

Gerfen, C. R., Engber, T.M., Mahan, L.C., Susel, Z., Chase, T.N., Monsma, F.J. Jr., and Sibley, D.R., 1990, D1 and D2 dopamine receptor-regulated gene expression of striatonigral and striatopallidal neurons, *Science,* 250:1429.

Gerfen, C. R., McGinty, J.F., and Young, W.S. III, 1991, Dopamine differentially regulates dynorphin, substance P, and enkephalin expression in striatal neurons: *In situ* hybridization histochemical analysis, *J. Neurosci.* 11:1016.

Gonya-Magee, T., and Anderson, M., 1983, An electrophysiological characterization of projections from the pedunculopontine area to entopeduncular nucleus and globus pallidus in the cat, *Exp. Brain Res.* 49:269.

Granata, A., and Kitai, S., 1991, Inhibitory substantia nigra inputs to the pedunculopontine neurons, *Exp. Brain Res.* 86:459.

Graybiel, A. M., 1990, Neurotransmitters and neuromodulators in the basal ganglia, *Trends Neurosci.* 13:244.

Grofova, I., and Zhou, M., 1993, Nigral innervation of cholinergic and non-cholinergic cells in the rat mesopontine tegmentum: A double label EM study, *Soc. Neurosci. Abstr.* 19:1433.

Grofova, I., and Zhou, M., 1998, Nigral innervation of cholinergic and glutamatergic cells in the rat mesopontine tegmentum: light and electron microscopic anterograde tracing and immunohistochemical studies, *J. Comp. Neurol.* 395:359.

Gunne, L.-M., Bachus, S. and Gale, K., 1988, Oral movements induced by interference with nigral GABA neurotransmission: Relationship to tardive dyskinesias, *Exp. Neurol.* 100:459.

Hallanger, A. E., Levey, A.I., Lee, H.J., Rye, D.B., and Wainer, B.H., 1987, The origins of cholinergic and other subcortical afferents to the thalamus in the rat, *J. Comp. Neurol.* 262:105.

Harnois, C., and Filion, M., 1982, Pallidofugal projections to thalamus and midbrain: A quantitative antidromic activation study in monkeys and cats, *Exp. Brain Res.* 47:277.

Hartmann-von Monakow, K., Akert, K., and Kunzle, H., 1979, Projections of precentral and premotor cortex to the red nucleus and other midbrain areas in macaca fasicularis, *Exp. Brain Res.* 34:91.

Hazrati, L.-N., and Parent, A., 1991, Contralateral pallidothalamic and pallidotegmental projections in primates: an anterograde and retrograde labeling study, *Brain Res.* 567:212.

Hirsch, E., Graybiel, A., Duyckaerts, C., and Jovoy-Agid, F., 1987, Neuronal loss in Parkinson's disease and in progressive supranucleur palsy, *Proc. Natl. Acad. Sci. USA,* 84:5976.

Iacono, R., Lonser, R., Mandybur, G., Morenski, J., Yamada, S., and Shima, F., 1994, Stereotactic pallidotomy results for Parkinson's exceed those of fetal graft, *The American Surgeon,* 60:777.

Inglis, W., Allen, L., Whitelaw, R., Latimer, M., Brace, H., and Winn, P., 1994a, An investigation into the role of the pedunculopontine tegmental nucleus in the mediation of locomotion and orofacial sterotypy induced by d-amphetamine and apomorphine in the rat, *Neuroscience,* 58:817.

Inglis, W., Dunbar, J., and Winn, P., 1994b, Outflow from the nucleus accumbens to the pedunculopontine tegmental nucleus: a dissociation between locomotor activity and the acquisition of responding for conditioned reinforcement stimulated by d-amphetamine, *Neuroscience,* 62:51.

Jellinger, K., 1988, The pedunculopontine nucleus in Parkinson's disease, progressive supranuclear palsy and Alzheimer's disease, *J. Neurol. Neurosurg. Psychiatr.* 51:540.

Jones, B., and Webster, H., 1988, Neurotoxic lesions of the dorsolateral pontomesencephalic tegmentum-cholinergic cell area in the cat. I. Effects upon the cholinergic innervation of the brain, *Brain Res.* 451:13.

Jones, B. E., 1991, Paradoxical sleep and its chemical/structural substrates in the brain, *Neuroscience,* 40:637.

Jones, B. E., and Cuello, A.C., 1989, Afferents to the basal forebrain cholinergic cell area from pontomesencephalic-catecholamine, serotonin, and acetylcholine-neurons, *Neuroscience,* 31:37.

Kang, Y., and Kitai, S., 1990, Electrophysiological properties of pedunculopontine neurons and their postsynaptic responses following stimulation of substantia nigra reticulata, *Brain Res.* 535:79-95.

Keating, G.L., and Rye, D.B., 1999, Pathways descending from the pedunculopontine region in the rat, *Soc. Neurosci. Abstr.* 25:1924.

Kelland, M., and Asdourian, D., 1989, Pedunculopontine tegmental nucleus-induced inhibition of muscle activity in the rat, *Behav. Brain Res.* 34:213-234.

Kimura, J., 1973, Disorder of interneurons in parkinsonism. The orbicularis oculi reflex to paired stimuli, *Brain,* 96:87.

Koch, M., Kungel, M., and Herbert, H., 1993, Cholinergic neurons in the pedunculopontine tegmental nucleus are involved in the mediation of prepulse inhibition of the acoustic startle response in the rat, *Exp. Brain Res.* 97:71.

Kojima, J., Yamaji, Y., Matsumura, M., Nambu, A., Inase, M., Tokuno, H., Takada, M., and Imai, H., 1997, Excitotoxic lesions of the pedunculopontine tegmental nucleus produce contralateral hemiparkinsonism in the monkey, *Neurosci. Lett.* 226:111.

Lai, Y., Clements, J., and Siegel, J., 1993, Glutamatergic and cholinergic projections to the pontine inhibitory area identified with horseradish peroxidase retrograde transport and immunohistochemistry, *J. Comp. Neurol.* 336:321.

Lai, Y., and Siegel, J., 1990, Muscle tone suppression and stepping produced by stimulation of midbrain and rostral pontine reticular formation, *J. Neurosci.* 10:2727.

Lavoie, B., Smith, Y., and Parent, A., 1989, Dopaminergic innervation of the basal ganglia in the squirrel monkey as revealed by tyrosine hydroxylase immunohistochemistry, *J. Comp. Neurol.* 289:36.

Lee, H. J., Rye, D.B., Hallanger, A.E., Levey, A.I., and Wainer, B.H., 1988, Cholinergic vs. noncholinergic efferents from the mesopontine tegmentum to the extrapyramidal motor system nuclei, *J. Comp. Neurol.* 275:469.

Leonard, C., and Llinas, R., 1994, Serotonergic and cholinergic inhibition of mesopontine cholinergic neurons controlling REM sleep: An *in vitro* electrophysiological study, *Neuroscience,* 59:309.

Lingenhohl, K., and Friauf, E., 1994, Giant neurons in the rat reticular formation: A sensorimotor interface in the elementary acoustic startle circuit, *J. Neurosci.* 14:1176.

Luebke, J., Greene, R., Semba, K., Kamondi, A., McCarley, R., and Reiner, P., 1992, Serotonin hyperpolarizes cholinergic low-threshold burst neurons in the rat laterodorsal tegmental nucleus *in vitro,* *Proc. Natl. Acad. Sci.* 89:743.

Mahowald, M., and Schenck, C., 1994, REM sleep behavior disorder, in: *Principles and Practices of Sleep Medicine,* M. Kryger, T. Roth, and W. Dement, eds., WB Saunders Company, Philadelphia.

Malin, A., Ciliax, B., and Rye, D., 1993, Organization of the mesopontine tegmental-striatal pathway in the rat, *Soc. Neurosci. Abstr.* 19:557.

Matsumura, M., Watanabe, K., and Ohye, C., 1997, Single-unit activity in the primate nucleus tegmenti pedunculopontinus related to voluntary arm movement, *Neurosci. Res.*28:155.

Mesulam, M.-M., Mufson, E.J., Levey, A.I., and Wainer, B.H., 1984, Atlas of cholinergic neurons in the forebrain and upper brainstem of the macaque based on monoclonal choline acetyltransferase immunohistochemistry and acetylcholinesterase histochemistry, *Neuroscience*, 12:669.

Mesulam, M.-M., Mufson, E. J., Wainer, B.H., and Levey, A.I., 1983, Central cholinergic pathways in the rat: An overview based on an alternative nomenclature (Ch1-Ch6), *Neuroscience*, 10:1185.

Milner, K., and Mogenson, G., 1988, Electrical and chemical activation of the mesencephalic and subthalamic locomotor regions in freely moving rats, *Brain Res.* 452:273.

Mitchell, I., Clarke, C.E., Boyce, S., Robertson, R.G., Peggs, D., Sambrook, M.A, and Crossman, A.R., 1989, Neural mechanisms underlying parkinsonian symptoms based upon regional uptake of 2-deoxyglucose in monkeys exposed to 1-methyl-4-phenyl-1,2,3,6-tetrahydropyridine, *Neuroscience*, 32:213.

Mogenson, G. J., and Wu, M., 1988, Differential effects on locomotor activity of injections of procaine into mediodorsal thalamus and pedunculopontine nucleus, *Brain Res. Bull.* 20:241.

Montplaisir, J., Bodbout, R., Poirier, G., and Bedard, M., 1986, Restless legs syndrome and periodic movements in sleep: Physiopathology and treament with L-Dopa, *Clin. Neuropharmacol.* 9:456.

Montplaisir, J., and Godbout, R., 1994, Restless legs syndrome and periodic movements during sleep, in: *Principles and Practice of Sleep Medicine*, M. Kryger, T. Roth, and W. Dement, eds., WB Saunders Company, Philadelphia.

Montplaisir, J., Godbout, R., Boghen, D., DeChamplain, J., Young, S., and Lapierre, G., 1985, Familial restless legs with periodic movements in sleep: electrophysiologic, biochemical, and pharmacologic study, *Neuro.* 35:130.

Mori, S., Sakamoto, T., Ohta, Y., Takakusaki, K., and Matsuyama, K., 1989, Site-specific postural and locomotor changes evoked in awake, freely moving intact cats by stimulating the brainstem, *Brain Res.* 505:66.

Morilak, D., and Ciaranello, R., 1993, 5-HT2 receptor immunoreactivity on cholinergic neurons of the pontomesencephalic tegmentum shown by double immunofluorescence, *Brain Res.* 627:49.

Morilak, D.A., Garlow, S.J., and Ciaranello, R.D., 1993, Immunocytochemical localization and description of neurons expressing serotonin2 receptors in the rat brain, *Neuroscience*, 54:701.

Mrzljak, L., Bergson, C., Pappy, M., Huff, R., Levenson, R. and Goldman-Rakic, P., 1996, Localization of dopamine D4 receptors in GABAergic neurons of the primate brain, *Nature*, 381:245.

Muhlethaler, M., Khateb, A., and Serafin, M., 1990, Effects of monoamines and opiates on pedunculopontine neurones, in: *The Diencephalon and Sleep*, M. Mancia and G. Marini, eds., Raven Press, New York.

Nakamura, Y., Tokuno, H., Moriizumi, T., Kitao, Y., and Kudo, M., 1989, Monosynpatic nigral inputs to the pedunculopontine tegmental nucleus neurons which send their axons to the medial reticular formation in the medulla oblongata. An electron microscopic study in the cat, *Neurosci. Lett.* 103:145.

Nakashima, K., Shimoyama, R., Yokoyama, Y., and Takahashi, K., 1993, Auditory effects on the electrically elicited blink reflex in patients with Parkinson's disease, *Electroenceph Clin. Neurophysiol.* 89:108.

Narabayashi, H., 1990, Surgical treatment in the levodopa era, in: *Parkinson's Disease*, G. Stem, ed., Chapman & Hall, London.

Noda, T., and Oka, H., 1986, Distribution and morphology of tegmental neurons receiving nigral inhibitory inputs in the cat: An intracellular HRP study, *J. Comp. Neurol.* 244:254.

Olszewski, J., and Baxter, D., 1954, *Cytoarchitecture of the Human Brain Stem*, JB Lippincott, Philadelphia.

Olszewski, J., and Baxter, D., 1982, *Cytoarchitecture of the Human Brain Stem*, S. Karger AG, Basel.

Parent, A., 1990, Extrinsic connections of the basal ganglia, *Trends Neurosci.* 13:254.

Parent, A., and DeBellefeuille, L., 1982, Organization of efferent projections from the internal segment of globus pallidus in primate as revealed by fluorescence retrograde labeling method, *Brain Res.* 245:201.

Parent, A., Pare, D., Smith, Y., and Steriade, M., 1988, Basal forebrain cholinergic and noncholinergic projections to the thalamus and brainstem in cats and monkeys, *J. Comp. Neurol.* 277:281.

Paxinos, G., and Watson, C., 1986, *The Rat Brain in Stereotaxic Coordinates*, Academic Press, Orlando.

Penders, C., and Delwaide, P., 1971, Blink reflex studies in patients with parkinsonism before and during therapy, *J. Neurol. Neurosurg. Psychiat.* 34:674.

Rye, D., 1997, Contributions of the pedunculopontine region to normal and altered REM sleep, *Sleep*, 20:757.

Rye, D., and Bliwise, D., 1997, Movement disorders specific to sleep and the nocturnal manifestations of waking movement disorders, in: *Movement Disorders: Neurologic Principles and Practice*, R. Watts and W. Koller, eds., McGraw-Hill, New York.

Rye, D., Lee, H., Saper, C., and Wainer, B., 1988, Medullary and spinal efferents of the pedunculopontine tegmental nucleus and adjacent mesopontine tegmentum in the rat, *J. Comp. Neurol.* 269:315.

Rye, D., Saper, C., Lee, H., and Wainer, B., 1987, Pedunculopontine tegmental nucleus of the rat: Cytoarchitecture, cytochemistry, and come extrapyramidal connections of the mesopontine tegmentum, *J. Comp. Neurol.* 259:483.

Rye, D., Thomas, J., and Levey, A., 1995a, Distribution of molecular muscarinic (m1-m4) receptor subtypes and choline acetyltransferase in the pontine reticular formation of man and non-human primates, *Sleep Res.* 24:59.

Rye, D., Turner, R., Vitek, J., Bakay, R., Crutcher, M., and DeLong, M., 1996, Anatomical investigations of the pallidotegmental pathway in monkey and man, in: *Basal Ganglia V,* H. Ohye, M. Kimura, and J. McKenzie, eds., Plenum Press, New York.

Rye, D., Vitek, J., Bakay, R., Kaneoke, Y., Hashimoto, T., Turner, R., Mirra, S., and DeLong, M., 1995b, Termination of pallidofugal pathways in man, *Soc. Neurosci. Abstr.* 21:676.

Sakai, K., 1980, Some anatomical and physiological properties of ponto-mesencephalic tegmental neurons with special reference to the PGO waves and postural atonia during paradoxical sleep in the cat, in: *The Reticular Formation Revisited: Specifying Function for a Nonspecific System: International Brain Research Organization Monograph Series,* J. A. Hobson, and M. A. B. Brazier, eds., Raven Press, New York.

Sakai, K., 1985, Anatomical and physiological basis of paradoxical sleep, in: *Brain Mechanisms of Sleep,* D. J. McGinty, R. Drucker-Colin, A. Morrison, and P. L. Parmeggiani, eds., Raven Press, New York.

Scarnati, E., Prioa, A., Loreto, S.D., and Pacitti, C., 1987, The reciprocal electrophysiological influence between the nucleus tegmenti pedunculopontinus and the substantia nigra in normal and decorticated rats, *Brain Res.* 423:116.

Schenck, C., Bundlie, S., Ettinger, M., and Mahowald, M., 1986, Chronic behavioral disorders of human REM sleep: A new category of parasomnia, *Sleep,* 9:293.

Schenck, C., Bundlie, S., and Mahowald, M., 1996, Delayed emergence of a parkinsonian disorder in 38% of 29 older men initially diagnosed with idiopathic rapid eye movement sleep behavior disorder, *Neurology,* 46:388.

Schenck, C., Bundlie, S., Patterson, A., Ettinger, M., and Mahowald, M., 1987a, Five years of clinical experinece with 21 patients having chronic REM sleep behavior disorder (RBD), *Sleep Res.* 16:424.

Schenck, C., Bundlie, S., Patterson, A., and Mahowald, M. 1987b, Rapid eye movement sleep behavior disorder, *JAMA,* 257:1786.

Schenck, C., Bundlie, S., Patterson, A., and Mahowald, M., 1987c, Rapid eye movement sleep behavior disorder: A treatable parasomnia affecting older adults, *JAMA,* 257:1786.

Schenck, C., Hopwood, J., Duncan, E., and Mahowald, M., 1992, Preservation and loss of REM-atonia in human idiopathic REM sleep behavior disorder (RBD): Quantitative polysomnographic (PSG) analyses in 17 patients, *Sleep Res.* 21:16.

Schenck, C., and Mahowald, M., 1992, Motor dyscontrol in narcolepsy: Rapid-eye-movement (REM) sleep without atonia and REM sleep behavior disorder, *Ann. Neurol.* 32:3.

Shimizu, T., Inami, Y., Sugita, Y., Iijima, S., Teshima, Y., Matsuo, R., Yasoshima, A., Egawa, I., Okawa, M., Tashiro, T., and Hishikawa, Y., 1990, REM sleep without muscle atonia (stage 1-REM) and its relation to delirious behavior during sleep in patients with degenerative diseases involving the brain stem, *Jap. J. Psychiatr. Neurol.* 44:681.

Shink, E., Sidibe, M., and Smith Y., 1997, Efferent connections of the internal globus pallidus in the squirrel monkey: II. Topography and synaptice organization of pallidal efferents to the pedunculopontine nucleus, *J. Comp. Neurol.* 382:348.

Smith, Y., and Shink, F., 1995, The pedunculopontine nucleus (PPN): A potential target for the convergence of information arising from different functional territories of the internal pallidum (GPi) in primates, *Soc. Neurosci. Abstr.* 21:677.

Spann, B.M., and Grofova, I., 1992, Cholinergic and non-cholinergic neurons in the rat pedunculopontine tegmental nucleus, *Anat.Embryol.* 186:215.

Spooren, W., Cuypers, E., and Cools, S., 1989, Oro-facial dyskinesia and the subcommissural part of the globus pallidus in the cat: Role of acetylcholine and its interaction with GABA. *Psychopharm.* 99:381.

Staedt, J., Stoppe, G., Kogler, A., Munz, D., Riemann, H., Emrich, D., and Ruther, E., 1993, Dopamine D2 receptor alteration in patients with periodic movements in sleep (nocturnal myoclonus), *J. Neural Trans. [GenSect]* 93:71.

Staedt, J., Stoppe, G., Kogler, A., Riemann, H., Hajak, G., Munz, D., Emrich, D., and Ruther, E., 1995, Nocturnal myoclonus syndrome (periodic movements in sleep) related to central dopamine D2-receptor alteration, *Eur. Arch. Psychiatr. Clin. Neurosci.* 245:8.

Steininger, T., Wainer, B., and Rye, D., 1997, Ultrastructural study of cholinergic and non-cholinergic neurons in the pars compacta of the rat pedunculopontine tegmental nucleus, *J. Comp. Neurol.* 382:285.

Steriade, M., 1992, Basic mechanisms of sleep generation, *Neurology,* 42:9.

Steriade, M., Datta, S., Pare, D., Oakson, G., and Dossi, R.C., 1990a, Neuronal activities in brain-stem cholinergic nuclei related to tonic activation processes in thalamocortical systems, *J. Neurosci.* 10:2541.

Steriade, M., and McCarley, R., 1990, *Brainstem Control of Wakefulness and Sleep,* Plenum Press, New York.

Steriade, M., Paré, D., Datta, S., Oakson, G., and Dossi, R.C., 1990b, Different cellular types in mesopontine cholinergic nuclei related to ponto-geniculo-occipital waves, *J. Neurosci.* 10:2560.

Swerdlow, N., and Geyer, M., 1993, Prepulse inhibition of acoustic startle in rats after lesions of the pedunculopontine tegmental nucleus, *Behav. Neurosci.* 107:104.

Vilaro, M. T., Palacios, J.M., and Mengod, G., 1994, Multiplicity of muscarinic autoreceptor subtypes? Comparison of the distribution of cholinergic cells and cells containing mRNA for five subtypes of muscarinic receptors in the rat brain, *Mol. Brain Res.* 21:30.

Vincent, S. R., and Satoh, K., 1984, Corticotropin-releasing factor (CRF) immunoreactivity in the dorsolateral pontine tegmentum: further studies on the micturition reflex system, *Brain Res.* 308:387.

Vincent, S.R., Satoh, K., Armstrong, D.M., and Fibiger, H.C., 1983a, NADPH-diaphorase: a selective histochemical marker for the cholinergic neurons of the pontine reticular formation, *Neurosci. Lett.* 43:31.

Vincent, S. R., Satoh, K., Armstrong, D.M., and Fibiger, H.C., 1983b, Substance P in the ascending cholinergic reticular system, *Nature*, 306:688.

Vitek, J., Kaneoke, Y., Turner, R., Baron, M., Bakay, R., and DeLong, M., 1993, Neuronal activity in the internal (GPi) and external (GPe) segments of the globus pallidus (GP) of parkinsonian patients is similar to that in the MPTP-treated primate model of parkinsonism, *Soc. Neurosci. Abstr.* 19:1584.

von Krosigk, M., Smith, Y., Bolam, J., and Smith, A., 1992, Synaptic organization of gabaergic inputs from the striatum and the globus pallidus onto neurons in the sustantia nigra and retrorubral field which project to the medullary reticular formation, *Neuroscience,* 50:531-549.

Wainer, B., and Mesulam, M.-M., 1990, Ascending cholinergic pathways in the rat brain, in: *Brain Cholinergic Systems*, M. Steriade, and D. Biesold, eds., Oxford University Press, New York.

Webster, H., and Jones, B., 1988, Neurotoxic lesions of the dorsolateral pontomesencephalic tegmentum-cholinergic cell area in the cat. II. Effects upon sleep-waking states, *Brain Res.* 458:285.

Wirtshafter, D., 1994, FOS-like-immunoreactivity in basal ganglia outputs following administration of dopamine agonists, *Soc. Neurosci. Abstr.* 20:1190.

Zweig, R., Hedreen, J., Jankel, W., Casanova, M., Whitehouse, P., and Price, D., 1988, Pathology in brainstem regions of individuals with primary dystonia, *Neurol.* 38:702.

Zweig, R., Whitehouse, P., Casanova, M., Walker, L., Jankel, W., and Price, D., 1987, Loss of pedunculopontine neurons in progressive supranuclear palsy, *Ann. Neurol.* 22:18.

FATAL FAMILIAL INSOMNIA: A DISEASE MODEL EMPHASIZING THE ROLE OF THE THALAMUS IN THE REGULATION OF THE SLEEP-WAKE CYCLE AND OTHER CIRCADIAN RHYTHMS

PATRIZIA AVONI, PIETRO CORTELLI, ELIO LUGARESI, AND PASQUALE MONTAGNA[*]

INTRODUCTION

Fatal Familial Insomnia (FFI) is a human hereditary prion disease due to a mutation at codon 178 of the prion protein gene located on chromosome 20, which cosegregates with a methionine polymorphism at codon 129 of the same gene on the mutated allele (Lugaresi et al., 1986; Medori et al., 1992). Prion diseases are characterized by the presence in the brain of an isoform of the prion protein, which is partially resistant to the action of proteases, so-called PrPres, and is transmissible. The brains of FFI patients indeed demonstrate accumulation of a 19 kDa (after deglycosylation) PrPres (Parchi et al 1998), and FFI has been transmitted to transgenic mice. FFI has been found in at least 21 pedigrees around the world, and probably represents the third most common inherited human prion disease. FFI is also characterized by a unique set of clinical symptoms and pathological findings, which give rise to peculiar clinico-pathological correlations and interesting physiopathological conclusions concerning the role of the thalamus, especially its anterior and mediodorsal nuclei, in the regulation of circadian activities such as the sleep-wake cycle and the attending hormonal and vegetative functions.

CLINICAL ASPECTS OF FATAL FAMILIAL INSOMNIA

FFI affects males and females equally and is transmitted as an autosomal dominant trait. FFI begins at a mean age of 51 years, running a uniformly fatal course and leading to death after 18.4 months as a mean. Duration of the disease, however, and other clinical features appear to be modified by the type of methionine/valine polymorphism, which is normally present at codon 129 of the prion protein gene on the non-mutated allele: methionine homozygote patients have a significantly shorter disease course than methionine-valine heterozygotes (9.1 vs 30.8 months) (Montagna et al., 1998). Patients

[*] PATRIZIA AVONI, PIETRO CORTELLI, ELIO LUGARESI, AND PASQUALE MONTAGNA • Institute of Clinical Neurology, University of Bologna Medical School, Bologna, Italy.

complain of sleep-wake disturbances, including insomnia with progressively worsening inability to fall asleep and to maintain nocturnal sleep, display oneiric stuporous episodes characterized by the intrusion upon wakefulness of stuporous states during which the patients actually move and gesture as if enacting their dream content, and show impaired autonomic functions such as increased blood pressure, irregular breathing and tachycardia, increased salivation, tearing and diaphoresis associated with irregular low-grade evening fever, all features suggesting abnormal sympathetic activation. Urinary and fecal incontinence and impotence in the males inevitably appear at some stage of the disease. Patients also display somatomotor manifestations, earlier and more evident in the 129 heterozygotes, consisting of diplopia, dysarthria and dysphagia, ataxia/abasia, dysmetria, and spontaneous and evoked myoclonus and spasticity. A few patients also develop sporadic tonic-clonic seizures.

From a neuropsychological point of view, patients remain cooperative and retain good intellectual funtioning until late in the disease course. Altered vigilance with frequent and spontaneous intrusions of dream-like states from which the patients can be awakened early in the disease course, are later superseded by lasting stupor and coma and, finally, death in an akinetic mutism state. Neuropsychological tests indicated defective working memory and impaired attention, especially upon sustained efforts.

NEURORADIOLOGICAL AND PATHOLOGICAL FEATURES

Standard CT and MRI scans were relatively aspecific and demonstrated diffuse cerebral and cerebellar atrophy. 18F-FDG PET scans performed at several stages of the disease in seven patients (Cortelli et al., 1997) showed severely decreased glucose utilization in the thalamus bilaterally in all patients, associated with a milder involvement of the cingulate gyrus. In 129 methionine homozygote patients this could represent the only PET abnormality found. In heterozygotes with a more prolonged disease course, additional hypometabolism could be seen in other cerebral regions, in particular the basal and lateral frontal, middle and inferior temporal and hippocampal areas, and the caudate and putaminal nuclei. The occipital cortex remained unaffected in all patients. Thus, hypometabolism of the thalamus and cingulate cortex may be considered the functional hallmark of FFI.

Histopathological post-mortem studies (Manetto et al., 1992) showed highly characteristic patterns, with a severe loss of neurons and reactive astrogliosis in the thalamus, especially the anterior and the mediodorsal nuclei; neuronal loss of over 90% could be observed. These findings were independent of the disease course. The inferior olives also showed neuronal loss and gliosis in all patients. The involvement of the cerebral cortex varied according to the duration of the disease, and was more pronounced in the 129 heterozygote patients. Spongiosis, the pathological hallmark of the prion diseases, was usually encountered in the entorhinal cortex, associated with gliosis, but was always absent in the thalamus. In the neocortex, focal spongiosis could be seen in the frontal, temporal, and parietal lobes, and minimally or not at all in the occipital lobe; these cortical areas were, however, found affected only in those patients who died after at least 12 months. Astrogliosis was also observed in the midbrain periaqueductal grey matter, the superior colliculus, and the hypothalamus. Overall, therefore, FFI may be pathologically defined as a preferential thalamic atrophy, with involvement of the cerebral cortex occurring as a late event in the disease course. Regional deposition of the PrPres protein was more widespread than the pathological changes: PrPres was detected only in the limbic regions and in the thalamus, hypothalamus, and brainstem in patients suffering earlier death. With progression of the disease, PrPres deposition became prominent in the cerebral cortex (Parchi et al., 1998). PrPres deposition correlated with the hypometabolism observed on Pet studies (Cortelli et al., 1997).

NEUROPHYSIOLOGICAL, AUTONOMIC AND ENDOCRINE STUDIES

Twenty-four hour polysomnographic studies documented the absence of normal sleep, even from the beginning of the disease in the 129 homozygote patients. NREM sleep was decreased or abolished and the EEG showed a pattern characterized by continuous oscillations between a diffuse alpha activity typical of relaxed wakefulness, and desynchronized theta activity. This mixture of alpha and theta activity resembled stage 1 NREM sleep. REM sleep episodes occurred in clusters, associated with the enactment of the dream content. Short REM episodes could abnormally emerge directly from wakefulness, recurring periodically and presenting abnormally retained muscle tone (absence of the physiological atonia) and intensified myoclonic jerks. Deep NREM sleep was strikingly reduced or completely absent, even upon spectral analysis of the EEG. A prominent feature was the progressive and sometimes early disappearance of the sleep spindles and K complexes, sleep figures which mark the transition from wakefulness to sleep and from one sleep stage to another. Circadian studies of blood pressure and heart rate run in association with the polysomnographic recordings demonstrated that the physiological nocturnal fall in blood pressure occurred early. Monitoring of the body core temperature showed persistently elevated values (Figure 1). Catecholamine (adrenaline and noradrenaline) levels were increased throughout the 24 hours, and their circadian rhythmicity was progressively lost. Likewise, cortisol levels were permanently increased in the face of normal or undetectable levels of ACTH. The circadian secretion of other hormones was also disrupted: melatonin showed a gradual decrease in the amplitude of its

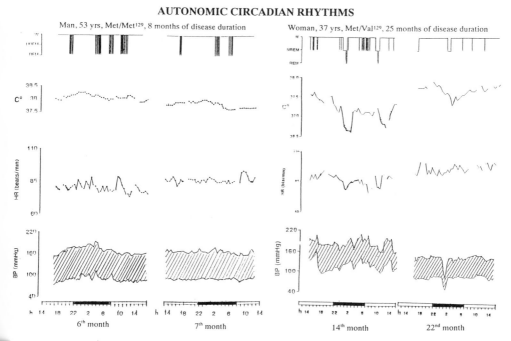

AUTONOMIC CIRCADIAN RHYTHMS

Man, 53 yrs, Met/Met129, 8 months of disease duration   Woman, 37 yrs, Met/Val129, 25 months of disease duration

6th month   7th month   14th month   22nd month

Figure 1. 24-hour sleep histogram and autonomic monitoring in an FFI patient demonstrating severe reduction of REM and total absence of NREM sleep, persistently elevated body core temperature, heart rate and blood pressure.

circadian fluctuations until a complete loss of rhythmicity, and GH showed reduced or absent rhythmicity, paralleling the loss of deep NREM sleep. Prolactin levels remained relatively unaffected. Autonomic investigations showed that the resting levels of blood pressure, heart rate and plasma noradrenaline were persistently elevated and rose disproportionately during a head-up tilt test. These autonomic investigations thus indicated sympathetic overactivity in FFI, increasing during the disease course and eventually leading to a loss of the physiological 24-hour circadian oscillations. Prolonged 24-hour actigraphic monitoring of motor activity demonstrated complete absence of any recognizable rest-activity circadian rhythm, and calorimetric determinations a strikingly increased (+60%) energy expenditure.

PATHOPHYSIOLOGICAL CONSIDERATIONS

Both the histopathological studies performed post-mortem and the Pet studies performed in vivo document that the principal change in FFI is a degeneration of the thalamus, especially its anterior and mediodorsal nuclei. Hence, the conclusion that the thalamus must be implicated in the genesis of the key features of the disease, e.g., the inability to produce and organize sleep activities and the disruption of vegetative and hormonal rhythms. The lack of sleep in FFI is indeed accompanied by ever-increasing sympathetic activation and energy expenditure. That the thalamus intervenes in the regulation of sleep, as a structure promoting sleep and rest activities, is a concept that dates back to the early studies of Hess (1944), later neglected. Other physiologists described how stimulation of limbic structures linked to the thalamus could induce a pre-sleep behaviour, or how lesions of the thalamus disrupt sleep (Parmeggiani, 1964; Villablanca and Salinas-Zebalos, 1972); more recently, Marini et al. (1988) provoked insomnia by bilateral lesions of the mediodorsal thalamic nuclei in the cat.

Thus, the thalamus must be considered a strategic structure in the central autonomic network which regulates the body homeostasis, and consequently wake-sleep and other circadian autonomic and hormonal functions. Anatomically, the anterior and mediodorsal nuclei constitute the limbic portion of the thalamus, interconnecting limbic and paralimbic areas to subcortical structures such as the hypothalamus. We suggest that damage to the thalamus in FFI releases these subcortical structures from corticolimbic control, shifting the metabolic balance to one of increased activitation and lack of sleep (Lugaresi et al. 1998).

REFERENCES

Cortelli, P., Perani, D., Parchi, P., Grassi, F., Montagna, P., De Martin, M., Castellani, R., Tinuper, P., Gambetti, P., Lugaresi, E., Fazio, F., 1997, Cerebral metabolism in fatal familial insomnia: relation to duration, neuropathology, and distribution of protease-resistent prion protein, *Neurology,* 49:126.

Hess, W.R., 1944, Das Schlaf syndrom als Folge diencephaler Reizung, *Helv. Physiol. Pharmacol. Acta,* 2:305.

Lugaresi, E., Medori, R., Montagna, P., Baruzzi, A., Cortelli, P., Lugaresi, A., Tinuper, P., Zucconi, M., Gambetti, P., 1986, Fatal familial insomnia and dysautonomia with selective degeneration of thalamic nuclei, *New Engl. J. Med.* 315:997.

Lugaresi, E., Tobler, I., Gambetti, P., Montagna, P., 1998, The pathophysiology of fatal familial insomnia, *Brain Pathol.* 8:521.

Manetto, V., Medori, R., Cortelli, P., Montagna, P., Baruzzi, A., Hauw, J.J., Rancurel, G., Vanderhaeghen, J.J., Mailleux, P., Bugiani, O., Tagliavini, F., Bouras, C., Rizzuto, N., Lugaresi, E., Gambetti, P., 1992, Fatal familial insomnia: clinical and pathological study of five new cases, *Neurology,* 42:312.

Marini, G., Imeri, L., Mancia, M., 1988, Changes in sleep-waking cycle induced by lesions of medialis dorsalis thalamic nuclei in the cat, *Neurosci. Lett.* 85:223.

Medori, R., Tritschler, H.J., LeBlanc, A., Villare, F., Manetto, V., Chen, H.Y., Xue, R., Leal, S., Montagna, P., Cortelli, P., Tinuper, P., Avoni, P., Mochi, M., Baruzzi, A., Hauw, J.J., Ott, J., Lugaresi, E., Autilio-

Gambetti, L., Gambetti, P.. 1992, Fatal familial insomnia is a prion disease with a mutation at codon 178 of the prion protein gene, *New Engl. J. Med.* 326:444.

Montagna, P., Cortelli, P., Avoni, P., Tinuper, P., Plazzi, G., Gallassi, R., Portaluppi, F., Julien, J., Vital, C., Delisle, C.B., Gambetti, P., Lugaresi, E., 1998, Clinical features of fatal familial insomnia: phenotypic variability in relation to a polymorphism at codon 129 of the prion protein gene, *Brain Pathol.* 8:515.

Parchi, P., Petersen, R.B., Chen, S.G., Autilio-Gambetti, L., Capellari, S., Monari, L., Cortelli, P., Montagna, P., Lugaresi, E., Gambetti, P., 1998, Molecular pathology of fatal familial insomnia, *Brain Pathol.* 8:539.

Parmeggiani, P.L., 1964, A study of the central representation of sleep behavior, Progr. Brain Res. 6:180.

Villablanca, J., and Salinas-Zebalos, M.E., 1972, Sleep-wakefulness, EEG, and behavioral studies of chronic cats without the thalamus:"athalamic", *Arch. Ital. Biol.* 110:383.

PART V. PLASTICITY IN MOVEMENT DISORDERS

PLASTICITY AND BASAL GANGLIA DISORDERS

MARK HALLETT[*]

INTRODUCTION

One of the important facts about the physiology of the brain that has emerged in recent years is that, even in adult life, the brain remains capable of changes in many different ways. These changes permit the brain to learn as well as respond to injury, and may be adaptive or non-adaptive. The brain is not only capable of reorganizing, it is constantly doing so.

There are a number of mechanisms for plasticity in humans. Their physiology has been studied in model systems, and we have only incomplete knowledge as to which mechanism applies to which phenomenon. First, a change in the balance of excitation and inhibition can happen very quickly. This process depends on the fact that neurons or neural pathways have a much larger region of anatomical connectivity than their usual territory of functional influence. Some zones may be kept in check by tonic inhibition. If the inhibition is removed, the region of influence can be quickly increased or unmasked, a process commonly called unmasking. For example, following the application of the GABA antagonist bicuculline to the forelimb area of the rat motor cortex, stimulation of the adjacent vibrissa area led to forelimb movements. This strongly suggests that GABAergic neurons are crucial to the maintenance of cortical motor representations (Jacobs and Donoghue, 1991).

The second process that can also be relatively fast is strengthening or weakening of existing synapses, in processes such as long-term potentiation (LTP) or long-term depression (LTD). LTP or LTD occur following specific patterns of synaptic activity and may last for long periods of time (Hess et al., 1996; Hess and Donoghue, 1996).

The third process is a change in neuronal membrane excitability. One possible mechanism for this process is an alteration of the Na^+-channels in neuronal membranes, which has been demonstrated to occur with operantly conditioned changes in the H-reflex (Halter et al., 1995).

The fourth process is anatomical change, which requires a longer period of time. Specific anatomical changes include sprouting of new axon terminals and formation of new synapses. For example, there can be increases in synaptic density that strengthen a pre-existing but previously weak connection. After long-term thalamic stimulation in adult cats,

[*] MARK HALLETT • Human Motor Control Section, National Institute of Neurological Disorders and Stroke, Bethesda, MD 20892.

synaptic proliferation can be identified in motor cortex (Keller et al., 1992). In extreme circumstances, there can be growth of new connections into new regions, as demonstrated by Pons et al. (1991) after long term deafferentation of a limb in adult monkeys.

EXAMPLES OF HUMAN PLASTICITY

In human physiology, the best evidence for plasticity is in the setting of peripheral deafferentation (such as amputation), motor learning, and stroke.

Peripheral Deafferentation

The pioneering work of Merzenich and colleagues in primates demonstrated that the organization of the sensory cortex changed after deafferentation of a limb (Merzenich et al., 1984; Merzenich and Jenkins, 1993). Similar observations were made by Donoghue et al. (1990) and Donoghue (1995) for the motor cortex. These studies demonstrated that the deafferented (or deefferented) cortex did not stay idle, but was taken over by body representations adjacent to the deafferented body part. Our group has investigated this situation in the motor system of humans (Hallett et al., 1993). We studied patients with traumatic, surgical, or congenital amputations of the arm at about the level of the elbow using transcranial magnetic stimulation (TMS) (Cohen et al., 1991a). Motor representation areas targeting muscles ipsilateral and immediately proximal to the stump were larger than those for muscles contralateral to the stump. These results are consistent with the idea that the motor cortex for the muscles proximal to the amputation had expanded into the territory of the amputated part.

We have done experiments to see how fast modulation of motor outputs could occur following reversible deafferentation of a limb segment in humans. Reversible deafferentation was accomplished by using a blood pressure cuff (Brasil-Neto et al., 1992, 1993). The amplitudes of motor-evoked potentials (MEPs) to TMS from muscles immediately proximal to the temporarily anesthetized forearm increased in minutes after the onset of anesthesia and returned to control values after the anesthesia subsided. Mano and colleagues showed that projections from the biceps region of the motor cortex can be directed to the spinal cord neurons of intercostal nerves in patients with brachial plexus avulsion after the intercostal nerve is anastomosed to the musculocutaneous nerve, but this process took a year or more to occur (Mano et al., 1995).

Motor Learning

Cortical changes result from changes in the patterns of behavior. Indeed, it appears that there is a constant battle for the control of each neuron among its various inputs. In order to investigate this issue, we performed detailed mapping of the motor cortical areas, targeting the first dorsal interosseus (FDI) and the adductor digiti minimi (ADM) bilaterally in Braille readers and in blind controls using focal TMS (Pascual-Leone et al., 1993). In the controls, motor representations of right and left FDI and ADM were not significantly different. However, in the proficient Braille readers, the representation of the FDI in the reading hand was significantly larger than that in the nonreading hand or in either hand of the controls. Conversely, the representation of the ADM in the reading hand was significantly smaller than that in the non-reading hand or in either hand of the controls. These results suggest that the cortical representation for the reading finger in proficient Braille readers is enlarged at the expense of the representation of other fingers.

Conversely, Leipert and colleagues studied cortical plasticity in patients who had a unilateral immobilization of the ankle joint without any peripheral nerve lesion (Liepert et al., 1995). The motor cortex area of the inactivated tibial anterior muscle diminished compared to the unaffected leg, without changes in spinal excitability or motor threshold.

As in the setting of amputation, we can look at how fast such changes can occur, and for this purpose we have studied motor learning. Pascual-Leone et al. (1995) looked at the motor cortical representation of the hand over a 5 day period in normal subjects as they learned a skilled hand task. As subjects became more skilled at a five finger exercise on a piano, the size of the motor representation of hand increased.

Stroke

After hemispherectomy, motor function in the limb contralateral to the excised hemisphere shows a substantial degree of recovery, particularly when surgery is performed at early age. To understand the mechanisms underlying this recovery of function, we studied patients with hemispherectomy. TMS of the remaining hemisphere induced bilateral activation of deltoid and biceps (Cohen et al., 1991b).

In accord with this finding, one theme for stroke recovery, with evidence from neuroimaging, has been recruitment of ipsilateral pathways (Chollet et al., 1991; Honda et al., 1997; Cramer et al., 1997). On the other hand, studies with TMS have not yet confirmed the utility of ipsilateral pathways for recovery when stroke occurs in adult age. Studies do show that ipsilateral MEPs are more likely, have lower thresholds, and have shorter latencies in patients with stroke than in normal subjects, but these responses are found more frequently in patients with poorer functional recovery (Turton et al., 1996).

It is the magnitude of the contralateral MEPs that correlate with good recovery. With recovery following stroke, Traversa et al. (1997) have demonstrated an enlarged TMS map area of the recovering muscles. This would seem to indicate the importance of plasticity mechanisms in recovery. Taub and colleagues have been advocating constraint-induced movement therapy that forces use of the hemiplegic limb, even in patients with chronic and apparently stable stroke (Taub et al., 1993; 1999). In a number of trials, there is behavioral improvement, and Liepert et al. (1998) have shown that TMS maps increase in size in this circumstance.

PLASTICITY IN PARKINSON'S DISEASE

Anatomical Compensation

Patients with Parkinson's disease have difficulty performing long sequences of movements, presumably as one result of their nigrostriatal loss of dopamine and its resultant effects on cortical function. We used PET to assess the rCBF associated with the performance of well-learned sequences of finger movements of varying length (Catalan et al., 1999). Sequential finger movements in the Parkinson's disease patients were associated with an activation pattern similar to that found in normal subjects, but they showed relative overactivity in the precuneus, premotor, and parietal cortices. Increasing the complexity of movements resulted in increased rCBF in the premotor and parietal cortices of normal subjects; the Parkinson's disease patients showed greater increases in these same regions and had additional significant increases in the anterior supplementary motor area (SMA)/cingulate. This last point has particular interest, since performance of self-selected movements induces significant activation of the anterior SMA/cingulate in normal subjects but not in Parkinson's disease patients. Thus, in Parkinson's disease patients, more cortical areas are recruited to perform sequential finger movements; this may be the result of increasing corticocortical activity to compensate for their striatal dysfunction.

Functional Compensation

In Parkinson's disease, the fundamental disturbance is bradykinesia. Some other aspects, such as rigidity, may arise as a reaction to that dysfunction rather than as a separate primary disturbance. The body generally tries to maintain itself in the same position. The smallest perturbations do not need any corrective action since they are blunted just by the inertia of the body. Slightly larger perturbations to the body can be dealt with by reflex action, the smaller ones by the monosynaptic reflex and larger ones by long-latency reflexes of various origins. The largest perturbations require voluntary correction. Each of these successive mechanisms is more powerful and more "intelligent," but at a longer latency. If voluntary correction is delayed, as it is in Parkinson's disease, then it might be helpful to increase the gain of reflexes. This will increase the stiffness of the body and help prevent disturbance from equilibrium, but, of course, such stiffness is not as effective as voluntary mechanisms in preventing falls. In Parkinson's disease patients, an important contributor to rigidity is the long-latency stretch reflex, which by this logic could be considered compensatory (Hallett, 1989).

Generally, greater postural sway while standing reflects instability. We examined postural stability in patients with Parkinson's disease and were surprised to find greater sway when on medication compared to when off medication (Panzer et al., 1990). When "off," patients were very stiff and swayed only little. Small perturbations did not affect them, but a larger perturbation caused them to fall. When "on," patients were less stiff and swayed more. This example shows how increased stiffness is functionally helpful.

DYSTONIA

Post-Hemiplegic Dystonia

Post-hemiplegic dystonia has long been known as a maladaptive consequence of stroke. There are a variety of similar syndromes in this category. Dystonia may also result from head trauma, hypoxia, cyanide poisoning, encephalitis, and carbon monoxide poisoning (Krauss et al., 1992; Choi et al., 1993; Borgohain et al., 1995; Scott and Jankovic, 1996; Apaydin et al., 1998; Choi and Cheon, 1999). The common thread in all these cases is damage to the basal ganglia, particularly the putamen, although the disorder can arise with basal ganglia damage sparing the putamen. One of these syndromes is called the 4-hemi syndrome, due to a combination of hemiplegia, hemidystonia, hemiatrophy, and hemi-seizures (Thajeb, 1996). In all of these circumstances there is a delay between the brain insult and the development of dystonia. The delay can be as short as one month and as long as many years; in the Scott and Jankovic series the average duration was 25 years (Scott and Jankovic, 1996). The latency was longer when the insult was at an earlier age.

The mechanism of post-hemiplegic dystonia is unknown, but it must reflect a slow process of plasticity, such as sprouting. That such a process is likely to be occuring has been demonstrated in a cat model. Kultas-Ilinsky et al. (1992) made kainic acid lesions in the substantia nigra pars reticularis and entopeduncular nucleus. At one year after the lesion, their findings in the ventral anterior thalamic nucleus (VA) included an increase in the appositional length of the dendrites of GABAergic local circuit neurons along projection neuron dendrites, accompanied by an increase in the number of dendrodendritic synapses. Pre-synaptic structures displayed some features of growth cones, such as an abundance of tubular and vesicular structures, many with electron-dense contents. These findings are consistent with continuing active remodeling of synaptic connections up to one year.

Focal Dystonia Of The Hand (Writer's Cramp)

The genesis of occupational cramps such as writer's cramp or pianist's cramp has been extensively studied. There is common agreement that these disorders are focal dystonias (Cohen and Hallett, 1988; Berardelli et al., 1998; Hallett, 1998a; 1998b). There are likely several causes or contributing factors, and repetitive activity may be one. A possible animal model of dystonia was created in nonhuman primates with synchronous, widespread sensory stimulation to the hand during a repetitive motor task (Byl et al., 1996; 1997). Over a period of months, the animals' motor performance deteriorated. After development of the movement disorder, the primary somatosensory cortex was mapped, and each cell was analyzed for the region of the body that activated the cell, its "receptive field." Receptive fields in area 3b were increased ten- to twenty-fold, often extending across the surface of two or more digits. The investigators suggested that synchronous sensory input over a large area of the hand can lead to remapping of the receptive fields and subsequently to a movement disorder. However, these tasks also involve repetitive movements, which can lead to remapping of the motor system directly.

If this situation holds in human dystonia, then the somatosensory cortex might also be abnormal. Specifically, if the receptive fields have enlarged, they should overlap more than in the normal situation. Then the centers of the sensory receptive field maps from the different fingers in patients with dystonia should be closer together. We mapped the human cortical hand somatosensory area of 6 patients with focal dystonia of the hand, using noninvasive, high-resolution electrophysiologic and magnetic resonance imaging techniques (Bara-Jimenez et al., 1998). The center of the receptive field from each finger was approximated by calculating the electrical dipole source for the somatosensory-evoked potential produced by cutaneous stimulation. We found degradation of the normal homuncular organization of the finger representations in the primary somatosensory cortex. The cortical finger representations in these patients were closer to each other than in normal subjects and often had an abnormal somatotopic arrangement. These findings support the concept that abnormal plasticity is involved in the development of dystonia. Our findings have been reproduced using MEG (Elbert et al., 1998). When doing thalamotomies in patients with dystonia, Lenz et al. (1999) and Lenz and Byl (1999) have found enlarged receptive fields in thalamic neurons. These results are also consistent.

Blepharospasm

A clever animal model of blepharospasm has been produced in the rat (Schicatano et al., 1997). Two mild alterations to the rat trigeminal reflex blink system are required. The first modification, a small striatal dopamine depletion, reduces the tonic inhibition of trigeminal reflex blink circuits. The second alteration, a slight weakening of the orbicularis oculi muscle, begins an adaptive increase in the drive on trigeminal sensory-motor blink circuits that initiates blepharospasm. By themselves, neither of these modifications causes spasms of lid closure, but combined they induce bilateral forceful blinking and spasms of lid closure. The authors suggest that their two-factor model may explain the development of benign essential blepharospasm in humans. The first factor, a subclinical loss of striatal dopamine, creates a permissive environment within the trigeminal blink circuits, and this might arise from a genetic defect. The second factor, an external ophthalmic insult, precipitates benign essential blepharospasm.

A similar situation has already been found as a possible human model of blepharospasm, proposed in a case report of blepharospasm-like symptoms developing contralateral to an eyelid weakened by facial nerve palsy (Chuke et al., 1996). With a weak lid, the central nervous system must increase the neural signal to the muscle in order to achieve eyelid closure. This is accomplished by increasing the output of the brainstem signal for a given command for eyelid closure. Such an increase in gain also affects the normal eyelid, since changes in gain are always bilateral. Thus, hyperexcitability of the normal eyelid might be a maladaptive consequence of the weakness of the affected lid. This theory is supported by the observation that the eyelid spasms were eliminated by the implantation of a gold weight to assist closure of the paretic eyelid. Subsequently, four additional patients were described with blepharospasm after Bell's palsy (Baker et al., 1997).

Excitability of the brainstem controller can be assessed using the blink reflex recovery cycle. In this test, the amplitude of a second blink reflex is compared with a first one at short intervals. Ordinarily, there is inhibition of the second blink reflex for a period lasting several seconds. We have added additional support to the concept of brainstem hyperexcitability in facial palsy by finding reduced inhibition in the blink reflex recovery cycle, similar to that seen in patients with blepharospasm (Syed et al., 1999).

There is reasonable evidence that many cases of focal dystonia are due, in part, to an autosomal dominant gene with low penetrance (Defazio et al., 1993). Perhaps just the right environmental exposure is needed to produce dystonia.

REFERENCES

Apaydin, H., Ozekmekci, S., and Yeni, N., 1998, Posthemiplegic focal limb dystonia: a report of two cases, *Clin. Neurol. Neurosurg.* 100:46.

Baker, R.S., Sun, W.S., Hasan, S.A., Rouholiman, B.R., Chuke, J.C., Cowen, D.E. et al., 1997, Maladaptive neural compensatory mechanisms in Bell's palsy-induced blepharospasm, *Neurology,* 49:223.

Bara-Jimenez, W., Catalan, M.J., Hallett, M., and Gerloff, C., 1998, Abnormal somatosensory homunculus in dystonia of the hand, *Ann. Neurol.* 44:828.

Berardelli, A., Rothwell, J.C., Hallett, M., Thompson, P.D., Manfredi, M., and Marsden, C.D., 1998, The pathophysiology of primary dystonia, *Brain,* 121:1195.

Borgohain, R., Singh, A.K., Radhakrishna, H., Rao, V.C., and Mohandas, S., 1995, Delayed onset generalized dystonia after cyanide poisoning, *Clin. Neurol. Neurosurg.* 97:213.

Brasil-Neto, J.P., Cohen, L.G., Pascual-Leone, A., Jabir, F.K., Wall, R.T., and Hallett, M., 1992, Rapid reversible modulation of human motor outputs after transient deafferentation of the forearm: A study with transcranial magnetic stimulation, *Neurology,* 42:1302.

Brasil-Neto, J.P., Valls-Solé, J., Pascual-Leone, A., Cammarota, A., Amassian, V.E., Cracco, R. et al., 1993, Rapid modulation of human cortical motor outputs following ischemic nerve block, *Brain,* 116:511.

Byl, N., Merzenich, M.M., and Jenkins, W.M., 1996, A primate genesis model of focal dystonia and repetitive strain injury: I. Learning-induced dedifferentiation of the representation of the hand in the primary somatosensory cortex in adult monkeys, *Neurology,* 47:508.

Byl, N.N., Merzenich, M.M., Cheung, S., Bedenbaugh, P., Nagarajan, S.S., and Jenkins, W.M., 1997, A primate model for studying focal dystonia and repetitive strain injury: effects on the primary somatosensory cortex, *Phys. Ther.* 77:269.

Catalan, M.J., Ishii, K., Honda, M., Samii, A., and Hallett, M., 1999, A PET study of sequential finger movements of varying length in patients with Parkinson's disease, *Brain,* 122:483.

Choi, I.S., and Cheon, H.Y., 1999, Delayed movement disorders after carbon monoxide poisoning, *Eur. Neurol.* 42:141.

Choi, Y.C., Lee, M.S., and Choi, I.S., 1993, Delayed-onset focal dystonia after diffuse cerebral hypoxia--two case reports, *J. Korean Med. Sci.* 8:476.

Chollet, F., DiPiero, V., Wise, R.J.S., Brooks, D.J., Dolan, R.J., and Frackowiak, R.S.J., 1991, The functional anatomy of motor recovery after stroke in humans. A study with positron emission tomography, *Ann. Neurol.* 29:63.

Chuke, J.C., Baker, R.S., and Porter, J.D., 1996, Bell's Palsy-associated blepharospasm relieved by aiding eyelid closure, *Ann. Neurol.* 39:263.

Cohen, L.G., Bandinelli, S., Findley, T.W., and Hallett, M., 1991a, Motor reorganization after upper limb amputation in man, *Brain,* 114:615.

Cohen, L.G., and Hallett, M., 1988, Hand cramps: clinical features and electromyographic patterns in a focal dystonia, *Neurology,* 38:1005.

Cohen, L.G., Roth, B.J., Wassermann, E.M., Topka, H., Fuhr, P., Schultz, J. et al., 1991b, Magnetic stimulation of the human cerebral cortex, an indicator of reorganization in motor pathways in certain pathological conditions, *J. Clin. Neurophysiol.* 8:56.

Cramer, S.C., Nelles, G., Benson, R.R., Kaplan, J.D., Parker, R.A., Kwong, K.K. et al., 1997, A functional MRI study of subjects recovered from hemiparetic stroke, *Stroke,* 28:2518.

Defazio, G., Livrea, P., Guanti, G., Lepore, U., and Ferrari, E., 1993, Genetic contribution to idiopathic adult-onset blepharospasm and cranial cervical dystonia, *Eur. Neurol.* 33:345.

Donoghue, J.P., 1995, Plasticity of adult sensorimotor representations, *Curr. Opin. Neurobiol.* 5:749.

Donoghue, J.P., Suner, S., and Sanes, J.N., 1990, Dynamic organization of primary motor cortex output to target muscles in adult rats. II. Rapid reorganization following motor nerve lesions, *Exp. Brain Res.* 79:492.

Elbert, T., Candia, V., Altenmuller, E., Rau, H., Sterr, A., Rockstroh, B. et al., 1998, Alteration of digital representations in somatosensory cortex in focal hand dystonia, *Neuroreport*, 9:3571.

Hallett, M., 1989, Long-latency reflexes, in: *Disorders of Movement,* N.P. Quinn and P.G. Jenner, eds., Academic Press, London.

Hallett, M., 1998a, The neurophysiology of dystonia, *Arch. Neurol.* 55:601.

Hallett, M., 1998b, Physiology of dystonia, in: *Dystonia 3. Advances in Neurology,* S. Fahn, C.D. Marsden, and M. DeLong, eds., Lippincott-Raven, Philadelphia.

Hallett, M., Cohen, L.G., Pascual-Leone, A., Brasil-Neto, J., Wassermann, E.M., and Cammarota, A.N., 1993, Plasticity of the human motor cortex, in: *Spasticity: Mechanisms and Management,* A.F. Thilmann, D.J. Burke, and W.Z. Rymer, eds., Springer-Verlag, Berlin.

Halter, J.A., Carp, J.S., and Wolpaw, J.R., 1995, Operantly conditioned motoneuron plasticity: possible role of sodium channels, *J. Neurophysiol.* 73:867.

Hess, G., Aizenman, C.D., and Donoghue, J.P., 1996, Conditions for the induction of long-term potentiation in layer II/III horizontal connections of the rat motor cortex, *J. Neurophysiol.* 75:1765.

Hess, G., and Donoghue, J.P., 1996, Long-term depression of horizontal connections in rat motor cortex, *Eur. J. Neurosci.* 8:658.

Honda, M., Nagamine, T., Fukuyama, H., Yonekura, Y., Kimura, J., and Shibasaki, H., 1997, Movement-related cortical potentials and regional cerebral blood flow change in patients with stroke after motor recovery, *J. Neurol. Sci.* 146:117.

Jacobs, K.M., and Donoghue, J.P., 1991, Reshaping the cortical motor map by unmasking latent intracortical connections, *Science,* 251:944.

Keller, A., Arissian, K., and Asanuma, H., 1992, Synaptic proliferation in the motor cortex of adult cats after long- term thalamic stimulation, *J. Neurophysiol.* 68:295.

Krauss, J.K., Mohadjer, M., Braus, D.F., Wakhloo, A.K., Nobbe, F., and Mundinger, F., 1992, Dystonia following head trauma: a report of nine patients and review of the literature, *Mov. Disord.* 7:263.

Kultas-Ilinsky, K., De Boom, T., and Ilinsky, I.A., 1992, Synaptic reorganization in the feline ventral anterior thalamic nucleus induced by lesions in the basal ganglia, *Exp. Neurol.* 116:312.

Lenz, F.A., and Byl, N.N., 1999, Reorganization in the cutaneous core of the human thalamic principal somatic sensory nucleus (Ventral caudal) in patients with dystonia, *J. Neurophysiol.* 82:3204.

Lenz, F.A., Jaeger, C.J., Seike, M.S., Lin, Y.C., Reich, S.G., DeLong, M.R. et al., 1999, Thalamic single neuron activity in patients with dystonia: dystonia-related activity and somatic sensory reorganization, *J. Neurophysiol.* 82:2372.

Liepert, J., Miltner, W.H., Bauder, H., Sommer, M., Dettmers, C., Taub, E. et al., 1998, Motor cortex plasticity during constraint-induced movement therapy in stroke patients, *Neurosci. Lett.* 250:5.

Liepert, J., Tegenthoff, M., and Malin, J.P., 1995, Changes of cortical motor area size during immobilization, *Electroencephalogr. Clin. Neurophysiol.* 97:382.

Mano, Y., Nakamuro, T., Tamura, R., Takayanagi, T., Kawanishi, K., Tamai, S. et al., 1995, Central motor reorganization after anastomosis of the musculocutaneous and intercostal nerves in patients with traumatic cervical root avulsion, *Ann. Neurol.* 38:15.

Merzenich, M.M., and Jenkins, W.M., 1993, Reorganization of cortical representations of the hand following alterations of skin inputs induced by nerve injury, skin island transfers, and experience, *J. Hand Ther.* 6:89.

Merzenich, M.M., Nelson, R.J., Stryker, M.P., Cynader, M.S., Schoppmann, A., and Zook, J.M., 1984, Somatosensory cortical map changes following digit amputation in adult monkeys, *J. Comp. Neurol.* 224:591.

Panzer, V.P., Zeffiro, T.A., and Hallett, M., 1990, Kinematics of standing posture associated with aging and Parkinson's disease, in: *Disorders of Posture and Gait 1990,* T. Brandt, W. Paulus, W. Bles, M. Dieterich, S. Krafczyk, and A. Straube, eds., Georg Thieme, Stuttgart.

Pascual-Leone, A., Cammarota, A., Wassermann, E.M., Brasil-Neto, J.P., Cohen, L.G., and Hallett, M., 1993, Modulation of motor cortical outputs to the reading hand of Braille readers, *Ann. Neurol.* 34:33.

Pascual-Leone, A., Dang, N., Cohen, L.G., Brasil-Neto, J.P., Cammarota, A., and Hallett, M., 1995, Modulation of muscle responses evoked by transcranial magnetic stimulation during the acquisition of new fine motor skills, *J. Neurophysiol.* 74:1037.

Pons, T.P., Garraghty, P.E., Ommaya, A.K., Kaas, J.H., Taub, E., and Mishkin, M., 1991, Massive cortical reorganization after sensory deafferentation in adult macaques, *Science,* 252:1857.

Schicatano, E.J., Basso, M.A., and Evinger, C., 1997, Animal model explains the origins of the cranial dystonia benign essential blepharospasm, *J. Neurophysiol.* 77:2842.

Scott, B.L., and Jankovic, J., 1996, Delayed-onset progressive movement disorders after static brain lesions, *Neurology*, 46:68.

Syed, N.A., Delgado, A., Sandbrink, F., Schulman, A.E., Hallett, M., and Floeter, M.K., 1999, Blink reflex recovery in facial weakness: an electrophysiologic study of adaptive changes, *Neurology*, 52:834.

Taub, E., Miller, N.E., Novack, T.A., Cook, E.W., Fleming, W.C., Nepomuceno, C.S. et al., 1993, Technique to improve chronic motor deficit after stroke, *Arch. Phys. Med. Rehabil.* 74:347.

Taub, E., Uswatte, G., and Pidikiti, R., 1999, Constraint-Induced Movement Therapy: a new family of techniques with broad application to physical rehabilitation--a clinical review, *J. Rehabil. Res. Dev.* 36:237.

Thajeb, P., 1996, The syndrome of delayed posthemiplegic hemidystonia, hemiatrophy, and partial seizure: clinical, neuroimaging, and motor-evoked potential studies, *Clin. Neurol. Neurosurg.* 98:207.

Traversa, R., Cicinelli, P., Bassi, A., Rossini, P.M., and Bernardi, G., 1997, Mapping of motor cortical reorganization after stroke. A brain stimulation study with focal magnetic pulses, *Stroke*, 28:110.

Turton, A., Wroe, S., Trepte, N., Fraser, C., and Lemon, R.N., 1996, Contralateral and ipsilateral EMG responses to transcranial magnetic stimulation during recovery of arm and hand function after stroke, *Electroencephalogr. Clin. Neurophysiol.* 101:316.

MECHANISMS OF SUBTHALAMIC DYSFUNCTION IN PARKINSON'S DISEASE

MARIE-FRANÇOISE CHESSELET, ARPESH MEHTA, AND PHILIPPE DE DEURWAERDÈRE[*]

INTRODUCTION

Changes in neuronal firing in the subthalamic nucleus have emerged as one of the cardinal features of Parkinson's disease. A central role for this region in Parkinson's disease can be inferred from the success of subthalamic lesions in non-human primates (Bergman et al., 1990) and of subthalamic deep brain stimulation (Kumar et al., 1998) in alleviating the symptoms of Parkinson's disease. Recordings in humans with Parkinson's disease during stereotaxic surgery have revealed that subthalamic neurons have a markedly increased rate of firing with increased bursting (Hutchinson et al., 1998). The principal mechanism leading to this effect has been traditionally considered to be a decreased inhibitory input from the external pallidum (Albin et al., 1989). However, several experimental data suggest that this explanation is not sufficient for understanding how dopamine depletion leads to subthalamic dysfunction (Chesselet and Delfs, 1996; Hassani et al., 1996; Levy et al., 1997). It is likely that changes in the activity of other subthalamic inputs, coupled with reactive changes in subthalamic receptors and/or transduction pathways, need to be taken into account to understand this key feature of Parkinson's disease. Increased activity of subthalamic neurons can be modeled in rats with a unilateral lesion of the nigrostriatal pathway (Kreiss et al., 1997; Hassani et al., 1996). Therefore, this model can be used to determine the cellular and molecular mechanisms induced in the subthalamic nucleus by the loss of dopaminergic neurons. Here we review recent studies conducted in our laboratory to examine the changes induced by chronic dopamine depletion in the subthalamic responses to local drug administration.

ALTERATIONS IN DOPAMINE-MEDIATED BEHAVIOR

We have shown several years ago that local administration of the non-selective dopaminergic agonist apomorphine into the subthalamic nucleus induced hyperkinetic behavior, which in the rat takes the form of an increase in orofacial dyskinesia (Parry et al., 1994). This effect was blocked by dopaminergic D1 but not D2 antagonists (Parry et al.,

* MARIE-FRANÇOISE CHESSELET, ARPESH MEHTA AND PHILIPPE DE DEURWAERDÈRE • Department of Neurology, University of California in Los Angeles, Los Angeles, CA 90095.

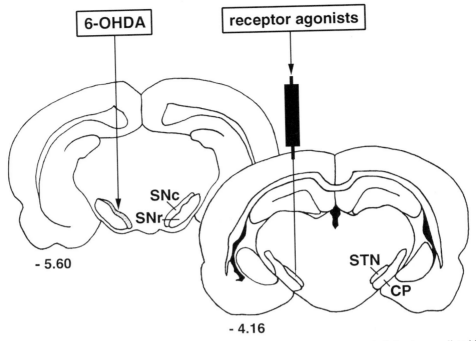

Figure 1. Schematic diagram of the experimental model used to study the changes in behaviors mediated by various subthalamic receptors following ipsilateral dopamine depletion. The receptor agonists were infused into the subthalamic nucleus via indwelling cannulae. The coordinates are in millimeters from bregma. SNc = substantia nigra pars compacta, SNr = substantia nigra pars reticulata, STN = subthalamic nucleus, CP = cerebral peduncle.

1994). We have recently confirmed the involvement of D1-like receptors in this effect by showing that it is reproduced by a D1 selective agonist, A77636, and blocked by a more selective D1 antagonist, SCH39166 (Mehta et al., 2000).

The existence of a dopaminergic input to the subthalamic nucleus has been highlighted in early anatomical studies (Lindvall and Bjorklund, 1978), and the presence of functional D1 receptors in this region was revealed in studies by Brown et al. (1979). However, the contribution of this input to the functioning of the subthalamic nucleus was challenged by the absence of conspicuous immunostaining for tyrosine hydroxylase in this region. Although the dopaminergic innervation of the subthalamic nucleus is probably minor, definite evidence of tyrosine-containing nerve terminals has been obtained in the primate (Lavoie et al., 1989; Herdeen, 1999). Furthermore, anatomical evidence suggests that these fibers are collaterals of nigrostriatal dopaminergic neurons originating in the substantia nigra pars compacta (Hassani et al., 1997). Whether this pathway is sufficient to provide a functionally significant innervation of the subthalamic nucleus remained uncertain.

To address this question, we sought to determine whether the effects of a locally applied dopaminergic agonist into the subthalamic nucleus were modified by prior lesion of dopaminergic cell bodies in the substantia nigra (Figure 1). Rats received a unilateral injection of 6-hydroxydopamine (8µg/4µl) into the substantia nigra one week prior to implantation of a guide cannula into the subthalamic nucleus. In these animals, local injection of apomorphine into the subthalamic nucleus 2-3 weeks after the dopaminergic lesion induced a markedly increased behavioral response compared to unlesioned rats (Figure 2). Interestingly, the elevated response persisted for at least one month when rats were tested every fourth day. The enhanced response was also decreased by selective D1 antagonists (Mehta et al., 2000). These data indicate that loss of dopaminergic inputs to the subthalamic nucleus resulted in an increased behavioral response. Thus, removal of the nigro-subthalamic dopaminergic pathway altered functional responses in the subthalamic nucleus, suggesting a functional role of endogenous dopamine in this region.

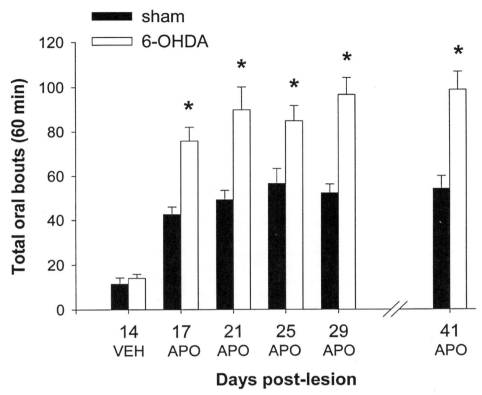

Figure 2. Effect of the dopaminergic agonist apomorphine (1.0 µg) administered locally in the subthalamic nucleus on orofacial dyskinesia in sham and dopamine depleted rats (n=6-8) tested every four days. Data are expressed as mean ± SEM of total oral bouts for the 60min period immediately following administration. *Refers to statistical difference when compared to corresponding sham data (Student's unpaired t-test). Reproduced from Mehta et al., 2000, with permission of Wiley Liss Inc., subsidiary of John Wiley and sons, Inc, copyright 2000.

The sustained increase in oral dyskinesia in response to administration of dopaminergic agonists after dopaminergic lesion suggests that the subthalamic nucleus could be directly involved in L-DOPA-induced dyskinesia. Furthermore, evidence for a functional dopaminergic nigro-subthalamic pathway suggests that, in Parkinson's disease, loss of dopaminergic inputs to the subthalamic nucleus could contribute to the increased firing and bursting activity seen in this region. The normal effect of dopamine on neuronal activity in the subthalamic nucleus remains unclear. Peripheral administration of the nonselective dopaminergic agonist apomorphine increases firing rates in the subthalamic nucleus (Kreiss et al., 1997), a finding at odds with our observation that dopaminergic agonists induce dyskinesia, a movement disorder usually associated with decreased activity in the subthalamic nucleus. However, other studies have found opposite effects of locally administered drugs (Hassani and Feger, 1999), and dopaminergic lesions seem to profoundly affect the cellular effects of dopaminergic drugs on neuronal activity in the subthalamic nucleus (Kreiss et al., 1997; Hassani and Feger, 1996).

One reason for these discrepancies may be that dopaminergic receptors located on different neuronal elements in the subthalamic nucleus may be involved, and their contribution to changes in cellular activity may be differentially appreciated depending on the experimental preparation. In situ hybridization studies failed to detect significant levels of D1 mRNA in the subthalamic nucleus (Fremeau et al., 1991), in contrast with the presence of binding sites, which when detected with tritiated ligands, reached levels similar to those observed in the cerebral cortex (Parry et al., 1994). In agreement with a predominantly presynaptic location of these receptors, kainic acid lesions of intrinsic subthalamic neurons did not significantly decrease D1 binding sites in the subthalamic nucleus (Parry et al., 1994). Presynaptic D1 receptors on cortical inputs to the subthalamic nucleus, for example, could alter cellular activity in this region by decreasing glutamatergic input from these neurons. However, this remains a matter of speculation in the absence of data on glutamate release in the subthalamic nucleus. More sensitive methods to detect mRNA, however, revealed the presence of D1 mRNA in the subthalamic nucleus, suggesting that a small number of receptors could be present on intrinsic subthalamic neurons (Flores et al., 1999).

GABA-ergic INPUTS TO THE SUBTHALAMIC NUCLEUS

The subthalamic nucleus receives a major GABA-ergic input from the external pallidal segment, or globus pallidus, in rats (Albin et al., 1989; Smith et al., 1998). In the subthalamic nucleus, axon terminals from these neurons form direct synapses with subthalamic output neurons (Smith et al., 1990), and GABA regulates their cellular activity by acting on $GABA_A$ receptors. Based on evidence that inhibition of subthalamic neurons induces hyperkinetic behavior (Hamada and Delong, 1992), one may expect that local stimulation of subthalamic $GABA_A$ receptors with an agonist would induce dyskinesia. We have indeed observed that local administration of muscimol, $0.01\mu g$ and $0.1\mu g$, into the subthalamic nucleus of freely moving rats induced a dose-dependent increase in orofacial dyskinesia.

Lesions of the subthalamic nucleus in intact rats induce a marked ipsilateral turning behavior when rats are challenged with a peripheral injection of apomorphine (Delfs et al., 1995a). While the mechanism of this effect is unknown, it is not due to a lesion-induced asymmetry in striatal dopaminergic receptors, a possible site of action of apomorphine (Parry and Chesselet, unpublished observation). More likely, the effect is due to the imbalance created by the increase in activity of the intact, contralateral subthalamic nucleus induced by peripheral administration of apomorphine (Kreiss et al., 1997). To determine whether local administration of muscimol into the subthalamic nucleus inhibits subthalamic activity in a way comparable to a lesion, we examined the effects of apomorphine administration after local administration of muscimol into the subthalamic nucleus. Indeed, an ipsilateral rotation was observed.

Dopaminergic lesions are known to profoundly affect the activity of neurons projecting from the external pallidum to the subthalamic nucleus. Firing rates in the external pallidum are decreased in rats (Pan and Walters, 1988) and monkeys (Filion et al., 1991) with a lesion of nigrostriatal dopaminergic neurons. This observation led to the hypothesis that a decrease in GABAergic tone was a major mechanism leading to the increased firing rate observed in the subthalamic nucleus (Albin et al., 1989). However, lesions of the external pallidum in rats do not induce the profound increase in firing rate observed in the subthalamic nucleus after dopaminergic lesions, suggesting that other mechanisms play a role (Hassani et al., 1996). Furthermore, although spontaneous firing rates are decreased in the globus pallidus after dopaminergic lesion, the pattern of firing is also altered, with some neurons showing a marked increase in burst firing (Pan and Walters, 1988). This type of firing pattern usually results in an increase of neurotransmitter release from axon terminals. Consistent with an increase in neurotransmitter release, numerous studies have shown an increase in the mRNA encoding glutamic acid decarboxylase, the enzyme of GABA synthesis, in the external pallidum of rats (Soghomonian and Chesselet, 1992) and monkeys (Soghomonian et al., 1994) with a lesion of dopaminergic neurons.

Currently, it is not technically possible to measure GABA release in the subthalamic nucleus of rats, and therefore it is not known whether GABA release is increased or decreased in the subthalamic nucleus after dopamine depletion. To indirectly assess

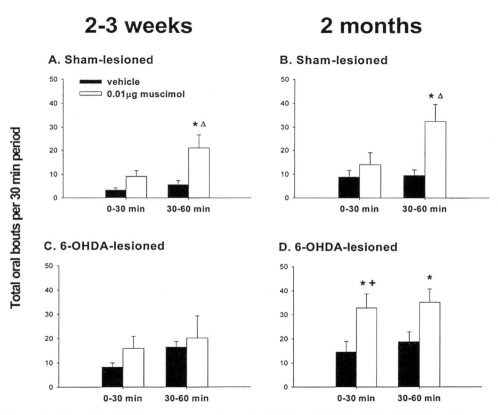

Figure 3. Effect of the GABA$_A$ receptor agonist muscimol (0.01 μg) administered locally in the subthalamic nucleus on orofacial dyskinesia in sham and dopamine depleted rats (n=6-10). One group of rats was tested 2-3 weeks after 6-OHDA lesions (A,C), and another group was tested 2 months after 6-OHDA lesions (B,D). Data are expressed as mean ± SEM of total oral bouts for the first two 30 min periods immediately following administration. * Refers to statistical difference when compared to corresponding vehicle data; Δ refers to statistical difference when compared to corresponding data from the previous 30 min time period; + refers to statistical difference when compared to corresponding data from the sham-lesioned group (Fisher's PLSD following significant one-way ANOVA, p<0.05 with Fisher PLSD).

GABAergic transmission in the subthalamic nucleus, we examined whether or not the behavioral effect of the GABA$_A$ agonist muscimol injected into the subthalamic nucleus was modified after lesions of the dopaminergic input. Three weeks after a 6-hydroxydopamine lesion, local administration of muscimol into the subthalamic nucleus no longer elicited oral dyskinesia, suggesting a decreased sensitivity of GABA responses (Figure 3). Experiments are in progress to determine whether this adaptation is the result of a change in receptor properties.

Previous studies with the dopaminergic antagonist haloperidol have shown that blockade of dopaminergic D2 receptors, a treatment which induces catalepsy in rats, also increases GAD mRNA in the external pallidum (Delfs et al., 1995b). However, this effect is transient: it is no longer observed after 4 weeks of chronic treatment (Mercugliano et al., 1992; Delfs et al., 1995b) and an opposite effect (i.e. a decreased GAD expression) is observed in rats treated for 8 months with the drug (Delfs et al., 1995b). We asked whether long term dopamine depletion also induced different effects than those observed shortly after lesion. Indeed, GAD mRNA was no longer increased in the external pallidum two months after 6-hydroxydopamine lesion (Mehta et al., in preparation). No evidence of recovery of dopaminergic innervation has been reported after this type of lesion, and dopamine transporter, measured with ligand binding autoradiography, remained profoundly decreased in the striatum of these animals. This suggests that the changes in GAD mRNA regulation in the globus pallidus are the result of functional adaptation rather than reinnervation of this region. Consistent with different changes in activity in the GABAergic globus pallidus-subthalamic nucleus pathway after short term and long term lesions, local injections of muscimol into the subthalamic nucleus induced a larger increase in orofacial dyskinesia two months after dopaminergic lesion than in controls. Specifically, the time course of the effect was altered, with a marked increase in the behavior during the first half hour following drug application (Figure 3).

These data indicate that prolonged dopamine depletion induces adaptive changes in GABAergic responses in the subthalamic nucleus. It is interesting to note that changes in GABAergic mechanisms in the globus pallidus and its output regions are tightly correlated with the occurrence of tardive dyskinesia after chronic neuroleptic treatment (Delfs et al., 1995b; Chesselet and Delfs, 1996). It is tempting to propose that changes in the responses to endogenous GABA could contribute to the development of dyskinesia after L-DOPA treatment in patients with chronic dopamine depletion. Current work in our laboratory examines this hypothesis.

When dopaminergic neurons are lesioned unilaterally, peripheral apomorphine injections induce a marked contralateral rotation, a behavior often used to assess the completeness of the lesion (Ungerstedt, 1971). Since silencing of the subthalamic nucleus itself induces an ipsiversive rotation in response to apomorphine (Delfs et al., 1995a), this test is not suitable to assess an antiparkinsonian effect of treatments that silence the subthalamic nucleus. Therefore, we examined the effects of muscimol administration into the subthalamic nucleus on other behaviors thought to reflect akinesia (Schallert and Tillerson, 1999). In agreement with a beneficial role of an inhibition of the subthalamic nucleus in patients with Parkinson's disease, local administration of muscimol into the subthalamic nucleus improved the deficit in contralateral limb use observed in a cylinder test after 6-hydroxydopamine lesions. Surprisingly, however, muscimol by itself induced a deficit in the forced step test, another measure of limb akinesia. Furthermore, muscimol potentiated the effect of a unilateral dopaminergic lesion in this test (Mehta and Chesselet, 2000). These data suggest that results from the forced step test must be interpreted with caution and may not accurately predict the antiakinetic properties of a drug.

EFFECTS OF DOPAMINERGIC LESIONS ON BEHAVIORAL AND CELLULAR RESPONSES TO 5-HT$_{2C}$ AGONISTS

The mRNA encoding the 5-HT$_{2C}$ subtype of the serotonergic receptors has a particularly interesting distribution in the basal ganglia. It is present in the striatum, particularly in its ventrolateral part, the internal pallidum, caudal part of the substantia nigra pars reticulata, and subthalamic nucleus (Eberle-Wang et al., 1997). It is absent from the external pallidum, both in rats (Eberle-Wang et al., 1997) and in humans (Pasqualetti et al., 1999). Although this distribution suggests a role for the 5-HT$_{2C}$ receptor in the control of movements, its functional role in the basal ganglia remains poorly understood. We have combined two approaches to examine the effects of 5-HT$_{2C}$ receptor stimulation in the basal ganglia and their modifications after dopaminergic lesions: measurement of orofacial dyskinesia after peripheral or local administration into the subthalamic nucleus and detection of the immediate early gene product Fos with immunohistochemical methods.

Peripheral administration of the nonselective 5-HT$_{2C}$ agonist m-CPP induced orofacial dyskinesia in rats in a dose-dependent manner (Eberle-Wang et al., 1996). A similar effect was observed when the drug was directly administered into the subthalamic nucleus (Eberle-Wang et al., 1996). In both cases, pharmacological analysis suggests that 5HT$_{2C}$ receptors played a principal role in this behavioral effect (Eberle-Wang et al., 1996). Furthermore, 5HT$_{2C}$ receptors in the subthalamic nucleus appear to be critical for the effect of peripherally administered drugs because local administration of the antagonist into the subthalamic nucleus markedly reduced the effect of peripherally administered m-CPP (Eberle-Wang et al., 1996).

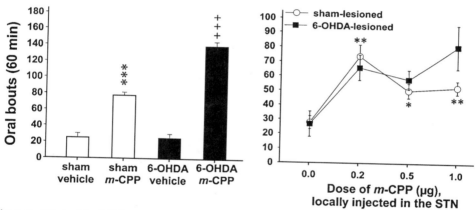

Figure 4. Effect of the 5-HT$_{2C}$ agonist m-CPP on orofacial dyskinesia in sham or dopamine depleted rats. m-CPP has been administered intraperitoneally (1 mg/kg; left panel) or locally in the subthalamic nucleus (right panel). Data are expressed as mean ± SEM of total oral bouts for the 60 min period immediately following administration. * and + Refer to statistical differences when compared to sham-lesioned animals receiving saline or m-CPP, respectively (Scheffe's test following significant one-way ANOVA). Reproduced with permission, from De Deurwaerdère and Chesselet, 2000 (copyright 2000 by the society for Neuroscience).

In rats with a unilateral lesion of the nigrostriatal pathway, the behavioral effect of peripherally administered m-CPP was markedly increased (De Deurwaerdère and Chesselet, 2000; Figure 4). However, the increased behavioral response does not seem to be mediated by the subthalamic nucleus. Indeed, no further increase in orofacial dyskinesia was observed when m-CPP was administered directly into the subthalamic nucleus of rats with lesions of the nigrostriatal dopaminergic pathway on the same side (Figure 4). Immunohistochemical detection of Fos labeled cells also supports the idea that subthalamic neurons do not increase their sensitivity to m-CPP after dopaminergic lesion. Indeed, the level of Fos induction observed in the subthalamic nucleus after peripheral administration of m-CPP was similar in intact and in lesioned rats. In contrast, the number of Fos immunoreactive cells induced by m-CPP was decreased in the medial striatum, and markedly increased in the entopeduncular nucleus of lesioned rats compared to controls (De Deurwaerdère and Chesselet, 2000). It remains unknown whether any of these sites are responsible for the increased behavioral response. However, the data clearly show that loss of dopamine induced profound alterations in the cellular response to 5HT$_{2C}$ agonists in discrete regions of the basal ganglia.

CONCLUSIONS

In summary, several conclusions emerge from these data. First, the mechanism of increased activity in subthalamic firing, a cardinal feature of Parkinson's disease, likely results from a combination of mechanisms rather than solely from a decreased GABAergic input. In particular, the loss of dopaminergic inputs from the substantia nigra may contribute significantly to this effect, but the specific mode of action of dopamine remains to be determined. In fact, the hypothesis that there is an increased release of GABA in the subthalamic nucleus at a time when subthalamic neurons show increased activity and rats display akinetic behavior is supported by a subsensitivity of subthalamic neurons to stimulation of GABA receptors. In rats with chronic dopaminergic lesions, a return of GAD mRNA levels in the globus pallidus to normal, together with an increased sensitivity to GABA agonists in the subthalamic nucleus, both suggest long term adaptive mechanisms, which may play a role in the decreased efficacy of L-DOPA, and perhaps in the appearance of side effects in patients with advanced disease. Finally, the loss of dopamine induces marked alterations in the response to serotonin in discrete regions of the basal ganglia, notably the internal pallidum. Clearly, adaptive mechanisms to changes in activity, not only of the dopaminergic system but of other neurotransmitter systems normally controlled by dopamine, are likely to play a critical role in the progression of symptoms and the response to treatment in this chronic disease.

REFERENCES

Albin, R.L., Young, A.B., and Penney, J.B., 1989, The functional anatomy of basal ganglia disorders, *Trends Neurosci.* 12:366.

Bergman, H., Wichmann, T., and DeLong, M.R., 1990, Reversal of experimental parkinsonism by lesions of the subthalamic nucleus, *Science.* 249:1436.

Brown, L.L., Markman, M.H., Wolfson, L.I., Dvorkin, B., Warner, C., and Katzman, R., 1979, A direct role of dopamine in the rat subthalamic nucleus and an adjacent intrapeduncular area, *Science.* 206:1416.

Chesselet, M.-F., and Delfs, J.M., 1996, Basal ganglia and movement disorders: an update, *Trends Neurosci.* 19:417.

De Deurwaerdère, P., and Chesselet, M.-F., 2000, Nigrostriatal lesions alter oral dyskinesia and c-fos expression induced by the serotonin agonist 1-(m-chlorophenyl)piperazine in adult rats. *J. Neurosci.* 20:5170.

Delfs, J.M., Ciaramitaro, V.M., Parry, T.J., and Chesselet, M.-F., 1995a, Subthalamic nucleus lesions: widespread effects on changes in gene expression induced by nigrostriatal dopamine depletion in rats, *J. Neurosci.* 15:6562.

Delfs, J.M., Ellison, G.D., Mercugliano, M., and Chesselet, M.-F., 1995b, Expression of glutamic acid decarboxylase mRNA in striatum and pallidum in an animal model of tardive dyskinesia, *Exp. Neurol.* 133:175.

Eberle-Wang, K., Lucki, I., and Chesselet, M.-F., 1996, A role for the subthalamic nucleus in 5-HT2C-induced oral dyskinesia, *Neuroscience.* 72:117.

Eberle-Wang, K., Mikeladze, Z., Uryu, K., and Chesselet, M.-F., 1997, Pattern of expression of the serotonin2C receptor messenger RNA in the basal ganglia of adult rats. *J. Comp. Neurol.* 384:233.

Filion, M., and Tremblay, L., 1991, Abnormal spontaneous activity of globus pallidus neurons in monkeys with MPTP-induced parkinsonism, *Brain Res.* 547:142.

Flores, G., Liang. J.J., Sierra, A., Martinez-Fong, D., Quiron, R., Aceves, J., and Srivastava, L.K., 1999, Expression of dopamine receptors in the subthalamic nucleus of the rat: characterization using reverse transcriptase-polymerase chain reaction and autoradiography, *Neuroscience*, 91:549.

Fremeau, R.T., Duncan, G.E., Fornaretto, M.G., Dearry, A., Gingrich, J.A., Breese, G.R., and Caron, M. G., 1991, Localization of D1 Dopamine receptor mRNA in brain supports a role in cognitive affective and neuroendocrine aspects of dopaminergic neurotransmission, *Proc. Natl. Acad. Sci. U.S.A.* 88:3772.

Hamada, I., and DeLong, M.R., 1992, Excitotoxic acid lesions of the primate subthalamic nucleus result in transient dyskinesias of the contralateral limbs. *J. Neurophysiol.* 68:1850.

Hassani, O.K., Mouroux, M., and Féger, J., 1996, Increased subthalamic neuronal activity after nigral dopaminergic lesion independent of disinhibition via the globus pallidus, *Neuroscience,* 72:105.

Hassani, O.K., François, C., Yelnik, J., and Feger, J., 1997, Evidence for a dopaminergic innervation of the subthalamic nucleus in the rat, *Brain Res.* 749:88.

Hassani, O.K., and Feger, J., 1999, Effects of intrasubthalamic injection of dopamine receptor agonists on subthalamic neurons in normal and 6-hydroxydopamine-lesioned rats: an electrophysiological and c-fos study, *Neuroscience,* 92:533.

Herdeen, J.C., 1999, Tyrosine hydroxylase-immunoreactive elements in the human globus pallidus and subthalamic nucleus, *J. Comp. Neurol.* 409:400-410.

Hutchinson, W.D., Allan, R.J., Opitz, H., Levy, R., Dostrovsky, J.O., Lang, A.E., and Lozano, A.M., 1998, Neurophysiological identification of the subthalamic nucleus in surgery for Parkinson's disease, *Ann. Neurol.* 44:622.

Kreiss, D.S., Mastropietro, C.W., Rawji, S.S., and Walters, J.R., 1997, The response of subthalamic nucleus neurons to dopamine receptor stimulation in a rodent model of Parkinson's disease, *J. Neurosci.* 17:6807.

Kumar, R., Lozano, A.M., Kim, Y.J., Hutchison, W.D., Sime, E., Halket, E., and Lang, A.E., 1998, Double-blind evaluation of subthalamic nucleus deep brain stimulation in advanced Parkinson's disease, *Neurology,* 51:850.

Lavoie, B., Smith, Y., and Parent, A., 1989, Dopaminergic innervation of the basal ganglia in the squirrel monkey as revealed by tyrosine hydroxylase immunohistochemistry, *J. Comp. Neurol.* 289:36.

Levy, R., Hazrati, L.N., Herrero, M.T., Vila, M., Hassani, O.K., Mouroux, M., Ruberg, M., Asensi, H., Agid, Y., Feger, J., Obeso, J.A., Parent, A., and Hirsch, E.C., 1997. Re-evaluation of the functional anatomy of the basal ganglia in normal and Parkinsonian states. *Neuroscience,* 76:335.

Lindvall, O., and Bjorklund, A., 1978, Anatomy of the dopaminergic neuron systems in the rat brain, *Adv. Biochem. Psychopharmacol.* 19:1.

Mehta, A., Thermos, K., and Chesselet, M.-F., 2000, Increased behavioral response to dopaminergic stimulation of the subthalamic nucleus after nigrostriatal lesions, *Synapse,* 37:298.

Mehta, A., and Chesselet, M.-F., 2000, Effects of GABA$_A$ receptor stimulation in the subthalamic nucleus on forelimb akinesia in the hemi-parkinsonian rat, *Soc. Neurosci. Abstr.* In press.

Mercugliano, M., Saller, C.F., Salama, A.I., U'Prichard, D.C., and Chesselet, M. F., 1992, Clozapine and haloperidol have differential effects on glutamic acid decarboxylase mRNA in the pallidal nuclei of the rat. *Neuropsychopharmacol.* 6:179.

Pan, H. S., and Walters, J.R., 1988. Unilateral lesion of the nigrostriatal pathway decreases the firing rate and alters the firing pattern of globus pallidus neurons in the rat, *Synapse.* 2:650.

Parry, T.J., Eberle-Wang, K., Lucki, I., and Chesselet, M.-F., 1994, Dopaminergic stimulation of subthalamic nucleus elicits oral dyskinesia in rats, *Exp. Neurol.* 128:181.

Pasqualetti, M., Ori, M., Castagna, M., Marazziti, D., Cassano, G.B., Nardi, I., 1999, Distribution and cellular localization of the serotonin type 2C receptor messenger RNA in human brain, *Neuroscience.* 92:601.

Schallert, T., and Tillerson, J.L., 1999, Intervention strategies for degeneration of dopamine neurons in parkinsonism; optimizing behavioral assessment of outcome, in: *Central Nervous System Diseases,* D.F. Emerich, R.L. Dean III and P.R. Sanberg, eds., Humana Press, Inc., Totowa, NJ.

Smith, Y., Bolam, J.P., and Von Krosigk, M., 1990, Topographical and synaptic organization of the GABAR-containing pallido-subthalamic projection in the rat, *Eur. J. Neurosci.* 2:500.

Smith Y., Bevan, M.D., Shink, E., and Bolam, J.P., 1998, Microcircuitry of the direct and indirect pathways of the basal ganglia, *Neuroscience,* 86:353.

Soghomonian, J.J., and Chesselet, M.-F., 1992, Effects of nigrostriatal lesions on the levels of messenger RNAs encoding two isoforms of glutamate decarboxylase in the globus pallidus and entopeduncular nucleus of the rat, *Synapse,* 11:124.

Soghomonian, J.J., Pedneault, S., Audet, G., and Parent, A., 1994, Increased glutamate decarboxylase mRNA levels in the striatum and pallidum of MPTP-treated primates. *J. Neurosci.* 14:6256.

Ungerstedt, U., 1971, Postsynaptic supersensitivity after 6-hydroxy-dopamine induced degeneration of the nigro-striatal dopamine system. *Acta. Physiol. Scand. Suppl.* 367:69.

NEUROTRANSMITTERS AND RECEPTORS IN THE PRIMATE MOTOR THALAMUS AND POTENTIAL FOR PLASTICITY

KRISTY KULTAS-ILINSKY AND IGOR A. ILINSKY[*]

INTRODUCTION

Unlike neuronal circuitry in the striatum, which is rich in a wide variety of neurotransmitters and modulators, the neurochemical composition of the motor thalamus[1] is relatively simple. The two major neurotransmitters in it are glutamate and GABA. Glutamate appears to be the neurotransmitter in the cerebellothalamic (Nieoullon et al., 1984) and corticothalamic synapses (Fonnum et al., 1981; Young et al., 1981, 1983; Deschenes and Hu, 1990), whereas GABA is the neurotransmitter released by axonal and dendritic synapses of local circuit neurons (LCN), and terminals of nigro-, pallido- and reticulo-thalamic afferents (DiChiara et al., 1979; Houser et al., 1980; Penney and Young, 1981; Ilinsky et al., 1997; Ilinsky et al., 1999). In addition, there is also a diffuse cholinergic input to the thalamus derived from the brainstem and the basal forebrain (Heckers et al., 1992), but its role in the primate motor thalamic nuclei has not been evaluated.

There are very few if any studies on the glutamatergic mechanisms in the motor thalamus. We know that all three types of ionotropic glutamate receptors (AMPA, KA and NMDA subtypes) are present and are distributed rather uniformly (Halpain et al., 1984; Monaghan and Cotman, 1985; Zuercher et al., 1987; Nielsen et al., 1990; Young et al., 1990). NMDA receptors mediate transmission at corticothalamic synapses as demonstrated by Deschenes and Hu (1990) in the cat VL. Little is known about metabotropic glutamate receptors and their functional role in the motor thalamus.

There is more information available about GABAergic mechanisms in the motor thalamus than about other neurotransmitter systems. However, this is not to say that we know a lot. The motor thalamus has not been subjected to such rigorous studies as the basal ganglia has, despite the fact that it is the major recipient of basal ganglia output and mediates practically all basal ganglia influences on behavior. Limited information on thalamic mechanisms involved in the processing of basal ganglia as well as cerebellum-derived information is the major deficiency in our overall understanding of the mechanisms

---

[*] KRISTY KULTAS-ILINSKY AND IGOR A ILINSKY • Department of Anatomy and Cell Biology, University of Iowa College of Medicine, Iowa City, Iowa 52242.
[1] motor thalamus is defined here as in the chapter by I.A. Ilinsky and Kultas-Ilinsky in this volume.

of normal motor control and movement disorders. One of the factors responsible for the slow progress in this field is the substantial interspecies differences in the organization of the motor thalamic circuitry, which is not entirely unexpected in view of the remarkable differences in movement repertoires of different species. Limited accessibility of the motor thalamic nuclei to experimental manipulations and confusion in the minds of investigators regarding thalamic delineations have also tempered enthusiasm for investigating this brain region. This chapter will summarize our current understanding of the GABAergic systems in the motor thalamus with emphasis on primates and their potential for lesion-induced plasticity.

NEUROANATOMICAL ORGANIZATION OF THE MOTOR THALAMIC NUCLEI IN PRIMATES AND RODENTS IS DIFFERENT

The difference between the two species is mainly due to the differences in the organization of GABAergic connections. There is a much greater variety of GABAergic synapses in the primate thalamus compared to the rat (Ilinsky et al., 1993). In the rat, aside from the GABAergic basal ganglia afferents, there is only one other source of GABAergic innervation: the reticular thalamic nucleus. In species with more complex behavior and especially in primates there is an added GABAergic component, namely local circuit neurons (LCN) or interneurons. However, even in the species that have LCN, the circuitry in the motor thalamus is not similar simply because the LCN connections differ in different species. For example, in the monkey, LCN seem to form synapses on other LCN (Kultas-Ilinsky and Ilinsky, 1991; Mason et al., 1996) a feature not described in the cat. Moreover, in the cat thalamus, there are very few, if any, synapses by reticular nucleus fibers on LCN (Liu et al., 1995). Yet, as our recent studies demonstrated, a minimum of 50 % of synapses formed by axons of the reticular thalamic nucleus in the monkey thalamus are on the LCN (Kultas-Ilinsky et al., 1995; Tai et al., 1995; Ilinsky et al., 1999). This makes the circuitry in the primate motor thalamus unique because unlike in the cat, the reticular nucleus in the monkey will exert not only a direct inhibitory effect on thalamocortical projection neurons but will also change the excitability of other pools of neurons via LCN mediated disinhibition. There are no electrophysiological studies in primates on relationships between the reticular nucleus and other thalamic nuclei. However, reticulothalamic mechanisms have been extensively studied in rodents and carnivores mainly in the sensory thalamic nuclei. It is interesting that biophysical properties of IPSPs generated in thalamocortical neurons by NRT stimulation are not universal but differ in different nuclei and species (see Huguenard, 1998, for review). The reasons for this variability are still debated, but we consider differences in the wiring to be the most likely culprit. It is not clear how different or similar the reticular nucleus interactions with the motor thalamic nuclei in primates will turn out to be compared, for example, to somatosensory nuclei adjacent to them.

Just as in the case of reticulothalamic connections, GABAergic pallidothalamic and nigrothalamic afferents in the cat and monkey also terminate on LCN, thereby exerting a dual influence on thalamocortical projection neurons (Kultas-Ilinsky et al., 1983; Ilinsky and Kultas-Ilinsky, 1990; Ilinsky et al., 1997). Again there are species differences in the organization of these inputs: pallidal synapses in the monkey participate in complex synaptic arrangements such as triads with three inhibitory synapses, which were not found in the cat.

Thus, in view of these species differences in neuroanatomical organization, the GABAergic regulation of the activity of thalamocortical projection neurons in the primate motor thalamus should differ substantially from that in rodents due to the fact that in primates a part of the afferent input is mediated by GABAergic LCN. It should also differ from that in other species containing LCN, such as cat, due to the higher number of GABAergic circuits involved in shaping the activity of thalamocortical projection neurons.

MOTOR THALAMIC NUCLEI IN THE SAME SPECIES DIFFER WITH RESPECT TO LCN CONNECTIVITY

If one utilizes the definition of the motor thalamus proposed by Ilinsky and Kultas-Ilinsky (see chapter in this volume), then in primates it will be comprised of a number of thalamic subdivisions, including such nuclei as ventral anterior, with its medial (VAmc) and lateral (VApc/VAdc) parts receiving nigrothalamic and pallidothalamic afferents, respectively; the ventral lateral nucleus (VL, recipient of cerebellothalamic afferents), mediodorsal nucleus (MD) receiving nigral, pallidal and even cerebellar efferents in its different subdivisions; centromedian (CM) and centrolateral nuclei (CL) receiving pallidothalamic and cerebellothalamic afferents, respectively; and some other nuclei. Again, the organization of neuronal circuits in these nuclei differs because of the peculiarities of LCN, i.e. GABAergic, connections. This is the case even when the extrinsic afferents to the nuclei are similar. For example, in the magnocellular part of the VAmc that receives GABAergic nigrothalamic afferents, LCN are very few (Ilinsky and Kultas-Ilinsky, 1990;

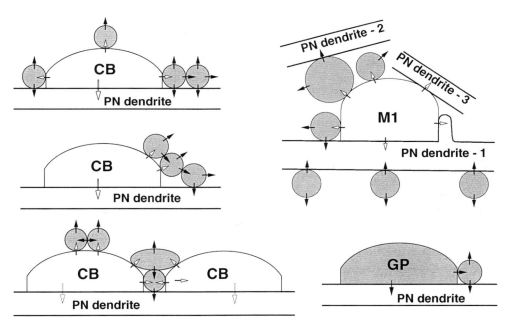

Figure 1. The diagrams illustrate some of the complex relationships of local circuit neuron dendrites (LCN) associated with some afferent inputs, such as cerebellar (CB), primary motor cortex (M1), and pallidal (GP) that contact projection neuron (PN) dendrites. Shaded ovals and circles represent GABAergic structures: LCN dendrites and a pallidal bouton. Black arrows indicate GABAergic synapses, white arrows - glutamatergic synapses. Two-headed arrow indicates a reciprocal synapse between two LCN dendrites. Free arrows pointing away from LCN indicate that these dendrites make additional synapses on PN and/or LCN dendrites. There is no direct evidence at this point to show whether different dendrites associated with one extrinsic afferent terminal belong to different LCN or a single one. Nonetheless, since each LCN interacts with more than one projection neuron, such relationships suggest that the input from a single afferent terminal will influence excitability of several projection neurons, even if mediated by one LCN. In cases where there are serial LCN to LCN synapses imposed between the afferent and projection neuron the net result will depend on the number and combination of synapses and neurons in these chains.

Kultas-Ilinsky and Ilinsky, 1990), whereas in the pars multiformis of the MD (MDmf), whose connections with the substantia nigra pars reticularis and prefrontal cortex are identical to those of the VAmc, the LCN are extremely numerous and form a multitude of synapses on thalamocortical projection neurons and in complex synaptic arrangements (Tai et al., 1996). The latter are practically absent in the VAmc (Ilinsky and Kultas-Ilinsky, 1990). There are also differences in the neuronal circuitry of the two pallidal afferent receiving nuclei, VAdc and CM (Ballercia et al., 1996; Kultas-Ilinsky et al., 1997).

In some motor thalamic nuclei of the monkey, LCN form synapses on one other. For example, these LCN to LCN synapses are very numerous in the VL, i.e., the cerebellar projection zone (Kultas-Ilinsky and Ilinsky, 1991) but not obvious in the basal ganglia afferent territory, i.e., VA. LCN to LCN synapses make the thalamic circuitry in primates more complex and perhaps also more versatile. It is important to keep in mind that thalamic LCN are special in that their dendrites, in addition to their axons, are presynaptic. It is not clear at this point whether LCN contact one another via their axons or whether their axons terminate only on projection neurons, but there is ample evidence showing that LCN in the monkey VL are connected with one another via unidirectional and reciprocal dendro-dendritic synapses (Harding, 1973; Harding and Powel, 1977; Kultas-Ilinsky and Ilinsky, 1991; Mason et al., 1996). Examples of such complex synapses with participation of LCN dendrites are illustrated in Figure 1.

If GABAergic synapses are classified by the origin of their pre- and postsynaptic components then it is possible to identify at least 10 different types in the motor thalamic nuclei. These will include: nigral synapses on projection neurons, pallidal synapses on projection neurons, reticular synapses on projection neurons, nigral synapses on LCN, pallidal synapses on LCN, reticular synapses on LCN, LCN dendrites on projection neurons, LCN axons on projection neurons, LCN dendrites on LCN, LCN axons on LCN.

However, this is a rather rough classification since it does not take into account whether synapses are on the dendrites, somata or axon initial segments. If these were taken into account then the number of combinations would at least double.

DO ANATOMICALLY DIFFERENT GABAERGIC SYNAPSES IN THE MOTOR THALAMUS DIFFER FUNCTIONALLY?

The answer to this question will most likely be affirmative, though at present direct evidence to support this claim is lacking. A recent study by Gupta et al. (2000) on identified cortical neurons in the rat convincingly demonstrated a great electrophysiological diversity of GABAergic synapses, even of synapses formed by axons of an identified cell type on different postsynaptic neurons. The exact mechanisms responsible for this diversity are not quite understood, but one of the contributing factors must be the presence of GABA receptors with different subunit composition at physiologically different GABAergic synapses. It has also been demonstrated that individual cortical neurons can express multiple receptor subtypes in functionally distinct synapses (Fritschy et al., 1998). All these findings in the cortex provide independent support for the idea expressed earlier that in the primate motor thalamus anatomically different types of GABAergic synapses utilize receptors with different subunit composition and hence differ functionally (Kultas-Ilinsky et al., 1998). Several lines of indirect evidence are available that point in this direction. These come from the studies on receptor binding autoradiography, *in situ* hybridization, and lesioning of extrinsic GABAergic inputs to the motor thalamus in monkeys and then comparing the results with the data available in other species. This evidence is summarized below.

GABA$_A$ receptor binding sites are differentially distributed in the nuclei of the monkey motor thalamus (Kultas-Ilinsky et al., 1993; Ambardekar et al., 2000) just as in the cat (Kultas-Ilinsky et al., 1988). The ratio of the high affinity GABA$_A$ binding sites to the low affinity $GABA_A$ or benzodiazepine binding sites is about five times lower in the monkey (Kultas-Ilinsky et al., 1993) compared to the rat (Olsen et al., 1990). This is due to the remarkably higher number of benzodiazepine binding sites in the monkey thalamus compared to the rat. It is interesting that the density of benzodiazepine binding sites in the cat motor thalamic nuclei is intermediate between the values for the rat and monkey (Kultas-Ilinsky et al., 1988; Kultas-Ilinsky et al., 1993). There is also a remarkably greater variety of $GABA_A$ receptor subunits expressed in the monkey motor thalamic nuclei. Eleven subunits of 11 studied so far are expressed in the motor thalamic nuclei of the monkey (Huntsman et al., 1996; Kultas-Ilinsky et al., 1998), compared to the rat, where only four were detected of 13 studied (Wisden et al., 1992). In the monkey, these subunits are differentially expressed in different nuclei and types of neurons. Even the nuclei that receive the same main subcortical GABAergic afferent input, like VAdc and CM, are characterized by remarkable differences in the expression level of $GABA_A$ receptor subunits and also in the binding density of both $GABA_A$ and $GABA_B$ receptor ligands (Kultas-Ilinsky et al., 1998; Ambardekar et al., 1999). The density of $GABA_B$ receptors is also several times higher in the monkey thalamus compared to that in the rat (Bowery et al., 1999; Ambardekar et al., 1999), although $GABA_B$ binding sites are distributed more uniformly throughout the thalamus compared to $GABA_A$ sites. These internuclear and interspecies differences in receptors may be directly related to the differences in the anatomical organization of GABAergic circuits in the motor thalamic nuclei of different species, as briefly reviewed above.

The data obtained from those studies as well as results from some lesioning experiments (Kultas-Ilinsky et al., 1990) allowed us to advance several hypotheses related to the composition of anatomically different GABAergic synapses in the primate motor thalamus (see Kultas-Ilinsky et al., 1998, for details). For example, we have predicted that the benzodiazepine binding sites, i.e., the low affinity $GABA_A$ sites, are definitely associated with the dendro-dendritic synapses formed by LCN, whereas reticulothalamic synapses, except for those in the intralaminar and midline nuclei, are rich in the high-affinity $GABA_A$ sites, with the most likely subunit composition of receptors at these synapses being $\alpha_1\alpha_4\beta_2\delta$. The data also suggest that $GABA_A$ receptors at the nigrothalamic and pallidothalamic synapses are different. The receptors at the nigrothalamic synapses are most likely enriched in benzodiazepine binding sites and contain γ_2 and β_1 subunits. Interestingly, $GABA_A$ receptors at the pallidothalamic synapses, which are derived from collaterals of the same axons in two different nuclei, VAdc and CM (Parent and De Bellefeuille, 1982), are likely to have significantly different subunit composition, and hence different functional characteristics (Kultas-Ilinsky et al., 1998). Finally, about 30% of the total number of $GABA_B$ receptors in the motor thalamic nuclei appear to be presynaptic on reticulothalamic synapses (Ambardekar et al., 1999), whereas some of the rest would be found on glutamatergic corticothalamic terminals. Obviously this information is by no means sufficient for any meaningful hypotheses about the functional properties of anatomically different GABAergic synapses, but it can be used as a starting point for the future quest to acquire such knowledge. A significant breakthrough could come from electrophysiological studies on the synaptic transmission between identified cell types in different nuclei of the primate motor thalamus, which would shed some light on the properties of GABAergic synapses in these regions, especially those formed by LCN. At the moment such data do not exist.

It is well known that the subunit composition determines the channel properties and pharmacology of the receptors. $GABA_A$ receptors, especially those that contain benzodiazepine binding sites, have especially intricate responses to a diverse range of compounds that bind to modulatory benzodiazepine sites and possess a variety of clinical properties (for review see Lüddens et al., 1995; Lüddens, 1998). A total of 19 $GABA_A$ receptor subunit genes have been cloned so far. This makes 800 subunit combinations, hence receptor subtypes, theoretically possible (Barnard, 1996). As noted above, mRNAs of

11 subunits have been found to be expressed in the primate motor thalamus. Thalamic tissue was also found to be enriched in recently cloned ε subunit mRNA, although its nuclear distribution is not clear as of yet (Davies et al., 1997). Even these incomplete data imply the existence of a substantial number of subunit combinations with very distinct pharmacological profiles in the primate thalamus. If and when the specific subtypes are localized to particular types of GABAergic synapses in the motor thalamus, a wide window of opportunities will open for designing drugs that selectively target specific parts of thalamic circuits. This, in turn, could be one of the means to restore balance to the malfunctioning thalamic circuits observed in various movement disorders.

PLASTICITY OF GABAERGIC SYSTEMS IN THE MOTOR THALAMUS

Basal ganglia afferent projection zones in the thalamus may be affected to different degrees in different types of movement disorders depending on the location of the primary lesion site. In conditions like multisystem degeneration or postencephalitic parkinsonism, where neuronal degeneration in the medial globus pallidus and/or substantia nigra pars reticularis takes place, the neurons in corresponding thalamic afferent zones become deafferented from their GABAergic basal ganglia inputs. In other instances, like idiopathic parkinsonism, dystonia, chorea and so on, direct degeneration might not occur, but the basal ganglia thalamic connections are functionally altered. In both instances, however, intact neurons and or axons in the area may attempt to compensate for the loss or malfunctioning of the inputs from the basal ganglia.

The classic concept derived from the animal lesioning experiments (Pan et al., 1983) is that if a brain region is deafferented, then the postsynaptic receptors associated with the lesioned pathways upregulate, whereas the presynaptic receptors decrease in number immediately after the lesion. At the same time, GABA receptors down regulate at the projection sites of deafferented neurons if these neurons are also GABAergic. In deafferentation experiments in the monkey that we performed recently, this concept was applicable only to some results. For example, we found a 20-30% decrease in the number of $GABA_B$ binding sites in the monkey motor thalamic nuclei one week after NRT lesions (Ambardekar et al., 1999), which is consistent with the presynaptic location of these receptors on reticulothalamic terminals suggested in physiological studies (Le Feuvre et al., 1997). On the other hand, we also found a statistically significant decrease of tritiated muscimol and flunitrazepam binding sites in the same nuclei after these lesions. It is well established that both $GABA_A$ and $GABA_B$ receptors mediate IPSPs on thalamocortical projection neurons generated by reticular input (Ulrich and Huguenard, 1996). There is little evidence for the presynaptic location of $GABA_A$ receptors on reticulothalamic terminals, but such a location has been suggested in studies on other systems (Churchill et al., 1991). Obviously, the presynaptic location of $GABA_A$ receptors on reticulothalamic synapses can not be entirely excluded at this point although other explanations for these results can also be found. One of the factors that may have contributed to the results is that half of the synapses of the lesioned pathway are on GABAergic LCN. These neurons may react to deafferentation differently from projection neurons. Results of experiments with denervation of somatosensory thalamus from the excitatory dorsal column input by Ralston et al. (1996) represent interest in this context. These investigators found a statistically significant decrease of GABAergic synapses formed by LCN and NRT that was paralleled by the loss of GABA immunoreactivity, while there were no signs of reactive synaptogenesis by intact excitatory inputs. Such a "withdrawal" reaction to denervation by LCN may be specific to somatosensory nuclei, but if it is a more common phenomenon it may account for the loss of at least some of the postsynaptic GABA receptors in our experiments.

Similar to the results of lesioning experiments in the reticular nucleus, kainic acid lesions of GPm also resulted in down regulation of muscimol binding sites in the pallidal territory of the thalamus a week after lesioning (Ambardekar et al., in preparation). In contrast, the number of flunitrazepam binding sites in these experiments was upregulated. Four months after similar lesions, the number of both muscimol and flunitrazepam sites slightly increased compared to the Bmax values at one week postlesion. It is interesting that in the two different lesioning experiments (NRT *vs.* GPm lesioning), the muscimol and benzodiazepine sites behaved differently. And it is even more interesting that in experiments with pallidal lesions the two sites displayed different time course of changes: the number of flunitrazepam binding sites continued to increase slowly over the time period studied, whereas the muscimol binding sites decreased initially but started to come back four months later. A similar trend was noted earlier in some deafferentation experiments in the cats (Kultas-Ilinsky et al., 1990). This suggests that flunitrazepam and muscimol binding sites that react to the lesioning of the globus pallidus are parts of different receptor complexes and most likely are located at anatomically different GABAergic synapses.

In our earlier EM study in the adult cat we made combined lesions in the two basal ganglia output structures, substantia nigra pars reticularis and entopeduncular nucleus, and analyzed the area of overlap of the two projections at two survival times - four days and one year (Kultas-Ilinsky et al., 1992). Quantitative EM analysis revealed that at four days postlesion a large proportion of boutons with symmetric synaptic contacts, a typical feature of the basal ganglia boutons in the thalamus, were removed from secondary dendrites, but 11% were still degenerating. On primary projection neuron dendrites, 49% of boutons with these features were degenerating and 33 % on tertiary dendrites. A year after lesioning, a few degenerating boutons were still observed on primary and tertiary dendrites of projection neurons, but none were seen on the secondary. Instead, one found unusual looking boutons that displayed features of growth cones at all three locations: 12% on primary, 22% on secondary, and 32% on tertiary dendrites. On the other hand, LCN displayed clear signs of hypertrophy at both light and electron microscopic levels accompanied by an increase in the immunostaining for glutamic acid decarboxylase. Moreover, the appositional length between the LCN and projection neuron dendrites increased, as did the number of dendrodendritic synapses formed by LCN on projection neurons. Nonetheless, the density of all symmetric synapses never returned to normal levels, except for those on tertiary dendrites where it was even slightly increased compared to the control. These findings demonstrated that after severe deafferentation of the motor thalamus from the extrinsic GABAergic inputs, LCN dendrites participated in synaptic remodeling along with some axon terminals of unidentified origin. Possible sources of these reactive boutons could be from contralateral basal ganglia afferents that are normally present in the cat thalamus in small numbers (Ilinsky et al., 1987), axons of LCN, or axons of the reticular thalamic nucleus. It is also noteworthy that degenerating terminals were still found a year after deafferentation. This suggests that lesion-induced remodeling may be an ongoing, lengthy process that also involves elimination of some sprouted terminals. Studies with shorter intervals between survival times are needed to determine whether different systems make re-innervation attempts at different postlesion survival times.

These findings clearly suggest the potential for plasticity in the motor thalamus on the part of GABAergic systems. Literature data suggest a variety of lesion-induced plasticity type changes in the thalamus in general. Regeneration of NRT axons in the adult rat thalamus has been suggested by Benfey et al. (1985). Reactive synaptogenesis on the part of cortical afferents to the rat VL was demonstrated after its deafferentation from the cerebellar input (Bromberg et al., 1987). Clinical studies in humans suggest that there is definitely a reorganization of connectivity in the motor and somatosensory thalamus in human patients with dystonia (see chapter by Lenz et al. in this volume). We suggest that some kind of slowly progressing reorganization also takes place in the motor thalamus in other movement disorders and, specifically, in Parkinson's disease.

The main conclusion of this review is that we still know very little about what is happening at the thalamic level in both the normally functioning brain and in movement disorders. Obviously, a considerable effort is needed to make progress. It is also obvious that the final answers can come only from studies on primates. On the other hand, since primate research is extremely complicated for a variety of reasons, unconventional and ingenious approaches must also be developed to make progress along these lines.

REFERENCES

Ambardekar, A.V., Ilinsky, I.A., Froestl, W., Bowery, N.G., and Kultas-Ilinsky K., 1999, Distribution and Properties of GABA$_B$ antagonist [^{3}H] CGP 62349 binding in the rhesus monkey thalamus and basal ganglia and the influence of lesions in the reticular thalamic nucleus, *Neuroscience*, 93:1339.

Ambardekar, A., Surin, A., Parts, K., Ilinsky, I.A., and Kultas-Ilinsky, K., 2000, Distribution and binding parameters of muscimol and flunitrazepam in the thalamic nuclei of *Macaca mulatta* and effects of lesioning in the globus pallidus and reticular thalamic nucleus, (*in preparation*)

Ballercia, G., Kultas-Ilinsky, K. Bentivoglio M., and Ilinsky I.A., 1996, Neuronal and synaptic organization of the centromedian nucleus of the monkey thalamus: a quantitative ultrastructural study, with tract tracing and immunohistochemical observations, *J. Neurocytol.* 25:267.

Barnard, E.A., 1996, The molecular biology of GABA$_A$ receptors and its applications, in: *GABA: Receptors, Transporters and Metabolism*, C. Tanaka and N.G. Bowery, eds., Bilkhauser, Basel.

Benfey, M., Bünger, U.R., Vidal-Sanz, M., Bray, G.M., and Aguayo, A.J., 1985, Axonal regeneration from GABAergic neurons in the adult rat thalamus, J. Neurocytol. 19:279.

Bowery, N.G., Parry, K., Goodrich, G., Ilinsky, I.A., and Kultas-Ilinsky, K., 1999, Distribution of GABA$_B$ binding sites in the thalamus and basal ganglia of the rhesus monkey (*Macaca mulatta*), *Neuropharmacol.* 38:1675.

Bromberg, M.B., Pamel, G., Stephenson, B.S., Young, A.B., and Penney, J.B., 1987, Evidence for reactive synaptogenesis in the ventrolateral thalamus and red nucleus of the rar, changes in high affinity glutamate uptake and numbers of coricofugal fibers, *Exp. Brain Res.* 69:53.

Churchill, L., Bourdelais, A., Austin, M., Zahm, D.S, Kaliwas, P.W., 1991, gamma-Aminobutyric acid and mu-opioid receptor localization and adaptation in the basal forebrain, *Adv. Exp. Biol. Med.* 295:101.

Davis, P.A., Hanna, M.C., Hales, T.G., and Kirkness, E.W., 1997, Insensitivity to anaesthetic agents conferred by a class of GABA$_A$ receptor subunit, *Nature,* 385:820.

Deschenes, M., and Hu, B., 1990, Electrophysiology and pharmacology of the corticothalamic input to the lateral thalamic nuclei: an intracellular study in the cat, *Eur. J. Neurosci.* 2:140.

DiChiara, G., Porceddu, M.L., Morelli, M., Mulas, M.L., and Gessa, M.L., 1979, Evidence for a GABAergic projection from substantia nigra to the ventromedial thalamus and to the superior colliculus of the rat, *Brain Res.* 176:273.

Fonnum, F., Storm-Mathisen, J., and Divac, I., 1981, Biochemical evidence for glutamate as a neurotransmitter in corticostriatal and corticothalamic fibres in rat brain, *Neurosci.* 6:863.

Fritschy, J.M., Weinmann, O., Wenzel, A. and Benke, D., 1998, Synapse specific localization of NMDA and GABA$_A$ receptor subunits revealed by antigen retrieval immunocytochemistry, *J. Comp. Neurol.* 390:194.

Gupta, A., Wang, Y., and Markram, H., 2000, Organizing principles for a diversity of GABAergic interneurons and synapses in the neocortex, *Science*, 287:273.

Halpain, S., Wieczorek, C.M., and Rainbow, T.C., 1984, Localization of glutamate receptors in rat brain by quantitative autoradiography, *J. Neurosci.* 4:2246.

Harding, B.N., 1973, An ultrastructural study of the termination of afferent fibres within the ventrolateral nuclei of the monkey thalamus, *Brain Res.* 54:341.

Harding, B.N., and Powell, T.P.S., 1977, An electron microscopic study of the centre-median and ventrolateral nuclei of the thalamus in the monkey, *Philos. Trans. R. Soc. Lond. (Biol)* 279:359.

Heckers, S., Geula, C., and Mesulam, M., 1992, Cholinergic innervation of the human thalamus: Dual origin and differential nuclear distribution, *J. Comp. Neurol.* 325:68.

Houser, C.R., Vaughn, J.E., Barber, R.P., and Roberts, E., 1980, GABA neurons are the major cell type of the nucleus reticularis thalami, *Brain Res.* 200:341.

Huguenard, J.R., 1998, Anatomical and physiological considerations in thalamic rhythm generation, *J. Sleep Res.* 7:24.

Huntsman, M.M., Leggio, M.G., and Jones, E.G., 1996, Nucleus-specific expression of GABA$_A$ receptor subunit mRNAs in monkey thalamus, *J. Neurosci.* 16:3576.

Ilinsky I.A., Ambardekar, A., and Kultas-Ilinsky, K., 1999, Organization of projections from the anterior pole of the nucleus reticularis thalami (NRT) to subdivisions of the motor thalamus: Light and electron microscopic studies in the rhesus monkey, *J. Comp. Neurol.* 409:369.

Ilinsky, I.A., and Kultas-Ilinsky, K., 1990, Fine structure of the ventral anterior thalamic nucleus (VAmc) of *Macaca mulatta*: I. Cell types and synaptology, *J. Comp. Neurol.* 294:455.

Ilinsky I.A., and Kultas-Ilinsky, K., 1997, Mode of termination of pallidal afferents in the thalamus: A light and Electron microscopic study with anterograde tracers and immunocytochemistry in *Macaca mulatta*, *J. Comp. Neurol.* 386:601.

Ilinsky, I.A., and Kultas-Ilinsky, K., Rosina, A., and Haddy, M., 1987, Quantitative evaluation of crossed and uncrossed projections from basal ganglia and cerebellum to the cat thalamus, *Neurosci.* 21:207.

Ilinsky I.A., Toga, A.W., and Kultas-Ilinsky, K., 1993, Anatomical organization of internal neuronal circuits in the motor thalamus, in: *Thalamic Networks for Relay and Modulation*, D. Minciacchi, M. Molinari, G. Macchi and E.G. Jones, eds., Pergamon Press, Oxford-New York.

Kultas-Ilinsky, K., DeBoom, T., and Ilinsky, I.A., 1992, Synaptic reorganization in the feline ventral anterior thalamic nucleus induced by lesions in the basal ganglia, *Exp. Neurol.* 116:312.

Kultas-Ilinsky, K., DeBoom, T., and Ilinsky, I.A., 1993, Interspecies comparison of expression of GABA/Benzodiazepine receptors and their subunits in the motor and limbic nuclei of the thalamus, in: *Thalamic Networks for Relay and Modulation*, D. Minciacchi, M. Molinari, G. Macchi and E.G. Jones, eds., Pergamon Press, Oxford-New York.

Kultas-Ilinsky, K., Fogarty, J.D., Hughes, B., and Ilinsky, I.A., 1988, Distribution and binding parameters of GABA and benzodiazepine receptors in the cat motor thalamus and adjacent nuclear groups, *Brain Res.* 459:1.

Kultas-Ilinsky, K., Hughes B., Fogarty, J.D., and Ilinsky, I.A., 1990, GABA and benzodiazepine receptors in the cat motor thalamus after lesioning of nigro and pallidothalamic pathways, *Brain Res.* 511:197.

Kultas-Ilinsky, K., and Ilinsky, I., 1990, Fine structure of the ventral anterior thalamic nucleus (VAmc) of *Macaca mulatta*: II. Organization of nigrothalamic afferents as revealed with EM autoradiography, *J. Comp. Neurol.* 294:479.

Kultas-Ilinsky, K., and Ilinsky, I., 1991, Fine structure of the ventral lateral nucleus (VL) of the *Macaca mulatta* thalamus: Cell types and synaptology, *J. Comp. Neurol.* 314:319.

Kultas-Ilinsky, K., Ilinsky, I., Warton S., and K.R. Smith, 1983, Fine structure of nigral and pallidal afferents in the thalamus: An EM study in the cat, *J. Comp. Neurol.* 216:390.

Kultas-Ilinsky, K., Kallsen, J., Tewfik, S., Freedman, L., and Ilinsky, I., 1989, Autradiography of GABA and benzodiazepine receptors in the monkey motor thalamus, *Soc. Neurosci. Abstr.* 15:289.

Kultas-Ilinsky, K., Leontiev, V., and Whiting P.J., 1998, Expression of 10 GABA$_A$ receptor subunit messenger RNAs in the motor-related thalamic nuclei and basal ganglia of *Macaca mulatta* studied with in situ hybridization histochemistry, *Neurosci.* 85:179.

Kultas-Ilinsky, K., Reising, L., Yi H., and Ilinsky, I., 1997, Pallidal afferent territory of the *Macaca mulatta* thalamus: neuronal and synaptic organization of the VAdc, *J. Comp. Neurol.* 386:573.

Kultas-Ilinsky, K., Yi, H., and Ilinsky, I., 1995, Nucleus reticularis thalami input to the anterior thalamic nuclei in the monkey: a light and electron microscopic study, *Neurosci. Lett.* 186:25.

Le Feuvre, Y., Fricker, D., Leresche, N., 1997, GABA$_A$ receptor-mediated IPSC in rat thalamic sensory nuclei: patterns of discharge and tonic modulation by GABA$_B$ autoreceptors. *J. Physiol.* 502:91.

Liu, X.B., Warren, R., and Jones, E.G., 1995 Synaptic distribution of afferents from reticular nucleus in ventroposterior nucleus of the cat thalamus, *J. Comp. Neurol.* 352:187.

Lüddens, H., 1998, The diversity of GABA$_A$ receptors. Pharmacological and electrophysiological properties of GABAA channel subtypes, *Mol. Neurobiol.* 18:35.

Lüddens, H., Korpi, E.R., and Seeburg, P.H., 1995, GABA$_A$/Benzodiazepine receptor heterogeneity: Neurophysiological implications, *Neuropharmacol.* 34:245.

Mason, A., Ilinsky, I.A., Beck, S., and Kultas-Ilinsky, K., 1996, Re-evaluation of synaptic relationships of cerebellar terminals in the ventral lateral thalamic nucleus of the rhesus monkey thalamus based on serial section analysis and three dimensional reconstruction, *Exp. Brain Res.* 109:219.

Monaghan, D.T., and Cotman, C.W., 1985, Distribution of N-methyl-D-aspartate-sensitive L-[^{3}H]glutamate-binding sites in the rat brain, *J. Neurosci.* 5:2909.

Nielsen, E.O., Drejer, J., Cha, J-H., Young, A.B., and Honore T., 1990, Autoradipographic characterization and localization of quisqualate binding sites in the rat brain using the antagonist [^{3}H]6-cyano-7-nitroquinoxaline-2,3-dione: Comparison with (*R,S*)-[^{3}H]α-amino-3-hydroxy-5-methyl-4-isoxazolepropionic acid binding sites, *J. Neurochem.* 54:686.

Nieoullon, A., Kerkerian, L., and Dusticier, N., 1984, High affinity glutamate uptake in the red nucleus and ventrolateral thalamus after lesion of the cerebellum in adult cat, *Exp. Brain Res.* 55:409.

Olsen, R.W., McCabe, R.T., and Wamsley, J.K., 1990, GABA$_A$ receptor subtypes: autoradiographic comparison of GABA, Benzodiazepine and convulsant binding sites in the rat central nervous system, *J. Chem. Neuroanat.* 3:59.

Pan, H.S., Frey, K.A., Young, A.B., and Penney, J.B., 1983, Changes in [^{3}H] muscimol binding in substantia nigra, entopeduncular nucleus, globus pallidus and thalamus after striatal lesions as demonstrated by quantitative receptor autoradiography, *J. Neurosci.* 3:1189.

Parent A., and De Bellefeuille L., 1982, Organization of efferent projections from the internal segment of the globus pallidus in primates as revealed by fluorescence retrograde labeling method, *Brain Res.* 245:201.

Penney, J.B., and Young, A.B., 1981, GABA as the pallidothalamic neurotransmitter: Implications for basal ganglia function, *Brain Res.* 207:195.

Ralston, H.J. III, Ohara, P.T., Meng, X.W., Wells, J., and Ralston, D.D., 1996, Transneuronal changes of the inhibitory circuitry in the macaque somatosensory thalamus following lesions of the dorsal column nuclei, *J. Comp. Neurol.* 371:325.

Tai, Y., Yi, H., Ilinsky, I.A., and Kultas-Ilinsky, K., 1995, Nucleus reticularis thalami connections with the mediodorsal thalamic nucleus: a light and electron microscopic study in the monkey, *Brain Res. Bull.* 38:475.

Tai, Y., Kultas-Ilinsky, K., and Ilinsky, I.A., 1996, Superior colliculus projections to the mediodorsal (MD) and ventral anterior (VAmc) thalamic nuclei in the monkey, *Soc. Neurosci. Abstr.* 22:3031.

Ulrich, D., and Huguenard, J.R., 1996, γ-Aminobutyric acid type B receptor dependent burst firing in thalamic neurons: A dynamic clamp study, *Proc. Natl. Acad. Sci. USA* 93:13245.

Wisden, W., Laurie, D.J., Monyer, H., and Seeburg P.H., 1992, The distribution of 13 $GABA_A$ receptor subunit mRNAs in the rat brain. 1. Telencephalon, diencephalon and mesencephalon, *J. Neurosci.* 12:1040.

Young, A.B., Bromberg, M.B., and Penney, J.B., 1981, Decreased glutamate uptake in subcortical areas deafferented by sensorimotor cortical ablation in the cat, *J. Neurosci.* 1:241.

Young, A.B., Dauth, G,W., Hollingworth, Z., Penney J.B., Kaatz, K., and Gilman S., 1990, Quisqualate- and NMDA-sensitive [^{3}H]glutamate binding in the primate brain, *J. Neurosci. Res.* 27:512.

Young, A.B., Penney J.B., Dauth, G.W., Bromberg, M., and Gilman S., 1983, Glutamate as a possible neurotransmitter of cerebral corticofugal fibers in the monkey, *Neurology*, 33:1513.

Zuercher, G., DeBoom, T., and Kultas-Ilinsky, K., 1987, Quantitative autoradiography of glutamate receptors in the feline thalamus, *Soc. Neurosci. Abstr.* 13:756.

MICROELECTRODE STUDIES OF BASAL GANGLIA AND VA, VL AND VP THALAMUS IN PATIENTS WITH DYSTONIA: DYSTONIA-RELATED ACTIVITY AND SOMATIC SENSORY REORGANIZATION

FREDERICK A. LENZ, NANCY N. BYL, IRA M. GARONZIK,
JUNG -I. LEE, AND SHERWIN E. HUA[*]

INTRODUCTION

Dystonia is a movement disorder characterized by sustained muscle contractions leading to twisting movements and abnormal postures (Fahn, 1988). It is usually assumed that dystonia is driven by abnormal activity arising in the basal ganglia (Marsden et al., 1985; Hallett, 1997; Vitek et al., 1998). Additionally, there is indirect evidence that neuronal activity in a cerebellar relay nucleus of the human thalamus (Ventral intermediate – Vim) and in a pallidal relay nucleus (Ventral oral posterior – Vop) is related to dystonic movements. Lesions in Vim and Vop relieve dystonia (Laitinen, 1970; Cooper, 1976; Gros et al., 1976; Andrew et al., 1983; Tasker et al., 1988b; Tasker et al., 1988a; Cardoso et al., 1995) and stimulation of Vim and Vop can increase (Tasker et al., 1982) or decrease dystonic movements (Benabid et al., 1996). The effects of thalamic lesioning and stimulation may be related to abnormalities in sensory processing observed in the central nervous system of patients with dystonia (Tempel and Perlmutter, 1990; Reilly et al., 1992; Tempel and Perlmutter, 1995).

During thalamotomy and pallidotomy for treatment of dystonia, the location of the target is first defined by radiologic studies (Bertrand and Lenz, 1995; Vitek and Lenz, 1997). Thereafter, microelectrode recordings may be used to confirm the target predicted by the radiologic studies. These physiologic studies provide a unique opportunity to examine pallidal and thalamic neuronal activity in patients with dystonia (dystonia patients). The present chapter reviews evidence from human microelectrode recordings that the thalamus and basal ganglia are involved in the mechanism of dystonia.

---

[*] FREDERICK A. LENZ • Department of Neurosurgery, Johns Hopkins University, Baltimore, Maryland 21287. NANCY N. BYL • Department of Physical Therapy, University of California at San Francisco, San Francisco, CA. IRA M. GARONZIK, JUNG –I. LEE, AND SHERWIN E. HUA • Department of Neurosurgery, Johns Hopkins University, Baltimore, Maryland 21287.

PALLIDAL STUDIES

We have carried out a detailed analysis of the activity in the globus pallidus of a single patient with hemidystonia (Lenz et al., 1998a). A woman with progressive, medically intractable right upper extremity dystonia underwent a pallidotomy with only transient improvement. During the procedure her dystonia became more severe as she repeatedly made a fist on command in order to provoke dystonia transiently (movement provoked dystonia). Comparisons within cells in the internal segment of the globus pallidus (GPi) revealed that the firing rate was the same at rest, with making a fist, and during movement provoked dystonia. However, the firing rate compared between cells decreased significantly throughout the procedure as the patient made a fist repeatedly. During the second half of the procedure the firing rate of cells in GPi was similar to that in hemiballismus.

If GPi is involved in the production of dystonic movements, then activity in GPi of dystonic patients might occur at low frequency and might be correlated with dystonia, as in the case of thalamic activity during dystonia or activity of GPi cells in hemiballismus (Suarez et al., 1997). Prevalence of low frequency activity correlated with dystonia has been reported in GPi by another study (Vitek et al., 1998). However, we found that the activity of cells in GPi was neither correlated with EMG activity during dystonia, nor dominated by dystonia frequency activity.

A higher proportion of cells in GPi of the patient with dystonia responded to somatosensory stimulation (53% - 9/17), than in hemiballismus (13% - 2/15, $P<0.05$ Fisher exact), Parkinson's 'off' (23/67 - 34%, $P>0.05$), and Parkinson's 'on' (6/19 - 32%, $P>0.05$). The proportion of GPi sensory cells was constant in the first (5/9) and second half (4/8) of the procedure in the patient with dystonia. In the two trajectories (one parasagittal plane) through GPi cells responding to sensory inputs consistently had RFs including the dystonic upper extremity (RFs on wrist or hand - 5). Similar consistency of RFs was never observed in a single parasagittal plane of 40 patients with Parkinson's disease. The proportion of cells in GPi that responded to sensory stimulation was significantly higher in dystonia. These results suggest that pallidal activity can correlate inversely with the severity of dystonia, perhaps due to activity-dependent changes in neuronal function resulting from repeated voluntary movement.

THALAMIC STUDIES

Indirect evidence suggests that the thalamus contributes to abnormal movements occurring in dystonia patients (Lenz et al., 1999). Thalamic neuronal activity was recorded as previously described (Lenz et al., 1988a). Standard techniques were used to record EMG activity in wrist flexors, wrist extensors, biceps, and triceps (Lenz et al., 1988a). EMG activity in deltoid was not monitored because dystonia, either at rest or during pointing, always involved elbow and wrist more than shoulder.

Tape recorded neuronal and EMG activities were examined postoperatively. The present report focuses on single neuron activity recorded in the region where cells exhibited activity related to active or passive movements of the upper extremity (Lenz et al., 1990). Action potentials were discriminated by amplitude (D-DISI, Bak Electronics, Rockville,

Figure 1. Thalamic spike train and EMG signals in a patient with dystonia. A. Five traces showing, from top, the times of occurrence of thalamic action potentials and the envelope of EMG activity in four arm muscles, as labeled. B. The raw spike power spectrum (resolution 0.01 Hz, left) and smoothed spike power spectra (resolution 0.78 Hz) after averaging the raw estimates in groups of 8 contiguous, non-overlapping frequencies (right). C. Raw and smoothed spectra for EMG channel 1 (wrist flexors). D. Coherence and phase functions for the cross-correlation function spike X EMG for wrist flexors. See text. (From Lenz et al., 1999, with permission.)

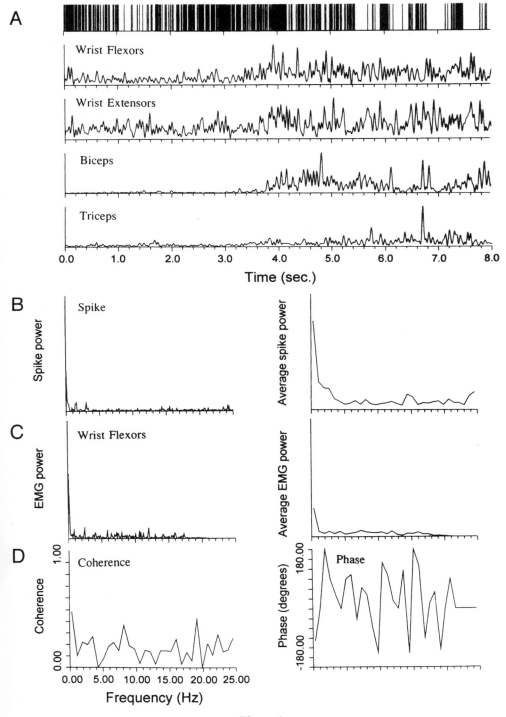

Figure 1

MD) and confirmed to arise from a single cell. The confirmation criterion was constant shape of the action potential, as verified by displaying the shape on a storage oscilloscope. Times of occurrence of discriminated action potentials were stored at a clock rate of 1000 Hz. EMG signals were digitized at a rate of 200 Hz on a digital computer (11/73, Digital Equipment Corp.). The data were analyzed on a workstation (DEC station 3100, Digital Equipment Corp.) with SAS-version 6.

Figure 1 shows an example of simultaneously recorded thalamic and EMG activity in a patient with dystonia. Both thalamic and EMG signals varied slowly at about the same frequency and increased at 4 seconds, and possibly at 6 and 7.5 seconds. Visual examination of the spike and EMG signals, however, was inadequate to assess either the composition of, or the correlation between, the spike and EMG signals. Therefore, thalamic and EMG signals were analyzed in the frequency domain. The spike train was converted into an equivalent analog signal by use of the French-Holden algorithm (French and Holden, 1971; French et al., 1972; Glaser and Ruchkin, 1976; Lenz et al., 1988a). The EMG signal was first processed to eliminate movement artifact by band-pass filtering (minus 6 Db below 20 Hz and above 120 Hz). The envelope of EMG activity was then produced by full-wave rectification and filtering (minus 6 Db above 20 Hz). The 10% cosine rule was then applied to eliminate low frequency components generated by finite sampling of the signals (Glaser and Ruchkin, 1976). The signal to noise ratio (SNR) was calculated as the spike power at dystonia frequency divided by the average power per component of the spike autopower spectrum. Thus the SNR is a measure of the extent to which power was concentrated at dystonia frequency.

Standard techniques used to take the spectrum of these two signals resulted in 256 estimates of the spectrum between 0 and 25 Hz (Oppenheim and Schafer, 1975; Bendat and Piersol, 1976; Glaser and Ruchkin, 1976; Lenz et al., 1988a). Cross-correlation spectral analysis (Figure 1B to D) was carried out to determine whether the two signals were correlated. The raw power spectra are shown in the left panels of Figure 1B and C (resolution – 0.1 Hz). In the right panels of Figure 1 B and C, eight contiguous, non-overlapping spectral estimates are averaged to produce smoothed spectral estimates (resolution - 0.78 Hz). Smoothed spectra give a more statistically reliable estimate, but at a cost of decreased resolution (Glaser and Ruchkin, 1976).

The coherence function was evaluated as a measure of the probability that two signals were linearly related. The coherence had a value of 0 if the spike and EMG signals were not linearly related and a value of 1 if the two had a perfect linear relationship (Oppenheim and Schafer, 1975; Bendat and Piersol, 1976; Glaser and Ruchkin, 1976; Lenz et al., 1988a). A coherence of greater than 0.42 at a given frequency indicated that the two signals were likely ($P<0.05$) to be linearly related at that frequency (Benignus, 1969). The signals in Figure 1 were coherent in the lowest frequency band as indicated by a coherence of 0.5. Phase was calculated by standard techniques (Oppenheim and Schafer, 1975; Bendat and Piersol, 1976; Glaser and Ruchkin, 1976; Lenz et al., 1988a) so that a negative phase for the cross-correlation function spike X EMG indicated that the spike signal was phase advanced on the EMG signal. The phase of the spike X EMG function at the frequency of dystonia in Figure 1 was minus 100°, which indicated that the spike train was phase advanced on the EMG signal.

The present study tested the hypothesis that thalamic activity contributed to the dystonic movements that occurred in such patients. During these movements, spectral analysis of EMG signals in flexor and extensor muscles of the wrist and elbow exhibited peak EMG power in the lowest frequency band (0 to 0.78 Hz, mean - 0.39 Hz - dystonia frequency) for 60-85% of epochs studied during a pointing task. Normal controls showed low frequency peaks for less than 16% of epochs during pointing. Among dystonia patients, simultaneous contraction of antagonistic muscles (co-contraction) at dystonia frequency during pointing was observed for muscles acting about the wrist (63% of epochs) and elbow (39%) but was not observed among normal controls during pointing.

The Thalamic Spike Signal: Populations Of Cells With Power Concentrated In The Lowest Frequency Band

Thalamic neuronal signals were recorded during thalamotomy for treatment of dystonia (Lenz et al., 1999) and were compared to those of control patients without motor abnormality who were undergoing thalamic procedures for treatment of chronic pain. Presumed nuclear boundaries of a human thalamic cerebellar relay nucleus (ventral intermediate - Vim) and a pallidal relay (ventral oral posterior - Vop) nucleus were estimated by aligning the atlas map (Schaltenbrand and Walker, 1982) of the anterior border of the principal sensory nucleus (ventral caudal - Vc) with the region where the majority of cells have cutaneous receptive fields (RFs).

Table 1 shows populations of cells having peak spectral power at dystonia frequency and a ratio of power at dystonia frequency to total spectral power (signal to noise ratio - SNR) greater than two. The proportion of cells with a dystonia frequency peak of $SNR \geq 2$ was significantly higher ($p<0.05$, Chi square) for patients with dystonia (37/83 - 44.6%) than for control patients with pain (15/59 – 25.4%). The proportion of nonsensory cells with a dystonia frequency peak of $SNR \geq 2$ was significantly higher in patients with dystonia than in controls ($p<0.01$, Chi-square). The proportion of sensory cells with these characteristics did not differ significantly ($p>0.05$, Chi-square) between these two populations.

Table 1. Cell populations with a concentration of power (signal noise ratio ≥ 2) in the spike power peak at dystonia frequency (< 0.79 Hz). Proportions were tested by Chi square or Fisher exact test, as appropriate. From Lenz et al., 1999, with permission.

	Cell Population	Dystonia	Control
A	All (P<0.05)	37/83 = 44.6%	15/59 = 25.4%
B	Sensory (NS)	6/25 = 24.0%	1/9 = 11.1%
	Nonsensory (P<0.01)	31/58 = 53.5%	14/50 = 28.0%
C	Presumed Vim (NS)	18/57 = 31.6%	15/49 = 30.6%
	Presumed Vop (P<0.001)	17/21 = 81.0%	0/9 = 0.0%
D	Sensory in presumed Vim (NS)	5/23 = 21.7%	1/9 = 11.1%
	Nonsensory in presumed Vim (NS)	13/34 = 38.2%	14/40 = 35.4%
E	Nonsensory in presumed Vop (P<0.001)	16/19 = 84.2%	0/9 = 0.0%

Nonsensory cells in presumed Vop of patients with dystonia (Table 1, Line E) were significantly more likely ($P<0.001$, Fisher exact test) to exhibit a peak in the lowest frequency band with a $SNR \geq 2$ (84%) than were such cells in controls (0%). In dystonia patients, the proportion of cells with a peak of activity ($SNR \geq 2$) in the lowest frequency band (Table 1, Line C) was significantly higher in presumed Vop (81%) than in presumed Vim (32% - $P<0.0005$, Chi square). The differences in the proportions of sensory and nonsensory cells in presumed Vim with dystonia frequency SNR>2 between dystonia patients and controls were not significant. These results demonstrate that peak spike power

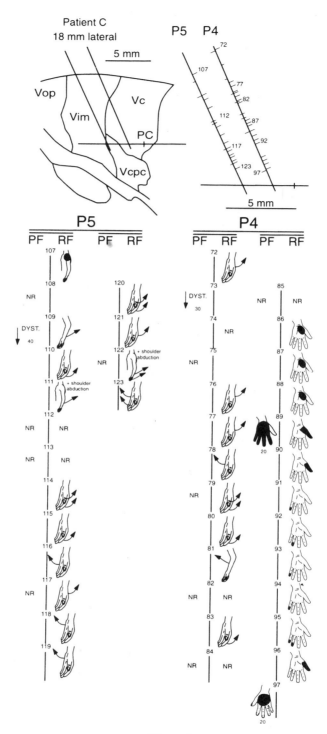

Figure 2

of nonsensory cells in Vop, but not in Vim, is concentrated at dystonia frequency more commonly in patients with dystonia than in controls.

The ratio of power at dystonia frequency to average spectral power was greater than 2 (P<0.001) for cells in presumed Vop often for dystonia patients (81%) but never for control patients. The percentage of such cells in presumed Vim of dystonia patients (32%) was not significantly different from that of controls (31%). Many cells in presumed Vop exhibited dystonia frequency activity that was correlated with and phase-advanced on EMG activity during dystonia, suggesting that this activity was related to dystonia.

Somatosensory Activity: Thalamic Reorganization In Dystonia

The proportion of deep sensory cells in presumed Vim was significantly higher (P < 0.01, Chi square) in dystonia patients (28%) than in controls (11%) (Lenz et al., 1999). Differences in presumed Vop of dystonia (11%) and controls (4%) followed the same trend, but did not reach significance (P < 0.15, Chi square). Presumed Vim contained a significantly greater (P<0.01, Chi square) proportion of deep sensory cells (28%) than did presumed Vop (11%) in dystonia patients, but not in controls (P>0.05, Fisher exact). No significant differences were found between the proportions of cells with deep somatic sensory responses in Vim or presumed Vop (Chi square, P>0.05) of individual patients with tremor of Parkinson's disease, essential tremor, cerebellar tremor, or chronic pain. The proportion of deep sensory cells was thus consistently higher in presumed Vim and presumed Vop of dystonia patients than in that of controls. Dystonia patients had a significantly higher proportion of deep sensory cells responding to movement of more than one joint (26%, 13/52) than did control patients (8%, 4/49).

In addition to the increased percentages of deep sensory cells, there was an apparent increase in the representation of individual joints in presumed Vim of patients with dystonia. For example, passive wrist movements were represented over a distance of 3.2 mm in trajectory P4 of patient 3 (see Figure 2, sites 72-78). The dimensions of the representation of deep structures could not be confirmed by detailed sensory mapping of 'motor thalamus' during surgery. Therefore, the maximum size of the representation of a part of the body was estimated from the lengths of trajectories along which all receptive fields (RFs) included a single joint. A previous study showed that the size of the thalamic representation of a body part varied with the maximum length of thalamic trajectories over which cellular RFs included that body part (Lenz et al., 1994). We restricted the analysis to the wrist or elbow representation because our exploration in patients with dystonia and

---

Figure 2. Map of neural activity in and anterior to nucleus Vc in Patient C. Upper panel: The locations of stimulation/recording sites relative to the anterior-posterior commisure (AC-PC) line indicated by the horizontal line. PC is located at the right end of the line. The electrode trajectories (P4 and P5) are shown by the two oblique lines. The locations of recording sites are indicated by ticks to the right of each line; stimulation sites are indicated by ticks to the left. All sites are numbered sequentially, and numbers are indicated for every fifth tick. Scale as indicated.

Panels P4 and P5 show the site number, effect of threshold microstimulation (TMIS), and receptive field (RF). The effect of TMIS at any site is listed below and to the left of the site number. If the evoked effect was a sensation, then a figurine indicating the location of the sensation (projected field – PF) is shown. The threshold current (in μA) is indicated below the PF figurine. NR indicates that no effect was evoked at that site by stimulation at up to 80 μA.

The RF is indicated to the right and below the site number. Note that many cells responded to movements of two joints. For example, cells 114 and 120 responded optimally to extension of both fingers and wrist. The shading on the figurine indicates the part of the body where mechanical stimulation-evoked cellular activity. All shaded figurines indicate cutaneous stimulation, except for neurons 86 and 107, at which cellular activity was evoked by manipulation of muscle but not by manipulation of the overlying skin. From Lenz et al., 1999, with permission.

tremor was directed toward the area where these joints were represented (Lenz et al., 1988b; Lenz et al., 1990). The explorations in the present study included this representation to a large degree, as indicated by RFs including the wrist or elbow joint for 79% of all deep RFs in the patients with dystonia.

Long trajectory lengths with cellular RFs including one joint might be compensated for shorter lengths along adjacent trajectories. However, the pairs of trajectories in Figure 1 had constant RFs over a substantial length of each trajectory. Furthermore, the differences between adjacent trajectories would be expected to be similar for dystonia and control patients since the mapping techniques were the same. Therefore, statistical comparison of these two populations (Figure 3) is valid.

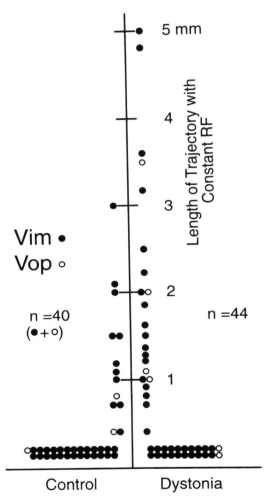

Figure 3. Lengths of electrode trajectories along which all cells had RFs for a single joint, either the wrist or the elbow. From Lenz et al., 1999, with permission.

In Figure 3 we have plotted trajectory lengths over which all RFs included a single joint, either wrist or elbow. Cells that had RFs covering more than one joint were excluded since this might increase trajectory lengths with a constant RF. Lengths of trajectories were measured by nucleus. If a trajectory traversed the border between Vim and Vop, then the lengths within either nucleus were measured separately and assigned to the appropriate nucleus. Lengths of trajectories through motor thalamus, including both Vim and Vop, with constant deep RFs were significantly longer in dystonic patients than in the controls (P<0.03, Kurskal Wallace ANOVA by ranks), and the same trend was observed for trajectories for Vim (P<0.055, Kurskal Wallace ANOVA). Figure 3 suggested that the same trend applied in Vop, although the number of data points was too small to test statistically. Thus, dystonia patients showed increased receptive fields and an increased thalamic representation of dystonic body parts. The motor activity of an individual sensory cell was related to the sensory activity of that cell by identification of the muscle apparently involved in the cell's receptive field. Specifically, we defined the effector muscle as the muscle that, by contraction, produced the joint movement associated with a thalamic neuronal sensory discharge when the examiner passively moved the joint. Spike X EMG correlation functions during dystonia indicated that thalamic cellular activity was less often related to EMG in effector muscles than to noneffector muscle contractions observed in dystonia. These observations are consistent with a model in which sensory input to Vim in dystonia is transmitted through altered sensory maps to activate multiple muscles in the periphery, producing the overflow of muscle activation that is characteristic of dystonia.

Studies of the Principal Sensory Nucleus of Human Thalamus

The map of the hand representation in the primary sensory cortex, area 3b, is altered in monkeys with dystonia-like movements resulting from overtraining in a gripping task (Byl et al., 1997). We have studied whether similar reorganization occurs in the somatic sensory thalamus of patients with dystonia (dystonia patients) (Lenz and Byl, 1999). Recordings of neuronal activity and microstimulation-evoked responses were studied in the cutaneous core of Vc of 11 dystonia patients who underwent stereotactic thalamotomy. Fifteen patients with essential tremor who underwent similar procedures were used as controls. The cutaneous core of Vc was defined as the part of the cellular thalamic region where the majority of cells had RFs to innocuous cutaneous stimuli. The proportion of RFs including multiple parts of the body was greater in dystonia patients (29%) than in patients with essential tremor (11%)(Lenz and Byl, 1999). Similarly, the percentage of projected fields (PFs) including multiple body parts was higher in dystonia patients (71%) than in patients with essential tremor (41%). A match at a thalamic site was said to occur if the RF and PF at that site included a body part in common. As illustrated in Figure 4, such matches were significantly less prevalent in dystonia patients (33%) than in patients with essential tremor (58%). The average length of the trajectory where the PF included a consistent, cutaneous RF was significantly longer in patients with dystonia than in control patients with essential tremor. The findings of sensory reorganization in Vc thalamus are congruent with those reported in the somatic sensory cortex of monkeys with dystonia-like movements resulting from overtraining in a gripping task.

CONCLUSIONS

The present data and previous studies suggest the mechanism by which reorganization may contribute to dystonia. Specifically, the mismatch between effector muscles and muscles correlated with dystonia (see Table 2) is reminiscent of findings in patients with spinal transection (Lenz et al., 1994) or amputation (Davis et al., 1998; Lenz et al., 1998b). In both of these groups of patients, i.e., spinal transection and amputation, the reorganization of the thalamic map of RFs leads to a mismatch between RFs (input) and the part of the body where thalamic microstimulation evokes sensation (projected field-output)

A

Matches: Neuronal Receptive Fields and Projected Fields
Controls

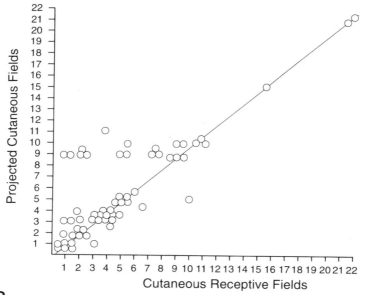

B

Patients with Dystonia

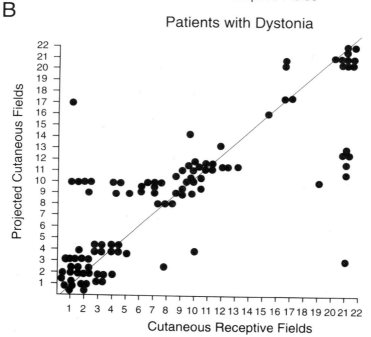

1	Intraoral
2	Perioral
3	Facial
4	D1
5	D2
6	D3
7	D4
8	D5
9	Multiple digits
10	Palm / Hand
11	Forearm
12	Whole arm
13	Upper arm
14	Wrist
15	Trunk
16	Leg
17	Upper leg
18	Lower leg
19	Pelvis
20	Ankle
21	Foot
22	Toes

Figure 4

Table 2. The incidence of dystonia-related activity in muscles that give rise to the movement that caused cellular firing when the movement is imposed by the examiner (effector muscle). Match describes the situation in which thalamic dystonia-related activity is correlated with EMG recorded in the effector muscle. Mismatch describes the situation in which thalamic dystonia-related activity is correlated with EMG recorded in a noneffector muscle. In this table, thalamic cell that has a negative spike X EMG phase with respect to a particular muscle is said to lead EMG in that muscle. From Lenz et al., 1999, with permission.

	Dystonia
Match: Thalamic cell correlated with EMG of the effector muscle	52% (11/21)
Thalamic cell correlated with and leads EMG of the effector muscle	38% (8/21)
Mismatch: Thalamic cell correlated with EMG of the noneffector muscle.	86% (18/21)
Thalamic cell correlated with and leads EMG of the noneffector	52% (11/21)

(Lenz et al., 1994; Lenz et al., 1998b). These patients have somatotopic reorganization of RF (input) maps but have less complete reorganization of the projected field map (output) (Lenz et al., 1998b). In dystonia, there may be a similar failure of the muscles correlated with thalamic spike activity to reflect the changes in the RF produced by shifts in the sensory map.

This mismatch between muscles in the RF and those correlated with dystonia may cause somatic sensory input to be transmitted to multiple muscles other than the effector muscles. This mismatch might produce the co-contraction and the overflow of muscle activation that is characteristic of dystonia (Rothwell et al., 1983; Hallett, 1993). A model of such a condition may be the stump jerks that occur in some patients with amputation (Marion et al., 1989). Stump jerks have features of dystonia (W. Olson - personal communication) and affect a population with demonstrated reorganization of thalamic maps (Lenz et al., 1998b). This type of reorganization may also produce focal dystonias that occur in patients, whose occupations (e.g. musicians) include repetitive motor tasks performed under high cognitive drive (Hochberg et al., 1983; Byl et al., 1996).

Figure 4. Matches in neuronal receptive fields and projected fields in tremor patients (panel A) and dystonia patients (panel B). Each of the neuronal receptive fields and projected fields at a site was coded according to body parts (inset, Figure 4), and all were plotted against each other (Figure 4). For this section of the analysis, all digits together (labeled 'multiple digits' in the inset) and the hand excluding digits (labeled 'palm/hand' in the inset) were each considered as separate parts of the body. The PF of a stimulation site where a cell was not recorded was paired with the RF of the closest recorded cell. For example, in Figure 2 stimulation at site 88 was paired with the RF for the cell recorded at site 89. For any RF/PF pair, we plotted against each other the part of the body where the RF of the recorded cell was closest to the part of the body involved in the PF (Figure 4A, essential tremor; 4B, dystonia). Thus, points on the 45 degree angle line indicate thalamic sites where the RF and PF matched or partially matched. The proportion of points on the line was significantly lower in dystonia than in essential tremor. From Lenz et al., 1999, with permission.

REFERENCES

Andrew, J., Fowler, C.J., and Harrison, J.G., 1983, Stereotaxic thalamotomy in 55 cases of dystonia, *Brain*, 106:981.

Benabid, A.L., Pollak, P., Gao, D., Hoffmann, D., Limousin, P., Gay, E., Payen, I., and Benazzouz, A., 1996, Chronic electrical stimulation of the ventralis intermedius nucleus of the thalamus as a treatment of movement disorders, *J. Neurosurg.* 84:203.

Bendat, J.S., and Piersol, A.G., 1976, *Random Data,* Wiley, New York.

Benignus, V.A., 1969, Estimation of the coherence spectrum and its confidence interval using the FFT, *IEEE Trans. Audio Electroacoustic.* 17:145.

Byl, N.N., Merzenich, M.M., Cheung, S., Bedenbaugh, P., Nagarajian, S.S., and Jenkins, W.M., 1997, A primate genesis model of focal dystonia and repetitive strain injury: II the effect of movement strategy on the de-differentiation of the hand representation in the primary somatosensory cortex of owl monkeys, *Soc. Neurosc. Abstr.* 22:1055.

Byl, N.N., Merzenich, M.M., and Jenkins, W.M., 1996, A primate genesis model of focal dystonia and repetitive strain injury: I. learning-induced de-differentiation of the representation of the hand in the primary somatosensory cortex in adult monkeys, *Neurology,* 47:508.

Cardoso, F., Jankovic, J., Grossman, R.G., and Hamilton, W., 1995, Outcome after stereotactic thalamotomy for dystonia and hemiballismus, *Neurosurg*ery, 36:501.

Cooper, I.S., 1976, 20-Year followup study of the neurosurgical treatment of dystonia musculorum deformans, in: *Advances in Neurology,* R.Eldridge and S.Fahn ,eds., Raven Press, New York.

Davis, K.D., Kiss, Z.H.T., Luo, L., Tasker, R.R., Lozano, A.M., and Dostrovsky, J.O., 1998, Phantom sensations generated by thalamic microstimulation, *Nature*, 391:385.

Fahn,S., 1988, Concept and classification of dystonia, in: *Advances in Neurology*, S. Fahn, ed., Raven Press, New York.

French, A.S., and Holden, A.V., 1971, Frequency domain analysis of neurophysiologic data, *Comp. Prog. in Biomed.* 1:219.

French, A.S., Holden, A.V., and Stein, R.B., 1972, The estimation of the frequency response function of a mechanoreceptor, *Kybernetic.* 11:15.

Glaser, E.M., and Ruchkin, D.S., 1976, *Principles of Neurobiological Signal Analysis,* Academic Press, New York.

Gros, C., Frerebeau, P., Perez-Dominquez, E., Bazin, M., and Privat, J.M., 1976, Long term results of stereotaxic surgery for infantile dystonia and dyskinesia, *Neurochirurgia.* 19:171.

Hallett, M., 1993, Physiology of basal ganglia disorders: an overview, *Canadian J. Neur. Sci.* 20: 177-183.

Hallett, M., 1997, Physiology of dystonia, *Adv. Neurol.* 78:11.

Hochberg, F.H., Leffert, R.D., Heller, M.D., and Merriman, L., 1983, Hand difficulties among musicians, *JAMA.* 249:1869.

Laitinen, L.V., 1970, Neurosurgery in Cerebral Palsy, *J. Neurol. Neurosurg. Psych.* 33:513.

Lenz, F.A., and Byl, N.N, 1999, Reorganization of the cutaneous core of the human thalamic principle thalamic somatosensory nucleus in patients with dystonia, *J. Neurophysiol.* 82:3204.

Lenz, F.A., Dostrovsky, J.O., Tasker, R.R., Yamashiro, K., Kwan, H.C., and Murphy, J.T., 1988b, Single-unit analysis of the human ventral thalamic nuclear group: somatosensory responses, *J. Neurophysiol.* 59:299.

Lenz, F.A., Jaeger, C.J., Seike, M., Lin, Y.C., Reich, S.G., DeLong, M.R., and Vitek, J.R., 1999, Single neuron analysis of the motor thalamus in patients with dystonia, *J. Neurophysiol.* 82:2372.

Lenz, F.A., Kwan, H. C., Dostrovsky, J.O., Tasker, R.R., Murphy, J.T., and Lenz, Y.E., 1990, Single unit analysis of the human ventral thalamic nuclear group: activity correlated with movement, *Brain,* 113:1795.

Lenz, F.A., Kwan, H.C., Martin, R., Tasker, R., Richardson, R.T., and Dostrovsky, J.O., 1994, Characteristics of somatotopic organization and spontaneous neuronal activity in the region of the thalamic principal sensory nucleus in patients with spinal cord transection, *J. Neurophysiol.* 72:1570.

Lenz, F.A., Suarez, J.L., Verhagen Metman, L., Reich, S.G., Hallett, M., Rowland, L.H., and Dougherty, P.M., 1998a, Pallidal activity during dystonia: somatosensory reorganization and changes with severity, *J. Neurol. Neurosurg. Psych.* 65:767.

Lenz, F.A., Tasker, R.R., Kwan, H.C., Schnider, S., Kwong, R., Murayama, Y., Dostrovsky, J.O., and Murphy, J.T., 1988a, Single unit analysis of the human ventral thalamic nuclear group: correlation of thalamic "tremor cells" with the 3-6 Hz component of parkinsonian tremor, *J. Neurosci.* 8:754.

Lenz, F.A., Zirh, A.T., Garonzik, I.M., and Dougherty, P.M., 1998b, Neuronal activity in the region of the principle sensory nucleus of human thalamus (ventralis caudalis) in patients with pain following amputations, *Neuroscience*, 86:1065.

Marion, M.H., Gledhill, R.F., and Thompson, P.D., 1989, Spasms of amputation stumps: a report of two cases, *Movement Dis.* 4:354.

Marsden, C.D., Obeso, J.A., Zarranz, J.J., and Lang, A.E., 1985, The anatomical basis of symptomatic hemidystonia, *Brain*, 108:463.

Oppenheim, A.V. and Schafer, R.W., 1975, *Digital Signal Processing,* Prentice-Hall, Englewood Heights, NJ.

Reilly, A.J., Hallett, M., Cohen, L.G., Tarkka, I.M., and Dang, N., 1992, The N30 component of somatosensory evoked potentials in patients with dystonia, *EEG Clin. Neurophysiol.* 84:243.

Rothwell, J.C., Obeso, J.A., Day, B.L., and Marsden, C.D., 1983, Pathophysiology of dystonias, in: *Motor Control Mechanisms in Health and Disease*, J.E. Desmedt, ed., Raven Press, New York.

Schaltenbrand, G., and Walker, A.E., 1982, *Stereotaxy of the Human Brain,* Thieme-Stratton, New York.

Suarez, J.L., Verhagen Metman, L., Reich, S.G., Dougherty, P.M., Hallett, M., and Lenz, F.A., 1997, Pallidotomy for hemiballismus: efficacy and characteristics of neuronal activity, *Ann. Neurol.* 42:807.

Tasker, R.R., Doorly, T., and Yamashiro, K., 1988a, Thalamotomy in generalized dystonia, in: *Advances in Neurology*, S. Fahn, ed., Raven Press, New York.

Tasker, R.R., Doorly, T., and Yamashiro, K., 1988b, Thalamotomy in generalized dystonia, *Adv. Neurol.* 50:615.

Tasker, R.R., Organ, L.W., and Hawrylyshyn, P., 1982, *The Thalamus and Midbrain in Man: A Physiologic Atlas Using Electrical Stimulation,* Thomas, Springfield, IL.

Tempel, L.W., and Perlmutter, J.S., 1995, Abnormal cortical responses in patients with writer's cramp, *Neurology*, 43:2252.

Tempel, L.W., and Perlmutter, J.S., 1990, Abnormal vibration-induced cerebral blood flow responses in idiopathic dystonia, *Brain*, 113:691.

Vitek, J.L., Zhang, J., Evatt, M., Mewes, K., DeLong, M.R., Hashimoto, T., Triche, S., and Bakay, R.A.E., 1998, GPi pallidotomy for dystonia: clinical outcome and neuronal activity, *Adv. Neurol.* 78:211.

PART VI. NEURONAL ACTIVITY IN MOVEMENT DISORDERS

PHYSIOLOGY OF GLOBUS PALLIDUS NEURONS IN MOVEMENT DISORDERS

MICHEL FILION[*]

INTRODUCTION

The goal of the present review is to determine the contribution of the globus pallidus to movement disorders. We will examine changes in the activity and reactivity of individual globus pallidus neurons in the monkey, as the normal animal is rendered hypokinetic and parkinsonian, following intoxication with the neurotoxin MPTP (1-methyl-4-phenyl-1,2,3,6-tetra-hydropyridine), and then hyperkinetic and dyskinetic, by pharmacological treatment with dopamine agonists. We will also examine data obtained in normal monkeys rendered transiently dyskinetic by intra-pallidal injection of an antagonist of the inhibitory neurotransmitter gamma-aminobutyric acid (GABA). We interpret the changes in the light of recent data from the literature.

TERRITORIES IN THE GLOBUS PALLIDUS

Physiology in the Normal Primate

A number of years ago, Parent et al. (1984) showed that the axons of caudate nucleus neurons travel and terminate dorsally in the globus pallidus, whereas those from the putamen do so more ventrally. Nearly simultaneously, Percheron et al. (1984) showed that such caudate and putamen territories were likely to overlap physiologically as a result of the wide dendrites of pallidal neurons crossing territorial limits, especially at the tip of the globus pallidus. The caudate territory is now called associative, because the caudate nucleus receives fibers mostly from associative cortical areas having cognitive functions. The putamen territory is now called sensorimotor, because the putamen receives somatotopic projections mainly from sensory and motor cortical areas.

To study the physiology of the two territories within the globus pallidus (Tremblay and Filion, 1989; Tremblay et al., 1989), we stimulated electrically one site in the caudate nucleus and two sites in the putamen bilaterally, and recorded the activity of individual globus pallidus neurons, in the lateral or external pallidal segment (GPe), and in the medial

---

[*] MICHEL FILION • Unite de Recherche en Neuroscience, Centre Hospitalier de l'Universite Laval, Sainte-Foy, Quebec G1V 4G2, Canada.

or internal pallidal segment (GPi) in normal monkeys. We combined several responses of the same neuron to make peristimulus histograms. In two-thirds of the cases, stimulation of either caudate or putamen produced a decrease followed by an increase in the activity of the pallidal neurons, an inhibition-excitation sequence.

As predicted by anatomical studies, neurons responding only to caudate stimulation were grouped in the dorsal part of the pallidum, in both GPe and GPi. Inversely, neurons responding only to putamen were located more ventrally in the pallidum. Moreover, there was a significant area of overlap between territories, a region where pallidal neurons responded to both caudate and putamen or to either one or the other.

Center-Surround Organization

The pallidal neurons in the center of each territory responded to stimulation by an inhibition-excitation sequence: the most frequent type of response. Surprisingly, however, at the periphery of each territory they responded with pure excitation. It was as if, in the center of the territory, the excitatory response limited the temporal duration of the initial inhibitory response, whereas at the periphery of the territory it restricted the spatial extent of the inhibitory response.

Such center-surround organization may be explained by surround or lateral inhibition, which allows neurons to emphasize their own message by repressing that of their neighbors. Within the striatum, lateral inhibition was thought to result mainly from the recurrent collaterals of the numerous striato-pallidal projection neurons, making up more than 90% of the total striatal neuronal population, and using GABA as an inhibitory neurotransmitter. However, neurophysiologists (Jaeger et al., 1994) tried hard, but failed to demonstrate that activation of the striatal projection neurons can result in lateral inhibition. Koós and Tepper (1999) have recently demonstrated that the small population of striatal interneurons using GABA as a neurotransmitter can exert powerful inhibitory control over striatal projection neurons.

GPe neurons are also GABAergic and inhibitory. Contrary to GPi neurons, they have recurrent collaterals (François et al., 1984; Shink and Smith, 1995) that may produce lateral inhibition. GPe neurons also send recurrent branches up-stream, into the striatum. Interestingly, Bolam et al. (1998) have recently shown that these branches target specifically inhibitory GABAergic interneurons, as if to terminate and take over the striatal lateral inhibition.

In summary, caudate and putamen each have their own territory in the globus pallidus, but the latter overlap at the neuronal level. There is, moreover, a center-surround organization of neuronal activity in the pallidum. Its role may be to contrast, to focus, or to select one representation or the other and possibly to switch from one to the other. We may now ask what happens to the representation and to the center-surround organization in movement disorders?

Pallidal Territories in Movement Disorders

After the monkeys had been rendered parkinsonian, the great majority of pallidal neurons responded excessively and unselectively to many stimulation sites (Filion et al., 1989; Tremblay et al., 1989). In particular, contralateral sites that were ineffective in the normal state often triggered responses in animals rendered parkinsonian. We then injected the monkeys with the dopamine agonist apomorphine. The signs of parkinsonism disappeared and the animals became transiently hyperactive and dyskinetic. The same pallidal neurons that responded excessively and unselectively in the parkinsonian state now responded to a single or a few sites with a short inhibition curtailed by excitation: a central type of response. Other pallidal neurons responded with excitation only: a peripheral type of response. Still others did not respond at all, as if they were now outside the territory of either the caudate or putamen. The center-surround mechanism for focusing and selection appeared to function as well as (or maybe even better than) in normal monkeys. When the effect of apomorphine terminated and the monkeys recovered their hypokinesia and parkinsonism, the pallidal neurons again responded excessively and unselectively, as if the focusing and selection mechanisms were again widely open or ineffective.

Let us now examine another representation, that of the body, in the sensorimotor territory of the globus pallidus, and see whether it appears to be controlled by mechanisms for selectivity.

Normal Somatotopy in the Globus Pallidus

Thirty years ago, DeLong (1971) showed that about 20% of globus pallidus neurons discharge in close relationship with movement. The relationship was generally with a single contralateral joint, suggesting the likelihood of somatotopy. Indeed, further studies (DeLong et al., 1985; Hamada et al., 1990) showed that the leg is represented dorsally in the globus pallidus, the face ventrally, and the arm in between. However, although pallidal neurons with similar activity or reactivity are grouped in clusters, the representation of the body is at times discontinuous (leg-arm-leg, arm-leg-arm, arm-leg-face) in given trajectories, suggesting complex or multiple representation.

Hoover and Strick (1993; 1999) used ingenious anatomical methods to study topological relationships between globus pallidus neurons and their afferent and efferent cortical projections. Thus, they injected a neurotropic virus in the primary motor cortex, the ventral premotor cortex, and the supplementary motor area. The virus infected the thalamic neurons projecting to each cortical area; and then, trans-neuronally infected the GPi neurons projecting to the latter thalamic neurons. Hoover and Strick (1999) were thus able to verify that the leg, the arm, and the face of the primary motor cortex are represented dorso-ventrally in GPi. They also showed (Hoover and Strick, 1993) that there are three separate representations of the arm in the GPi: a dorsal one related to supplementary motor area, a ventral one related to ventral premotor cortex, and a middle one related to the primary motor cortex. Interestingly, the representation of primary motor cortex is centrally located, whereas the others are peripheral. Moreover, the GPi representation of the leg of the primary motor cortex is likely to overlap or to interdigitate with the arm of the supplementary motor area, while the face of the primary motor cortex is likely to overlap or to interdigitate with the arm of the premotor cortex.

It is important to realize here that the primary motor cortex is specialized mainly for the execution of movements, a basic sensorimotor function, tightly related to sensory receptors and motoneurons. On the other hand, the supplementary motor area and the premotor cortex are specialized for the immediate preparation of movement, which can be considered a low-level associative function. Thus, the premotor cortex is thought to follow external cues orienting movement, whereas the supplementary motor area is thought to follow internal cues, learned and stored in memory (Middleton and Strick, 2000). More frontal cortical areas are represented in the caudate nucleus and more dorsally in the globus pallidus. They are thought to be related to high-level associative and cognitive functions.

To illustrate this point further, let us recall the electrophysiological recordings of Nambu et al. (1990) in monkeys trained to execute movements. The investigators identified two types of pallidal neurons. The first type responds to stimulation of premotor cortex and supplementary motor area, and is active mostly after the instruction signal and before the go-signal and subsequent movement. This type of pallidal neuron is therefore active mainly during preparation for movement. The second type of pallidal neuron responds to stimulation of primary motor cortex and is active only after the go-signal, just before and mainly during execution of movement.

In summary, globus pallidus neurons do not represent only individual joints or limb segments, but also represent whether each segment is involved in preparation or execution of movement or both or alternately one and the other. Let us now see what happens to such body representation in movement disorders.

Pallidal Somatotopy in Movement Disorders

In parkinsonian monkeys, an abnormally large number of neurons respond to passive movement. Moreover, instead of responding to movement of a single contralateral joint in one direction only, they often respond with excessive amplitude to more than one joint, in more than one direction, and in both ipsi- and contralateral limbs (Filion et al., 1988). This has now been demonstrated in human parkinsonians during neurosurgery (Sterio et al., 1994; Taha et al., 1996). Parkinson's patients thus appear to have a big selection problem, both between limb segments and between execution and preparation of movement. Whether, inversely, hyperkinetic patients focus excessively on the central, executive pallidal neurons at the expense of peripheral, preparatory pallidal neurons is a possibility that we will examine later.

Let us now explore briefly the representation of yet another neurological concept in the globus pallidus: the direct and indirect striato-pallidal pathways.

DIRECT AND INDIRECT STRIATO-PALLIDAL PATHWAYS

Functional Model of the Basal Ganglia

The current functional model of the basal ganglia proposes that disorders of movement result from imbalance between the direct and indirect striato-pallidal pathways. Thus, about 50% of striatal neurons influence the output of the basal ganglia (the GPi and the substantia nigra pars reticulata) directly. Activation of this pathway results in disinhibition of the motor thalamus and cortex, and therefore facilitates movement. The model predicts that the pathway is hypoactive in parkinsonism, and hyperactive in hyperkinetic disorders. The other 50% of striatal neurons are intermingled with the first and influence the output of the basal ganglia indirectly, through the GPe and the subthalamic nucleus. Activation of this pathway results in increased inhibition of the motor thalamus and cortex, and therefore represses movement. The model predicts that the pathway is hyperactive in parkinsonism and hypoactive in hyperkinesia.

Abnormal Pallidal Firing Rates in Movement Disorders

We collaborated in obtaining data supporting the above model, in particular, by recording neuronal activity in the globus pallidus of intact and parkinsonian monkeys, with and without dopaminergic treatment (Filion and Tremblay, 1991; Filion et al., 1991). Thus, in parkinsonian monkeys compared to normal, the mean firing rate of GPi neurons increases, whereas that of GPe neurons decreases. The reverse occurs when the dopamine agonist apomorphine is used to abolish parkinsonian signs and to render the animal hyperkinetic and dyskinetic. The latter changes have now been observed in human parkinsonians during neurosurgery (Hutchison et al., 1997).

Nevertheless, the decrease in GPe activity in parkinsonism has been refuted by a number of neuroscientists. In particular, it does not agree with results of molecular biology showing that the messenger ribonucleic acid for glutamate decarboxylase, the enzyme for the synthesis of GABA, is either unchanged or increased in the GPe of parkinsonians, instead of being decreased (Soghomonian and Chesselet, 1992). If changes in mean firing rates do not tell the whole story, what about changes in firing patterns?

Abnormal Pallidal Firing Patterns in Parkinsonism

Twenty years ago, we claimed that bursting activities, at times synchronous with limb tremor, were the major abnormality of both GPe and GPi neurons in monkeys rendered parkinsonian by electrolytic midbrain lesion or injection of dopamine antagonists (Filion, 1979). Since then, such bursting activities have been recorded repeatedly in monkeys rendered parkinsonian by MPTP (Miller and DeLong, 1986; Filion and Tremblay, 1991; Bergman et al., 1994) and in parkinsonian patients (Hutchison et al., 1994; Sterio et al., 1994; Taha et al., 1996). Bergman et al. (1994) even claimed that the pallidal neurons showing abnormal firing rates in parkinsonian subjects are those that also show abnormal bursting patterns. Finally, Bergman et al. (1998) used auto- and cross-correlation analyses to demonstrate that pallidal neurons discharge irregularly and independently in the intact monkey, but that the discharge becomes oscillatory and synchronized between neurons in the parkinsonian. They concluded that mechanisms for independent parallel processing of information in the pallidum become ineffective in parkinsonism.

To try to understand such changes in firing patterns, we measured and analyzed distributions of intervals between spikes (Filion and Tremblay, 1991). Thousands of intervals were used to make histograms for individual neurons. The histograms were then normalized in percent and averaged for populations of neurons in GPi and GPe, in normal and parkinsonian states. In the parkinsonian subjects compared to normal, the mode of the population histograms (the more frequent interval) decreases by half, from 8 to 4 ms, in both the GPi and the GPe. Shorter intervals between spikes indeed occur in the abnormal bursts of discharges in the parkinsonian subjects. They necessarily result from increased neuronal depolarization or excitation. Such excessive excitation is probably imposed directly by the subthalamic nucleus, which is overactive in parkinsonism (Miller and DeLong, 1986; Bergman et al., 1994), and projects to both GPe and GPi neurons (Shink et al., 1996). But how can one explain the decrease in mean firing rate of GPe neurons if they are more excited? The population histograms show that this is very likely due to an increase in long intervals that occurs in GPe but not in GPi in parkinsonism. Only a few long intervals can greatly reduce mean firing rate, giving the false impression that the activity of GPe neurons is reduced, whereas, as a group, the neurons are simultaneously over-excited and over-inhibited in parkinsonism. We also observed such co-existence of hyper- and hypo-activity within each pallidal segment in monkeys with hyperkinesia.

Abnormal Pallidal Firing Patterns in Hyperkinesia

We produced transient hyperkinesia and dyskinesia in normal monkeys by injection of the GABA antagonist bicuculline locally in the GPe (Matsumura et al., 1995). GPe neurons located close to the injection site greatly increased their activity just before the occurrence of dyskinesia, as predicted by the model. Similarly, GPi neurons, located grossly in register with the GPe injection site, decreased their activity, again as predicted by the model. In contrast, however, GPe neurons located far from the bicuculline injection site decreased their activity, and GPi neurons out of register with the injection site increased their activity, results counter to the model's prediction. This is a clear expression of lateral inhibition within GPe, likely projected (inverted) upon GPi.

Further study of this phenomenon by repeated bicuculline injections at the same site in GPe while recording sequentially from different GPi neurons showed that there is a center-surround organization of activity during dyskinesia, with hypoactivity in the center and hyperactivity at the periphery. Remember that GPi neurons located centrally in the nucleus project to the primary motor cortex and control movement execution, while peripheral GPi neurons project to pre-motor cortex and supplementary motor area and control preparation of movement. According to this scheme and the center-surround organization of GPi discharge during bicuculline-induced dyskinesia, cortical executive neurons would be disinhibited, whereas cortical preparatory neurons would be inhibited, facilitating execution of unprepared movement.

Trans-Cortical and Sub-Cortical Loops Through the Basal Ganglia

Recent anatomical and physiological studies in the primate suggest that cortical neurons mostly target the striatal neurons of the indirect pathway (Strick et al., 1995; Parthasarathy and Graybiel, 1997), whereas thalamic intralaminar neurons preferentially target the striatal neurons of the direct pathway (Sidibé and Smith, 1996). Half the striatal projection neurons would therefore be involved in a mainly trans-cortical loop, while the other half would participate in a mainly sub-cortical, trans-thalamic loop.

The fact that cortico-striatal terminals are located specifically on the head of dendritic spines, whereas thalamo-striatal axons end on dendritic shafts (Smith et al., 1994), suggests that the trans-cortical loop is more specific, and the sub-cortical loop more diffuse. Moreover, surround inhibition, focusing, and selection processes appear to occur mostly within the trans-cortical loop, including the indirect striato-pallidal pathway, and the GPe. Thus, tampering with the indirect striato-pallidal pathway always results in excessive and inappropriate motor activity. For example, degeneration of the striato-pallidal neurons of the indirect pathway in early stages of Huntington's disease gives rise to choreic dyskinesia (Albin et al., 1990). Disinhibition of GPe neurons with local injection of bicuculline also gives rise to choreic dyskinesia, as we have seen above. Cytotoxic destruction of the GPe exaggerates levodopa-induced dyskinesia in parkinsonian monkeys (Blanchet et al., 1994). Finally, lesion of the subthalamic nucleus also produces excess movement, even in parkinsonian subjects (Guridi et al., 1996).

Lesion of the GPi, on the other hand, decreases both the excess and the lack of movement, as demonstrated repeatedly by pallidotomy in parkinsonian patients. The role of the GPi therefore appears to be mainly the expression of the selection of movement already made up-stream in the basal ganglia.

In summary, unaltered functioning of the trans-cortical loop appears to be essential for adequate selection, preparation, and execution of movement, whereas the sub-cortical, trans-thalamic loop is more likely to modulate the gains of the system in relation to attentional mechanisms. In this context, the imbalance between direct and indirect striato-pallidal pathways, repeatedly demonstrated in movement disorders, reflects an imbalance between selection and attentional mechanisms. Much research has been focused on the trans-cortical loop. We should now pay more attention to the direct influence of the thalamus on the basal ganglia.

SUMMARY

Recordings from single neurons suggest that functional territories, including somatotopy, are maintained in the globus pallidus of the normal monkey by selection mechanisms comprising surround inhibition. In movement disorders, the territories overlap excessively or are otherwise abnormal, and the center-surround organization of neuronal responses to stimulation is correspondingly abnormal. Motor disorders have also been attributed to imbalance between the direct and the indirect striato-pallidal pathways. Recent data suggest that the direct pathway is related mainly to thalamic attentional mechanisms, whereas the indirect pathway is related mainly to cortical selection mechanisms.

REFERENCES

Albin, R.L., Reiner, A., Anderson, K.D., Penney, J.B., and Young, A.B., 1990, Striatal and nigral neuron subpopulations in rigid Huntington's disease: implications for the functional anatomy of chorea and rigid-akinesia, *Ann. Neurol.* 27:357.

Bevan, M.D., Booth, P.A.C., Eaton, S.A., and Bolam, J.P., 1998, Selective innervation of neostriatal interneurons by a subclass of neurons in the globus pallidus of the rat, *J. Neurosci.* 18:9438.

Bergman, H., Wichmann, T., Karmon, B., and DeLong, M.R., 1994, The primate subthalamic nucleus. II. Neuronal activity in the MPTP model of parkinsonism, *J. Neurophysiol.* 72:507.

Bergman, H., Feingold, A., Nini, A., Raz, A., Slovin, H., Abeles, M., and Vaadia, E., 1998, Physiological aspects of information processing in the basal ganglia of normal and parkinsonian primates, *Trends Neurosci.* 21:32.

Blanchet, P.J., Boucher, R., and Bédard, P.J., 1994, Excitotoxic lateral pallidotomy does not relieve L-dopa-induced dyskinesia in MPTP parkinsonian monkeys, *Brain Res.* 650:32.

DeLong, M.R., 1971, Activity of pallidal neurons during movement. *J. Neurophysiol.* 34:414.

DeLong, M.R., Crutcher, M.D., Georgopoulos, A.P., 1985, Primate globus pallidus and subthalamic nucleus: functional organization. *J. Neurophysiol.* 53:530.

Filion, M., and Tremblay, L., 1991, Abnormal spontaneous activity of globus pallidus neurons in monkeys with MPTP-induced parkinsonism, *Brain Res.* 547:142.

Filion, M., Tremblay, L., and Bédard, P.J., 1988, Abnormal influences of passive limb movement on the activity of globus pallidus neurons in parkinsonian monkeys, *Brain Res.* 444:165.

Filion, M., Tremblay. L., and Bédard, P.J., 1989, Excessive and unselective responses of medial pallidal neurons to both passive movement and striatal stimulation in monkeys with MPTP-induced parkinsonism, in: *Neural Mechanisms in Disorders of Movement,* A.R. Crossman and M.A. Sambrook, eds., Libbey, London.

Filion, M., Tremblay, L., and Bédard, P.J., 1991, Effects of dopamine agonists on the spontaneous activity of globus pallidus neurons in monkeys with MPTP-induced parkinsonism, *Brain Res.* 547:152.

François, C., Percheron, G., Yelnik, J., and Heyner, S., 1984, A Golgi analysis of the primate globus pallidus. I. Inconstant processes of large neurons, other neuronal types and efferent axons, *J. Comp. Neurol.* 227:182.

Guridi, J., Herrero, M.T., Luquin, M.R., Guillen, J., Ruberg, M., Laguna, J., Vila, M., Javoy-Agid, F., Agid, Y., Hirsch, E., and Obeso, J.A., 1996, Subthalamotomy in parkinsonian monkeys. Behavioral and biochemical analysis, *Brain,* 119:1717.

Hamada, I., DeLong, M.R., and Mano, N., 1990, Activity of identified wrist-related pallidal neurons during step and ramp wrist movements in the monkey, *J. Neurophysiol.* 64:1892.

Hoover, J.E., and Strick, P.L., 1993, Multiple output channels in the basal ganglia, *Science,* 239:819.

Hoover, J.E., and Strick, P.L., 1999, The organization of cerebello- and pallido-thalamic projections to primary motor cortex: an investigation employing retrograde transneuronal transport of herpes simplex virus type I, *J. Neurosci.* 19:1446.

Hutchison, W.D., Levy, R., Dostrovsky, J.O., Lozano, A.M., and Lang, A.E., 1997, Effects of apomorphine on globus pallidus neurons in parkinsonian patients, *Ann. Neurol.* 42:767.

Hutchison, W.D., Lozano, A.M., Davis, K.D., Saint-Cyr, J.A., Lang, A.E., and Dostrovsky, J.O., 1994, Differential neuronal activity in segments of globus pallidus in Parkinson's disease patients, *Neuroreport,* 5:1533.

Jaeger, D., Kita, H., and Wilson, C., 1994, Surround inhibition among projection neurons is weak or nonexistent in the rat neostriatum, *J. Neurophysiol.* 72:2555.

Koós, T., and Tepper, J.M., 1999, Inhibitory control of neostriatal projection neurons by GABAergic interneurons, *Nature Neurosci.* 2:467.

Matsumura, M., Tremblay, L., Richard, H., and Filion, M., 1995, Activity of pallidal neurons in the monkey during dyskinesia induced by injection of bicuculline in the external pallidum, *Neuroscience,* 65:59.

Middleton, F.A., and Strick, P.L., 2000, Basal ganglia and cerebellar loops: motor and cognitive circuits, *Brain Res. Rev.* 31:236.

Miller, W.C., and DeLong, M.R., 1986, Altered tonic activity of neurons in the globus pallidus and subthalamic nucleus in the primate MPTP model of parkinsonism, in: *The Basal Ganglia II, as Advances in Behavioral Biology, Vol 32,* M.B. Carpenter and A. Jayaraman, eds., Plenum Press, New York.

Nambu, A., Yoshida, S., Jinnai, K., 1990, Discharge patterns of pallidal neurons with input from various cortical areas during movement in the monkey, *Brain Res.* 519:183.

Parent, A., Bouchard, C., and Smith, Y., 1984, The striatopallidal and striatonigral projections: two distinct fiber systems in primate, *Brain Res.* 303:385.

Parthasarathy, H.B., and Graybiel, A.M., 1997, Cortically driven immediate-early gene expression reflects modular influence of sensorimotor cortex on identified striatal neurons in the squirrel monkey, *J. Neurosci.* 17:2477.

Percheron, G., Yelnik, J., and François, C., 1984, A Golgi analysis of the primate globus pallidus. III. Spatial organization of the striato-pallidal complex, *J. Comp. Neurol.* 227:214.

Shink, E., and Smith, Y., 1995, Differential synaptic innervation of neurons in the internal and external segments of the globus pallidus by the GABA- and glutamate-containing terminals in the squirrel monkey, *J. Comp. Neurol.* 358:119.

Shink, E., Bevan, M.D., Bolam, J.P., and Smith, Y., 1996, The subthalamic nucleus and the external pallidum: two tightly interconnected structures that control the output of the basal ganglia in the monkey, *Neuroscience,* 73:335.

Sidibé, M., and Smith, Y., 1996, Differential synaptic innervation of striatofugal neurones projecting to the internal or external segments of the globus pallidus by thalamic afferents in the squirrel monkey, *J. Comp. Neurol.* 365:445.

Smith, Y., Bennett, B.D., Bolam, J.P., Parent, A., and Sadikot, A.F., 1994, Synaptic relationships between dopaminergic afferents and cortical or thalamic input in the sensorimotor territory of the striatum, *J. Comp. Neurol.* 344:1.

Soghomonian, J.J., and Chesselet, M.F., 1992, Effects of nigrostriatal lesions on the levels of messenger RNAs encoding two isoforms of glutamate decarboxylase in the globus pallidus end entopeduncular nucleus in the rat, *Synapse,* 11:124.

Sterio, D., Beric, A., Dogali, M., Fazzini, E., Alfaro, G., and Devinsky, O., 1994, Neurophysiological properties of pallidal neurons in Parkinson's disease, *Ann. Neurol.* 35:586.

Strick, P.L., Dum, R.P., and Picard, N., 1995, Macro-organization of the circuits connecting the basal ganglia with the cortical motor areas, in: *Models of Information Processing in the Basal Ganglia,* J.C. Houk, J.L. Davis, and D.G. Beiser, eds., MIT Press, Cambridge.

Taha, J.M., Favre, J., Baumann, T.K., and Burchiel, K.J., 1996, Characteristics and somatotopic organization of kinesthetic cells in the globus pallidus of patients with Parkinson's disease, *J. Neurosurg.* 85:1005.

Tremblay, L., and Filion, M., 1989, Responses of pallidal neurons to striatal stimulation in intact waking monkeys, *Brain Res.* 498:1.

Tremblay, L., Filion, M., and Bédard, P.J., 1989, Responses of pallidal neurons to striatal stimulation in monkeys with MPTP-induced parkinsonism, *Brain Res.* 498:17.

PHYSIOLOGY OF SUBTHALAMIC NUCLEUS NEURONS IN ANIMAL MODELS OF PARKINSON'S DISEASE

ABDELHAMID BENAZZOUZ, ZHONGGE NI, RABIA BOUALI-BENAZZOUZ, SORIN BREIT, DONGMING GAO, ADNAN KOUSIE, PIERRE POLLAK, PAUL KRACK, VALERIE FRAIX, ELENA MORO, AND ALIM-LOUIS BENABID[*]

INTRODUCTION

The subthalamic nucleus (STN) has been reported to play an important role in the control of movement. It is described as a "driving force of the basal ganglia" because it exerts a glutamatergic excitatory influence on the output structures of the system: the pars reticulata of substantia nigra (SNr) and the internal part of globus pallidus (GPi, the equivalent of the entopeduncular nucleus (EP) in rodents) (Kitai and Kita, 1987; Robledo and Feger, 1990). From the review of Parent and Hazrati (1995), it appears that the STN, like the striatum, is a major structure through which cortical signals are transmitted to the output nuclei of the basal ganglia. Moreover, in addition to the direct cortico-STN pathway, the cortex influences the STN through an indirect pathway involving the striatum and the external part of globus pallidus (GPe, an equivalent of GP in rodents).

The STN has been shown to play an important role in the control of movement. It was commonly related to hemiballismus, the manifestation of abnormal involuntary movements (AIM) resulting from a vascular accident in the region including the STN (Martin, 1927; Whitter, 1947; Meyers et al., 1968; Baruma and Lakke, 1986). These AIMs can be reproduced in non-human primates by inhibiting STN activity after *in situ* injection of GABA agonists or after lesioning of the nucleus (Whittier, 1947; Whittier and Mettler,

[*] ABDELHAMID BENAZZOUZ, ZHONGGE NI, RABIA BOUALI-BENAZZOUZ, AND SORIN BREIT • Laboratoire de Neurosciences Précliniques, INSERM U.318, Université Joseph Fourier, CHRU-Pavillon B, BP217, 38043 Grenoble Cedex 09, France. DONGMING GAO • Department of Physiology, Jinzhou Medical College, Jinzhou, P.R. China. ADNAN KOUSIE • Laboratoire de Neurosciences Précliniques, INSERM U.318, Université Joseph Fourier, CHRU-Pavillon B, BP217, 38043 Grenoble Cedex 09, France. PIERRE POLLAK • Department of Physiology, Université Joseph Fourier, CHRU-Pavillon B, BP217, 38043 Grenoble Cedex 09, France. PAUL KRACK • Department of Neurosciences, University of Grenoble, France; and Neurology Department, University of Kiel, 24105 Kiel, Germany. VALERIE FRAIX, ELENA MORO, AND ALIM-LOUIS BENABID • Laboratoire de Neurosciences Précliniques, INSERM U.318, Université Joseph Fourier, CHRU-Pavillon B, BP217, 38043 Grenoble Cedex 09, France.

1949; Hammond et al., 1979; Crossman et al., 1984; Hamada and DeLong, 1992). Recently, several studies demonstrated that the STN is implicated in the pathophysiology of Parkinson's disease. The experimental animal models of Parkinson's disease have allowed extensive investigation of the functional consequences of the degeneration of the dopaminergic nigrostriatal pathway. The intracerebral administration of 6-hydroxydopamine (6-OHDA) to rodents or the systemic administration of 1-methyl-4-phenyl-1,2,3,6-tetrahydropyridine (MPTP) to primates induces the loss of dopamine cells in the substantia nigra pars compacta (SNc), leading to the depletion of dopamine in the striatum. The consequence of this dopamine depletion is impairment in the information processing in the basal ganglia nuclei. The most significant alterations include increases in the neuronal firing rate (Bergman et al., 1994; Miller and DeLong, 1987), glucose metabolism (Mitchell et al., 1989), and mitochondrial enzyme activity in the subthalamic nucleus and its efferent structures (Porter et al., 1994; Vila et al., 1996; 1999).

THE SUBTHALAMIC NUCLEUS IN MPTP-TREATED MONKEYS

In monkeys rendered parkinsonian by systemic injections of MPTP, abnormal activity of STN neurons is registered (Bergman, et al., 1994; Miller and DeLong,, 1987). Electrophysiological studies in the *Macaca mulata* monkey (Miller and DeLong, 1987), in which MPTP induced only akinesia and rigidity without tremor at rest, showed an increase in the firing rate of STN neurons. However, in African green monkeys, in which MPTP induces akinesia, rigidity, and tremor at rest, Bergman et al. (1994) have shown a significant increase in the firing rate of STN neurons from 19±10 spikes/s before to 26±15 spikes/s after MPTP treatment. These authors also reported significant changes in the firing pattern after MPTP-treatment: STN cells displayed a high tendency to fire in bursts. In addition, 16% of the cells exhibiting burst activity showed periodic oscillatory bursting with a frequency close to the tremor frequency (4-6 Hz), and 10% of neurons displayed periodic oscillatory activity at higher frequencies. The changes reported in electrophysiological studies are in good agreement with the changes in metabolic activity. In MPTP-treated monkeys, several studies have found a significant reduction in 2-deoxyglucose uptake (Mitchell et al., 1989) and a significant increase in cytochrome oxidase mRNA and GAD mRNA expression at the STN level (Guridi et al., 1996; Vila et al., 1997). Similar results were obtained in STN efferent structures (GPi and SNr), showing also that these abnormalities can be reversed by levodopa treatment or STN lesion (Guridi, et al., 1996; Herrero, et al., 1996). These findings enabled investigators to postulate that the STN plays a critical role in the pathophysiology of Parkinson's disease. Recently, Bezard et al. (1999) demonstrated that the STN also plays a central role in the compensatory mechanism that sustains the dopaminergic function in the pre-symptomatic phase of Parkinson's disease. It is thought that parkinsonian motor signs appear when dopaminergic neuronal death exceeds a critical threshold: 70-80% of striatal nerve terminals and 50-60% of SNc perikarya (Agid, 1991; Poewe, 1993). This gradual appearance of clinical signs is due to the existence of compensatory mechanisms corresponding to an increased efficiency of residual dopaminergic neurons, which results in increased release of dopamine at the striatal level (Zigmond, et al., 1990; Bezard, et al., 1998). It has been demonstrated that SNc excitatory glutamatergic afferents can activate the remaining dopaminergic neurons and contribute to the compensatory effects, which mask the clinical manifestation of experimental parkinsonism in the pre-symptomatic period (Bezard and Gross, 1998). The same authors demonstrated that temporary blockade of these SNc glutamatergic inputs elicited parkinsonian motor abnormalities in asymptomatic monkeys treated with MPTP (Bezard et al., 1997a; 1997b). More recently, these authors showed that the compensatory increase in the glutamatergic input to the SNc originates from the STN (Bezard et al., 1999). Using electrophysiological recordings, they have shown that STN neurons enhanced their activity during the course of MPTP-treatment, even before the first appearance of clinical signs.

THE SUBTHALAMIC NUCLEUS IN 6-OHDA RAT MODEL OF PARKINSONISM

Results from different studies of the spontaneous activity of STN neurons in the 6-hydroxydopamine (6-OHDA) rat model of parkinsonism are inconsistent. Two of these studies Hassani, et al., 1996; Kreiss, et al., 1997) reported an increase in the firing rate of STN neuronal activity after SNc lesioning, but their results differed with respect to the changes in the firing patterns. In contrast, Hollerman and Grace (1992) demonstrated important changes in the firing pattern without modifications in the firing rate. Recently, we studied the time-course of changes in the firing rate and firing pattern of STN neuronal activity in the rat model of parkinsonism (Ni et al., 1999). Electrophysiological recordings were done in normal rats and in four groups of rats at different time points after 6-OHDA microinjection into the SNc. Results showed a significant decrease in firing rate during the first and second weeks post-lesion compared to normal rats. However, from the 3rd week after 6-OHDA injection the firing rates returned to baseline. With regard to firing pattern, the majority of STN cells discharged regularly or slightly irregularly in normal animals. Only a few cells exhibited burst or mixed patterns of activity. After SNc-lesion the percentage of cells exhibiting burst and mixed patterns increased progressively (Ni et al., 1999). Our data are in good agreement with the results of Hollerman and Grace (1992). These findings provide evidence for a transient change in the firing rate of STN neuronal activity while a progressive and stable change in the firing pattern was taking place. This suggests that modifications in firing patterns are more important than changes in firing rates in STN neurons in this 6-OHDA rat model of Parkinson's disease. More recent data (Plenz and Kitai, 1999) from an *in vitro* model, in which STN and GP were co-cultured with the cortex and the striatum without dopamine inputs (a model resembling a dopamine-deficient basal ganglia system), supported the finding that dopamine depletion increases burst activity in the STN. In addition, these investigators reported that the mean firing rate of STN neurons in this model was 13±2 spikes/s, which is close to the firing rate observed in our control rats.

The presence of burst cells in parkinsonism is apparently the most constant finding among published data on the spontaneous activity of the STN and other basal ganglia nuclei. Burst activity has been reported in SNr (Burbaud et al., 1995; Murer et al., 1997) and GP neurons of rats with 6-OHDA lesions (Hassani et al., 1996). According to the current model of basal ganglia circuitry (Albin et al., 1989), it can be postulated that the change in the firing pattern of STN neurons is mediated by the indirect pathway. However, it has been shown that burst activity in rats with 6-OHDA lesions is not modified by ibotenic GP lesion (Hassani et al., 1996). It is also possible that the change in the firing pattern in GP neurons of rats with 6-OHDA lesions could originate in the STN. We demonstrated in a recent study (Ni et al., 2000) that GP neurons in normal rats exhibited three different firing patterns: regular, irregular, and bursting, and that STN lesion in normal rats induced a ceasing of burst cell activity. In addition, we found that in rats with 6-OHDA SNc lesions, GP neurons showed a significant increase in the number of cells exhibiting bursting activity, and that STN lesions in rats with previous SNc lesions induced normalization of the firing patterns. These results demonstrate that the STN plays an important role in the modulation of the pattern of activity of GP neurons. From other recent evidence, it appears that the STN also participates in the genesis of the burst pattern of activity of SNr neurons in rats with 6-OHDA lesions, and that STN lesions can reverse this abnormal spontaneous firingpattern (Burbaud et al., 1995; Murer et al., 1997).

Since the burst discharge pattern is correlated with physiological and pathological phenomena, neuronal bursting would exert specific effects on target cells. The physiological relevance of burst firing and modulation of regularity is further supported by the results that show specific increases in neurotransmitter release in response to burst stimulation as compared to regular stimulation (Dutton and Dyball, 1979; Gonon, 1988; Lundberg et al., 1986). From a functional viewpoint, the burst firing pattern is more effective than the regular one at the same firing frequency (Bergman et al., 1994). From this evidence, one could consider a bursting distribution of spikes within a normal firing pattern range after dopamine depletion to be evidence of a state functionally equivalent to hyperactive activity, as compared to the regular and continuous firing pattern characteristic of the normal state.

THE SUBTHALAMIC NUCLEUS AS A TARGET FOR THE THERAPY OF PARKINSON'S DISEASE

From the above review, it is clear that the subthalamic nucleus plays a critical role in the pathophysiology of Parkinson's disease. From a functional view point and in accordance with the accepted model of basal ganglia organization, the abnormal activity observed in the STN provokes an increase in firing rate and a change in firing pattern of the main output nuclei of basal ganglia: GPi (Bergman et al., 1994; Bezard et al., 1999; Filion and Tremblay, 1991; Boraud et al., 1996) and SNr (Wichmann et al., 1999) in MPTP-treated monkeys. This change in the firing activity of output structures of the system is supposed to induce an increase in the tonic inhibitory influence exerted by these structures on the activity of motor thalamic nuclei, resulting in deactivation of motor cortical areas. Finally, the cortical hypoactivity results in the manifestation of parkinsonian motor symptoms. STN lesions have been shown to induce an improvement in parkinsonian motor symptoms in monkeys rendered parkinsonian by MPTP (Guridi et al., 1996; Aziz et al., 1991; Bergman, et al., 1990) and to abolish the rotational response to apomorphine of rats with 6-OHDA lesions of the SNc (Burbaud et al., 1995; Blandini et al., 1997). Functionally, STN ablation reverses the increases in mitochondrial enzyme activity found in the entopeduncular nucleus and SNr of 6-OHDA lesioned rats (Blandini et al., 1997). In addition, it normalizes the firing rate and discharge pattern of SNr neurons (Burbaud et al., 1995; Murer et al., 1997) and prevents changes in gene expression in the entopeduncular nucleus and globus pallidus of rats with 6-OHDA SNc lesions (Delfs et al., 1995).

Since the beneficial effect of STN lesions is accompanied by the appearance of abnormal movements (Aziz et al., 1991; Bergman et al., 1990) and multiple deficits in attentional tasks (Baunez et al., 1995; Baunez and Robbins, 1997), it was difficult to propose this surgical procedure for treatment of Parkinson's disease. To avoid the side effects, we have replaced ablative lesions with high frequency stimulation (HFS), which had been used in the ventral-intermedial (Vim) nucleus of the thalamus to replace thalamotomy in the treatment of tremor (Benabid et al., 1987; 1991; 1996). HFS of the subthalamic nucleus has been shown to reverse the three cardinal motor symptoms, akinesia, rigidity, and tremor, in primate models of parkinsonism (Benazzouz et al., 1993; 1996; Gao et al., 1999) and in severe parkinsonian patients (Limousin et al., 1995; 1998; Krack et al., 1998). Moreover, the threshold of current intensity inducing positive effects does not induce side effects (Benazzouz et al., 1993; 1996; Limousin et al., 1996; 1998). In addition, we have found that when the stimulation is stopped, the positive effect can be maintained for seconds to minutes, but it is always reversible (Benazzouz et al., 1993; 1996). The advantage of this technique is that the effect is reversible and easily graded by changing electrical parameters. From the results of these studies it appears that STN-HFS is functionally equivalent to STN-lesion, but its precise mechanism of function remains unclear. In previous studies (Benazzouz et al., 1995; Burbaud et al., 1994), we reported that STN-HFS induced a significant decrease of neuronal activity in the output structures of basal ganglia, SNr and EP, and a net increase of activity in GP neurons. From these results it was unclear whether the inhibition of activity in the output structures of basal ganglia is due to a direct inhibition of STN neurons or to the activation of GP neurons, which are known to project to these nuclei using GABA as a neurotransmitter. In order to investigate the role of GP as a potential step in inducing SNr inhibition, we studied the responses of SNr neurons to STN-HFS in rats with ibotenic acid lesion of the GP. The results from this group of animals show that STN-HFS still inhibits the activity of SNr neurons but induces a dramatic effect on the duration and recovery from the inhibition after the stimulation is stopped (Benazzouz et al., 2000) The inhibitory effect observed in the output structures of basal ganglia could be the consequence of the inhibition of STN neuronal activity induced by STN-HFS. In fact, as revealed in our study (Benazzouz et al., 2000), when STN-HFS is given, the firing rate of STN neurons decreases. From this evidence, we can postulate that HFS induces a direct inhibitory influence on the stimulated neurons, through a still unknown mechanism. Recently, Beurrier et al. (personal communication) found that STN-HFS applied to slices containing the STN without its afferent structures induced an immediate and clear inhibition of its neuronal activity. All these findings fit with the notion that high frequency stimulation of the STN is inhibitory to STN neurons, thus reducing excitatory output from the STN, which, in turn, results in deactivation of the output nuclei of the basal ganglia. According to the current model of cortico-basal ganglia-thalamo-cortical circuits, the reduction in tonic inhibitory drive of SNr and EP neurons in response to STN-HFS induces a disinhibition of activity in motor thalamic nuclei, which results in activation of the motor cortical system. This hypothesis is reinforced by other published data showing that STN-HFS induces a significant increase in the firing rate of the ventral-lateral and the ventral-medial nuclei of the rat thalamus (Benazzouz et al., 2000; Gao et al., 1997). So far, in the absence of data demonstrating the existence of direct projections from the STN to these thalamic nuclei, the effects of STN-HFS on these nuclei should be mediated by SNr and EP GABAergic inhibitory projections. In addition, Limousin et al. (1997) demonstrated that STN-HFS, which produced significant improvement in movement performance, was accompanied by an increase in cortical activity of the supplementary motor area, dorso-lateral prefrontal cortex, and cingulate. In another recent study, we have found that STN-HFS also influences the activity of the SNc dopaminergic neurons by increasing their firing rate (Benazzouz et al., 2000b). This result is consistent with recent biochemical studies using microdialysis technique (Bruet et al., 1999; Paul et al., 2000) demonstrating that STN-HFS with the same parameters induced an increase in the level of dopamine and/or metabolites in the striatum of normal rats. In conclusion, we can postulate that STN-HFS could induce an increase of activity of residual dopaminergic neurons in parkinsonian patients resulting in an increase of dopamine production at the striatal level. Nevertheless as a large number of SNc dopaminergic neurons are damaged in severe Parkinson's disease patients, the improvement of motor symptoms in response to STN-HFS could not solely be due to this phenomenon.

REFERENCES

Agid, Y., 1991, Parkinson's disease: pathophysiology, *Lancet*, 337:1321.

Albin, R.L., Young, A.B., and Penney, J.B., 1989, The functional anatomy of basal ganglia disorders, *Trends Neurosci.* 12:366.

Aziz, T.Z., Peggs, D., Sambrook, M.A., and Crossman, A.R., 1991, Lesion of the subthalamic nucleus for the alleviation of 1-methyl-4- phenyl-1,2,3,6-tetrahydropyridine (MPTP)-induced parkinsonism in the primate, *Mov. Disord.* 6:288.

Baunez, C., and Robbins, T.W., 1997, Bilateral lesions of the subthalamic nucleus induce multiple deficits in an attentional task in rats, *Eur. J. Neurosci.* 9:2086.

Baunez, C., Nieoullon, A., and Amalric, M., 1995, In a rat model of parkinsonism, lesions of the subthalamic nucleus reverse increases of reaction time but induce a dramatic premature responding deficit, *J. Neuroscience,* 15:6531.

Benabid, A.L., Pollak, P., Gao, D., Hoffmann, D., Limousin, P., Gay, E., Payen, I., and Benazzouz, A., 1996, Chronic electrical stimulation of the ventralis intermedius nucleus of the thalamus as a treatment of movement disorders, *J. Neurosurg.* 84:203.

Benabid, A.L., Pollak, P., Gervason, C., Hoffmann, D., Gao, D.M., Hommel, M., Perret, J.E., and de Rougemont, J., 1991, Long-term suppression of tremor by chronic stimulation of the ventral intermediate thalamic nucleus, *Lancet*, 337:403.

Benabid, A.L., Pollak, P., Louveau, A., Henry, S., and de Rougemont, J., 1987, Combined (thalamotomy and stimulation) stereotactic surgery of the VIM thalamic nucleus for bilateral Parkinson disease, *Appl. Neurophysiol.* 50:344.

Benazzouz, A., Boraud, T., Feger, J., Burbaud, P., Bioulac, B., and Gross, C., 1996, Alleviation of experimental hemiparkinsonism by high-frequency stimulation of the subthalamic nucleus in primates: a comparison with L-Dopa treatment, *Mov. Disord.* 11:627.

Benazzouz, A., Gao, D., Ni, Z., and Benabid, A.L., 2000, High frequency stimulation of the STN influences the activity of dopamine neurons in the rat, *Neuroreport*, 11:1593.

Benazzouz, A., Gao, D.M., Ni, Z.G., Piallat, B., Bouali-Benazzouz, R., and Benabid AL., 2000b, Effect of high-frequency stimulation of the subthalamic nucleus on the neuronal activities of the substantia nigra pars reticulata and ventrolateral nucleus of the thalamus in the rat, *Neuroscience*, 99:289.

Benazzouz, A., Gross, C., Feger, J., Boraud, T., and Bioulac, B., 1993, Reversal of rigidity and improvement in motor performance by subthalamic high-frequency stimulation in MPTP-treated monkeys, *Eur. J. Neurosci.* 5:382.

Benazzouz, A., Piallat, B., Pollak, P., and Benabid, A.L., 1995, Responses of substantia nigra pars reticulata and globus pallidus complex to high frequency stimulation of the subthalamic nucleus in rats: electrophysiological data, *Neurosci. Lett.* 189:77.

Bergman, H., Wichmann, T., and DeLong, M.R., 1990, Reversal of experimental parkinsonism by lesions of the subthalamic nucleus, *Science*, 249:1436.

Bergman, H., Wichmann, T., Karmon, B., and DeLong, M.R., The primate subthalamic nucleus. II. Neuronal activity in the MPTP model of parkinsonism, *J. Neurophysiol.* 72:507.

Bezard, E., Bioulac, B., and Gross, C.E., 1998, Glutamatergic compensatory mechanisms in experimental parkinsonism, *Prog. Neuropsychopharmacol. Biol. Psychiatry*, 22:609.

Bezard, E., Boraud, T., Bioulac, B., and Gross, C.E., 1997a, Presymptomatic revelation of experimental parkinsonism, *Neuroreport*, 8:435.

Bezard, E., Boraud, T., Bioulac, B., and Gross, C.E., 1999, Involvement of the subthalamic nucleus in glutamatergic compensatory mechanisms, *Eur. J. Neurosci.* 11:2167.

Bezard, E., and Gross, C.E., 1998, Compensatory mechanisms in experimental and human parkinsonism: towards a dynamic approach, *Prog. Neurobiol.* 55:93.

Bezard, E., Imbert, C., Deloire, X., Bioulac, B., and Gross, C.E., 1997b, A chronic MPTP model reproducing the slow evolution of Parkinson's disease: evolution of motor symptoms in the monkey, *Brain Res.* 766:107.

Blandini, F., Garcia-Osuna, M., and Greenamyre, J.T., 1997, Subthalamic ablation reverses changes in basal ganglia oxidative metabolism and motor response to apomorphine induced by nigrostriatal lesion in rats, *Eur. J. Neurosci.* 9:1407.

Boraud, T., Bezard, E., Bioulac, B., and Gross, C., 1996, High frequency stimulation of the internal Globus Pallidus (GPi) simultaneously improves parkinsonian symptoms and reduces the firing frequency of GPi neurons in the MPTP-treated monkey, *Neurosci. Lett.* 215:17.

Bruet, N., Windels, F., Bertrand, A., Chatellard-Causse, C., Feuerstein, C., Benabid, A,L, and Savasta, M., 1999, Modification of striatal dopamine release after electrical stimulation of the subthalamic nucleus in normal and partially dopaminergic denervated rats, *Soc. Neurosci. Abstr.* 541:18.

Burbaud, P., Gross, C., and Bioulac, B., 1994, Effect of subthalamic high frequency stimulation on substantia nigra pars reticulata and globus pallidus neurons in normal rats, *J. Physiol.(Paris)* 88:359.

Burbaud, P., Gross, C., Benazzouz, A., Coussemacq, M., and Bioulac, B., 1995, Reduction of apomorphine-induced rotational behaviour by subthalamic lesion in 6-OHDA lesioned rats is associated with a normalization of firing rate and discharge pattern of pars reticulata neurons, *Exp. Brain Res.* 105:48.

Buruma, O.J.S., and Lakke, J.P.W.F., 1986, Ballism, in: *Handbook of Clinical Neurology*, P.J.Vinken, G.W. Bruyn, and H.L. Klawans, eds., Elsevier, Amsterdam.

Crossman, A.R., Sambrook, M.A., and Jackson, A., 1984, Experimental hemichorea/hemiballismus in the monkey. Studies on the intracerebral site of action in a drug-induced dyskinesia, *Brain,* 107:579.

Delfs, J.M., Ciaramitaro, V.M., Parry, T.J., and Chesselet, M.F., 1995, Subthalamic nucleus lesions: widespread effects on changes in gene expression induced by nigrostriatal dopamine depletion in rats, *J. Neurosci.* 15:6562.

Dutton, A., and Dyball, R,E. 1979, Phasic firing enhances vasopressin release from the rat neurohypophysis, *J. Physiol. (Lond.)* 290:433.

Filion, M., and Tremblay, L., 1991, Abnormal spontaneous activity of globus pallidus neurons in monkeys with MPTP-induced parkinsonism, *Brain Res.* 547:142.

Gao, D., Benazzouz, A., Bressand, K., Piallat, B., and Benabid, A.L., 1997, Roles of GABA, glutamate, acetylcholine and STN stimulation on thalamic VM in rats, *Neuroreport,* 8:2601.

Gao, D.M., Benazzouz, A., Piallat, B., Bress, K., Ilinsky, I.A., Kultas-Ilinsky, K., and Benabid, A.L., 1999, High-frequency stimulation of the subthalamic nucleus suppresses experimental resting tremor in the monkey, *Neuroscience,* 88:201.

Gonon, F.G., 1988, Nonlinear relationship between impulse flow and dopamine released by rat midbrain dopaminergic neurons as studied by in vivo electrochemistry, *Neuroscience,* 24:19.

Guridi, J., Herrero, M.T., Luquin, M.R., Guillen, J., Ruberg, M., Laguna, J., Vila, M., Javoy-Agid, F., Agid, Y., Hirsch, E., and Obeso, J.A., 1996, Subthalamotomy in parkinsonian monkeys. Behavioural and biochemical analysis, *Brain,* 119:1717.

Hamada, I., and DeLong, M.R., 1992, Excitotoxic acid lesions of the primate subthalamic nucleus result in transient dyskinesias of the contralateral limbs, *J. Neurophysiol.* 68:1850.

Hammond, C., Feger, J., Bioulac, B., and Souteyrand, J.P., 1979, Experimental hemiballism in the monkey produced by unilateral kainic acid lesion in corpus Luysii, *Brain Res.* 171:577.

Hassani, O.K., Mouroux, M., and Feger, J., 1996, Increased subthalamic neuronal activity after nigral dopaminergic lesion independent of disinhibition via the globus pallidus, *Neuroscience,* 72:105.

Herrero, M.T., Levy, R., Ruberg, M., Luquin, M.R., Villares, J., Guillen, J., Faucheux, B., Javoy-Agid, F., Guridi, J., Agid, Y., Obeso, J.A., and Hirsch, E.C., 1996, Consequence of nigrostriatal denervation and L-dopa therapy on the expression of glutamic acid decarboxylase messenger RNA in the pallidum, *Neurology,* 47:219.

Hollerman, J.R., and Grace, A.A., 1992, Subthalamic nucleus cell firing in the 6-OHDA-treated rat: basal activity and response to haloperidol, *Brain Res.* 590:291.

Kitai, S.T., and Kita H., 1987, Anatomy and physiology of the subthalamic nucleus: a driving force of the basal ganglia, in: *The Basal Ganglia II: Structure and Function,* M.B. Carpenter and A. Jayaraman, eds., Plenum Press, New York.

Krack, P., Pollak, P., Limousin, P., Hoffmann, D., Xie, J., Benazzouz, A., and Benabid, A.L., 1998, Subthalamic nucleus or internal pallidal stimulation in young onset Parkinson's disease, *Brain,* 121:451.

Kreiss, D.S., Mastropietro, C.W., Rawji, S.S., Walters, J.R. 1997, The response of subthalamic nucleus neurons to dopamine receptor stimulation in a rodent model of Parkinson's disease, *J Neurosci.* 17:6807.

Limousin, P., Greene, J., Pollak, P., Rothwell, J., Benabid, A.L., and Frackowiak, R., 1997, Changes in cerebral activity pattern due to subthalamic nucleus or internal pallidum stimulation in Parkinson's disease, *Ann. Neurol.* 42:283.

Limousin, P., Krack, P., Pollak, P., Benazzouz, A., Ardouin, C., Hoffmann, D., and Benabid, A.L., 1998, Electrical stimulation of the subthalamic nucleus in advanced Parkinson's disease, *N. Engl. J. Med.* 339:1105.

Limousin, P., Pollak, P., Benazzouz, A., Hoffmann, D., Broussolle, E., Perret, J.E., and Benabid, AL., 1995, Bilateral subthalamic nucleus stimulation for severe Parkinson's disease, *Mov. Disord.* 10:672.

Limousin, P., Pollak, P., Hoffmann, D., Benazzouz, A., Perret, J.E., and Benabid, A.L., 1996, Abnormal involuntary movements induced by subthalamic nucleus stimulation in parkinsonian patients, *Mov. Disord.* 11:231.

Lundberg, J.M., Rudehill, A., Sollevi, A., Theodorsson-Norheim, E., and Hamberger, B., 1986, Frequency- and reserpine-dependent chemical coding of sympathetic transmission: differential release of noradrenaline and neuropeptide Y from pig spleen, *Neurosci. Lett.* 63:96.

Martin, J.P., 1927, Hemichorea resulting from a local lesion of the brain, *Brain,* 50:637.

Meyers, R., 1968, Ballismus, in: *The Handbook of Clinical Neurology,* P.J. Vinken and G.W. Bruyn, eds., North-Holland Press, Amsterdam.

Miller, W.C., and DeLong, M.R., 1987, Altered tonic activity of neurons in the globus pallidus and subthalamic nucleus in the primate MPTP model of parkinsonism, in: *The Basal Ganglia II,* M.B. Carpenter and A. Jayaraman, eds., Plenum Press, New York.

Mitchell, I.J., Jackson, A., Sambrook, M.A., and Crossman, A.R., 1989, The role of the subthalamic nucleus in experimental chorea. Evidence from 2-deoxyglucose metabolic mapping and horseradish peroxidase tracing studies, *Brain,* 112:1533.

Murer, M.G., Riquelme, L.A., Tseng, K.Y., and Pazo, J.H., 1997, Substantia nigra pars reticulata single unit activity in normal and 6-OHDA-lesioned rats: effects of intrastriatal apomorphine and subthalamic lesions, *Synapse,* 27:278.

Ni, Z.G., Bouali-Benazzouz, R., Gao, D.M., Benabid, A.L., and Benazzouz, A., 1999, Time-course of changes in neuronal activity of subthalamic nucleus after 6-OHDA induced dopamine depletion in rats, *Soc. Neurosci. Abstr.* 29:2116.

Ni, Z.G., Bouali-Benazzouz, R., Gao, D.M., Benabid, A.L., and Benazzouz, A., 2000, Changes in the firing pattern of globus pallidus neurons after the degeneration of nigrostriatal pathway are mediated by the subthalamic nucleus in the rat, *Mov. Disord.* 31:278.

Pan, H.S., and Walters, J.R., 1988, Unilateral lesion of the nigrostriatal pathway decreases the firing rate and alters the firing pattern of globus pallidus neurons in the rat, *Synapse,* 2:650.

Parent, A., and Hazrati, L.N., 1995, Functional anatomy of the basal ganglia. II. The place of subthalamic nucleus and external pallidum in basal ganglia circuitry, *Brain Res. Rev.* 20:128.

Paul, G., Reum, T., Meissner, W., Marburger, A., Sohr, R., Morgenstern, R., and Kupsch, A., 2000, High frequency stimulation of the subthalamic nucleus influences striatal dopaminergic metabolism in the naive rat, *Neuroreport,* 11:441.

Plenz, D., and Kital, S.T., 1999, A basal ganglia pacemaker formed by the subthalamic nucleus and external globus pallidus, *Nature,* 400:677.

Poewe, W., 1993, Clinical features, diagnosis, and imaging of parkinsonian syndromes, *Curr. Opin. Neurol. Neurosurg.* 6:333.

Porter, R.H., Greene, J.G., Higgins, D.S., Jr., and Greenamyre, J.T., 1994, Polysynaptic regulation of glutamate receptors and mitochondrial enzyme activities in the basal ganglia of rats with unilateral dopamine depletion, *J. Neurosci.* 14:7192.

Robledo, P., and Feger, J., 1990, Excitatory influence of rat subthalamic nucleus to substantia nigra pars reticulata and the pallidal complex: electrophysiological data, *Brain Res.* 518:47.

Vila, M., Levy, R., Herrero, M.T., Faucheux, B., Obeso, J.A., Agid, Y., and Hirsch, E.C., 1996, Metabolic activity of the basal ganglia in parkinsonian syndromes in human and non-human primates: a cytochrome oxidase histochemistry study, *Neuroscience,* 71:903.

Vila, M., Levy, R., Herrero, M.T., Ruberg, M., Faucheux, B., Obeso, J.A., Agid, Y., and Hirsch, E.C., 1997, Consequences of nigrostriatal denervation on the functioning of the basal ganglia in human and nonhuman primates: an in situ hybridization study of cytochrome oxidase subunit I mRNA, *J. Neurosci.* 17:765.

Vila, M., Marin, C., Ruberg, M., Jimenez, A., Raisman-Vozari, R., Agid, Y., Tolosa, E., and Hirsch, E.C., 1999, Systemic administration of NMDA and AMPA receptor antagonists reverses the neurochemical changes induced by nigrostriatal denervation in basal ganglia, *J. Neurochem.* 73:344.

Whittier, J.R., 1947, Ballism and the subthalamic nucleus (nucleus hypothalamicus, corpus Luysi), *Arch. Neurol. Psych.* 58:672.

Whittier, J.R., and Mettler, F.A., 1949, Studies of the subthalamus of the rhesus monkey. II. Hyperkinesia and other physiologic effects of subthalamic lesions, with speicial reference to the subthalamic nucleus of Luys, *J. Comp. Neurol.* 90:319.

Wichmann, T., Bergman, H., Starr, P.A., Subramanian, T., Watts, R.L., and DeLong, M.R., 1999, Comparison of MPTP-induced changes in spontaneous neuronal discharge in the internal pallidal segment and in the substantia nigra pars reticulata in primates, *Exp. Brain Res.* 125:397.

Zigmond, M.J., Abercrombie, E.D., Berger, T.W., Grace, A.A., and Stricker, E.M.,1990, Compensations after lesions of central dopaminergic neurons: some clinical and basic implications, *Trends Neurosci.* 13:290.

THE MOTOR THALAMUS: ALTERATION OF NEURONAL ACTIVITY IN THE PARKINSONIAN STATE

CHRISTOPHER M. ELDER AND JERROLD L. VITEK[*]

INTRODUCTION

Derivations of the term thalamus have been used in various ways since the time of Virgil and Ovid around the beginning of the first millennium. Historians attributed the first clear anatomical use of the term thalamus to Galen in the 2nd century A.D. It appears, however, that Galen was not describing the structure we recognize today as the thalamus, but instead used the word to describe a funnel-like region representing the inferior horn of the lateral ventricle. Galen called this area thalamus, since it resembled the inner chamber of a ship's galley, or "thalami." The first descriptions of the structure we currently recognize as the thalamus arose with Mondinus in the 14th century. Although Vesalius continued to lay the foundation of our understanding by providing visual representations of the thalamus, corpus striatum, and internal capsule, descriptions remained incomplete, since they lacked named structures. Thomas Willis provided the first clear representations of the thalamus to modern anatomists. Nuclear subdivisions of the thalamus, however, would not be described for another two centuries until Burdach did so in 1822 (Jones, 1985).

The thalamus is among the largest organized structures in the brain, measuring approximately 15,000 mm^3, or 1 percent of the total brain volume of the average human being. The thalamus contains approximately ten million neurons (Jones, 1985) and is anatomically positioned in such a way that it may influence neuronal activity throughout the cerebral cortex. It is, in fact, often referred to as the gateway to the cortex and contains over 50 anatomically and functionally distinct subnuclei that contribute to the regulation of many motor as well as nonmotor processes. Although the thalamus has long been considered important in motor control, a majority of the investigations into the function of the thalamus have focused on the sensory thalamus (Poggio and Mountcastle, 1963; Friedman and Jones, 1980; Jones, 1981; Jones et al., 1982; Jones and Friedman, 1982; Jones, 1991; Morrow and Casey, 1992). Relatively few studies in comparison have been conducted on the motor thalamus (Strick, 1976; Horne and Porter, 1980; Macpherson, 1980; Anner-Baratti et al., 1986; Anderson and Turner, 1991; Butler, 1992; Vitek et al.,

[*] CHRISTOPHER M. ELDER AND JERROLD L. VITEK • Department of Neurology, Emory University School of Medicine, Atlanta, Georgia 30322.

1994 and 1996). These studies, however, have led to an improved understanding of its functional organization.

FUNCTIONAL ORGANIZATION OF THE MOTOR THALAMUS

The primate motor thalamus is composed of regions receiving projections from the basal ganglia and deep cerebellar nuclei and projecting primarily to motor and premotor cortical areas. In the terminology of Olszewski, those areas of the thalamus receiving basal ganglia input include nucleus ventralis lateralis, pars oralis (VLo), and nucleus ventralis anterior (VA). Those areas of the thalamus receiving cerebellar inputs include nucleus ventralis lateralis, pars caudalis (VLc), nucleus ventralis posterior lateralis, pars oralis (VPLo), and Area X (Strick, 1976; Kunzle, 1976; Kievet and Kuypers, 1977; Schell and Strick, 1984; Anner-Baratti et al., 1986; Holsapple et al., 1991).

Early studies of thalamic function suggested that the motor thalamus served as an integrator of neural activity, processing converging afferent information from the basal ganglia and cerebellum and relaying this "processed" information to the cerebral cortex. The idea of the thalamus as an integrator became less favorable, however, as it became clear that afferent projections from the basal ganglia and cerebellum terminated in largely segregated subnuclei of the motor thalamus (Stanton, 1980; Asanuma et al., 1983; Schell and Strick, 1984; Ilinsky and Kultas-Ilinsky, 1987; Marolli, 1989; Holsapple, 1991). These observations led to the consideration that thalamic subnuclei may differ in their physiological characteristics and play differential roles in motor control.

More recent studies of the physiological characteristics of neurons in the motor thalamus have revealed that somatosensory responses of neurons in basal ganglia-receiving and cerebellar-receiving areas are limb, joint, and directionally specific. It has been demonstrated that a single neuron in the motor thalamus responds predominately to movement of a single joint in the contralateral limb in one direction (Vitek et al., 1994). Although specific in their receptive field properties, the relative incidence of cells responding to active or passive movement varies greatly between subnuclei in the motor thalamus. Neurons within VPLo and VLc respond more frequently to passive limb movement than those within VLo, VA, and Area X, which respond preferentially during active movement. Anatomic studies have demonstrated direct or indirect spinal afferent projections through the pretectum or cerebellum to VPLo and VLc, but not to VLo or VA. Area X, similar to VPLo and VLc, receives projections from the cerebellum. Unlike VPLo and VLc, however, which receive input from the interposed nucleus, a nucleus with dense peripheral afferent input, Area X receives its input from the dentate nucleus, a nucleus with little peripheral afferent input from the spinal cord. Thus, those subnuclei with a greater incidence of responses to passive manipulation receive projections from the spinal cord either directly or indirectly through the cerebellum or brainstem.

Neurons within subnuclei of the motor thalamus also have different mean discharge rates. Although the distribution of mean discharge rates of neurons within these nuclei is varied and overlapping, VPLo has a significantly higher mean discharge rate compared to VLo, VLc, VA, and Area X. The observed differences in mean discharge rates among subnuclei can be explained in part by differences in afferent projections to these areas. VLo and VA receive tonic inhibitory projections from the internal segment of the globus pallidus, while VPLo receives tonic excitatory projections from deep cerebellar nuclei and spinothalamic pathways. Area X receives little afferent input from the spinal cord, either directly or indirectly, which may account for the lower mean discharge rates observed in this subnucleus. The lower mean discharge rates observed in VLc are more difficult to explain. The influence of peripheral afferent input on neuronal activity in VLc, however, does not appear to be as strong as that to VPLo, given the relatively longer latency and weaker response of VLc neurons to torque-induced perturbations of the limb (Vitek et al.,

1994). The relatively weaker effect of peripheral afferent activity to this subnucleus may account for its lower mean discharge rate. Consistent with this hypothesis, neurons in the lemniscal-receiving area, ventralis posterior lateralis (VPLc), had the shortest latency and most robust response to torque-induced limb perturbations and the highest mean discharge rate (Vitek et al., 1994).

Although subcortical areas projecting to, and cortical areas receiving projections from, the motor thalamus were known to be somatotopically organized (Muakkassa and Strick, 1979; Brooks and Thach, 1981; DeLong et al., 1985; Orioli and Strick, 1989; Hoover and Strick, 1993), a detailed somatotopic organization of the motor thalamus had not been determined until recently (Vitek et al., 1994). A separate somatotopy has now been described for both cerebellar (VPLo)-receiving and pallidal (VLo)-receiving areas (Vitek et al., 1994). The somatotopic organization of VPLo and VLo is similar to that described for the sensory thalamus, with different body parts arranged in a system of concentric lamellae. In this organization, leg areas are located lateral and rostral to arm areas, which are in turn located lateral and anterior to face areas (Jones et al., 1982; Jones and Friedman, 1982; Vitek et al., 1994;).

Neuronal activity changes in response to task-related movements have been observed in both the pallidal (VLo)-receiving and cerebellar (VPLo)-receiving areas of the motor thalamus. Mean discharge rates of neurons within both VLo and VPLo may be modulated prior to or during the execution of goal-directed movement (Strick, 1976; Anderson, Horne and Porter, 1980; Macpherson et al., 1980; Vitek et al., 1989; Nambu et al., 1991; Turner, 1991; Butler et al., 1992; Anderson et al., 1993; Jinnai et al., 1993). Similar to neuronal activity changes in response to passive manipulation of the limbs, neurons in both the pallidal- and cerebellar-receiving areas respond to limb movement in a preferred direction during task-related movements (Strick, 1976; Horne and Porter, 1980; Macpherson et al., 1980; Vitek et al., 1989). Many task-related neurons in the motor thalamus respond in a reciprocal fashion to movement in opposite directions, with increased activity during movement in one direction, and decreased activity during movement in the opposite direction (Vitek et al., 1989). The onset of neuronal activity relative to the start of movement is similar in VLo and VPLo, with neuronal activity preceding the onset of movement in both subnuclei with an average latency of approximately 50 ms (Vitek, unpublished observations). Unlike VLo, however, where responses to kinesthetic stimuli are generally weak and occur at long latencies (mean 47 ms), responses to kinesthetic stimuli within the VPLo are robust and occur at latencies as early as 7 ms (Vitek et al., 1994). The presence of short-latency responses to kinesthetic stimuli in the posterolateral portions of VPLo suggest that this subnucleus could subserve an important role in the regulation of ongoing movement, while VLo, with its preponderance of responses to active movement, may be more directly related to aspects of movement that do not require online assessment of movement, such as the planning or preparation for movement. Given its strong projections to motor areas of the cerebral cortex, the motor thalamus may play a significant role in the modulation of cortical neuronal activity and the control of movement. It is reasonable, therefore, to assume that alterations in neuronal activity at the level of the thalamus could disrupt signal processing at the level of the cortex and play a significant role in the development of the altered movement associated with hypokinetic and hyperkinetic movement disorders (Vitek and Giroux, 2000).

ALTERED PHYSIOLOGICAL ACTIVITY IN THE MOTOR THALAMUS IN THE PARKINSONIAN STATE

The 1-methyl-4-phenyl-1,2,3,6-tetrahydopyridine (MPTP) model of parkinsonism produced in nonhuman primates has been used to investigate changes in neuronal activity that occur throughout the pallido-thalamocortical circuit in the parkinsonian state. This induced state resembles parkinsonism in humans and is characterized by bradykinesia/akinesia, rigidity, tremor, gait, and postural disorders. Previous studies in parkinsonian primates have demonstrated increased mean discharge rates in the subthalamic nucleus (STN) and internal segment of the globus pallidus (GPi) (Bergman et al., 1984; Filion et al., 1985; Miller and DeLong, 1987; Filion and Tremblay, 1991; Wichmann et al., 1994b). Blockade of this excessive increase in neuronal activity in the STN or GPi, either by lesioning or by injections of the GABA agonist muscimol, has resulted in improvement in parkinsonian motor signs (Bergman et al., 1990; Aziz et al., 1991; Wichmann et al., 1994a, 1994b). These observations led to the hypothesis that excessive inhibitory output from GPi resulted in inhibition of thalamocortical activity, which was, in turn, responsible for the development of parkinsonian motor signs (DeLong, 1990). Although there is significant support for this hypothesis at the level of the basal ganglia, there is little support for this hypothesis at the level of the thalamus. Furthermore, based on this model, lesions in the motor thalamus would be expected to worsen parkinsonian motor signs. In fact, thalamotomy has been used for decades as a treatment for parkinsonian tremor and may also improve rigidity. An added paradox is that lesions in cerebellar-receiving, not pallidal-receiving, areas of the motor thalamus improve parkinsonian tremor.

To further investigate the role of the thalamus in the development of parkinsonian motor signs, we examined the physiological changes that occurred in thalamic neurons by recording spontaneous neuronal activity and characterizing the response of individual neurons to passive movement following induction of the parkinsonian state using the neurotoxin MPTP.

Following treatment with MPTP and the development of parkinsonian motor signs, the number of neurons in VLo that responded to passive manipulation of the limbs was significantly increased. In addition, receptive fields were less specific than those found in the normal state, with cells responding to passive manipulation of multiple joints, in multiple directions, and in multiple body segments. Furthermore, manipulation of the ipsilateral limb produced a response in over 20% of the cells, a response that was never seen in the normal animal (Vitek et al., 1990; Vitek et al., 1994b). Overall, more than half of the population of neurons examined in VLo in the parkinsonian monkey responded aberrantly to passive movement.

In addition to broadened, less specific receptive fields, mean discharge rates and neuronal discharge patterns were also different from those observed in the normal animal. The mean discharge rate of cells in VLo was decreased (from 16 Hz prior to MPTP treatment to 11.5 Hz subsequent to induction of the parkinsonian state), and there was an increased incidence of bursting activity (from four percent of cells in the normal monkey to 22 percent of cells in the parkinsonian monkey) (Figure 1). Analysis of the intraburst spike pattern revealed two patterns. Some cells exhibited a pattern similar to that seen in sleeping animals (i.e., each successive spike in the burst train occurring at a longer interval than the preceding one), while others exhibited a random pattern of spike occurrences within the burst (Vitek et al., 1994). Interestingly, the changes in somatosensory responses, discharge rates, and patterns were observed not only in the basal ganglia-receiving area VLo, but also in the cerebellar-receiving area VPLo (Vitek et al., 1994). Within the VPLo, the mean discharge rates following administration of MPTP decreased from 22 Hz to 15 Hz and the incidence of bursting increased from 2 percent to 14 percent after induction of the parkinsonian state (Figure 1). More recently, a decrease in the mean discharge rate of neurons in VPLc in the parkinsonian state has also been observed (Vitek and Zhang, unpublished observations). Thus, in the parkinsonian state we observed a reduction in mean discharge rate, an increased incidence of bursting, and a loss of specificity in the receptive field properties of neurons in both pallidal- and cerebellar-receiving areas. Furthermore, based on our initial observations of reduced mean discharge rates in VPLc, changes in neuronal activity observed in the parkinsonian state may not be restricted to the motor thalamus, but may also involve nonmotor areas as well.

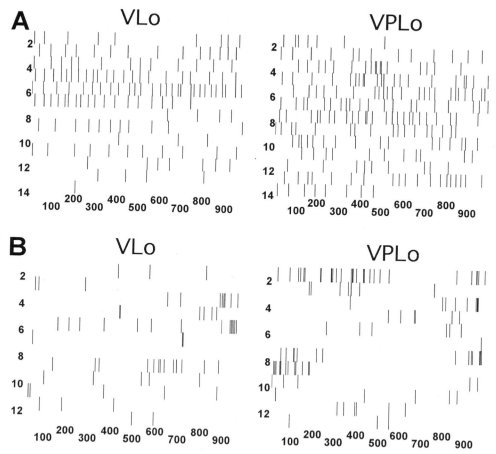

Figure 1. Raster diagrams of the firing pattern in VPLo and VLo in the normal (A) and parkinsonian (B) states in the MPTP model of parkinsonism in nonhuman primates. Values along the ordinate axis represent the elapsed time in seconds. Values along the abscissa represent the elapsed time in milliseconds within each one-second raster line display.

ROLE OF ALTERED PATTERNS OF THALAMIC ACTIVITY IN THE DEVELOPMENT OF THE PARKINSONIAN STATE

During sleep, the increased bursting that occurs throughout the thalamus acts to disconnect cortical to cortical signal transmission. A disruption in cortical signal processing of this type could occur in the parkinsonian state as a result of the "sleep-like" bursting pattern of thalamic activity and could contribute to the altered movement that occurs in this condition (Llinas, 1984; Llinas, 1988; Buzsaki et al., 1990; Steriade and McCarley, 1990; Steriade, 1992). The relative contribution of sleep-like versus random intraburst patterns to the disturbance in movement is unclear. However, both may be destructive to cortical signal processing, introducing an increasingly "noisy" signal to the cortex, disrupting the normal spatial temporal pattern(s) of cortical to cortical signal transmission, and perhaps even contributing to the development of sleep disorders so commonly seen in this patient population.

ALTERNATIVE MODEL FOR PARKINSON'S DISEASE

Taken together, the present observations suggest that current models of movement disorders must incorporate altered patterns of neuronal activity and loss of specificity in receptive field properties of neurons in both pallidal- and cerebellar-receiving areas of the motor thalamus. This model (Figure 2) proposes that aberrant activity in the motor thalamus disrupts normal signal processing in cortical motor areas that subserve movement, leading to the disordered movement that occurs in the parkinsonian state. Furthermore, the presence of similar changes in the cerebellar-receiving area VPLo offers an explanation for the observation that lesions in the cerebellar-receiving, not the pallidal-receiving, area of the motor thalamus improves parkinsonian tremor, a paradox that has heretofore been unexplained.

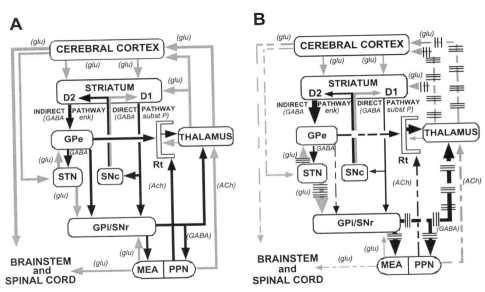

Figure 2. Schematic representation of the basal ganglia-thalamocortical "motor" circuit in the normal (A) and parkinsonian (B) models, including those projections from the PPN and GPe to Rt and thalamus. Wider lines represent a relative increase while thinner lines represent a relative decrease in neuronal activity. Gray lines represent excitatory connections while black lines represent inhibitory connections. GPe, globus pallidus pars externa; GPi, globus pallidus pars interna; STN, subthalamic nucleus; SNr, subtantia nigra pars reticulata; SNc, substantia nigra pars compacta; PPN, pedunculopontine nucleus; MEA, midbrain extrapyramidal area; Rt, reticularis; glu, glutamate; GABA, γ–aminobutyric acid; enk, enkephalin; subst P, substance P. Dopamine (DA) receptor subtypes 1 and 2 are represented as D1 and D2, respectively.

Although the development of aberrant neuronal activity in the motor thalamus in the parkinsonian state provides a ready explanation for the amelioration in parkinsonian motor signs following thalamotomy, the lack of improvement in bradykinesia suggests that a site other than VLo or VPLo may subserve this motor function. In early studies by Hassler, it was argued that lesions encompassing more anterior portions of the pallidal-receiving area were effective in improving bradykinesia, suggesting there may be a different motor subcircuit that subserves bradykinesia (Hassler et al., 1979). Since pallidotomy is effective in improving bradykinesia, it has been hypothesized that pallido-brainstem projections may subserve this function. However, it is compelling to suggest that bradykinesia, rigidity, and tremor may occur as the result of alterations in neuronal activity in different portions of the motor thalamus, with anterior pallidal-receiving areas subserving bradykinesia, posterior pallidal-receiving areas subserving rigidity, and the cerebellar-receiving area VPLo (analogous to Vim in the human) subserving the development of tremor (Vitek et al., 1994).

The pathophysiological basis for the development of aberrant neuronal activity in the motor thalamus in Parkinson's disease remains unclear. One possible explanation is that altered cellular activity is transmitted from the GPi to the motor thalamus. Consistent with this hypothesis, increased mean discharge rates, altered patterns, and loss of specificity in somatosensory response properties of GPi neurons have been described (Miller and DeLong, 1987; Filion et al., 1988; Filion and Tremblay, 1991). This explanation, however, is insufficient to address the widespread changes in thalamic activity observed. Alternate pathways, including projections from the external segment of the globus pallidus to the reticular nucleus of the thalamus (Asanuma, 1994; Hazrati and Parent, 1991), or projections from the pedunculopontine nucleus (PPN) to the reticular nucleus as well as to other motor and nonmotor subnuclei of the thalamus (Hallanger et al., 1987; Levey et al., 1987; Pare et al., 1988; Steriade et al., 1988), may provide an explanation for the present observations. Given the extensive projections from the GPi to the PPN, together with the diffuse projection from the PPN to the thalamus, the PPN is positioned to play a significant role in mediating the observed changes in thalamic nuclear activity.

CONCLUSION

Given the present observations of changes in neuronal activity throughout the motor thalamus, alternative models must be considered. In addition to changes in firing rate, these models must incorporate the observed changes in firing pattern and somatosensory responses that occur within and across subnuclei of the motor thalamus. They must also take into consideration the potential influence of pathways other than that from GPi to the motor thalamus, including those from the PPN or the external segment of the globus pallidus. Finally, the role of changes in nonmotor portions of the thalamus must also be taken into account. Numerous studies have reported altered sensory processing in patients with Parkinson's disease (Schneider et al., 1987; Sathian, et al, 1997; Teo et al., 1997; Boecker et al., 1999). Until now, there have been few explanations for those observations. It is certainly possible, however, that much of the disturbance in sensory processing as well as many of the nonmotor features of Parkinson's disease could occur as a result of the widespread changes that occur across different areas of the thalamus.

ACKNOWLEDGMENTS: We would like to thank Dr. Michael Crutcher for his helpful review of the manuscript, Rachel Miller for excellent administrative support, and Dr. Jian-Yu Zhang for his assistance in the preparation of Figure 1.

REFERENCES

Anderson, M., Inase, M., Buford, J., and Turner, R., 1993, Movement and preparatory activity of neurons in pallidal-receiving areas of the monkey thalamus, in: *Role of the Cerebellum and Basal Ganglia in Voluntary Movement*, N. Mamo, I. Hamada, and M. DeLong, eds., Elsevier Science, New York.

Anderson, M., and Turner, R., 1991, Activity of neurons in cerebellar-receiving and pallidal-receiving areas of the thalamus of the behaving monkey, *J. Neurophysiol.* 66:879.

Anner-Baratti, R., Allum, J., and Hepp-Raymond, M., 1986, Neural correlates of isometric force in the "motor" thalamus, *Exp. Brain Res.* 63:567.

Asanuma, C., 1994, Organization of the external pallidal projection upon the thalamic reticular nucleus in squirrel monkeys, *Soc. Neurosci. Abstr.* 20:332.

Asanuma, C., Thach, W., and Jones, E., 1983, Anatomical evidence for segregated focal grouping of efferent cells and their terminal ramifications in the cerebellothalamic pathway of the monkey, *Brain Res. Rev.* 5:267.

Asanuma, C., Thach, W., and Jones, E., 1983, Brainstem and spinal projections of the deep cerebellar nuclei in the monkey, with observations on the brainstem projections of the dorsal column nuclei, *Brain Res. Rev.* 5:299.

Asanuma, C., Thach, W., and Jones, E., 1983, Cytoarchitectonic delineation of the ventral lateral thalamic region in the monkey, *Brain Res. Rev.* 5:219.

Asanuma, C., Thach, W., and Jones, E., 1983, Distribution of cerebellar terminations in the ventral lateral thalamic region of the monkey, *Brain Res. Rev.* 5:237.

Aziz, T., Pegs, D., Sambrook, M., and Crossman, A., 1991, Lesion of the subthalamic nucleus for the alleviation of 10methyl-4-phenyl-1,2,3,6-tetrahydrapyridine (MPTP)-induced parkinsonism in the primate, *Mov. Disord.* 6:288.

Bergman, H., Wichmann, T., and Delong, M., 1990, Reversal of experimental parkinsonism by lesions of the subthalamic nucleus, *Science*, 249:1436.

Bergman, H., Wichmann, T., Karmon, B., and DeLong, M., 1984, The primate subthalamic nucleus II. Neuronal activity in the MPTP model of parkinsonism, *J. Neurophysiol.* 72:507.

Brooks, V., and Thach, W., 1981, Cerebellar control of posture and movement, in: *Handbook of Physiology. Section 1: The Nervous System. Volume II. Motor Control, Part 2*, J. Brookhart, V. Mountcastle, and V. Brooks, eds., American Physiological Society, Bethesda.

Butler, E., Horne, M., and Rawson, J., 1992, Sensory characteristics of monkey thalamic and motor cortex neurones, *J. Physiol. (Lond.)* 445:1.

Buzsaki, G., Smith, A., Berger, S., Fisher, L., and Gage, F., 1990, Petit mal epilepsy and parkinsonian tremor: Hypothesis of a common pacemaker, *Neuroscience*, 36:1.

DeLong, M., 1990, Primate models of movement disorders of basal ganglia origin, *Trends Neurosci.* 13:281.

DeLong, M., Crutcher, M., and Georgopoulos, A., 1985, Primate globus pallidus and subthalamic nucleus: Functional organization, *J. Neurophysiol.* 53:530.

Feingold, A., Nini, A., Raz, A., Slovin, H., Zelenskaya, V., Vaadia, E., Abeles, M., and Bergman, H., 1996, Neuronal oscillations and correlations in the basal ganglia of normal and MPTP-treated monkeys, *Soc. Neurosci. Abstr.* 22:414.

Filion, M., Boucher, R., and Bedard, P., 1985, Globus pallidus unit activity in the monkey during induction of parkinsonism by 1-methyl-4-phenyl-1,2,3,6-tetrahydropyridine (MPTP), *Soc. Neurosci. Abstr.* 11:1160.

Filion, M., and Tremblay, L., 1991, Abnormal spontaneous activity of globus pallidus neurons in monkeys with MPTP-induced parkinsonism, *Brain Res.* 547:142.

Friedman, D., and Jones, E., 1980, Focal projection of electrophysiologically defined groupings of thalamic cells on the monkey somatic sensory cortex, *Brain Res.* 191:249.

Hallanger, A., Levey, A., Lee, H., Rye, D., and Wainer, B., 1987, The origins of cholinergic and other subcortical afferents to the thalamus in the rat, *J. Comp. Neurol.* 262:105.

Hassler, R., Mundinger, F., and Reichert, T., 1979, *Stereotaxis in Parkinson Syndrome: Clinical-Anatomical Contributions to Its Pathophysiology*, Springer-Verlag, New York.

Hazrati, L., and Parent, A., 1991, Projection from the external pallidum to the reticular thalamic nucleus in the squirrel monkey, *Brain*, 550:142.

Holsapple, J., Preston, J., and Strick, P., 1991, The origin of thalamic inputs to the "hand" representation in the primary motor cortex, *J. Neurosci.* 11:2644.

Hoover, J., and Strick, P., 1993, Multiple output channels in the basal ganglia, *Science*, 25:819.

Horne, M., and Porter, R., 1980, The afferents and projections of ventroposterolateral thalamus in the monkey, *J. Physiol. (Lond.)* 304:349.

Jinnai, K., Nambu, A., Yoshida, S., and Tanibuchi, I., 1993, The two separate neuron circuits through the basal ganglia concerning the preparatory or execution processes of motor control, in: *Role of the Cerebellum and Basal Ganglia in Voluntary Movement*, N. Mano, I. Hamada, and M. DeLong, eds., Elsevier Science, New York.

Jones, E., 1981, Functional subdivision and synaptic organization of the mammalian thalamus, *Int. Rev. Physiol.* 25:173.

Jones, E., 1985, *The Thalamus*, Plenum Press, New York.

Jones, E., 1991, The anatomy of sensory relay functions in the thalamus, *Prog. Brain Res.* 87:29.

Jones, E., and Friedman, D., 1982, Projection pattern of functional components of thalamic ventrobasal complex on monkey somatosensory cortex, *J. Neurophysiol.* 48:521.

Jones, E., Friedman, D., and Hendry, H., 1982, Thalamic basis of place- and modality-specific columns in the monkey somatosensory cortex: A correlative anatomical and physiological study, *J. Neurophysiol.* 48:545.

Kievit, J. and Kuypers, H., 1977, Organization of the thalamo-cortical connexions to the frontal lobe in the rhesus monkey, *Exp. Brain Res.* 29:299.

Kunzle, H., 1976, Thalamic projections form the precentral motor cortex in *Macaca fascicularis*, *Brain Res.* 105:253.

Levey, A., Hallanger, A., and Wainer, B., 1987, Cholinergic nucleus basalis neurons may influence the cortex via the thalamus, *Neurosci. Lett.* b74:7.

Llinas, R., 1984, Rebound excitation as the physiological basis for tremor: a biophysical study of the oscillatory properties of mammalian central neurones in vitro, in: *Movement Disorders: Tremor*, L. J. Findley and R. Capildea, eds., Macmillan, London.

Llinas, R., 1988, The intrinsic electrophysiological properties of mammalian neurons: Insights into central nervous system function, *Science*, 242:1654.

Macpherson, J., Rasmusson, D., and Murphy, J., 1980, Activities of neurons in "motor" thalamus during control of limb movement in the primate, *J. Neurophysiol.* 44:11.

Miller, W.C., and DeLong, M.R., 1987, Altered tonic activity of neurons in the globus pallidus and subthalamic nucleus in primate MPTP of parkinsonism, in: *The Basal Ganglia II*, M. Carpenter and A. Jayaraman, eds., Plenum Press, New York.

Muakkassa, K., and Strick, P., 1979, Frontal lobe inputs to primate motor cortex: Evidence for four somatotopically organized "premotor" areas, *Brain Res.* 177:176.

Nambu, A., Yoshida, S, and Jinnai, K., 1991, Movement-related activity of thalamic neurons with input from the globus pallidus and projection to the motor cortex in the monkey, *Exp. Brain Res.* 84:279.

Orioli, P., and Strick, P., 1989, Cerebellar connections with the motor cortex and the arcuate premotor area: An analysis employing retrograde transneural transport of WGA-HRP, *J. Comp. Neurol.* 288:612.

Pare, D., Smith, Y., Parent, A., and Steriade, M., 1988, Projections of brainstem core cholinergic and non-cholinergic neurons of cat to intralaminar and reticular thalamic nuclei, *Neuroscience,* 25:69.

Poggio, G., and Mountcastle, V., 1963, The functional properties of ventrobasal thalamic neurons studied in unanesthetized monkeys, *J. Neurophysiol.* 26:775.

Riehle, A., Grun, S., Diesmann, M., and Aertsen, A., 1997, Spike synchronization and rate modulation differentially involved in motor cortical function, *Science*, 278:1950.

Sathian, K., Zangaladze, A., Green, J., Vitek, J.L., and DeLong, MR., 1997, Tactile spatial acuity and roughness discrimination: Impairments due to aging and Parkinson's disease, *Neurology,* 49:168.

Schell, G., and Strick, P., 1984, The origin of thalamic inputs to the arcuate premotor and supplementary motor areas, *J. Neurosci.* 4:539.

Singer, W., 1994, A new job for the thalamus, *Nature*, 369:444.

Steriade, M., 1992, Basic mechanisms of sleep generation, *Neurology*, 42 (Suppl 6) :9.

Steriade, M., and McCarley, R., 1990, *Brainstem control of wakefulness and sleep*, Plenum Press, New York.

Steriade, M., Pare, D., Parent, A., and Smith, Y., 1988, Projections of cholinergic and non-cholinergic neurons of the brainstem core to relay and associational thalamic nuclei in the cat and macaque monkey, *Neuroscience,* 25:47.

Strick, P., 1976, Activity of ventrolateral thalamic neurons during arm movement, *J. Neurophysiol.* 39:1032.

Strick, P., 1976, The influence of motor preparation on the response of cerebellar neurons to limb displacements, *Soc. Neurosci. Abstr.* 3:2007.

Vitek, J., Ashe, J., DeLong, M., and Alexander, G., 1990, Altered somatosensory response properties of neurons in the 'motor' thalamus of MPTP treated parkinsonian monkeys, *Soc. Neurosci. Abstr.* 16:425.

Vitek, J., Ashe, J., DeLong, M., and Alexander, G., 1989, Organization of primate ventrolateral thalamus: Neuronal relations to active and passive limb movements, *Soc. Neurosci. Abstr.* 15:287.

Vitek, J., Ashe, J., DeLong, M., and Alexander, G., 1994, Physiologic properties and somatotopic organization of the primate motor thalamus, *J. Neurophysiol.* 71:1498.

Vitek, J., Ashe, J., and Kaneoke, Y., 1994, Spontaneous neuronal activity in the motor thalamus: Alteration in pattern and rate in parkinsonism, *Soc. Neurosci. Abstr.* 20:561.

Vitek, J., and Giroux, M., 2000, Physiology of Hypokinetic and Hyperkinetic Movement Disorders: Model for Dyskinesia, *Ann. Neurol.* 47 (Suppl 1):S131.

Vitek, J., Ashe, J., DeLong, M., and Alexander, G., 1990, Altered somatosensory response properties of neurons in the "motor" thalamus of MPTP treated parkinsonian monkeys, *Soc. Neurosci. Abstr.* 16:425.

Wichmann, T., Bergman, H., and DeLong, M., 1994a, The primate subthalamic nucleus. III. Changes in motor behavior and neuronal activity in the internal pallidum induced by subthalamic inactivation in the MPTP model of parkinsonism, *J. Neurophysiol.* 72:521.

Wichmann, T., Baron, M., and DeLong, M., 1994b, Local inactivation of the sensorimotor territories of the internal segment of the globus pallidus and the subthalamic nucleus alleviates parkinsonian motor signs in MPTP treated monkeys, in: *The Basal Ganglia IV: New Ideas and Data on Structure and Function*, J. Percheron, J. McKenzie, and J. Feger, eds., Plenum Press, New York.

NEURONAL ACTIVITY IN MOTOR THALAMUS OF PARKINSON'S DISEASE PATIENTS

JONATHAN O. DOSTROVSKY, GREGORY F. MOLNAR, ANDREW
PILLIAR, WILLIAM D. HUTCHISON, KAREN D. DAVIS, AND
ANDRES M. LOZANO[*]

INTRODUCTION

Current models of basal ganglia dysfunction in Parkinson's disease (PD) attribute the motor disturbances largely to hyperactivity of the internal segment of the globus pallidus (GPi) (Delong, 1990; Alexander and Crutcher, 1991). Since the GPi neurons are inhibitory, their hyperactivity is believed to produce excessive inhibition of the GPi target neurons in the motor thalamus. This in turn reduces the activity of the cortical targets of these neurons, thereby depressing motor performance and giving rise to akinesia and bradykinesia, and possibly to tremor. When thalamic neurons are hyperpolarized, as occurs during sleep, their firing pattern is dramatically altered into a bursting pattern, consisting of bursts with a very characteristic intraburst firing pattern caused by a low threshold calcium spike (Steriade et al., 1997). It is possible that in PD the increased inhibition in thalamus resulting from the GABAergic inputs from GPi could lead to such bursting, and it has been proposed that such activity may lead to the generation of tremor (Pare et al., 1990).

Electrophysiological recordings from GPi neurons in PD patients and in monkeys treated with 1-methyl-4-phenyl-1, 2, 3, 6-tetrahydropyridine (MPTP) confirm the hyperactivity of GPi neurons (Filion and Tremblay, 1991; Hutchison et al., 1994). In addition, lesions in GPi that would result in decreased GPi output improve PD symptoms (Lozano et al., 1995). Functional imaging studies in PD patients have revealed that treatments which improve the PD symptoms also lead to increased activation of motor-related cortical areas (Jenkins et al., 1992; Limousin et al., 1997; Davis et al., 1997). Recent recordings in thalamus in MPTP treated monkeys show a decrease in spontaneous firing rate and an

[*] JONATHAN O. DOSTROVSKY • Department of Physiology, University of Toronto, Toronto, Ontario M5S 1A8, Canada. GREGORY F. MOLNAR • Institute of Medical Science, University of Toronto, Toronto, Ontario M5S 1A8, Canada. ANDREW PILLIAR • Department of Physiology, University of Toronto, Toronto, Ontario M5S 1A8, Canada. WILLIAM D. HUTCHISON • Department of Physiology and Department of Surgery, University of Toronto, Toronto, Ontario M5S 1A8, Canada; Toronto Western Research Institute, Toronto Western Hospital, 399 Bathurst St., Toronto, Ontario M5T 2S8, Canada. KAREN D. DAVIS AND ANDRES M. LOZANO • Institute of Medical Science and Department of Surgery, University of Toronto, Toronto, Ontario M5S 1A8, Canada; Toronto Western Research Institute, Toronto Western Hospital, 399 Bathurst St., Toronto, Ontario M5T 2S8, Canada.

increased tendency for irregular firing patterns; surprisingly, though, these changes are not limited to pallidal-receiving areas (Vitek et al., 1994b). However, there has been no confirmation in humans that the spontaneous firing rates of thalamic neurons receiving pallidal inputs are depressed in PD patients.

The aim of this study was to test the hypothesis that neurons in the pallidal-receiving area of thalamus in PD patients have lower firing rates than normal. Therefore, this study determined the mean firing rate and discharge pattern of neurons in motor thalamus in PD patients and compared these data with those obtained in other patients in whom GPi is assumed to be normal.

METHODS

Recordings of neuronal activity were obtained from 122 neurons in motor thalamus in 5 PD, 10 essential tremor (ET), and 6 pain patients (chronic pain; pain group) during microelectrode-guided functional neurosurgery. Since we assumed the motor areas of thalamus in pain patients to be 'normal,' they served as controls. Detailed descriptions of the techniques used in this study have been previously published (Lenz et al., 1988; Tasker et al., 1998; Dostrovsky, 1999) and will only be described briefly. A Leksell stereotactic frame was attached to the patient's head under local anesthesia, and MRI or CT scans were obtained to determine the 3 dimensional coordinates of the anterior commissure (AC) and posterior commissure (PC). A computer program was used to fit the patient's AC and PC points to a standard sagittal map of the human thalamus and to plot the predicted electrode track (see Figure 1). A small hole was then made in the skull under local anesthesia to permit insertion of the microelectrode. A 25µm exposed-tip, parylene-insulated tungsten microelectrode was driven through the thalamus with the aid of a hydraulic microdrive. Neuronal recordings were made along several parasagittal trajectories targeting the ventral thalamus, but only a subset of these, those at the laterality of hand sensory representation (i.e. usually 15mm lateral), were selected for the present study. Thalamic neurons were tested for responses to passive and active joint movements or cutaneous and deep stimulation. Active and passive movements of the hand and wrist including clenching of the fist, wrist flexion and extension, and finger movements were used to determine movement-related neurons. Stimulation of the body surface by brushing, stroking, or squeezing was used to identify somatosensory-responsive neurons in the tactile relay nucleus, the ventral caudal nucleus (Vc).

Neurons that responded only or preferentially to voluntary movements were classified as voluntary cells (Vol), while those that responded preferentially to movements of a joint were classified as kinesthetic cells (Ki). The positions of these cells along the electrode track were reconstructed and those responding to tactile stimulation were assumed to be in Vc. The kinesthetic cells were assumed to be in ventralis intermedius (Vim) and those responding to voluntary movements to be in ventralis oralis posterior or anterior (Vop/Voa) (Figure 1). Data were included in the analysis only from electrode tracks that traversed the hand cutaneous representation. In addition, only well-isolated neurons were studied and then only if there was a period of spontaneous firing at a time when there were no active or passive movements and no tremor (see examples in Figure 2).

Mean spontaneous firing rates were determined from spontaneous neuronal activity (usually 20 s) when the patient was at rest and not performing a task. A customized Spike 2 (Cambridge Electronic Design Limited, Cambridge, England) script was used to quantify the regularity of firing patterns. The script compared the pattern of firing (i.e. digitized action potentials) of a single neuron over the sampling period to that generated by a poisson distribution (mean = 1) over the same period. A value called 'variance' of '1' indicated completely random firing, a value '<1' indicated a pattern with increasing regularity, and a value '>1' indicated increased irregularity. Cells firing in a bursting pattern thought to be due to an underlying low threshold calcium spike were initially identified qualitatively by visual and aural analysis of the firing pattern and then subjected to quantitative analysis as described previously (Radhakrishnan et al., 1999; Tsoukatos et al., 1997).

The mean firing rate and variance values for each neuron were pooled in either one of two groups in each of the three patient groups: 'Vol' for voluntary-responsive neurons and 'Ki' for kinesthetic-responsive neurons. The firing rate and variance-pooled values were statistically analyzed within and between patient groups using a one-way ANOVA and the Newman-Keuls Multiple Comparison post-hoc test, with a level of significance of p = 0.05.

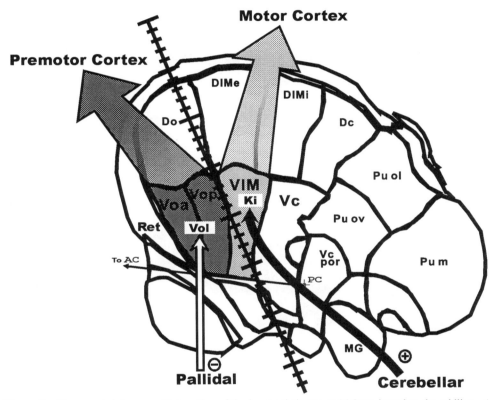

Figure 1. Diagram showing a sagittal section of the human thalamus at 14.5mm lateral to the midline. A typical electrode track is shown superimposed on the section (1 mm scale). Also shown are the presumed cerebellar- and pallidal-receiving areas of thalamus (Hirai and Jones, 1989; Macchi and Jones, 1997). Abbreviations: AC = anterior commissure; Dc = dorsal caudal nucleus; Dim.e = dorsal intermediate external nucleus; Dim.i = dorsal intermediate internal nucleus; Do = dorsal oral nucleus; MG = medial geniculate nucleus; PC = posterior commissure; Pu = pulvinar (m = medial; ol = oral lateral; ov = oral ventral) nucleus; Ret = reticular nucleus; Vc = ventral caudal nucleus; Vc.por = ventral caudal portae nucleus; Vim = ventral intermediate nucleus; Voa = ventral oral anterior nucleus; Vop = ventral oral posterior nucleus.

The stereotactic locations of the recorded neurons were plotted relative to the AC- PC line (abscissa) and a line perpendicular to the AC-PC line at the location of the first tactile neuron in that same trajectory (ordinate) (i.e. anterior Vc border). This was done because the Vc border is a well-defined physiological landmark. The difference in mean anterior-posterior locations of Vol and Ki neurons was compared for each patient group (one-way ANOVA with Newman-Keuls Multiple Comparison Post-hoc test, p = 0.05). Mean spontaneous firing rates of neurons within or beyond 2mm of the Vc border were also compared for each patient group. The approximate width of Vim is 2mm based on anatomy (Hassler, 1959; Schaltenbrand and Wahren, 1977) and physiological recordings obtained by our group. This analysis provides an alternate, anatomical-based method of differentiating neurons in the pallidal- and cerebellar-receiving areas of motor thalamus, since neurons located >2mm from the Vc border would likely be in Vop/Voa.

A   Vol neuron PD

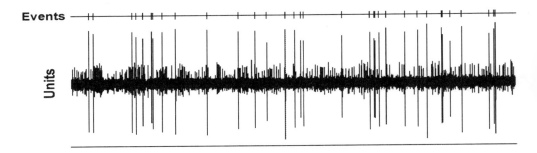

B   Vol neuron Pain

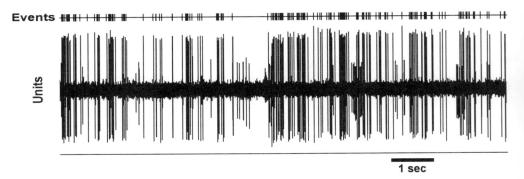

1 sec

Figure 2. Two examples showing the spontaneous firing activity of a 'voluntary' neuron in the motor thalamus of a PD patient and a pain patient. The bottom trace in each section is the raw recording of neuronal activity. In each case the large unit was digitized by template matching and the events shown on the top trace.

RESULTS

The mean spontaneous firing rate of the Vol neurons was significantly lower (p<0.01) in the PD patients than in the ET and pain patients (see Figure 3). In contrast, the mean spontaneous firing rate of Ki cells was similar in PD and pain patients but significantly lower (p<0.05) than in ET patients. When the neurons were classified according to whether they were located at a distance of more or less than 2mm anterior to the tactile border, the resulting mean firing rates of the two cell groups were very similar to those obtained when the cells were classified according to physiology (i.e. <2mm groups were similar to Ki cells and >2mm to Vol cells). Analysis of firing patterns (poisson surprise method) did not reveal any significant differences between PD and ET patients for Vol or Ki neurons or between

Vol and Ki neurons in PD patients. Analysis of the incidence of bursting cells (with firing characteristics presumed to result from a low threshold calcium spike) revealed that none of the Vol or Ki neurons in any of the three patient groups fired in such bursts. Furthermore, for the few calcium spike-type bursting cells that were found in these patients there was no significant difference (p>0.05) in incidence between PD patients and ET patients.

The locations of the Ki cells were predominantly in the region designated as Vim, whereas the Vol cells were predominantly in Vop with a few in Voa. The mean location of the Vol cells along the AC - PC axis was significantly more anterior (p<0.001) than the location of the Ki cells.

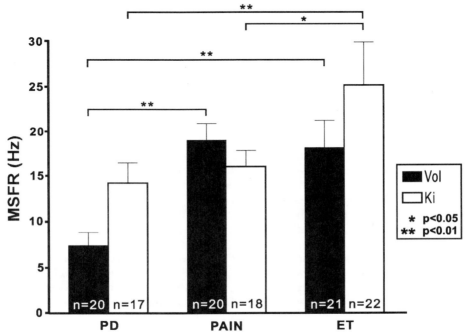

Figure 3. Bar graph showing the mean spontaneous firing rates (MSFR) for Vol and Ki neurons in PD, ET, and pain patients.

DISCUSSION

This study has revealed that the mean spontaneous firing rate of thalamic voluntary neurons in PD patients was significantly lower than those in ET and pain patients. In contrast, there was no difference in mean firing rate of Ki cells in PD and pain patients. Unexpectedly, the Ki neurons in ET patients had a significantly higher firing rate than those in PD and pain patients. These findings support the hypothesis that the hyperactivity of GPi neurons in PD patients leads to decreased firing of thalamocortical neurons in pallidal-receiving areas of motor thalamus (see below).

There is good reason to believe that the voluntary neurons are predominantly located in the pallidal-receiving area of motor thalamus and the Ki neurons in the cerebellar-receiving area of thalamus. It has been shown in monkeys that neurons in nucleus ventralis lateralis pars oralis (VLo) and ventralis anterior (VA) respond primarily to voluntary movements, whereas those in nucleus ventralis posterior lateralis, pars oralis (VPLo) respond to kinesthetic inputs but not to active movements (Vitek et al., 1994a). In the monkey, VLo and VA receive inputs from the globus pallidus, whereas VPLo receives cerebellar inputs (Buford et al., 1996). Furthermore, the anatomical studies of Jones and colleagues indicate that the human Vim is equivalent to the monkey VPLo, and the human Vop and Voa are equivalent to monkey VLo (Hirai and Jones, 1989; Macchi and Jones, 1997). It is thus generally assumed that Vim is the cerebellar-receiving area of motor thalamus and Voa and Vop the pallidal-receiving areas (e.g. Zirh et al., 1997) of human motor thalamus.

Further validation of our characterization of voluntary and kinesthetic cells, and our assumption that they are located in Voa/Vop and Vim respectively, arises from our findings that Ki cells are usually located posterior to Vol cells and usually are located in Vim according to the reconstructions. Furthermore, when the cells were categorized according to their location anterior or posterior to a line passing 2 mm anterior to the tactile border (the border between Vim and Vc), similar findings were obtained with respect to mean firing rates in the three groups of patients as were obtained when the cells were grouped according to whether they were Vol or Ki cells.

This study is the first to report firing rates of Ki and Vol neurons in motor thalamus of PD patients. However, there has been one brief report of firing rates in motor thalamus of the parkinsonian (MPTP treated) monkey (Vitek et al., 1994b). That study found a significant reduction in the mean spontaneous firing rate and increased irregularity in the firing pattern in the pallidal-receiving area (VLo; 16 ± 8Hz to 11.5 ± 7Hz, $p < 0.001$) and also in the cerebellar-receiving area (VPLo; 22 ± 8 Hz to 15 ± 8 Hz, $p < 0.001$). These results for VLo are consistent with our findings for Voa/Vop in PD patients in comparison with the ET and pain patients. However, it is not clear why in the MPTP treated monkeys there is also a reduction in firing rate of neurons in the cerebellar-receiving areas of thalamus. The mean firing rates of neurons in Vop and Vim in the pain (control) patients in our study are similar to those reported in normal monkey motor thalamus (Vitek et al., 1994a) and similar to those reported by Lenz et al. (1999) for neurons in Vim in pain patients.

The finding of this study that cells in the presumed pallidal receiving area of motor thalamus in PD patients have lower spontaneous firing rates than those in ET and pain patients is consistent with predictions of the GPi hyperactivity model of the pathophysiology of PD. Studies in human PD patients (Hutchison et al., 1994) and in MPTP monkeys (Filion and Tremblay, 1991) have indeed confirmed that GPi is hyperactive. Furthermore, injections of apomorphine, a dopamine agonist which alleviates the bradykinesia and akinesia symptoms, have been shown to reduce the firing rate of GPi neurons in both human PD patients and MPTP treated monkeys (Filion et al., 1991; Hutchison et al., 1997). It has also been shown that electrical stimulation (single pulses) of GPi results in inhibition of pallidal receiving neurons in monkey thalamus (Buford et al., 1996). The depressed firing in Voa/Vop of PD patients revealed in the present study indicates that even following long-term GPi hyperactivity these neurons remain depressed. This is also consistent with cortical imaging studies. Basal ganglia loops through Voa/Vop of thalamus have large inputs to the supplementary motor area (SMA) of the cortex (Macchi and Jones, 1997). Functional imaging studies have demonstrated reduced SMA activation in PD patients in off-treatment states during a motor task, which is reversed by apomorphine-induced improvement in akinesia (Jenkins et al., 1992), and increased SMA activity resulting from alleviation of akinesia by deep brain stimulation of GPi (Davis et al., 1997; Limousin et al., 1997). Our observations are consistent with the findings of these studies, as they predict that reducing the inhibitory input from GPi should increase (normalize) the firing of neurons in Vop/Voa and thus result in increased cortical activation.

The lack of a significant increase in bursting activity in pallidal receiving areas in PD patients indicates that the increased inhibition from GPi is not sufficient to alter the mode of firing of these neurons. This finding suggests that this type of bursting is unlikely to be an important factor in the etiology of parkinsonian tremor. This is also consistent with our own unpublished observations and those in a recent study by Zirh et al. (1997) showing that the firing pattern of tremor cells in PD patients is not due to an underlying low threshold calcium spike.

The increased firing of Ki neurons in Vim in ET patients suggests that this increased activity might be related to the pathophysiology of ET. Although little is known concerning the etiology of ET, several imaging studies have reported findings suggesting increased cerebellar output. Functional imaging studies in ET patients have reported increased regional blood flow in cerebellum and deep cerebellar nuclei during both tremor and non-tremor states and in contralateral thalamus during tremor (Bucher et al., 1997; Jenkins et al., 1993). Thus, it is possible that cerebellar hyperactivity could account for the increase in the spontaneous firing rates of Ki neurons in ET patients revealed using the microelectrode techniques of the present study. This would be consistent with our findings, since the cerebellar projections to thalamus are excitatory. The increased excitability of Vim neurons might lead to increased kinesthetic feedback and contribute to the generation of the tremor.

In summary, this study has shown that changes in PD patients at the level of the motor thalamus include a reduction in the mean spontaneous firing rate of presumed Voa/Vop neurons probably as a result of increased inhibitory output from the basal ganglia. Increased activity of neurons in the presumed Vim of patients with ET suggests that the changes may be related to the pathophysiology and may provide a possible explanation for the effectiveness of lesions in Vim in treating ET.

ACKNOWLEDGMENTS: This work was supported by grants from The Parkinson Foundation of Canada, US NINCDS (NS 36824), Brain Injury Community Re-Entry (Niagara) Inc. (to G.F.M.), and an Ontario Graduate Scholarship (to G.F.M).

REFERENCES

Alexander, G.E., and Crutcher, M.D., 1991, Functional architecture of basal ganglia circuits: neural substrates of parallel processing, *Trends Neurosci.* 13:266.

Bucher, S.F., Seelos, K.C., Dodel, R.C., Reiser, M., and Oertel, W.H., 1997, Activation mapping in essential tremor with functional magnetic resonance imaging, *Ann. Neurol.* 41:32.

Buford, J.A., Inase, M., and Anderson, M.E., 1996, Contrasting locations of pallidal-receiving neurons and microexcitable zones in primate thalamus, *J. Neurophysiol.* 75:1105.

Davis, K.D., Taub, E., Houle, S., Lang, A.E., Dostrovsky, J.O., Tasker, R.R., and Lozano, A.M., 1997, Globus pallidus stimulation activates the cortical motor system during alleviation of parkinsonian symptoms, *Nature Med.* 3:671.

DeLong, M.R., 1990, Primate models of movement disorders of basal ganglia origin, *Trends Neurosci.* 13:281.

Dostrovsky, J.O., 1999, Invasive techniques in humans: microelectrode recordings and microstimulation, in: *Modern Techniques in Neuroscience Research*, U. Windhorst and H. Johansson, eds., Springer-Verlag, Berlin.

Filion, M. and Tremblay, L., 1991, Abnormal spontaneous activity of globus pallidus neurons in monkeys with MPTP-induced parkinsonism, *Brain Res.* 547:142.

Filion, M., Tremblay, L., and Bedard, P.J., 1991, Effects of dopamine agonists on the spontaneous activity of globus pallidus neurons in monkeys with MPTP-induced parkinsonism, *Brain Res.* 547:152.

Hassler, R., 1959, Anatomy of the thalamus, in: *An Introduction to Stereotaxis with an Atlas of the Human Brain*, G. Schaltenbrand and P. Bailey, eds., Thieme, Stuttgart.

Hirai, T., and Jones, E.G., 1989, A new parcellation of the human thalamus on the basis of histochemical staining, *Brain Res.* Rev. 14:1.

Hutchison, W.D., Levy, R., Dostrovsky, J.O., Lozano, A.M., and Lang, A.E., 1997, Effects of apomorphine on globus pallidus neurons in parkinsonian patients, *Ann. Neurol.* 42:767.

Hutchison, W.D., Lozano, A.M., Davis, K.D., Saint-Cyr, J.A., Lang, A.E., and Dostrovsky, J.O., 1994, Differential neuronal activity in segments of globus pallidus in Parkinson's disease patients, *NeuroReport*, 5:1533.

Jenkins, I.H., Fernandez, W., Playford, E.D., Lees, A.J., Frackowiak, R.S., Passingham, R.E., and Brooks, D.J., 1992, Impaired activation of the supplementary motor area in Parkinson's disease is reversed when akinesia is treated with apomorphine, *Ann. Neurol.* 32:749.

Jenkins, I.H., Bain, P.G., Colebatch, J.G., Thompson, P.D., Findley, L.J., Frackowiak, R.S., Marsden, C.D., and Brooks, D.J., 1993, A positron emission tomography study of essential tremor: evidence for overactivity of cerebellar connections, *Ann. Neurol.* 34:82.

Lenz, F.A., Dostrovsky, J.O., Kwan, H.C., Tasker, R.R., Yamashiro, K., and Murphy, J.T., 1988, Methods for microstimulation and recording of single neurons and evoked potentials in the human central nervous system, *J. Neurosurg.* 68:630.

Lenz, F.A., Jaeger, C.J., Seike, M.S., Lin, Y.C., Reich, S.G., Delong, M.R., and Vitek, J.L., 1999, Thalamic single neuron activity in patients with dystonia: dystonia-related activity and somatic sensory reorganization, *J. Neurophysiol.* 82:2372.

Limousin, P., Greene, J., Pollak, P., Rothwell, J., Benabid, A.L., and Frackowiak, R., 1997, Changes in cerebral activity pattern due to subthalamic nucleus or internal pallidum stimulation in Parkinson's disease, *Ann. Neurol.* 42:283.

Lozano, A.M., Lang, A.E., Galvez-Jimenez, N., Miyasaki, J., Duff, J., Hutchison, W.D., and Dostrovsky, J.O., 1995, Effects of GPi pallidotomy on motor function in Parkinson's disease, *Lancet*, 346:1383.

Macchi, G. and Jones, E.G., 1997, Toward an agreement on terminology of nuclear and subnuclear divisions of the motor thalamus, *J. Neurosurg.* 86:670.

Pare, D., Curro'Dossi, R., and Steriade, M., 1990, Neuronal basis of the parkinsonian resting tremor:a hypothesis and its implications for treatment, *Neuroscience*, 35:217.

Radhakrishnan, V., Tsoukatos, J., Davis, K.D., Tasker, R.R., Lozano, A.M., and Dostrovsky, J.O., 1999, A comparison of the burst activity of lateral thalamic neurons in chronic pain and non-pain patients, *Pain*, 80:567.

Schaltenbrand, G., and Wahren, W., 1977, *Atlas for Stereotaxy of the Human Brain*, Thieme, Stuttgart.

Steriade, M., Jones, E.G., and McCormick, D.A., 1997, *Thalamus*, Elsevier Science, Oxford.

Tasker, R.R., Davis, K.D., Hutchison, W.D., and Dostrovsky, J.O., 1998, Subcortical and thalamic mapping in functional neurosurgery, in: *Textbook of Stereotactic and Functional neurosurgery*, P.L. Gildenberg and R.R. Tasker, eds., McGraw-Hill, New York.

Tsoukatos, J., Kiss, Z.H., Davis, K.D., Tasker, R.R., and Dostrovsky, J.O., 1997, Patterns of neuronal firing in the human lateral thalamus during sleep and wakefulness, *Exp. Brain Res.* 113:273.

Vitek, J.L., Ashe, J., Delong, M.R., and Alexander, G.E., 1994a, Physiologic properties and somatotopic organization of the primate motor thalamus, *J. Neurophysiol.* 71:1498.

Vitek, J. L., Ashe, J., and Kaneoke, Y. 1994b, Spontaneous neuronal activity in the motor thalamus: Alteration in the pattern and rate in parkinsonism, *Soc. Neurosci. Abstr.* 20:561.

Zirh, T.A., Lenz, F.A., Reich, S.G., and Dougherty, P.M., 1997, Patterns of bursting occurring in thalamic cells during parkinsonian tremor, *Neuroscience*, 83:107.

ELECTROPHYSIOLOGICAL INSIGHTS INTO THE MOTOR CONTROL SYSTEM IN PARKINSONISM

SVETLANA RAEVA[*]

INTRODUCTION

The results of microrecording and of macro- and microstimulation in the human ventral lateral (VL; Voa-Vop nuclei according to Hassler's nomenclature, 1982) and ventral intermediate (Vim) thalamic nuclei in parkinsonism offer an opportunity to study mechanisms involved in motor control (Jasper and Bertrand, 1966, Crowell et al., 1968; Raeva, 1972, 1977, 1986, 1993; Tasker et al., 1987; Alexander et al., 1986, 1990; Lenz et al., 1990; Graybiel, 1994; Raeva et al., 1999 a,b) and generation of parkinsonian tremor (Albe-Fessard et al., 1962, 1966; Jasper and Bertrand, 1966; Ohye and Narabayashi, 1972; Ohye et al., 1974; Ohye and Albe-Fessard, 1978; Raeva et al., 1986; Tasker et al., 1987; Lenz et al., 1993, 1994; Zirh et al., 1998).

Our previous studies on patients with and without tremor (rigid and akinetic forms of parkinsonism) described the existence of neurons in the human VL (consisting of Voa and Vop nuclei in Hassler's nomenclature) that had a latent rhythmogenic 5 ± 1 Hz tendency, which plays a significant role in both the genesis of pathological tremor and in the control of normal voluntary movements (Raeva, 1972, 1977; Raeva et al., 1986). More recently, we also described two functionally different cell types in the Voa nucleus: neurons with tonic transfer function (named A-units) and bursting Ca^{2+} oscillatory-mode neurons (named B-units). The two types of cells reacted differently during voluntary movements and somatosensory stimulation (Raeva et al., 1999 a,b). These studies also suggested that the two cell types play functionally different roles in the mechanisms of motor signal transmission and parkinsonian tremor.

The aim of the present study was: (1) to analyze in detail peculiarities in firing patterns of the human ventralis intermedius (Vim) neurons related to motor and tremor activity and to compare these data with alterations of Vop cell activity previously described (Raeva et al., 1999 a,b); (2) to elucidate the role of the human motor thalamus and the non-specific and associative thalamic nuclei in the motor control system and in tremorogenic mechanisms; and (3) to evaluate the significance of some factors contributing to the

---

[*] SVETLANA RAEVA • Institute of Chemical Physics, Russian Academy of Sciences, and Burdenko Neurosurgery Institute, Russian Academy of Medical Sciences, Moscow 117977, Russia.

development of the central pathological oscillatory thalamic mechanisms involved in parkinsonian tremor.

MATERIAL AND METHODS

The present data are based on results of microrecording in various thalamic nuclei, mainly in the Voa, Vop, and Vim, obtained during 166 stereotactic operations in parkinsonian patients, in collaboration with neurosurgeons in the Burdenko Neurosurgery Institute (A. Kadin, N. Vasin, V. Shabalov, A. Shtock) over the last decades. Microrecording techniques and data analysis were basically the same as those described in our previous studies (Raeva 1972, 1986, 1999a,b). Single-unit activity was recorded extracellularly with tungsten microelectrodes (resistance 1-3 MΩ). Microelectrode trajectories were monitored with X-ray or computer tomography. Several functional tests were used, including various forms of active voluntary movements such as clenching and unclenching a fist, lifting and lowering the extremities, closing and opening the eyes, and so forth.

According to elaborate methodology of verbally ordered functional cell testing (Raeva, 1972, 1977, 1986; Raeva and Lukashev, 1993; Raeva et al., 1999b), these movements were performed in response to verbal commands such as: "Clench your hand!", "Unclench!", "Raise your hand!", "Lower your hand!", etc. Changes in neural and electromyogram (EMG) activities were analyzed with computerized programs at different functional phases of the verbally ordered voluntary movement as follows: (i) phase before the stimulus presentation; (ii) preparation phase, from the moment of verbal command presentation to the beginning of initial EMG changes; (iii) phase of movement initiation and execution, indicated by the EMG changes; and (iv) the final phase of the movement termination and its after-effect. Passive joint movements (flexion-extension of the wrist, elbow, leg etc.) and light touch were applied as well. The surface electroencephalogram (EEG), the mechanogram, and phonogram of verbal stimuli were recorded simultaneously with unit and EMG activities.

Spike discharges from single units, or from two to three adjacent neurons recorded simultaneously, were distinguished by their amplitudes and patterns. The following functions were estimated: firing rate frequency; interspike and interburst interval histograms; peristimulus rasters of spikes or bursts; peristimulus firing rate histograms of units with averaged EMG responses and phonogram signals; auto- and cross-correlation histograms between activity of units and/or pathological tremor and spectral power functions (interval = 20–200 ms). The interneuronal connections between neighboring units and relationships between neural and EMG changes were estimated with the cross-correlation and cross-spectral power functions and the joint peristimulus scatter diagrams (Gerstein and Perkel, 1972). Each of these traditional statistical functions was determined in the course of functional tests; their alterations characterized the pattern of evoked activity and temporal relations between different units, on the one hand, and units and tremor, on the other.

Taking into account the dynamics and variability of unit reactions of the human brain, principal component analysis (PCA) (Donchin and Heffley, 1978; Raeva and Lukashev, 1993) and correlation techniques were used to investigate the structure of response patterns in relation to both the temporal processing and the dynamics of reactions during different functional stages of the identical voluntary motor test performance. The quantitative evaluation of functional differences in response patterns during voluntary movements and some somatosensory stimulation, as well as the evaluation of the features of spontaneous activity of units, were the main criteria for identification of records of different thalamic nuclei. The data presented concern the precise quantitative analysis of a total of 1100 thalamic neurons, of which 155 were located in the Voa, 80 - in the Vop, 220 - in the Vim, 426 - in the Rt, 104- in the MD, and 115 - in the CM.

RESULTS AND DISCUSSION

The present results illustrate the firing pattern of movement-related Vim neurons in relation to active movements and tremorogenic mechanisms. The response of Vim neurons during movement performance was characterized by a single discharge pattern. The essential features of this firing pattern included: (1) an absence of anticipatory changes during the preparatory phase of voluntary movements (140 out of 140 reacted, 100%); (2) the uniform biphasic activatory configuration of response activity correlating with muscle or joint influences during the onset and the termination of movements (140 out of 220 studied, 69%); (3) the ability of neurons to react during movements by transformation of their initial irregular non-rhythmic activity into the rhythmic one at a frequency of 5 ± 1 Hz (72 out of 140 reacted, 51%); (4) the very prolonged movement after-effect, closely correlated with the peripheral tremor activation (90 out of 140 reacted, 64%). The latter peculiarities of the response pattern of movement-related units were observed also in the Vop nucleus. The Vop cells responded similarly during the muscle contraction and/or its after-effect by a transient transformation of the unitary discharge pattern into a 5 ± 1 Hz rhythmic one (30 out of 69 reacted, 43%) and by a rebound late activation (24 out of 69 reacted, 35 %).

Cells displaying activity related to active movements and to somatosensory stimulation have been described previously in human motor thalamus by several authors (Albe-Fessard 1962, 1966; Jasper and Bertrand, 1966; Ohye and Narabayashi, 1972; Raeva, 1986, 1993; Tasker et al., 1987; Lenz et al., 1988, 1990).

The comparative analysis of Vop and Vim neurons performed in the present study showed similarities between unit activities in these two nuclei in parkinsonian patients with tremor. These similarities may be a result of altered, abnormal Vim and Vop activity possibly connected with Parkinson's disease pathology. This pattern was manifested as: (1) altered hyper-excitability of unit activity with abnormal prolonged late activation after voluntary movements determined by hyper- and/or hypokinetic state of patients (35% in the Vop; 64% in the Vim); (2) the "latent" rhythmogenic 5 ± 1 Hz tendency, which may be accentuated either by afferent influences or motor cortical stimulation during voluntary movements (43% in the Vop; 51% in the Vim); (3) a high degree of synchronization of oscillatory rhythmic discharges in many cell populations functioning as local oscillators, arising between neighboring cells during the rebound post-inhibitory activation and accentuated by feedback afferent influences (63% in the Vop; 68% in the Vim). Our results are in general agreement with results by Lenz et al. (1994), from the human ventral thalamic nuclear group, and by Vitek et al. (1990), from the MPTP model of parkinsonism in monkeys. In view of such functional homogeneity of the human Vop and Vim nuclei, many tremor-related and tremorogenic systems-linked cell populations seem to operate as self-maintained centers of oscillatory synchronized hyperexcitable activity, correlated with peripheral tremor.

At present, two general hypotheses explaining the mechanisms of parkinsonian tremor are being considered: (1) the intrinsic pacemaker as the central oscillator, possibly located in the basal ganglia, or (2) the peripheral feedback circuit (Tatton and Lee, 1975; Lamarre and Joffroy, 1979; Stein and Lee, 1981; Llinas, 1984; Steriade and Deschénes, 1984; Marsden, 1984; Schnider et al., 1989; Elble and Koller, 1990; Paré et al., 1990; Lenz et al., 1993, 1994; Zirch et al., 1998). We have assessed these two hypotheses by detailed analysis of the characteristics of single-spike (named A-type) and bursting calcium spike (named B-type) neurons in the motor thalamus and in the non-specific and associative thalamic nuclei.

Table 1 presents the general statistical evaluation of the percentage of single-spike A-neurons and bursting Ca^{2+} B-neurons located in the different thalamic nuclei in parkinsonian patients - in the Vim, Vop, Voa, Rt, CM, and MD. The data show significant differences between the percentage of discharges of unitary A-units located in the Vim and Vop nuclei of the motor thalamus, on the one hand, and bursting B-units located mainly in the nonspecific (Rt, CM) and associative (MD) thalamic nuclei, on the other. These data suggest that bursting calcium spike neurons located in the human non-specific Rt and CM thalamic structures play an important role in the mechanisms of central thalamic oscillators involved in tremor. A question of interest is what roll these nonspecific thalamic nuclei might play in the central burst rhythmicity involved in the tremorogenic oscillatory mechanisms.

Table 1. Statistical evaluation of location and percentage of single-spike A- and bursting calcium spike B-neurons in different thalamic nuclei in parkinsonian patients

Thalamic nuclei	Percentage of single-spike A-units	Percentage of bursting calcium spike B-units
Vim	98	2
Vop	95	5
Voa	63	37
Rt	51	42
CM	31	66
MD	18	79

According to numerous animal and human studies, the reticular thalamic nucleus (Rt) plays a very important role in the thalamo-reticulo-cortical interaction (Andersen et al., 1967; Steriade et al., 1985, 1987, 1997; Huguenard and Prince, 1992; Destexhe et al., 1998; Timofeev et al., 1998) during information processing (Raeva and Lukashev, 1993) and, especially, in the generation of thalamic rhythmicity (Steriade et al., 1987, 1997; Destexhe, 1998). It is believed that the GABAergic Rt neurons may modulate and synchronize the thalamo-cortical neurons via an inhibitory control.

In our previous study (Raeva et al., 1991, 1999a) no direct correlation between central 3-6 Hz rhythmicity of bursting B-units and the peripheral tremor was ever found for Rt and Voa neuronal activity. However, our present results using multiparametric and factor (principal components) analysis demonstrate the existence in the Rt of certain indirect correlations between the increase in the average frequency with stabilization of discharges of bursting B-type units and the intensification of parkinsonian tremor. (Coefficient of correlation K = 0.47, $P < 0.05$). No relationship was found, however, between average parameters of single-spike A-units and the tremor intensity in the Rt nucleus (K = 0.23, $P < 0.05$). Figure 1 illustrates the dynamics of average frequency with stabilization of discharges of bursting calcium spike B-unit and single-spike A-unit activities in relation to parkinsonian tremor intensity (from I to III stages) in the Rt and the Voa thalamic nuclei. As seen in Figure 1, the firing rate of A- and B-units with stabilization of discharges in the Voa increased with tremor intensification at similar rates (for A-units K = 0.46, $P < 0.05$; for B-units K = 0.49, $P < 0.01$); These findings suggest that the bursting calcium spike B-units can play an important role in the formation of local thalamic oscillators functioning at a frequency of about 5 Hz, and in the synchronizing mechanisms of the central thalamic oscillator involved in the generation of parkinsonian tremor.

These observations are in agreement with conclusions arrived at earlier by other authors (Paré et al., 1990; Lenz et al., 1993, Destexhe et al., 1996; Jeanmonod et al., 1996; Steriade et al., 1997; Tsoukatos et al., 1997; Hutchison et al., 1998; Zirh et al., 1998; Raeva et al., 1999a; Radhakrishnan et al., 1999). Our own earlier results in the human Rt nucleus, indicating the significance of bursting calcium spike neurons in the thalamo-reticulo-cortical interaction during verbally-ordered voluntary movements (Raeva and Lukashev,

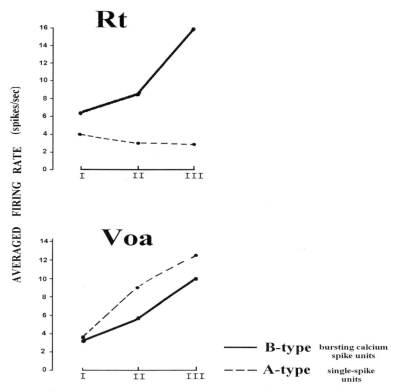

Figure 1. Relationship between mean average frequency of single-spike A-units and bursting calcium spike B-units and tremor intensity (from I to III stages) in the Rt and the Voa thalamic nuclei in parkinsonian patients.

1993) also support this. According to experimental data, inhibitory GABAergic Rt influences seem to be a powerful source of non-specific thalamic recruiting for generation of rhythmic oscillations and rebound spike bursts (Llinas, 1984; Huguenard and McCormick, 1992; Destexhe et al., 1998; Steriade et al., 1999).

Comparative statistical analysis of neural changes between the non-specific centrum medianum (CM) and motor Voa-Vop thalamic nuclei during identical voluntary motor test performances reveals important factors contributing to formation of the central oscillatory thalamic mechanisms involved in both voluntary movements and parkinsonian tremor.

Figure 2 presents an example of a general statistical evaluation of response pattern dynamics in single-spike Voa-Vop and bursting calcium-spike CM neurons before, during, and after verbally ordered voluntary movements in parkinsonian patients. These data provide evidence for some internuclear relationships between Voa-Vop and CM neurons that emerge during repeated voluntary movements. Along with differences between response patterns during the movement execution, there are also fundamental similarities between neighboring unit response patterns during the movement after-effect. The latter are characterized by an appearance of transient positive synergic rhythmicity at a frequency of about 4-5 Hz and synchronization of oscillations correlating with EMG tremor activation (39 out of 57 pairs studied, 68%, in the Voa-Vop; 22 out of 28 pairs studied, 79%, in CM).

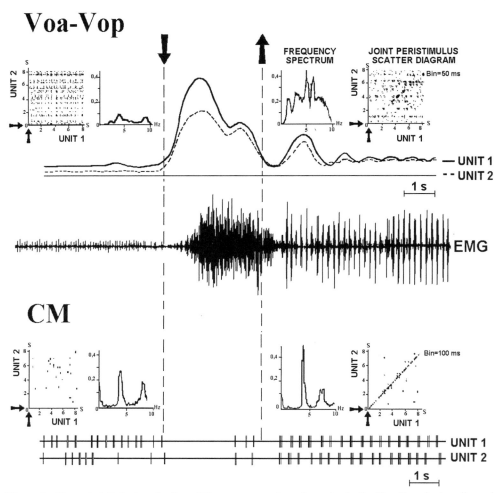

Figure 2. General statistical evaluation of the response pattern dynamics in the Voa-Vop single-spike and CM bursting calcium spike neurons during verbally ordered voluntary movements in a parkinsonian patient. The arrows mark the time of verbal command presentation: "Clench your fist!" (downward arrow); and "Unclench!" (upward arrow).

The results presented here, as well as those of earlier studies in patients with and without tremor (Raeva, 1972, 1986, 1993; Raeva et al., 1999a,b), provide some new insights into motor control and tremorogenic mechanisms. These studies suggest that the central 5 Hz rhythmic phenomenon may manifest a plausible common mechanism, related not only to the generation of parkinsonian tremor, but also to the motor signal transmission in response to motor stimulation during normal natural states in man. Two key factors seem to underlie this mechanism: (1) ability of Vop and Vim neurons to react during voluntary movements by transforming their unitary activity into transient, rhythmic 5 Hz burst-like discharges, triggered by cortical and/or afferent feedback influences; and (2) critical significance of bursting calcium spike neurons and of inhibitory GABAergic Rt influences in these transformations, supporting the generation of rhythmic oscillations and the rebound spike burst in thalamo-cortical neurons (Llinas, 1984; Huguenard and McCormick, 1992; Destexhe et al., 1998; Steriade, 1999; Raeva et al., 1999a).

In the resting tremor of Parkinson's disease, the central thalamic rhythmic 5 Hz phenomenon described by many authors (Albe-Fessard et al., 1962, 1966; Jasper and Bertrand, 1966; Ohye and Narabayashi, 1972; Ohye et al., 1974; Raeva et al., 1986; Tasker et al., 1987; Lenz et al., 1994) seems to transform into a pathological system functioning as a stable, self-maintained oscillatory mechanism driving synchronized unified oscillations of many cell populations. Our results suggest that this pathological oscillatory mechanism, linking the thalamus to parkinsonian tremor, involves both the central local oscillators and the peripheral feedback processes, which are closely correlated. This mechanism may be a result of the endogenous inhibitory deficit in Parkinsonism (Riclan and Levita, 1969; Velasco and Velasco, 1979; Scherman et al., 1989; Cooper et al., 1994), caused by a loss of dopaminergic neurons of substantia nigra (Biorklund and Lindvall, 1975; Hornykiewicz and Kish, 1986; Lavoie et al., 1989) and stable exogenous afferent influences from trembling limbs.

Our new findings support the view that the oscillatory mechanism, linking the thalamus to parkinsonian tremor, is a model of a pathological system according to the principles of the Ukhtomsky dominanta organization, the set of general principles underlying neural center behavior (Ukhtomsky, 1923, 1966). These principles are summarized in Table 2.

Table 2. Comparison of basic parameters of principles of the Ukhtomsky dominanta and obtained properties of the neural pathological oscillatory thalamic mechanism involved in parkinsonian tremor

Principles of the Uhktomsky dominanta	Properties of the neural pathological oscillatory thalamic mechanism involved in parkinsonian tremor
1. Increased excitability	1. Increased hyperexcitability determined by altered abnormal activities of Vop and Vim neurons
2. Persistence of excitation 3. Forced synchronization of many local oscillators composing the dominanta state	2. Higher degree of synchronization of oscillatory discharges in many ensembles functioning as local thalamic oscillators
4. Inertia of dominanta functioning	3. Stable inertial character of local oscillators self-maintained by afferent inflows and/or cortical motor stimulation during voluntary movements

Comparison of basic parameters of essential properties of the cellular pathological oscillatory mechanism obtained in this study with the Ukhtomsky principles reveals striking similarities.

The functional approach based on the Ukhtomsky dominanta principles seems to offer some strategies for Parkinson's disease treatment. According to Ukhtomsky, suppression of the pathological hyperactive dominanta may be reached in two main ways: by direct derangement of its central link, or by creation of another competitive dominanta as an antagonistic "antisystem," which might lead to suppression of the pathological system. Ukhtomsky suggests that direct derangement of the pathological dominanta might be achieved by two different methods. The first is the radical destruction of the central link of the dominanta, an approach widely used by ablative neurosurgery as a method for Parkinson's disease treatment. The second is the destabilization/transformation of the dominanta by modulation of the level of hyperactivity in its central link. This approach is realized in the method of deep brain stimulation.

Our functional model of a pathological oscillatory thalamic mechanism further develops Ukhtomsky principles and suggests strategies for further research of combined methods of treatment, including electrophysiological, biochemical, local, pharmacological, etc. Such approaches should be directed towards affecting the central link of the oscillatory thalamic mechanism through changing the state of the neuromodulatory thalamic system and/or the system-forming factors of oscillatory and synchronizing mechanisms linked to parkinsonian tremor. This functional approach might possibly lead to improvement of most parkinsonian motor signs.

ACKNOWLEDGMENTS: The author thanks her colleagues N. Vainberg, A. Lukashev, V. Pavlenko, and W. Dubinin for their great assistance with the statistical analysis and clinical physiological investigations; Yu. Tikhonov, A. Lashin, and A. Bogoljub for technical assistance; and R. Pavlov for improving the English text and typing the manuscript.

REFERENCES

Albe-Fessard, D., Arfel, G., Guiot, G., Hardy, J., Vourc'h, G., Hertzog, E., Aleonard, P., and Derome, P., 1962, Dérivations d'activités spontanées et evoquées dans les structures cérébrales profondes de l'homme, *Rev. Neurol.* 106:89.

Albe-Fessard, D., Guiot, G., Lamarre, Y., and Arfel, G., 1966, Activation ofthalamocortical projections related to tremorogenic processes, in: *The Thalamus*, D.P. Purpura and M.D. Yahr, eds., Columbia University Press, New York.

Alexander, G.E., Crutcher, M.D., and DeLong, M.R., 1990, Basal ganglia-thalamocortical circuits: parallel substrates for motor, oculomotor, "prefrontal" and "limbic" functions, *Prog. Brain Res.* 85:119.

Alexander, G.E., DeLong, M.R., and Strick, P.L., 1986, Parallel organization of functionally segregated circuits linking basal ganglia and cortex, *Annu. Rev. Neurosci.* 9:357.

Andersen, P., Anderson, S.A., and Lomo, T., 1967, Nature of thalamocortical relations during spontaneous barbiturate spindle activity, *J. Physiol. (Lond).* 192:283.

Biorklund, A., and Lindvall, O., 1975, Dopamine in dendrites of the substantia nigra neurons: suggestions for a role in dendritic terminals, *Brain Res.* 83:531.

Cooper, J.A., Sagar, H.J., Tidswell, Ph., and Jordan, N., 1994, Slowed central processing in simple and go/no go reaction time tasks in parkinson's disease. *Brain,* 117:517.

Crowell, R.M., Perret, E., Siegfried, J., and Villoz, JP., 1968, 'Movement units' and 'tremor phasic units' in the human thalamus, *Brain Res.* 11:481.

Destexhe, A., Bal, T., McCormick, D.A., and Sejnowski, T.J., 1996, Ionic mechanisms underlying synchronized oscillations and propagating waves in a model of ferret thalamic slices, *J. Neurophysiol.* 76:2049.

Destexhe, A., Contreras, D., and Steriade, M., 1998, Mechanisms underlying the synchronizing action of corticothalamic feedback through inhibition of thalamic relay cells, *J. Neurophysiol.* 79:999.

Donchin, E., and Heffley, E., 1978, Multivariate analysis of event-related potential data: a tutorial review, in: *Multidisciplinary Perspectives in Event-Related Brain Potential Research,* D. Otto ed., U.S. Gov. Printing Office, Washington, DC.

Elble, R.J., and Koller, W., 1990, *Tremor,* Johns Hopkins University Press, Baltimore.

Graybiel, A.M., Aosaki, T., Flaherty, A.W., and Kimura, M., 1994, The basal ganglia and adaptive motor control, *Science,* 265:1826.

Gerstein, G.L., and Perkel, D.H., 1972, Mutual temporal relationships among neuronal spike trains. Statistical techniques for display and analysis, *Biophys. J.* 12:453.

Hassler, R.,1982, Architectonic organization of the thalamic nuclei, in: *Stereotaxy of the Human Brain. Anatomical, Physiological and Clinical Applications,* G. Shaltenbrand and A. Walker, eds., Thieme Stratton, New York.

Hornykiewicz, O., and Kish, S.J., 1986, Biochemical pathophysiology of parkinson's disease, *Adv. Neurol.* 45:19.

Huguenard, J.R., and McCormick, D.A., 1992, Simulation of the currents involved in rhythmic oscillations in thalamic relay neurons, *J. Neurophysiol.* 68:1373.

Huguenard, J.R., and Prince, D.A., 1992, A novel T-type current underlies prolonged Ca^{2+}-dependent burst firing in GABAergic neurons of rat thalamic reticular nucleus, *J. Neurosci.* 12:3804.

Hutchison, W.D., Allan, R.J, Opitz, H., Levy, R., Dostrovsky, J.O., Lang, A.E., and Lozano, A.M., 1998, Neurophysiological identification of the subthalamic nucleus in surgery for parkinson's disease, *Ann. Neurol.* 44:622.

Jasper, H., and Bertrand, G., 1966, Thalamic units involved in somatic sensation and voluntary and involuntary movements in man, in: *The Thalamus*, D.P. Purpura and M.D. Yahr, eds., Columbia University Press, New York.

Jeanmonod, D., Magnin, M., and Morel, A., 1996, Low-threshold calcium spike bursts in the human thalamus. Common physiology for sensory, motor and limbic positive symptoms, *Brain,* 119:363.

Lamarre, Y., Joffroy, A.J., 1979, Experimental tremor in monkeys: Activity of thalamic and precentral cortical neurons in the absence of peripheral feedback, *Adv. Neurol.* 24:109.

Lavoie B., Smith, Y., and Parent, A., 1989, Dopaminergic innervation of the basal ganglia in the squirrel monkey as revealed by tyrosine hydroxylase immunohistochemistry, *J. Comp. Neurol.* 289:36.

Lenz, F.A., Dostrovsky, J.O, Tasker, R.R, Yamashiro, K., Kwan, H.C., and Murphy, J.T., 1988, Single-unit analysis of the human ventral thalamic nuclear group: somatosensory responses, *J. Neurophysiol.* 59:299.

Lenz, F.A., Kwan, H.C., Dostrovsky, J.O., Tasker, R.R., Murphy, J.T., and Lenz, Y.E., 1990, Single unit analysis of the human ventral thalamic nuclear group. Activity correlated with movement, *Brain,* 113:1795.

Lenz, F.A., Kwan, H.C., Martin, R.L., Tasker, R.R., Dostrovsky, J.O., and Lenz, Y.E., 1994, Single unit analysis of the human ventral thalamic nuclear group. Tremor-related activity in functionally identified cells, *Brain,* 117:531.

Lenz, F.A., Vitek, J.L., and DeLong, M.R., 1993, Role of the thalamus in parkinsonian tremor: evidence from studies in patients and primate models, *Stereotac. Funct. Neurosurg.* 60:94.

Llinas, R.R., 1984, Rebound excitation as a physiological basis for tremor: a biophysical study of oscillatory properties of mammalian central neurons in vitro, in: *Movement Disorders Tremor*, L.J. Findley and R. Capildeo, eds., Macmillan, London.

Marsden, C.D., 1984, Origins of normal and pathological tremor, in: *Movement Disorders Tremor*, L.J. Findley and R. Capildeo, eds., Macmillan, London.

Ohye, C., and Narabayashi, H., 1972, Activity of thalamic neurons and their receptive fields in different functional states in man, in: *Neurophysiology Studied in Man*, G.G. Somjen, ed., Excerpta Medica, Amsterdam.

Ohye, C., and Albe-Fessard, D., 1978, Rhythmic discharges related to tremor in human and monkey, in: *Abnormal Neuronal Discharges,* N. Chalazonitis, ed., Raven Press, New York.

Ohye, C., Saito, U., Fukamachi, A., and Narabayashi, H., 1974, An analysis of the spontaneous rhythmic and non-rhythmic burst discharges in the human thalamus, *J. Neurol. Sci.* 22:245.

Paré, D., CurróDossi, R., and Steriade, M., 1990, Neuronal basis of the parkinsonian resting tremor: a hypothesis and its implications for treatment, *Neuroscience,* 35:217.

Radhakrishnan, V., Tsoukatos, J., Davis, KD., Tasker, R.R., Lozano, A.M., and Dostrovsky, J.O., 1999, A comparison of the burst activity of lateral thalamic neurons in chronic pain and non-pain patients, *Pain,* 80:567.

Raeva, S.N., 1972, Unit activity of some deep nuclear structure of the human brain during voluntary movements, in: *Neurophysiology Studied in Man*, Somjen, ed., Exerta Medica, Amsterdam.

Raeva, S.N., 1977, *Microelectrode Investigations of Unit Activity of the Human Brain*, Nauka, Moscow (In Russian).

Raeva, S.N., 1986, Localization in human thalamus of units triggered during 'verbal commands,' voluntary movements and tremor, *Electroencephal. Clin. Neurophysiol.* 63:160.

Raeva, S.N., 1993, Unit activity of nucleus ventralis lateralis of human thalamus during voluntary movements, *Sereotact. Func. Neurosurg.* 60:86.

Raeva, S.N., Lukashev, A., and Lashin, A., 1991, Unit activity in human thalamic reticularis nucleus. I. Spontaneous activity, *Electroenceph. Clin. Neurophysiol.* 79:133.

Raeva, S.N., and Lukashev, A., 1993, Unit activity in human thalamic reticularis neurons. II. Activity evoked by significant and non-significant verbal or sensory stimuli, *Electroenceph. Clin. Neurophysiol.* 86:110.

Raeva, S.N., Vainberg, N.A., and Dubinin, V.A., 1999a, Analysis of spontaneous activity patterns of human thalamic ventrolateral neurons and their modifications due to functional brain changes, *Neuroscience,* 88:365.

Raeva, S.N., Vainberg, N.A., Tikhonov, Yu.N., and Tsetlin, I.M., 1999b, Analysis of evoked activity patterns of human thalamic ventrolateral neurons during verbally ordered voluntary movements, *Neuroscience,* 88:377.

Riclan, M., and Levita, E., 1969, *Subcortical correlates of human behavior: a psychological study of thalamus and basal ganglia surgery,* Williams and Wilkins, Baltimore.

Scherman, D., Desnos, C., Darchen, F., Pollak, P., Javoy-Agid, F., and Agid, Y., 1989, Striatal dopamine deficiency in Parkinson's disease: role of aging, *Ann. Neurol.* 26:551.

Schnider, S.M., Kwong, R.H., Lenz, F.A., and Kwan, H.C., 1989, Detection of feedback in the central nervous system using system identification techniques, *Biol. Cybern.* 60:203.

Stein, R.B., and Lee, R.G., 1981, Tremor and Clonus, in: *Handbook of Physiology*, V.B. Brooks, ed., American Physiological Society, Bethesda, MD.

Steriade, M., and Deschénes, M., 1984, The thalamus as a neuronal oscillator, *Brain Res.* 320:1.

Steriade, M., Deschénes, M., Domich, L., and Mulle, C., 1985, Abolition of spindle oscillations in thalamic neurons disconnected from nucleus reticularis thalami, *J. Neurophysiol.* 54:1473.

Steriade, M., Domich, L., Oakson, G., and Deschénes, M., 1987, The deafferented reticular thalamic nucleus generates spindle rhythmicity, *J. Neurophysiol.* 57:260.

Steriade, M., Jones, E.G., and McCormick, D.A., 1997, *Thalamus Organization and Function. Vol. 1,* Elsevier, Oxford.

Steriade, M., 1999, Coherent oscillations and short-term plasticity in corticothalamic networks, *Trends Neurosci.*, 22:337.

Tasker, R.R., Lenz, F.A., Dostrovsky, J.O., Yamashito, K., Chodakiewitz, G., and Albe Fessard, D., 1987, The physiological basis of Vim thalamotomy for involuntary movement disorders, in: *Clinical Aspects of Sensory and Motor Integration,* Struppler A. and Weindl A., eds., Springer-Verlag, Berlin/ Heidelberg.

Tatton, W.G., and Lee, R.G., 1975, Evidence for abnormal long-loop reflexes in rigid parkinsonian patients, *Brain Res.* 100:671.

Timofeev, I., Grenier, F., and Steriade, M., 1998, Spike-wave complexes and fast components of cortically generated seizures. IV. Paroxysmal fast runs in cortical and thalamic neurons, *J. Neurophysiol.* 80:1495.

Tsoukatos, J., Kiss, Z.H., Davis, K.D., Tasker, R.R., and Dostrovsky, J.O., 1997, Patterns of neuronal firing in the human lateral thalamus during sleep and wakefulness, *Exp. Brain Res.* 113:273.

Ukhtomsky, A.A., 1923, Dominanta as a working principle of nervous centers, in: *Complete Works, Vol. 1,* Leningrad University Press, Leningrad (In Russian).

Ukhtomsky, A.A., 1966, *The Dominanta,* Nauka, Moscow (In Russian).

Velasco, F., and Velasco, M., 1979, A reticulothalamic system mediating proprioceptive attention and tremor in man, *Neurosurgery,* 4:30.

Vitek, J.L., Asher, J., DeLong, M.R., and Alexander, G.E., 1990, Altered somatosensory response properties of neurons in the "motor" thalamus of MPTP treated parkinsonian monkeys, *Soc. Neurosci. Abstr.* 6:425.

Zirh, T.A., Lenz, F.A., Reich, S.G., and Dougherty, P.M., 1998, Patterns of bursting occurring in thalamic cells during parkinsonian tremor, *Neuroscience,* 83:107.

BEHAVIOR OF THALAMIC NEURONS IN THE MOVEMENT DISORDERS - TREMOR AND DYSTONIA

CHIHIRO OHYE AND TOHRU SHIBAZAKI[*]

INTRODUCTION

In the course of stereotacic thalamotomy for treatment of movement disorders such as Parkinson's disease, essential tremor, post-traumatic tremor, other kinds of tremor, dystonia, choreic movement, etc., we always use microrecording to determine the precise target to be coagulated (Ohye, 1994; 1996). While depth recording is useful for targeting, it also provides a unique opportunity to gain insight into the ongoing pathophysiological changes of neuronal activity in the human thalamus (Ohye et al., 1989; Ohye, 1997).

In our series of some 600 cases of stereotatic operations, we have been interested mostly in the tremor-related thalamic activity, as first described by Albe-Fessard et al. (1963), and later by Ohye (1994;1997) and Ohye et al. (1989). Although dystonia cases were not numerous in our patient population, we have operated on patients with cerebral palsy and post-traumatic or post-CVD movement disorders, in which a dystonic element was, more or less, involved. In this chapter, we describe briefly the tremor-related thalamic activity and then discuss several particular findings in the cases with dystonia.

SUBJECTS AND METHODS

Principal methods of stereotactic thalamotomy have already been described (Ohye, 1998a; 1998b). In brief, the patient is operated on under local anesthesia. Leksell's stereotactic apparatus is fixed on the skull. In our earlier series we performed ventriculography to determine a tentative target point but recently we depend on MRI and Surgiplan to obtain its 3D coordinates.

The recording electrode is of a bipolar concentric needle type, with outer diameter of 0.4 mm, tip length of about 10 μm, interpolar distance of about 0.1-0.2 mm, and electrical resistance of about 100 Kohm. Since Leksell's stereotactic apparatus uses two coagulation needles, the recording electrode takes the place of one coagulation needle. Conveniently,

---
[*] CHIHIRO OHYE AND TOHRU SHIBAZAKI • Functional and Gamma Knife Surgery Center, Hidaka Hospital, Takasaki, Gunma, Japan.

the recording electrode just enters the position of the needle stylet. Usually, the center position of the guide hole is used for the reference electrode (anterior electrode, oriented to the tentative target of zero point, at the lower border of the thalamic ventral intermediate nucleus (Vim) of Hassler's nomenclature in Shaltenbrand and Warren atlas (1997). The posterior hole (3 mm apart) is used for the posterior electrode. For the sake of patient safety and time economy, however, only the posterior track is used for microrecording. In fact, earlier study revealed that almost 80% of the kinesthetic neurons including tremor time-locked rhythmic discharges were recorded along the posterior track (Ohye et al., 1989). After positioning of the electrode, it is introduced slowly into the brain using a micromanipulator driven by a step motor and controlled by a remote switch. In our newly designed device, which we've used for electrode introduction in more recent cases, the speed and step are chosen as required, and the distance from the zero (tentative target) to the tip of the electrode is automatically and continuously displayed in micron steps on a digital counter. Spontaneous electrical activity is recorded continuously from the cortical surface down to the tentative target point or to a deeper point if necessary. Depth electrical activity is lead to an oscilloscope, and recorded with related EMG on both running paper by thermal pen and on a cassette tape recorder.

In addition to the spontaneous background activity mentioned above, the unitary or distinguishable multi-unitary spike discharge(s) are also selected with the aid of the micromanipulator. Whenever isolated stable spike discharge is found, passive and/or active limb movements are performed, and several natural stimuli are given to test whether the pattern of spike discharges is modified as a result. Light touch, tap, compression, muscle stretch, and voluntary contractions of the contra- and ipsi-lateral extremities or face area are tested. This paper presents results obtained mainly in the presumed ventral oral (Vo) and ventral intermediate (Vim) nuclei.

A day before the operation, all operative procedures, and especially the process and significance of microrecording, are fully explained to the patient and his (or her) family to get an informed consent.

RESULTS

Tremor

As has already been reported, in cases with tremor of any kind, we almost always find the tremor time-locked rhythmic discharges in the area of the lateral part of the Vim (Ohye et al., 1989; Ohye, 1996). In cases with spontaneous tremor, the activity is also responsive to passive movement or compression of the contralateral body part, but it is never responsive to light cutaneous stimuli. Therefore, these cells are classified as kinesthetic neurons. An example of typical kinesthetic neuron behavior is shown in Figure 1. Careful observation revealed that, in most of the cases, the thalamic group discharge is directly related to the passive stretch phase of the contralateral limb muscle reflected in the tremor grouping displayed by EMG. For example, in the case illustrated, wrist extension (passive stretching of the flexor muscle) induced a thalamic response and the natural tremor showed alternative contraction (grouping of EMG discharges around the wrist joint), then thalamic grouped discharge occurred at the silent phase of the wrist flexor muscle, due to the natural stretch of the flexor muscle by the antagonist (in this situation, wrist extensor muscle). This allows us to conclude that the tremor rhythm of a Vim neuron is the result of a passive stretch of the responsible muscle by active contraction of its antagonist muscle by tremor.

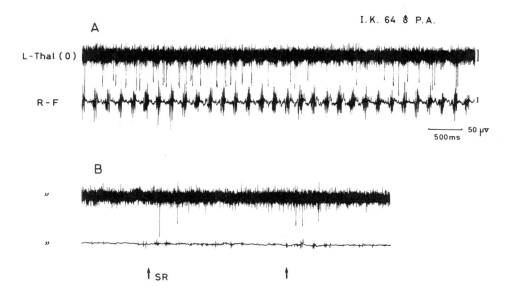

Figure 1. (A) Tremor time-locked thalamic Vim unit discharge (upper trace) is shown with the corresponding peripheral tremor demonstrated by EMG (lower trace). Note that in this case, thalamic discharge is at the silent (passive stretch) phase of the forearm flexor muscle. (B) The same thalamic unit responded to passive movement (extension of the wrist in this case). Approximate moment of the passive movement is indicated by arrow labeled SR.

Dystonia

It is not quite clear whether there is a dystonia-specific change in the ventral thalamus, but several interesting neuronal patterns or responses have been observed and will be presented here. In general, the spontaneous activity of ventral thalamic neurons (probably in Vo and Vim nuclei) is lower than that in Parkinson's disease. At a single or distinguishable multi-unitary spike level, we often encountered different types of irregular burst discharges, either as a grouped discharge of about 3-5 spikes of a short train or as a long train of bursts, up to 1 s. In one case we found a large number of successive burst discharges along the descending trajectory, as if the patient were in deep sleep. In roughly one third of the cases with cerebral palsy, increased burst discharges are found. Figure 2 shows two different patterns of burst discharges in the same thalamic unit, recorded from a case with partial dystonia in cerebral palsy. In this case, an irregular burst consisting of a couple of spikes was seen during a relatively calm, silent period of muscle tone and movement (Figure 2A), but it changed into a long-lasting burst discharge of more than 10 spikes, starting with a high amplitude spike during the period of increased muscle tone and movement indicated by irregular oscillation of the contralateral hand muscles (Figure 2B).

In the case of a 42 year old man with mixed types of involuntary movements, including dystonic extensor thrust in his left upper limb after severe head trauma at four years of age, MRI showed a large, irregular, wedge-shaped brain tissue deficit in the right parietal lobe extending into the ventral posterior thalamic area. His involuntary movement consisted of intentional tremor, ballistic movement, dystonia of the distal part of the arm (hand, fingers), and irregular jerky movement with hypesthesia. Microrecording toward the anterior part of the old damaged area (estimated as being thalamic Vo or Vim nuclei) revealed that the spontaneous background activity was relatively low compared to that in Parkinson's disease. However, an increased general background activity was occasionally recorded during his hand's dystonic contraction phase at three different thalamic points. One such increase of tonic discharge is shown in Figure 3. When dystonia with jerky arm-

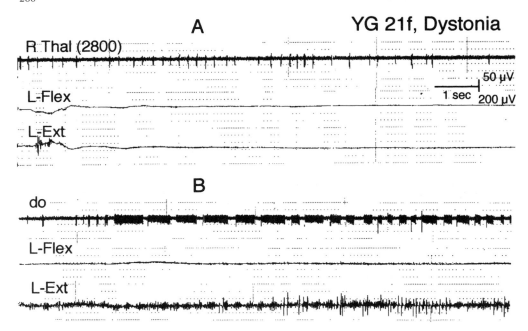

Figure 2. An example of different types of irregular burst discharges recorded in the Vim nucleus (at 2800 microns above zero) shown with the contralateral EMG from the forearm flexor (L-Flex) and extensor (L-Ext) muscles. (A) Without dystonic movement. No EMG discharge. (B) Irregular burst of long duration with periphral dystonia, more marked in the forearm extensor muscle.

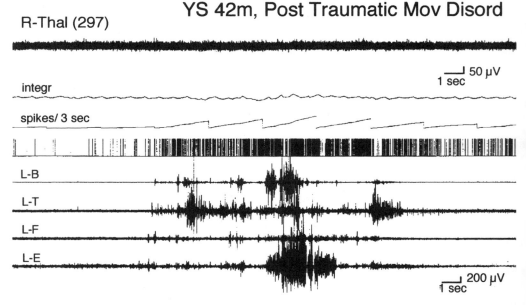

Figure 3. Concomitant increase of the thalamic discharge with the contralateral dystonic movement in a case with post-traumatic movement disorder. The first trace: thalamic discharge with small spikes, which is more clear after transformation into pulse in the fourth trace. Lower four traces: EMG from the contralateral upper limb muscles. B: biceps brachii, T: tricepsbrachii, F: forearmflexor, E: forearm extensor.

hand movement manifested, the thalamic activity increased concomitantly. Although time relation was not precisely measured, the abnormal muscle contraction seemed to precede the thalamic discharge.

In the case of a 14 year old girl with an early onset severe torsion dystonia linked to chromosome 9q34, traditionally called Oppenheim's dystonia (DYT 1), the operation was performed under general anesthesia. It was not surprising that the irregular burst discharges were very exaggerated in the anesthetized condition while kinesthetic responses were difficult to detect. Only two ambiguous responses and one clear response to passive arm movement (arm raising) were recorded (Figure 4). In this case, after physiological identification of the Vim nucleus, mainly Vo with Vim, thalamotomy reduced her dystonic posture considerably, an improvement great enough to enable her to enter a rehabilitation program.

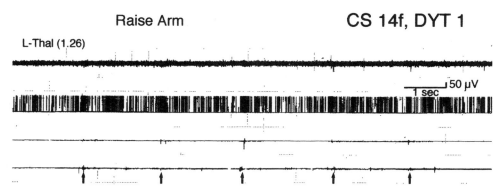

Figure 4. Thalamic response to passive arm raising movement (indicated by arrows under the lowest trace of EMG), recorded in the case with torsion dystonia of DYT1, under general anesthesia. The first trace: thalamic discharge, which is converted into pulse in the second trace.

As this case shows, dystonia often involves the trunkal muscles, but Vim neurons responding to trunkal stimulation have not been reported. Therefore, in the 200 most recent thalamotomy cases, mainly Parkinson's disease and essential tremor patients, we analyzed trunkal responses. In 12 cases we found 15 such responses. Of these, six responded to the compression of deltoid muscle, three to trapezius muscle, two each to SCM and chest, and one each to pectoralis major and back muscle. A typical response to stimulation (compression) of back muscle is shown in Figure 5. In Figure 6, all the recorded points of these trunkal neurons were plotted. Their distribution is almost within the range of typical kinesthetic neurons of Vim or dorsal to it. In order to determine their relationship to typical kinesthetic neurons responding to stimulation of extremities, sequential distribution patterns along each trajectory were examined. Five trunkal neurons were found segregated or almost segregated from kinesthetic neurons, whereas the other seven neurons were distributed among other kinesthetic neurons. In the latter cases, the trunkal neuron was often found between a lower limb and an upper limb responsive area. For example, along the recording trajectory the first kinesthetic response to the contralateral stimulus was related to the lower limb, then came trunkal response, followed by upper limb response. This sequential distribution pattern indicates that the recording trajectory was inclined slightly more laterally to traverse from lower limb to upper limb zone with the trunkal neuron in between.

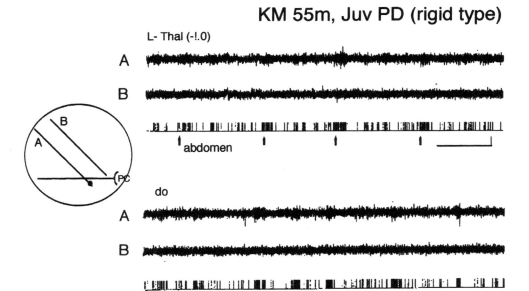

Figure 5. An example of the trunkal response. The thalamic units responded both to compression of abdomen and back muscle. The latter response was more sensitive in this case. The circle on the left shows the thalamic point where this response was recorded. The point is plotted on the lateral view of recording along electrode (A) trajectory (A and B, a pair of recording electrodes). PC: posterior commissure.

DISCUSSION

Earlier we discussed in detail the nature of kinesthetic tremor-related neurons in the Vim (Ohye et al., 1989). We suggest that activating stimuli probably originate from muscle spindles and reach the lateral part of the Vim via fast-conducting spinothalamic tract fibers. Therefore, when tremor appears at the phase of stretching by the antagonist contraction, the signal from the responsible muscle arrives at the responsive Vim neuron with a relatively short delay of 11~12 ms. It is then relayed to cortical area 3A, according to our experimental sudy in the monkey (Ohye, 1987). The descending pathway is not necessarily the corticospinal tract. In fact, it's likely to be the reticulospinal tract on the ipsilateral side to the peripheral tremor. Thus, the tremor rhythm comes back to the trembling muscle itself, making a long closed loop. There are still ambiguous or missing links between the cortex and brain stem, but this is our working hypothesis on the tremor maintenance system.

Activity of the thalamic neurons related to dystonia has not been throughly studied. There are definitely several different types of dystonia (Fahn et al., 1998). In our series of patients undergoing stereotactic surgery, we treated only partial dystonia of cerebral palsy, dystonia combined with other types of movement disorders caused by stroke, head injury, or, rarely, a special type of dystonia such as DYT1. Although no specific feature common to all dystonic cases was identified, less active spontaneous discharges in the ventral thalamus (Vo and Vim nuclei), irregular burst discharges, and tonic increase of discharge during dystonic movement were observed. At this time, we can not predict what type of abnormal pattern a given dystonic case will display.

Distribution of Trunkal Neurons

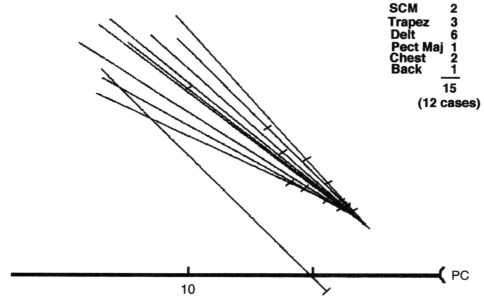

Figure 6. Distribution of the trunkal responses within the Vim nucleus. Lateral views of individual trajectories are shown by oblique lines, on which each point of trunkal response is plotted. The horizontal line is the intercommissural line. PC: posterior commissure. Origin of the response and respective numbers are shown in the upper right corner.

Lenz et al. (1992, 1999a, 1999b) and Zirh et al. (1998) also reported decreased spontaneous activity in the ventral thalamus in cases with dystonia. They analyzed the correlation between thalamic unit activity and corresponding EMG and concluded that the thalamic unit activity led EMG activity in dystonia. Our experience is not sufficient to arrive at a definite conclusion on this point. Lenz et al. (1992, 1999a, 1999b) also emphasized the sensory-motor mismatch in dystonia patients in the sense that the receptive fields of the thalamic neuron and thalamic stimulation-evoked peripheral muscles do not always coincide. We did not try to stimulate to test their findings.

The basal ganglia have been considered to be responsible for the pathophysiology of dystonia (Hallet, 1998; Berardelli et al., 1998). Vitek et al. (1998) reported neuronal activity of the pallidal neurons recorded from the cases with dystonia during the course of stereotactic pallidotomy. It is interesting to note that the neuronal behavior of the internal segment of the globus pallidus is similar to that found, probably in Vo nucleus, in the present study. They found somewhat reduced spontaneous activity compared with that in Parkinson's disease and the irregular grouped discharges.

The trunkal neurons, which are often involved in dystonic posture, have never been reported systematically, probably because they are rarely recorded. The present study revealed that they are located in between lower limb and upper limb areas, when recording is made by laterally inclined trajectory, being found in the rather dorsal part of the Vim nucleus.

SUMMARY

During the course of stereotactic thalamotomy for cases with movement disorders such as Parkinson's disease, essential tremor, dystonia associated with head trauma, stroke, cerebral palsy or congental disorder, we always perform microrecording to determine the exact target point to be coagulated. In cases with tremor of different origin, rhythmic discharges time-locked with the peripheral tremor or kinesthetic responses are recorded in the lateral part of the Vim nucleus in the thalamus. A small coagulation at this point arrests the tremor in most of the cases (up to 95 %) on a long-term basis.

Although the specific thalamic neuronal behavior was not determined in dystonia cases, decreased spontaneous activity, tonic increase of thalamic discharge during dystonic movement, and increased irregular burst discharge were observed. This activity was different from that in cases with tremor, although specific patterns or pattern combinations typical for dystonias of different origin could not be identified.

REFERENCES

Albe-Fessard, D., Arfel, G., Guiot, G., 1963, Activites electriques caracteristiques de quelques structures cerebral chez l'homme, *Ann. Chir.* 17:1185.
Berardelli, A., Rothwell, J.C., Hallet, M., et al., 1998, The pathophysiology of primary dystonia, *Brain*, 121:1195.
Fahn, S., Bressman, S.B., Marsden, C.D., 1998, Classification of dystonia in dystonia 3, in: *Advances in Neurology, Vol. 78,* S. Fahn, C.D. Marsden, M.R. DeLong, eds., Lippincott-Raven Publishers, Philadelphia.
Hallet, M., 1998, Physiology of dystonia, in: *Dystonia 3: Advances in Neurology, Vol.78*, S. Fahn, C.D. Marsden, M.R. DeLong, eds., Lippincott-Raven Publishers, Philadelphia.
Hassler, R., 1997, Architechtonic organization of the thalamic nuclei, in: *Atlas for Stereotaxy of The Human Brain*, G. Schaltenbrand, W. Wahren, eds., G. Thieme, Stuttgart.
Lenz, F.A., and Byl, N.N., 1999, Reorganization in the cutaneous core of the human thalamic principal somatic sensory nucleus (ventral caudal) in patients with dystonia, *J. Neurophysiol.* 82:3204.
Lenz, F.A., Jaeger, C.J., Seike, M.S., et al., 1999, Thalamic single neuron activity in patients with dystonia: dystonia-related activity and somatic sensory reorganization, *J. Neurophysiol.* 82:2372.
Lenz, F.A., Seike, M.S., Jaeger, C.J., et al., 1992, Single unit analysis of thalamus in patients with dystonia, *Movement Disorders 7, Supp.* 1:126.
Ohye, C., 1987, Neural circuit involved in Parkinsonian motor disturbances studied in monkey, Eur. Neurol. 26:41-46.
Ohye, C., 1997, Functional organization of the human thalamus, in: *Thalamus*, M. Steriade, E.G. Jones, D.A. McCormick, eds., Elsevier, Amsterdam
Ohye, C., 1998a, Thalamotomy for Parkinson's disease and other types of tremor. Part 1, Historical background and technique, in: *Textbook of Stereotactic and Functional Neurosurgery*, P.L. Gildenberg, R.R. Tasker, eds., McGraw Hill, New York.
Ohye, C., 1998b, Neural noise recording in functional neurosurgery, in: *Textbook of Stereotactic and Functional Neurosurgery*, P.L. Gildenberg, R.R. Tasker, eds., McGraw Hill, New York.
Ohye, C., Shibazaki, T., Hirai, T., et al., 1989, Further physiological observations on the ventralis intermedius neurons in the human thalamus, *J. Neurophysiol.* 61:488.
Vitek, J.L., Zhang, J., Evatt, M., et al., 1998, GPi pallidotomy for dystonia: Clinical outcome and neuronal activity, in: *Dystonia 3: Advances in Neurology, Vol.78*, S. Fahn, C.D. Marsden, M.R. DeLong, eds., Lippincott-Raven Publishers, Philadelphia.
Zirh, T.A., Reich, S.G., Perry, V., et al., 1998, Thalamic single neuron and electromyographic activities in patients with dystonia, in: *Dystonia 3: Advances in Neurology, Vol.78*, S. Fahn, C.D. Marsden, M.R. DeLong, eds., Lippincott-Raven Publishers, Philadelphia.

THE MOTOR CONTROL OUTPUT FORMING IN HEALTHY SUBJECTS AND PARKINSON'S DISEASE PATIENTS

SERGEY P. ROMANOV AND MICHAEL G. PCHELIN[*]

INTRODUCTION

The motor control system is considered to be divided into the central and peripheral parts. Whereas the anatomy and function of the spinal or segmental levels of the nervous system have been studied quite thoroughly, less is known about the interaction of the higher brain centers in the programming and execution of movements. The role of the cerebellum in the co-ordination of complex movements (Ito, 1984), of motor cortex in the initiation of motor activity (Evarts and Thach, 1969), and of subcortical structures in learning motor skills and in programming movements (Evarts et al., 1984; Asanuma, 1995) has been demonstrated in neurophysiological studies. Occurrence of uncontrollable or involuntary forced movements (hyperkineses), such as in the case of Parkinson's disease, is caused by pathological changes in the basal ganglia structures (Contreras-Vidal and Stelmach, 1995). Ability of neural centers at different CNS levels to maintain cyclic intrinsic activity was demonstrated by Grillner (1996) in the study of fictitious locomotion.

Simulation (Romanov, 1974, 1976, 1996) of the neuronal network at the segmental level controlling muscle contraction showed that the steady generation of contractions (periodicity of fluctuations of muscle contraction or a repeatability of impulse burst patterns in both afferent fibers and electronic interneuron analogues) has supported the concept of loop structures that are closed through proprioceptive connections, where the impulse flows are homeostated (Ashby, 1952; Pribram, 1963) by the properties of excitable elements. The periodicity of cyclic activity is determined by total delays in transmission of impulses in the pathways involved and transformation of excitation into the action potentials in the output of each neuron or receptive organ. The existence of loop circuits in the central brain regions is well established (Lukowiak, 1991; Buchanan and McPherson, 1995; Grillner, 1996). It is logical to assume that activity of the cerebral cortex and subcortical regions leaves specific traces of its influences directly on the executors of motor commands: discharges of motoneurons and the visible movements. We believe that the attributes of relationships between brain regions at the higher levels of the motor control system can be reflected in the parameters of tremors as well as in the properties of interactions of neuronal networks at

[*] SERGEY P. ROMANOV AND MICHAEL G. PCHELIN • Pavlov Institute of Physiology, Russian Academy of Sciences, St.-Petersburg, Russia.

the spinal level. We suggest that the cyclic activity in the closed structures that have various "extensions" should form a specific spectral frequency of fluctuations at the motor control system output and manifest in oscillations of jointly moving body parts. In fact, it is known that the significant spectral power of oscillations in the range of 8-12 Hz corresponds to the normal physiological tremor formed by the synchronised discharges of the motor units at segmental levels. In Parkinson's disease patients, the central motor control mechanisms are damaged, and the oscillations arise in a range of 4-6 Hz. Oscillations in the motor cortex that modulate descending corticospinal pathways might result in a similar pattern of modulation of muscle electromyographic activity and movement in the form of tremulous oscillations (McAuley et al., 1997). One strategy to study these mechanisms is to focus directly on the motor output, looking for any kinds of manifestations of the central rhythmic activity at the peripheral level. The purpose of the present work was to study the characteristics of the involuntary component of movement, in which the activities at different levels of the motor control system in both normal and various central pathological states are reflected.

MATERIAL AND METHODS

The analysis of movement parameters is commonly used to study motor control mechanisms in the normal central nervous system and to diagnose pathological states. We believe that for these purposes it is more suitable to investigate the parameters of muscle isometric efforts that reflect most fully the structure (voluntary and involuntary components) of motor commands in accordance with the principle of the final pathway or Sherrington's funnel. In the testing paradigm used in these studies, the subject was instructed to keep a constant level of voluntary effort by pressing his/her fingers on a force plate. Absolute values of the isometric efforts and the involuntary (tremor) force components of both hands were transformed into electrical signals that were entered enter into a computer program for subsequent statistical processing. Strain gauge signals were amplified, digitized with 12-bit resolution, and sampled at 100 Hz for analysis. Details of the device and methods used have been described previously (Romanov et al., 1996).

The study was performed on 43 normal volunteers and 345 patients with Parkinson's syndrome (from 14 to 88 years of age). This allowed us to reveal the features of the isometric-registered tremor shapes. Subjects had an opportunity to track the values of their own efforts separately for the right and left hand by viewing indicator marks on a monitor screen. In the standard test procedure we used two levels of effort: minimal (near 1 N) and near-maximal voluntary force that could be steadily maintained by the subject during 30s. The examinee sat in front of the monitor screen. By pressing the ends of the fingers of each hand on two separate force plates, he/she maintained the requested value of effort by keeping the marks on the screen scale at approximately stationary level, equal for the right and left hand during 30-60 s. The efforts were registered with an accuracy of 0.1 N in a range from 0 up to 50 N. The involuntary component of the effort was designated as physiological or pathological tremor (Romanov et al., 1997). For modulation of the activity of the motor control system we used the γ-test ('Jendrassic' manoeuvre) and the maintenance of isometric effort without a visual feedback. To reveal the peculiarities of strength oscillations the statistical and time series analyses, correlation analysis and fast Fourier transformations (standard procedure of "Statistics for Windows") were carried out.

The recording of tremor in the isometric regime shortens each trial by 30-60 s and does not require special preparation of patients. Standardisation of diagnostic methods enables evaluation and comparison of objective alterations of tremor parameters resulting from drug therapy, and allows for observation of patients during long treatment courses.

ISOMETRIC RECORDED TREMOR CHANGES IN THE PROCESS OF INCREASING VOLUNTARY EFFORT BY HEALTHY SUBJECTS

In this study tremor is interpreted as an involuntary component of motor control that reflects a functional state of the central nervous system. Since we introduce here a new procedure for registration of tremor, it is expedient to demonstrate at first how the isometric recorded tremor in healthy subjects changes during the process of increasing voluntary effort (Figure 1).

The top traces on Figure 1 demonstrate the relatively stable force control levels maintained by a healthy subject lightly touching the measuring plates by the tips of his fingers. In this case the subject develops a small voluntary effort in the fingers of each hand. Fluctuations of this effort correspond to rest tremor oscillations that are usually measured by accelerometers. Tensograms (measured by a strain gauge) of physiological tremor are characterized by low amplitudes and irregularity of "shapeless" fluctuations with occasional peaks arising from increased effort. With an increase of voluntary effort, the amplitude of low-frequency fluctuations increases. Under close to maximal voluntary effort it increases considerably. If the strength of pressing is more than 2.5 kg, the regular fluctuations of effort are appreciable in the 8 Hz range. These peaks may result from synchronization of motor unit activities subserved by normal, functioning mechanisms of loop regulation of the motoneuron activity through the proprioceptive connections at segmental levels. The amplitude of the isometric recorded tremor increases approximately linearly along with the increase in voluntary effort and in the physiological norm does not usually exceed 4-5 % of the voluntary isometric force. Note that the tensotremorogram records in Figure 1 display some differences between the shapes of isometric-registered tremors for the left and right hands.

FEATURES OF PATHOLOGICAL TREMORS

Pathological tremors (e.g., subcortical hyperkineses or essential and idiopathic tremors) are characterized primarily by an increase in the amplitude of the involuntary

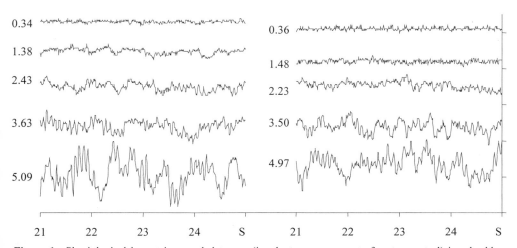

Figure 1. Physiological isometric recorded tremor (involuntary component of motor control) in a healthy subject. Four-second (starting from 21s) fragments of tensotremorograms of the left (on the left side) and right (on the right side) arms. Records were produced while maintaining varying levels (kg) of voluntary isometric effort (numerated at the left next to the curves) simultaneously by two hands. Distance between points on the right scale corresponds to 200 g.

component of an effort, and by the occurrence of characteristic periodic oscillations, (spindle shaped and complex patterns of the sine waveforms), which presumably can be closely related to specific impairments in some central brain regions. The typical oscillation patterns of isometric effort (involuntary component of motor control) from one control subject and three patients are demonstrated in Figure 2.

In healthy subjects (Figure 2A) the involuntary component (physiological tremor) of the voluntary effort displays relatively low amplitude fluctuations of irregular frequency and is represented here for comparison with the involuntary component of isometric efforts of patients with Parkinson's disease. In patients diagnosed with the tremor form of Parkinson's disease (Figure 2B), the isometric tremor represents irregular fluctuations of force, which considerably exceed the amplitude of physiological tremor and are more expressed in the right hand (patient's dominant hand). In patients with diagnosis of symptomatic tremor (Figure 2C), the involuntary fluctuations of effort have a slightly smaller amplitude but higher steady frequency, more pronounced in the left hand. In patients (Figure 2D) with a rigid akinetic form of idiopathic parkinsonism, the amplitude and pattern of effort fluctuations is similar to those of involuntary effort in healthy subjects, but unlike the normal physiological tremor, high-frequency components are not expressed.

Systematic study was carried out to determine how different muscle contraction strengths influence the distribution of frequency peaks in the power spectra. Fourier analysis appears to be the most informative in isometric registration of efforts. Figure 3 demonstrates the results of spectral analysis of the voluntary components of the effort of one healthy subject and three patients. It is important to note that in healthy subjects the distribution of spectral density is continuous in the range of 1-15 Hz, and the spectral power in this range grows linearly (in logarithmic scale) with the increase of strength of voluntary effort.

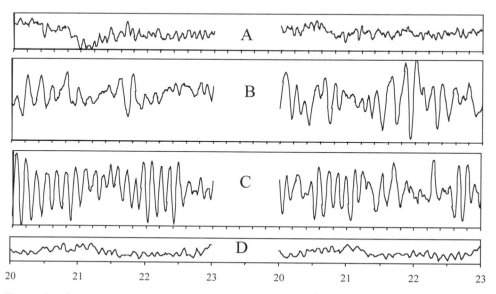

Figure 2. Comparison of various forms of tremor in four subjects (A-D). Tremors are registered simultaneously for the right and left arm at the voluntary effort of about 3 kg (see explanation in Table 1). The three-second fragments of isometric tremor for the left hand (on the left) and for the right hand (on the right) are shown for the same magnitude of effort.

Table 1. Comparison of averages of isometric strength and standard deviations of effort and tremor of the four subjects whose tremor records are illustrated in Figure 2

Subject (Fig. 2)	Clinical diagnosis	Age (years)	Hand	Isometric strength (kg, mean/std.dv.)	Tremor (std. dv.)
A (man)	Healthy	50	Left	3,27±0,09	±0,07
			Right	2,91±0,06	±0,04
B (man)	Parkinson's disease	70	L	3,10±0,15	±0,12
			R	3,11±0,13	±0,14
C (man)	Symptomatic	49	L	3,03±0,10	±0,17
			R	3,01±0,11	±0,19
D (woman)	Rigid-akinetic form	50	L	2,93±0,08	±0.05
			R	2,73±0,10	±0.05

When the voluntary effort increases to the level of near maximal contraction strength that could be maintained during 60s without fatigue, the power of spectral components increases in the range of 8-13 Hz where the power spectrum exceeds that in adjacent regions. The range 0-1 Hz represents an area of special interest, since it most likely characterizes voluntary effort control capabilities. In this range, increase of voluntary effort is accompanied by growth of spectral power and by expansion of the spectrum in the range of 0-3 Hz. Usually, increase of spectral power does not occur in the range of 4-7 Hz in comparison with adjacent parts of the spectrum (Figure 3A).

Entirely different spectral power distribution is seen in pathological states. If the range of 0.01-0.001 on the scale of spectral density is defined as the range of normal activity of the motor control system, then in the cases of the tremor form of Parkinson's disease and of symptomatic tremor (patients B and C, respectively), the spectral density considerably exceeds this level above the 4-7 Hz range with characteristic peaks at frequencies of 4.5 Hz and 6 Hz for patient B and at 6 Hz and 7.5 Hz for patient C. In the rigid akinetic form (Figure 3D) the spectral density is concentrated at the 0.001 level and limited by a highest frequency of 10 Hz. Spectral density in the range of 0-1 Hz remains approximately at the level of 1-0.1, a level in line with the force of voluntary effort. The basic spectral power at isometric effort regime registration is concentrated in the range of 0-15 Hz, though in some patients insignificant single peaks in the range of 13-20 Hz are found, as in patients B and C, or in the frequencies of 30 Hz and 40 Hz, e.g., as discussed by McAuley et al. (1997).

EFFECT OF DRUG THERAPY ON TREMOR PARAMETERS

Parkinson's disease provides an opportunity to study central mechanisms of the organization and execution of movements. To understand the factors involved in the shaping of the motor output we investigated parameters of isometric tremor while changing the voluntary effort in patients undergoing drug therapy. Although the pathological tremor of each patient has its own frequency, characteristic amplitude, and reactions to the tests, especially under different drug therapy protocols, there are still typical features common to all. An example of a typical result of data processing from one patient is presented in Figure 4. In this patient the tremor registered at the minimal voluntary efforts and had steady fluctuations (persistent tremor) characteristic of tremor at rest in diagnosis of Parkinson's disease.

Although only unilateral tremor was diagnosed clinically in this case, the simultaneous records of efforts from both hands revealed the presence of bilateral pathological tremor (Figure 4A), which was just more pronounced in the left hand. Increasing the voluntary

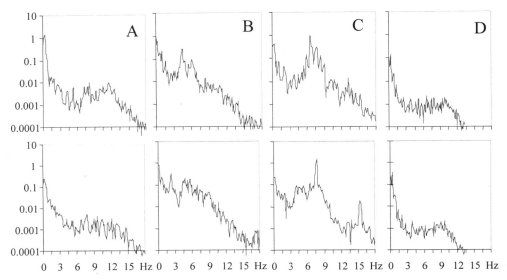

Figure 3. Estimation of the spectral density of records of isometric registered forces made during the maintenance of close to maximal voluntary effort. The data were assembled from fast Fourier transformation of 3000 points at sampling frequency of 100 Hz. The spectral density in the top row corresponds to the voluntary effort of the left hand and in the bottom row - to the right hand. A-D - subjects characterized in Table 1.

effort by approximately 8 times (see Table 2), the tremor essentially did not change with respect to its amplitude but became less regular. The changes in the movement control system required to increase voluntary effort were reflected in the spectral characteristics of the effort (Figure 4C-D). With an increase of voluntary effort, the amplitude of the spectral components in the range of 0-3 Hz had increased. The amplitude of peaks at 6 Hz frequency decreased, and the peaks themselves were slightly displaced towards the band of higher frequencies. The curves of the isometric tremor and distribution of the spectral density of voluntary effort obtained from the same patient against the background of drug therapy, are also shown in Figure 4. It is obvious that pathological tremor is present neither at rest, nor at the end of the maximal voluntary effort.

The spectra of voluntary efforts became similar for the right and left hand and decreased in the wide range of 0-15Hz. They became displaced upward with the increase of voluntary effort. (compare D to C in Figure 4). Unfortunately, this process is reversible and the patient is compelled to apply the supporting therapy continuously. When the patients achieve a satisfactory condition after the drug treatment or perhaps even the "normal" state of the motor control system, the basic power spectrum is concentrated in the range of 0-1 Hz, which, in our opinion, corresponds to voluntary control of efforts during the tracking. The higher frequency bands of 1-4 Hz reflect the states of centers of involuntary regulation, including supraspinal and exteroceptive pathways, such as tactile, visual, etc.

The method of registration of the involuntary component (tremor) of the effort during voluntary control of isometric tension of both hands in the task of tracking the feedback signals provides an opportunity to study the interaction of the central motor commands that can be directed to each hand or are separated at the spinal or segmental level. Cross-correlation analysis allowed us to reveal some peculiarities in the parameters of the motor output during different functional states of the motor control system (Figure 5). Here we represent the results of the data processing of the patient illustrated in Figure 4. These data

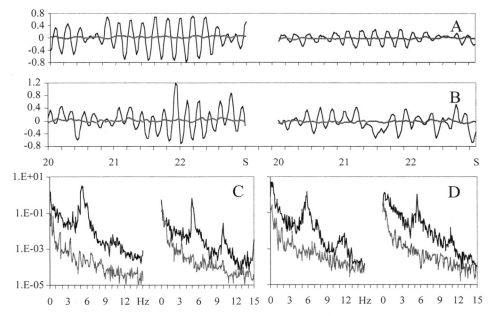

Figure 4. (A-B) Fluctuations of isometric registered involuntary (tremor) component of the hand efforts under minimal (A) and near maximal (B) voluntary force in Parkinson's disease patient (see parameters in Table 2). The three-second fragments of isometric tremor for the left hand (on the left) and for the right hand (on the right). The middle traces are the isometric recordings of tremor under the same testing conditions against the background of therapeutic action of L-Dopa. The scale of force (y-axes) is in kg. (C-D) Power spectral estimates of the above records accordingly for (A) and (B). The top traces reflect the state of patient withdrawal treatment. The lower traces are computed for this patient while under L-Dopa therapy.

Table 2. Comparison of average values of isometric strength and standard deviations of the effort and tremor in Parkinson's disease patient (age 53 years) before and during drug therapy

Date	Conditions of testing	L-dopa nakom	Hand	Isometric effort (kg, mean ± SD)	Tremor (SD)	Largest periodogram peaks (value/frequency)
16 March	Minimal voluntary effort	no	Left	0,36±0,147	±0,353	37,43/5,23; 29,58/5,13
			Right	0,27±0,061	±0,111	6,28/5,00; 2,57/5,17
6 April	-"-	yes	L	0,36±0,021	±0,028	0,35/0,60; 0,28/0,30
			R	0,37±0,027	±0,024	0,10/0,30; 0,90/0,67
16 March	Near maximal voluntary effort	no	L	2,38±0,182	±0,273	14,33/5,87; 11,25/5,73
			R	2,46±0,119	±0,191	9,66/5,40; 4,26/1,23
6 April	-"-	yes	L	2,12±0,08	±0,032	0,34/0,90; 0,25/0,60
			R	2,15±0,10	±0,034	0,23/0,53; 0,17/0,33

demonstrate the differences in isometric force and tremor between the pathological state (at the left side of Figure 5) and a condition close to physiologically normal (in the right part of Figure 5) reached by the patient under L-Dopa therapy. As described in the Material and Methods section, in these tests the patient sits in front of the monitor and, watching the marks on a screen, maintains the requested levels of identical right and left hand voluntary efforts.

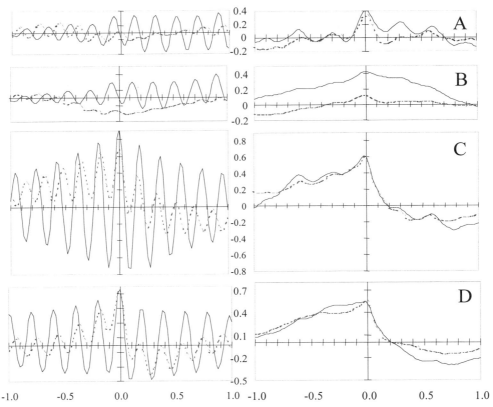

Figure 5. Cross-correlation of interrelations between left and right hand isometric force and tremor before (on the left) and during (on the right) drug therapy in a Parkinson's disease patient. (A) Cross-correlation functions between left (L) and right (R) hand tremor. (B) Cross-correlation functions between L and R isometric voluntary efforts. (C) Cross-correlation between tremor and voluntary effort of left hand. (D) The same as C but for the right hand. Thin lines correspond to the minimal voluntary efforts. Interrupted lines correspond to the near-maximal voluntary efforts. Y-axis is the coefficient cross-correlation. X-axis is the lag (s) between two temporary sequences or processes (variable on time series). The positive lag interval corresponds to a delay of the second process relative to the first.

The cross-correlation functions between involuntary components (tremor) in the right and left hand are shown in Figure 5A. Note that the cross-correlation functions are asymmetrical as related to the zero point of shift of the two processes analyzed. The maximal values of the cross-correlation functions and the displacement of these values on the time axis are the parameters of linear communication between the registered processes. In a pathological state of the central nervous system (the left side of Figure 5A), we see that the involuntary components (tremors) in the right and left hands correlate in the interval of minus 60-70 ms at the minimal voluntary efforts, when the right hand is ahead of the left hand up to an interval +1.5 s. Where the cross-correlation coefficient reaches the maximal of 0.3 (standard error 0.018) the right hand lags behind the left. If the level of the voluntary effort is increased then the values of the cross-correlation function are decreased and reflect the diminution of causal relationship between tremors in the right and left hands. The relationship between voluntary efforts of the two hands (Figure 5B) submits to the same principle, only the right hand outstrips the left hand by 70-80 ms. The cross-correlations between tremor and voluntary effort are of an oscillatory type in the form of sine waves

(Figure 5C,D). The association between these parameters is steadily maintained at an interval of ±1.5s of cross-correlation function for both hands, reflecting the large amplitude of oscillations of pathological tremor in the left hand. With the increase of voluntary efforts, the frequency of the tremor oscillations increases also, but the period of oscillations is decreased by 30 ms in the left hand, and by 18ms in the right hand during approximately equal to 200 ms duration periods of pathological oscillations in the right and left hand when the voluntary efforts are kept to a minimum.

If a Parkinson's disease patient is undergoing L-Dopa therapy, the above-described relationships become normalized (the right half of Figure 5) and are similar in shape to these relationships in healthy subjects, except for cross-correlation functions of the voluntary components between the left and right hand (Figure 5B). In the normal state of the motor control system, the latter are usually supported at the high level at an interval of larger than ±2 s cross-correlation function. It is interesting that for normal states of the motor control system the cross-correlation functions between the involuntary (tremor) component and voluntary effort (Figure 5C,D) remain similar to each other independently of the strength of voluntary effort for both hands. The higher correlation coefficients reflecting the functioning of the segmental mechanisms are registered in the interval up to ±20 ms from the zero point of the cross-correlation function. These are independent from the level of voluntary effort in the normal state of motor control, since they are equally high at different levels of voluntary strength applied.

THE EFFECT OF SENSORY FEEDBACK

The influence of a visual feedback also manifests in the shaping of motor commands and the regulation of effort. In these studies the subjects performed four consecutive tests. In the first test as well as prior to it the subject pressed on the force plate, keeping the voluntary effort under visual control and correcting the level of his effort by moving marks on a screen. In the second test, the marks remained motionless on the screen, and the subject tried to maintain the same degree of effort from memory. We suppose that in this case tactile sensitivity (touch) and the program control played basic roles in the maintenance of effort. In the third and fourth tests, the marks for the left and right hand were switched off separately, offering an opportunity to displace the marks either for the right or left hand. Changes in the visual feedback are reflected in changes of the distribution of spectral densities in the select ranges (Figure 6).

If the visual feedback for both hands is interrupted in healthy subjects, then the oscillations are increased only in the range of 0-1 Hz for the left and right hand. A typical result from one subject (man, age 60 years) is shown in Figure 6A. If one mark is excluded from the feedback then the spectral power is increased in the same 0-1 Hz region in the hand unilateral to the stop mark. Such changes of spectrum power were found in all healthy volunteers investigated.

The effect of visual feedback interruption on tremor in the symptomatic tremor patient (woman, age 36 years) was somewhat different. Results of the fast Fourier transformation of voluntary effort in this case are shown in Figure 6B. The pattern of distribution of the power spectra in the accentuated ranges in the right hand corresponded to that in the healthy subject. However, in the power spectra of the left hand effort inadequate reactions are visible during the fourth test: in the left hand the tremor is increased in the range of 6-7 Hz and 10-11 Hz; in the right hand the physiological tremor is increased in the range of 8-9 Hz. In this case, the spectral power of both voluntary and involuntary oscillations is higher especially in the range of 8-15 Hz relative to the level of voluntary strength when compared to the healthy subject. In a Parkinson's disease patient (man, age 57 years, bilateral tremor at 5.62 Hz, higher in the right hand) the involuntary component of oscillations had maximum spectral power in the range of 4-8 Hz (Figure 6C).

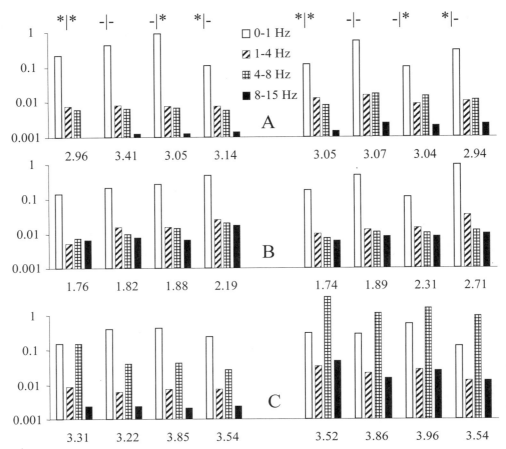

Figure 6. Distribution of average spectral density values of the voluntary component of effort at the frequency bands of 0-1 Hz, 1-4 Hz, 4-8 Hz and 8-15 Hz for the left hand (on the left) and right hand (on the right) for the healthy subject (A) and patients (B and C). Y axis is the spectral power. Under each group of columns there are average values of isometric force (kg) in each test during 30 s intervals of registration of effort. The symbolical designations at the top correspond to the conditions of the experiment. (*) pertains to mark controlled by the subject; (--) signifies that the external visual feedback is interrupted for the left or right hand.

Although the voluntary component of the effort reacted to the feedback signals in the spectral power range of 0-1 Hz, the involuntary component in the range of 4-8 Hz exceeded the spectral power of the voluntary component in the range of 0-1 Hz (significantly in the right hand). The testing showed that the components of a spectrum are subject to modulating influences from exteroceptive (visual and tactile, in our case) pathways, but these are insignificant in comparison with the modulating influence of the center of injury in subcortical structures (systems of extrapyramidal tracts) at segmental levels.

DISCUSSION

The main result of this study is a demonstration of the independence of the systems of voluntary control and program control for isometric effort, reflected in the changes of predicted spectral power of effort in different frequency ranges of tremor. The hierarchical structures of regulation of motoneuron activity via numerous parallel closed-loop pathways may be considered a homeostatic mechanism (using term offered by W.R. Ashby, 1952). In line with this idea we believe that changes of the spectral density in the 0-1 Hz range correspond to the transmission of signals in the circuits which are closed through exteroceptive (in our case visual or tactile feedback) inputs and motor output (feedback from the environment). The changes in the 1-4 Hz range may correspond to voluntary control (or a program) utilizing the internal long latent regulatory loops generating cyclic activity through the ascending spino-thalamo-cortical pathways. The activity generated at the segmental level is reflected in the range of 7-15 Hz. In healthy subjects, spectral power in the range of 4-7 Hz is suppressed. But an increase of the spectral power in this "forbidden" range of 4-7 Hz seems to be characteristic of impairments of the central servo-loop mechanisms for involuntary control of movements.

Obviously, the discussion about interactions between concrete brain structures based on the material presented can be only speculative. Nonetheless, we would like to conclude that the active control of isometric force is carried out by two independent systems. The first system carries out voluntary control by the strength of the force and intervenes in the process of strength maintenance by correcting the task performance. The second system provides a program control by generating descending flow to the segmental level from multiple closed-loop regulatory centers via basal ganglia and thalamic mechanisms controlling attention. The third, segmental level of control, represents the mechanism of homeostatic regulation of the motoneuron activity (Romanov, 1996) via proprioceptive pathways in accordance with the descending influences on it. Parkinson's disease, in which the pathways between substantia nigra and striatum are damaged, has been investigated to gain insight into the function of basal ganglia in movement control. Contreras-Vidal and Stelmach (1995) showed in a model that impairment of the interaction of thalamocortical pathways with segments of globus pallidum results in creation of local excitation centers that modulate initiation and performance of movements. Horak and Frank (1996) determined whether the effects of parkinsonism and the effects of L-Dopa treatment were similar on anticipatory, reactive, and tonic postural activity. They suggested that parkinsonism impairs the force control in three different postural systems: centrally initiated, which adjusts the body displacements associated with voluntary movement, peripherally triggered postural activity, which reacts to external perturbations, and background tonic system, which provides regulation of muscle stiffness and through which both centrally initiated and peripherally triggered postural activity must act.

The cross-correlation analysis between the efforts of both hands allows for interpretation of the interaction of exteroceptive inputs and their influence at the segmental level. We tested a group of healthy (right-handed) volunteers by registering the effort at a sampling frequency of 400 Hz. Cross-correlation estimates showed that the voluntary component outstripped the involuntary by 4-5 ms for both the right and the left hand under all conditions of manipulation with visual feedback. The voluntary and involuntary components of the right hand followed those of the left hand by 15 ms under conditions of tracking both marks, by 6-9 ms at the switched off visual feedback, and by 8-10 ms or 3-4 ms when the left or right mark was switched off, respectively. These data emphasize the asymmetry of contours of signal passage to the right and left hand and confirm the independence of regulatory mechanisms at the segmental level from the supraspinal systems of motor control.

However, the delay of activation of one hand relative to the other by 40-60 ms in Parkinson's disease patients may attest to the impairment of unilateral thalamocortical interactions with short-loop circuits through striatum and globus pallidus that can influence the thalamus and the midbrain directly. The delays of 8-10 ms are characteristic of the passage of signals through the thalamus during performance of programmed movements. In our case, the 8 ms delay is manifested in the interaction of both hands under conditions of the switched on and switched off visual feedback. It is likely that there are tactile inputs to the motor thalamus forming closed-loop circuits of involuntary regulation and correction of posture or effort in accordance with the hypothesis that sensory input may also be conveyed directly to the motor cortex via thalamocotrical pathways (Evarts et al., 1984; see also chapter by Mackel et al. in this volume).

When investigating principles of impulse flow transformation in the electronic model of a neural network, we demonstrated that the multiple-level neural network in layered brain structures can generate output based on anticipated meaning of the input signal (Romanov, 2000). The structure of the network executing the input signal extrapolations can be compared with the relationships of neurons in the cerebellar cortex. It has been suggested that described above relationships of neurons and the function executed by the cerebellar network are inherent in all brain structures, including neocortex. There is a regular redundancy in different levels of this network. The number of levels correlates with the accuracy of the prediction and the form of the input signal changes. The degree of excitation of the structure is automatically determined by the velocity and acceleration changes in the input signal.

We assume that the distortion of the normal functioning of movement organization is a consequence of structural changes in the central nervous system. Confirmation of this is provided in Figure 5, which shows how the parameters of the motor output are restored under the influence of L-Dopa treatment. Three basic structural mechanisms can be responsible for the impairments that cause tremor at the output of the motor control system: (1) decreased activity in the anticipatory structures, primarily compensating delays of signal conduction in descending and ascending pathways. This mechanism is subserved by the structures of feed-forward inhibition; (2) decreased activity in the structures responsible for lateral inhibition that results in synchronization of activity in the parallel projection pathways; (3) demyelination of tracts, which changes the delay of impulse spread.

The fluctuation patterns of visually observable free movements, on the basis of which the differential diagnostics of different forms of movement disorders are made (Bain et al., 1994; Deuschl et al., 1994), do not coincide with the objectively registered patterns of change of effort in the isometric mode. The comparison of results of the analysis of isometrical-registered effort with the established diagnoses of diseases of CNS, as listed by Marsden and Obeso, (1994), allows for presentation of a more detailed hierarchy of interactions of the subcortical structures in the organization of movements. Physiological and pathological tremors are usually classified into two broad groups: (1) those produced by oscillation in sensorimotor loops, so-called mechanical-reflex tremors, and (2) those produced by the oscillatory properties of central neuronal networks. Elble (1996) provided a contemporary perspective of tremor pathophysiology but acknowledged that no form of tremor is understood completely, because the origin of oscillations in most forms of tremor is undefined, and in many instances the underlying pathology is unknown. From our experiments it is possible to assume that the regulation of maintenance of stationary effort comes through the closed loops of automatic control (highest levels of the reflex - thalamus and basal ganglia) without participation of voluntary (cerebral cortex?) motor control, and the spectral power estimates of oscillations of involuntary effort in different frequency ranges may reflect the state of these brain structures.

CONCLUSIONS

Analysis of the spectral density of pathological tremors and of cross-correlation functions of bilateral relationships between tremor and voluntary effort reveals manifestations of the central rhythmic activity at the peripheral level in three different frequency bands. This information can be useful in developing new insights about the interactions of basal ganglia and thalamic structures in normal and pathological states, thus advancing our knowledge of the role of different levels of CNS in the organization of behavior.

REFERENCES

Asanuma, H., 1995, Neural basis of motor learning in mammals, in: *Abstr Fourth IBRO World Congress of Neurosci.,* Rapid Comm. Oxford.

Ashby, W.R., 1952, *Design for a Brain*, Chapman and Hall, London.

Bain, P.G., Findley, L.J., Thompson, P.D., Gresty, M.A., Rothwell, J.C., Harding, A.E., and Marsden, C.D., 1994, A study of hereditary essential tremor, *Brain,* 117:805.

Buchanan, J.T., and McPherson, D.R., 1995, The neuronal network for locomotion in the lamprey spinal cord: evidence for the involvement of commissural interneurons, *J. Physiol.* 89:221.

Contreras-Vidal, J.L., Stelmach, G.E., 1995, A neural model of basal ganglia-thalamocortical relations in normal and parkinsonian movement, *Bio. Cybernet.* 73:467.

Deuschl, G., Toro, C., Valls-Sole, J., Zeffiro, T., Zee, D.S., and Hallett, M., 1994, Symptomatic and essential palatal tremor. 1. Clinical, physiological and MRI analysis, *Brain,* 117:775.

Elble, R.J., 1996, Central Mechanisms of tremor, *J. Clin. Neurophysiol.* 13:133.

Evarts, E.V., and Thach, W.T., 1969, Motor mechanisms of the CNS: cerebro-cerebellar interrelations, *Ann. Rev. Physiol.* 31:451.

Evarts, E.V., Shunoda, Y., and Wise, S.P., 1984, *Neurophysiological Approaches to Higher Brain Functions,* John Wiley & Sons, New York.

Grillner, S., 1996, Neural networks for vertebrate locomotion, *Scient. Amer.* 274:48.

Horak, F., and Frank, J., 1996, Three separate postural systems affected in parkinsonism, in: *Motor Control VII. Proceedings of the VIIth International Symposium on Motor Control, held June 21-25, 1993 in Borovets, Bulgaria,* D.G., Stuart, ed., Motor Control Press, Tucson, AZ.

Ito, M., 1984, *The Cerebellum and Neural Control*, Raven Press, New York.

Lukowiak, K., 1991, Central pattern generators: some principles learned from invertebrate model systems, *J. Physiol.* (Fr.), 85:63.

Marsden, C.D., and Obeso, J.A., 1994, The functions of the basal ganglia and the paradox of stereotaxic surgery in Parkinson's disease, *Brain,* 117:877.

McAuley, J.H., Rothwell, J.C., and Marsden, C.D., 1997, Frequency peaks of tremor, muscle vibration and electromyographic activity at 10 Hz, 20 Hz and 40 Hz during human finger muscle contraction may reflect rhythmicities of central neural firing, *J. Exp. Brain Res.* 114:525.

Pribram, K., 1963, Control systems and behavior, in: *Brain and Behavior, Vol. II: The Internal Environment and Alimentary Behavior,* M.A.B. Brazier, ed., Am. Ins. Biol. Sci., Washington, D. C.

Romanov, S.P., 1974, Modelling of spinal level mechanisms of muscle contraction control, *Physiol. J. USSR* (In Russian), 60:1508.

Romanov, S.P., 1976, Study on a model of a role of Renshaw cells in regulation of motoneuron discharges, *Physiol. J. of USSR* (In Russian), 62:528.

Romanov, S.P., 1996, The inhibited feedback in regulation of motoneuron discharges, *Sechenov Physiol. J.* (In Russian). 82:33.

Romanov, S.P., 2000, The conceptual approaches to revealing a function of the structural organization of a neuron network, *J. HNA* (In Russian), 2:320.

Romanov, S.P., Yakimovsky, A.F., and Pchelin, M.G., 1996, A tensometry technique of quantitative estimation of tremor, *Sechenov Physiol. J.* (In Russian), 82:118.

Romanov, S.P., Pchelin, M.G., and Yakimovsky, A.F., 1997, Characteristics of isometrically recorded tremor in lesions of extrapyramidal system, *Russian J. Physiol.* 83:133.

PART VII. MECHANISMS AND EFFICIENCY OF NOVEL TREATMENTS FOR MOVEMENT DISORDERS

HOW ARE WE INHIBITING FUNCTIONAL TARGETS WITH HIGH FREQUENCY STIMULATION?

ALIM L. BENABID, ZHONGGE NI, STEPHAN CHABARDES,
ABDELHAMID BENAZZOUZ AND PIERRE POLLAK[*]

INTRODUCTION

It is commonly accepted that high frequency stimulation (HFS) inhibits neuronal somatic structures. HFS has been used to treat a variety of conditions, but its primary use has been to treat movement disorders by targeting different basal ganglia and thalamic nuclei. The basal ganglia circuitry is didactically described by the parallel pathway model (Alexander and Crutcher, 1990; Albin et al., 1989; DeLong, 1990). Three principal areas of the basal ganglia are targeted for therapeutic purposes: the thalamic ventralis intermedialis nucleus (Vim), the internal pallidum (GPi), and the subthalamic nucleus (STN). The substantia nigra lies below these three nuclei within the stereotactic coordinate system. Degeneration of the pars compacta of substantia nigra (SNc) is the cause of Parkinson's Disease, while the substantia nigra reticulata (SNr), which is more anterior and dorsal to it, is considered an equivalent of GPi in the model, because these two nuclei receive common afferents from both the STN and direct pathway, and the efferent output of both nuclei is GABAergic.

The statement: « High frequency stimulation inhibits neuronal somatic structures » raises two questions: Is this really an inhibition? If yes, what is its mechanism?

WHAT ARE THE FACTS?

1. High frequency stimulation (HFS) above 50 Hz (ideally between 130 and 185 Hz) mimics the effect of ablative procedures considered to be inhibitory in neuronal soma and dendrites (from here on referred to as neuronal somatic structures) in such nuclei as:

---

[*] ALIM L. BENABID • Department of Neurosurgery, Joseph Fourier University, Grenoble 38043, France. ZHONGGE NI • INSERM U318, Joseph Fourier University, Grenoble 38043, France. STEPHAN CHABARDES • Department of Neurosurgery, Joseph Fourier University, Grenoble 38043, France. ABDELHAMID BENAZZOUZ • INSERM U318, Joseph Fourier University, Grenoble 38043, France. P IERRE POLLAK • Department of Neurology, Joseph Fourier University, Grenoble 38043, France.

Thalamus: Vim, centromedian, and parafascicularis nuclei (CM-Pf); Basal Ganglia: GPi and STN; and Hypothalamus: VMH.

In contrast, HFS, as well as low frequency stimulation (LFS), excite axons in fiber bundles such as: optic track (flashes), pyramidal track (contractions), lemniscus medialis (paresthesias), and third nerve (ocular deviations).

2. HFS of Vim mimics thalamotomy, which is effective on tremor (Guiot et al., 1967, 1968; Ohye et al., 1982; Nagaseki et al., 1986; Ohye, 1986; Narabayashi, 1989; Benabid et al., 1989; 1991, 1996, 1999; Goldman et al., 1992). Observations during surgery in an awake patient with tremor (rest tremor in Parkinson's Disease or essential tremor) show that at the onset of the stimulation the tremor stops almost immediately. This can be demonstrated using accelerometric recording of the tremor, which shows that the delay between the onset of stimulation and the total tremor arrest is very short. Actually it can even be shortened by increasing the voltage of the stimulation. Conversely, when the stimulation is stopped, the after effect lasts only a few seconds, but it can also be prolonged by increasing the intensity of stimulation. Microrecordings of single cells in the thalamus have identified typical kinesthetic cells that fire in bursts synchronous to tremor (Albe-Fessard et al., 1961; Raeva et al., 1986; Ohye et al., 1989). Videorecordings of patients under HFS demonstrate how arrest of the tremor functionally improves the patients' lives, allowing them to write again, to handle a glass filled with water without spilling it, etc. This improvement is as immediately reversible as the arrest of tremor at the onset of stimulation.

3. The group in Lille (Blond et al., 1992) have reported that HFS of CM-Pf alleviated tremor and levodopa-induced dyskinesias (LIDS), effects that are similar to HFS effects in Vim and GPi, respectively. However, the group in Grenoble (Benabid et al., 1991) had not seen such effects. A comparative study of the positions of electrodes, performed by Caparros-Lefebvre et al. (1999), revealed that the positions of the clusters of electrodes placed by the two teams differed significantly, by as much as a few millimeters. This explained the reported differences in HFS effects. The comparison of the positions of the two clusters of electrodes with the atlas of Schaltenbrand and Bailey (1959) showed that the electrodes of the Grenoble group were more closely situated in Vim than those of the group in Lille, which were instead situated at the base of the CM-Pf or at the top of the prelemniscal radiations.

4. It has been well documented that HFS in GPi alleviates dyskinesias. Therefore, the interest in the pallidal target is mainly based on its efficiency on LIDs. HFS in GPi also has an effect on the other symptoms of Parkinson's Disease, but the conclusions from a comparative study of results of HFS in STN and GPi is not in favour of GPi. We have reported that in two matched groups of young onset patients, in the medication "off" condition, the improvement during STN stimulation is higher than 60 % on all items, while it is lower than 60 % in the GPi group (Krack et al., 1998).

5. HFS of STN reduces the triad of cardinal symptoms of Parkinson's disease. The improvement is highly significant, including a 60 % improvement for rigidity, 60 % improvement for akinesia, and 80 % improvement for tremor, when the benefits at 12 months in the "off-drug on-stim" condition are compared to the pre-operative "off-drug" condition (Limousin et al., 1995a, 1995b, 1998.) In STN, it is also possible to record single units firing in bursts synchronous to the tremor similar to those recorded in Vim, CM-Pf, and also in GPi. When patients in the pre-operative stage are unable to take care of themselves, for example, when they are unable to stand up from a chair or to walk without aid due to akinesia, rigidity, or "off" dystonia, the post-operative improvement is striking. Most of the patients under stimulation recover stable neurological status, which can be quantitatively compared to the level reached even for a few minutes at the best "on" condition during the pre-operative stage. HFS of STN does induce dyskinesias, but these are not actually complications because they are induced at higher voltages than those needed to obtain the therapeutic improvements. Videorecordings demonstrate that in patients who are "off-drug" the stimulation at a few volts above the threshold for obtaining

the improvement of the Parkinsonian symptoms induces huge abnormal movements, which are choreoballic and closely resemble LIDs or, at their extreme, ballistic movements (Limousin et al., 1996). This has suggested to us that there is a continuum, both pathophysiological and clinical, between the chorea, dystonic, and ballic dyskinesias induced by levodopa and the ballism induced by inhibition of STN. Actually, the occurrence of STN-induced dyskinesias is strengthened by simultaneous administration of dopaminergic drugs, mainly levodopa. Fortunately, the strong clinical improvement achieved by STN stimulation allows for significant reduction in drug dosage - an average decrease of 70 % at one year - while 30 % of these patients become drug free. It is interesting to note that the inducibility of dyskinesias by HFS in STN, as well as the tendency to have LIDs, decrease with time when the patients are submitted to continuous STN stimulation. This resembles what is observed in patients with LIDs when treated by subcontinuous infusion of apomorphine or lisuride (Colzi et al., 1998). This effect could be interpreted as desensitization of the post synaptic striatal receptors, which are no longer submitted to the oscillatory changes of the systemic concentration of levodopa and can thus return to their normal sensitivity characteristics. This is also consistent with the frequent observation that ballism induced by lesions in the STN tends to decrease with time. Therefore, it seems that the indirect improvement of LIDs by continuous STN-HFS is most likely due to the decrease in medication and not to a direct specific effect of HFS, as is the case in HFS of GPi (Krack et al., 1997).

6. HFS in the ventromedial hypothalamus (VMH) induces hyperphagia. The experiments in rats implanted with chronic electrodes in the VMH show that after the animals have been fed and additional food is presented, they will consume just a few more grams if there is no stimulation in the VMH. On the contrary, if in the same situation the stimulation of VMH is turned on at a high frequency, the behaviour of the rats changes, and even though they have been previously fed, they may take in a significant amount of food. Stimulation at a low frequency induces the exact opposite effect. This suggests that the stimulation is mimicking the excitatory-like properties in these nuclei when it is delivered at a low frequency, and mimicking the effects of lesions when it is delivered at a high frequency.

WHAT IS THE MECHANISM?

At the moment there is no clear understanding of the mechanism of action of HFS. Moreover, it has not even been demonstrated that this mechanism is unique and should be the same in different targets. It is certain, however, that the high frequency is the key factor responsible for effects.

The intensity-frequency relationship curve (Limousin et al, 1995a) shows that the threshold intensity drops dramatically at 100 Hz and remains quite stable up to 2500 Hz, which means that making stimulators with variable frequency is no longer necessary. It is interesting to note that this curve is even quantitatively superimposable on the threshold intensity-frequency relationship curve, which was obtained from axons of the motor nerve of crab leg muscles, and has since been presented frequently in handbooks of neurophysiology (Coppée, 1934). This raises the question of the neural elements that are affected by stimulation. All current data available are not helping to prove or disprove any of the hypotheses.

Based on all experimental and clinical observations, one can come up with a list of properties/criteria that should be met by any putative mechanism of HFS effect to be seriously considered. We could define these as has been done for the list of criteria for putative neurotransmitters and putative receptors. The profile of a putative mechanism must: (P1) - mimic effects of ablative procedures (except in fiber bundles); (P2) - be applicable to all structures inhibited by HFS; (P3) - be reversible and follow time course of

HFS; (P4) - depend on frequency; (P5) - excite fibers at all frequencies; (P6) - be coherent with at least some of the known physiology. The mechanism could also involve an additional structure instead of being intrinsic to the stimulated nucleus, for example, and extra nucleus, fibers of passage, extra network, etc.

A - For instance, in the case of STN stimulation an extra nucleus could be GPe (GP in rats). This mechanism could be coherent with (P1) (equals ablation), (P3) (equals reversible) and (P6) (physiology). In the model (Albin et al.,1989; Alexander and Crutcher, 1990; DeLong, 1990) it is shown that GPe not only projects its GABAergic inhibitory output onto STN, but also onto the GPi and SNr, which already receive an inhibitory influence from the striatum. To investigate this possibility, we set up a rat experiment in which we destroyed GP. According to the simplified approach of the model, this should be responsible for increasing STN activity, a condition normally observed in parkinsonian patients. The results showed that HFS of STN induced a strong but transitory inhibition of SNr cells and, at the same time, increased the activity of GP cells. This excitation of GP could be due to the backfiring of the GP to STN axons or could result from a cascade of events that finally led to excitation of GP (Benazzouz et al., 1995). Therefore, with GP being a potent inhibitory GABAergic structure, one might wonder whether the observed SNr inhibition induced by STN stimulation could not in fact be due to the increased GABAergic output of GP onto SNr. We therefore repeated the experiment with kainic acid lesions of GP, but this did not prevent the induction of inhibition in SNr cells when HFS was applied to STN. The results were still compatible with the hypothesis that the glutamatergic output of STN was shut down directly by the HFS of STN rather than indirectly by backfiring that shut off the GABAergic output of GP on the entopeduncular nucleus (EP, rat equivalent of human GPi) and SNr. The conclusion from the above data that the GPe does not play the main role in the induction of the pathophysiological mechanism leading to Parkinson's disease status, which was implied from the model, raises the question of the source of the high frequency firing rate observed in this disease. One must remember that the GABAergic output of GP onto STN, although well documented, is much less important than the strong glutamatergic input coming onto STN from the cortex as well as from the CM-Pf (Canteras et al., 1990; Fujimoto et al., 1993; Feger et al., 1994). The inhibition induced in SNr by STN stimulation after lesion of GP is comparable to that observed with the GP intact. However, one may note that the duration and intensity of the strong inhibitory periods after STN stimulation are slightly weaker when GP is missing as compared to when GP is intact. Therefore, backfiring of GP by STN stimulation could to some extent contribute to the inhibition of SNr but by no means can it be the major mechanism. For the latter, HFS induced shutdown of the glutamatergic output of STN is probably the best candidate at this point.

B - Fibers of passage could be excited by the stimulation, but this is not applicable because it goes against (P1): STN lesion can not excite fibres and thus be responsible for the effect.

C - The effect of STN stimulation could also be due to the increased production of dopamine by excited SNc. It has been observed in normal rats that STN HFS would induce a post-discharge in SNc cells, which happen to be excited. The mechanism of the SNc excitation is not understood. It could be due to backfiring, since SNc is known to project to STN. It could also be due to a disinhibition of SNc by the shutdown of STN hyperactivity, which in turn would decrease the activity of EP and SNr, which are GABAergic and project at least in part to SNc (Ni and Benazzouz, submitted). This cannot actually be considered a good candidate for the mechanism, as it runs counter to (P6): common knowledge in physiology. In advanced parkinsonian patients, SNc is strongly depleted, and it is unlikely that even its increased activity could reinduce a normal and sustained level of dopamine (DA) production. This hypothesis of increased production of DA also runs counter to (P1), since the lesion of STN could not by itself induce the hyperactivity of SNc cells.

D - An extra network? The jamming of a feedback loop by "zero filling" gaps between bursts would make the bursting or oscillatory (such as in tremor) message meaningless. If one considers the network involved in Vim stimulation, one may schematically see Vim as receiving proprioceptive inputs from the muscles through the posterior horn and the columns of the spinal cord. Vim itself projects to the motor cortex, which, in turn, excites the muscles via the pyramidal tract and lower motor neurons. This would constitute a feedback loop and could be one of the mechanisms of induction of tremor if the gain is not properly set in the loop. In this case, the contraction of the muscles would induce volleys of proprioceptive inputs, reaching the cortex through Vim and resulting in contraction of a muscle and so on. Such oscillatory activity could be the basis of tremor. Kinesthetic cells with bursting activity have been recorded in Vim (Ohye et al., 1989; see also chapter by Ohye and Shibasaki in this volume). HFS corresponds to the adjunction of continuous activity, which would superimpose on the bursting pattern observed in Vim and would thus fill up the gaps between these bursts. This would lead to a non-bursty pattern of firing, where periodic or pseudo periodic patterns would not be recognized, and therefore the system would not get engaged in the periodic activity leading to tremor. However, this jamming hypothesis needs an oscillatory background, which is against (P2): a principle requiring that the mechanism could be applied in all cases. This might be an acceptable hypothesis for tremor but burst patterns could also be erased by blank noise and lead to various cut-offs depending on the system considered.

So the next questions are, if this is an intrastructure mechanism: a) What leads to inhibition? and b) What prevents neurons from firing?

Whatever the internal mechanism is, it should lead to an apparent inhibition of the neuronal cellular structures in the nucleus. A progressive lesion induced by chronic HFS has been suggested as one of the mechanisms leading to the inhibition of the stimulated structure. However, this is against the (P3) principle (reversibility). If HFS induces progressive lesions, then the effect would not disappear when the stimulation is turned off. In fact, even twelve years after implantation, the arrest of stimulation in Vim consistently leads to recurrence of the tremor.

Another mechanism could be preferential activation of large myelinated GABAergic terminals. This has been proposed by Holsheimer et al. (2000) and Dostrovsky et al. (2000). This hypothesis is compatible with (P2) (all structures), and it seems to work in GPi, as well as in STN, which receives numerous GABAergic terminals from GP projections. From the recent results of the group in Toronto, it does not look like it is working perfectly in Vim.

Is there a structure that is excited by LFS and inhibited by HFS? This is typically the case in the VMH where, as stated above, HFS induces a hyperphagic behaviour but LFS promotes decreased food intake in the fed rat when additional food is presented. In our opinion this situation is the best example of the opposite effects of HFS and LFS.

Are the delayed stimulation effects compatible with an acute inhibition? It is known that dystonia is improved progressively both by pallidotomy and GPi HFS, but the time course of these effects is not comparable to what is observed in Parkinson's Disease or essential tremor. The longer time course of the former might be due to the induction of a "second step" phenomenon (*plasticity?*). There are now numerous reports of the progressive and spectacular improvement of dystonic children in whom inhibition of the GPi results in progressive disappearance of uncontrollable dystonic movements, thus enabling them to resume normal childhood activities.

CONCLUSIONS

In conclusion, "stimulation means excitation" is no longer a dogma, since it is not always (or never?) true at high frequency in cell clusters. "HFS is inhibitory" still needs direct evidence to become a theorem. But since it is verified in all tested situations, so far it can be considered an axiom.

ACKNOWLEDGMENTS: The authors' work discussed here was performed in cooperation with P. Krack, D.M. Gao, S. Breit, P. Limousin-Dowsey, E. Moro, E. Caputo, V. Fraix, C. Ardouin, A. Koudsie, D. Hoffmann, A. Ashraf, M. Dias, R. Villavicencio, B. Piallat, K. Bressand, M. Dematteis, R. Bouali-Benazzouz, L. Vercueil.

REFERENCES

Albe-Fessard, D., Arfel, G., Guiot, G., Hardy, J., Vourc'h, G., Hertzog, E., and Alèonard, P., 1961, Identification et délimitation précise de certaines structures sous-corticales de l'homme par l'électrophysiologie. Son interêt dans la chirurgie stéréotaxique des dyskinesies. C. R.. Acad. Sci. Paris 253:2412.

Alexander, G.E., and Crutcher, M.D., 1990, Functional architecture of the basal ganglia circuits: neural substrates of parallel processing, Trends Neurosci. 13:266.

Albin, R.L., Young, A.B., and Penney, J.B., 1989, The functional anatomy of basal ganglia disorders, Trends Neurosci. 12:366.

Benabid, A.L., Pollak, P., Hommel, M., Gaio, J.M., de Rougemont, J., and Perret, J., 1989, Traitement du tremblement parkinsonien par stimulation chronique du noyau ventral intermédiaire du thalamus, Rev. Neurol. (Paris) 145:320.

Benabid, A.L., Pollak, P., Gervason, C., Hoffmann, D., Gao, D.M., Hommel, M., Perret, J.E., and de Rougemont, J., 1991, Long-term suppression of tremor by chronic stimulation of the ventral intermediate thalamic nucleus, Lancet, 337:403.

Benabid, A.L., Pollak, P., Gao, D.M., Hoffmann, D., Limousin, P., Gay, E., Payen, I., and Benazzouz, A., 1996, Long-term suppression of tremor by chronic electrical stimulation of the ventralis intermedius nucleus of the thalamus as a treatment of movement disorders, J. Neurosurg. 84:203.

Benabid, A.L., Benazzouz, A., Hoffmann, D., Gao, D.M., Limousin, P., Koudsie. A., Krack P., and Pollak P., 1999, Chronic electrical stimulation of the ventralis intermedius nucleus of the thalamus and of other nuclei as a treatment for Parkinson's disease, Techn. in Neurosurg. 5:5.

Benazzouz, A., Piallat, B., Pollak, P., and Benabid, A.L., 1995, Responses of substantial nigra reticulata and globus pallidus complex to high frequency stimulation of the subthalamic nucleus in rats: electrophysiological data, Neurosci. Lett. 189:77.

Blond, S., Caparros-Lefebvre, D., Parker, F., Assaker, R., Petit, H., Guieu, J.D., Christiaens, J.L., 1992, Control of tremor and involuntary movement disorders by chronic stereotactic stimulation of the ventral intermediate thalamic nucleus, J. Neurosurg. 77:62.

Canteras, N.S., Shammah-Lagnado, S.J., Silva, B.A., and Ricardo, J.A., 1990, Afferent connections of the subthalamic nucleus: a combined retrograde and anterograde horseradish peroxidase study in the rat, Brain Res. 513:43.

Caparros-Lefebvre, D., Blond, S., Feltin, M.P., Pollak, P., and Benabid, A.L. , 1999, Improvement of levodopa induced dyskinesias by thalamic deep brain stimulation is related to slight variation in electrode placement: possible involvement of the centre median and parafascicularis complex, J. Neurol. Neurosurg. Psychiat. 67:306.

Colzi, A., Turner, K., and Lees, A.J., 1998, Continuous subcutaneous waking day apomorphine in the long-term treatment of the levodopa induced interdose dyskinesias in Parkinson's disease, J. Neurol. Neurosurg. Psychiat. 64:573.

Coppée, G., 1934, La pararésonance dans l'excitation par les courants sinusoïdaux, Arch. Inst. Physiol. Biochim. Biophys. 40:1.

DeLong, M.R., 1990, Primate models of movement disorders of basal ganglia origin, Trends Neurosci. 13:281.

Dostrovsky, J.O., Levy, R., Wu, J.P., Hutchison, W.D., Tasker, R.R., and Loazano, A.M., 2000, Micro-stimulation-induced inhibition of neuronal firing in human globus pallidus, J. Neurophysiol. 84, in press.

Feger, J., Bevan, M., and Crossman, A.R., 1994, The projections from the parafascicular thalamic nucleus to the subthalamic nucleus and the striatum arise from separate neuronal populations, a comparison with the corticostriatal and corticosubthalamic efferents in a retrograde fluorescent double-labeling study, Neuroscience, 60:125.

Fujimoto, K., and Kita, H., 1993, Response characteristics of subthalamic neurons to the stimulation of the sensorimotor cortex in the rat, Brain Res. 609:185.

Goldman, M.S., Ahlskog, J.E., and Kelly, P.J., 1992, The symptomatic and functional outcome of stereotactic thalamotomy for medically intractable essential tremor, *J. Neurosurg.* 76:924.

Guiot, G., Derome P., and Trigo, J.C., 1967, Le tremblement d'attitude. Indication la meilleure de la chirurgie stéréotaxique, *Presse Méd.* 75:2513.

Guiot, G., Arfel, G., and Derôme, P., 1968, La chirurgie stéréotaxique des tremblements de repos et d'attitude. *Gaz. Méd. France*, 75:4029.

Holsheimer, J., Dijkstra, E.A., Demeulemester, H., and Nuttin, B., 2000, Chronaxie calculated from current-duration and voltage duration data, *J. Neurosci. Meth.* 97:45.

Krack, P., Limousin, P., Benabid, A.L., and Pollak, P., 1997 Chronic stimulation of the subthalamic nucleus improves levodopa-induced dyskinesias in Parkinson's disease, *Lancet*, 350:1676.

Krack, P., Pollak, P., Limousin, P., Hoffmann, D., Xie, J., Benazzouz, A., and Benabid, A.L., 1998, Subthalamic nucleus and internal pallidal stimulation in young onset Parkinson's disease, *Brain*, 121:45.

Limousin, P., Pollak, P., Benazzouz, A., Hoffmann, D., Le Bas, J.F., Broussolle, E., Perret, J.E., Benabid, A.L., 1995a, Effect on parkinsonian signs and symptoms of bilateral subthalamic nucleus stimulation, *Lancet*, 345:91.

Limousin, P., Pollak, P., Benazzouz, A., Hoffmann, D., Broussolle, E., Perret, J.E., Benabid, A.L., 1995, Bilateral subthalamic nucleus stimulation for severe Parkinson's disease, *Mov. Dis.* 10:672.

Limousin, P., Pollak, P., Hoffmann, D., Benazzouz, A., Perret, J.E., and Benabid, A.L., 1996, Abnormal involuntary movements induced by subthalamic nucleus stimulation in parkinsonian patients, *Mov. Dis.* 11:231.

Limousin, P., Krack, P., Pollak, P., Benazzouz, A., Ardouin, D., Hoffmann, D., and Benabid, A.L., 1998, Electrical stimulation of the subthalamic nucleus in advanced Parkinson's disease, *N. Engl. J. Med.* 339:1105.

Nagaseki, Y., Shibazaki, T., Hirai, T., Kawashima, Y., Hirato, M., Wada, H., Miyazaki, M., and Ohye, C., 1986, Long term follow-up of selective Vim-thalamotomy, *J. Neurosurg.* 65:296.

Narabayashi, H., 1989, Stereotaxic Vim thalamotomy for treatment of tremor, *Eur. Neurol.* 29:29.

Ohye, C., Hirai, T., Miyazaki, M., Shibazaki, T., Nakajima, H., 1982, VIM thalamotomy for the treatment of various kinds of tremor, *Appl. Neurophysiol.* 45:275.

Ohye, C., 1986, Rôle des noyaux thalamiques dans l'hypertonie et le tremblement de la maladie de Parkinson. *Rev. Neurol. (Paris)* 142:362.

Ohye, C., Shibazaki, T., Hirai, T., Wada, H., Hirato, M., and Kawashima, Y., 1989, Further physiological observations on the ventralis intermedius neurons in the human thalamus, *J. Neurophysiol.* 61:488.

Raeva, S., 1986, Localization in human thalamus of units triggered during verbal commands, voluntary movements and tremor, *Electroencephal. Clin. Neurophysiol.* 63:160.

Shaltenbrand, G., and Bailey, P., 1959, *Introduction to Stereotaxic Operations with an Atlas of the Human Brain*, Thieme, Stuttgart.

IS THERE A SINGLE BEST SURGICAL PROCEDURE FOR THE ALLEVIATION OF PARKINSON'S DISEASE?

TIPU AZIZ, SIMON PARKIN, CAROLE JOINT, RALPH GREGORY, PETER BAIN, JOHN F. STEIN, AND RICHARD SCOTT[*]

INTRODUCTION

The last few years have seen a renewal of interest in stereotactic surgery as a result of both a deeper understanding of the neural mechanisms underlying parkinsonism, based largely on studies in the MPTP-exposed primate model (Mitchell et al., 1989), and a disenchantment with the long term effects of L-DOPA therapy. Since the publication by Laitinen et al. (1992) of the clinical effects of a postero-ventral pallidotomy (PVP), there has been a massive resurgence of interest in this procedure. Despite the wealth of publications, few were of a quality to assess the suitability of the procedure (Lozano, 1995; Vitek et al., 1997), even from a basic neurological viewpoint (Shima, 1994; Iacono et al., 1995), and as a result, most groups could only look to their own results for outcome assessment. Furthermore, most publications were reporting the outcomes of unilateral procedures (Dogali et al., 1995; Baron et al., 1996; Lang et al., 1997; Subramanian et al., 1997; Alterman and Kelly, 1998; Giller et al., 1998; Ondo et al., 1998; Samuel et al., 1998; Shannon et al., 1998; Uitti et al., 1998; Dalvi et al., 1999; De Bie et al. 1999; Jankovic et al., 1999) despite the need for bilateral symptomatic benefit. Bilateral pallidotomy was generally abandoned by most groups due to concern over unacceptable side effects, particularly cognitive deficit (Ghika et al., 1999). We remained optimistic about the future of lesional procedures until such time as our series could guide us to a rational management of the condition.

The development of neurostimulation as a therapy has led many groups to explore bilateral pallidal and subthalamic deep brain stimulation (DBS). Subthalamic stimulation

---

[*] TIPU AZIZ • Department of Neurosurgery, Radcliffe Infirmary, Oxford, OX2 6HE, U.K.; and Imperial College of Medicine, Charing Cross Hospital, London, U.K. SIMON PARKIN • Department of Neurology, Radcliffe Infirmary, Oxford, OX2 6HE, U.K.; and Imperial College of Medicine, Charing Cross Hospital, London, U.K. CAROLE JOINT • Department of Neurosurgery, Radcliffe Infirmary, Oxford, OX2 6HE, U.K. RALPH GREGORY • Department of Neurology, Radcliffe Infirmary, Oxford, OX2 6HE, U.K. PETER BAIN • Department of Neurology and Imperial College of Medicine, Charing Cross Hospital, London, U.K. JOHN F. STEIN • Department of Physiology, University of Oxford, Oxford U.K. RICHARD SCOTT • Department of Neuropsychology, Radcliffe Infirmary, Oxford, OX2 6HE, U.K.

appears to have greater potential for improving symptoms in the "off" medication state, and subsequent reduction in medication can alleviate dyskinesia. Reducing medication holds many benefits for patients and physicians, including simplification of regime, freedom from drug dependency, reduced drug expenditure, and avoidance of both side-effects and the hypothetical neurotoxicity of L-dopa. However, in some patients the benefits of STN stimulation may not match the range or degree of symptomatic improvement seen with medical therapy. In these cases it may be preferable to improve the side effects of drug therapy (such as dyskinesia, dystonia, and motor fluctuations) with pallidal stimulation or lesioning.

Despite remarkable results from a few centers (Kumar et al., 1998; Limousin et al., 1998; Volkmann et al., 1998; Moro et al., 1999; Houeto et al., 2000), the benefits of subthalamic and pallidal stimulation remain to be confirmed by a larger series with long term follow-up.

Given this situation we have continued to pursue all modalities of surgical therapy. It has also been our concern that outcomes be monitored not only from a neurological standpoint, but also from a standpoint based on 'quality of life' and cognitive data (Scott et al., 1998). As Hoehn and Yahr made it clear in 1969, a good surgical outcome may not benefit all patients functionally.

Large series allow dedicated groups to predict what a certain procedure (Duff and Sime, 1997; Alterman and Kelly, 1998; Bronstein et al., 1999; Limousin-Dowsey et al., 1999) can or cannot do, and, therefore, more recently we have begun to target the subthalamic nucleus in sub-groups of patients with Parkinson's disease (PD). We hope to demonstrate the cost effectiveness of such procedures, particularly implantation of deep brain stimulators. These are expensive devices requiring a large amount of clinical input per patient. In a resource-limited health service, cost benefit is of prime importance.

We present our experience over the last few years in movement disorder surgery at Radcliffe Infirmary, Oxford and Charing Cross Hospital, London. The technique used has been described previously (Papanastassiou et al., 1998), using the Radionics Image Fusion and Stereoplan for target acquisition and macrostimulation, and impedance monitoring for lesion placement, and more recently, recording of field potentials from the Medtronics DBS electrode when placing deep brain stimulators.

THALAMOTOMY

The best indication for thalamic surgery in movement disorder surgery is tremor. In Parkinson's disease there is a beneficial effect on rigidity, but this requires a single large lesion with potential cognitive risks. We performed 90 thalamotomies between January 1994 and March 1998 for the following conditions: (1) Parkinsonian tremor- 44 (40pts); (2) BET – 10 (10 pts); (3) MS – 17 (12pts); (4) Dystonia – 11 (7pts); and (5) Others (post-traumatic, post CVA, tumour) – 8.

Parkinson's Disease

In the absence of any medical or cognitive contraindications, patients were offered a VIM thalamotomy for drug resistant tremor. Patients had a long history of tremor predominant symptoms with drug responsive rigidity and bradykinesia.

In addition to a full neurological assessment and video recording of tremor, patients also undergo full neuropsychological and quality of life assessments. It has been our aim to evaluate the nature and extent of any gains that might accrue from thalamotomy and also to identify any cognitive sequelae that might be associated with a good neurological outcome.

Tremor was abolished or rendered minimal in 33 of the 34 patients studied for up to 60 months (mean 31 months). 4 patients required extension of the thalamotomy for tremor recurrence within one year of the procedure. Generally we found that if tremor was absent at three months it remained suppressed. There was one case of a hemiparesis and two of marked dysarthria. There were no deaths or operative haemorrhages.

There was no significant postoperative deterioration in either verbal or non-verbal IQ; speech articulation rates were reduced (N.S.), as was phonemic verbal fluency with lesions in the dominant hemisphere. Interestingly, 'Quality of life' outcomes, assessed by questionnaires, showed very few significant changes on 'generic' measures (i.e., SF-36, FLP), in marked contrast, for example, to the multidimensional improvements observed following pallidal lesions. On the SF-36, the only gain was in the patients' perception of 'improvements in health over the last 12 months;' there were no significant changes on the FLP (i.e., UK version of SIP), although a non-significant trend toward a deterioration in 'communication' was observed. On the Parkinson's disease-specific PDQ-39 there were, however, significant (p<.01) reductions in 'stigma,' 'bodily discomfort,' and the frequency and severity of tremor. These findings suggest that either the chosen questionnaires are insensitive to the specific functional gains following tremor suppression (i.e., ability to hold a glass of water *without* spillage), or that functional gains following thalamotomy are largely 'cosmetic' with psychosocial consequences. The work of Laitinen suggested that thalamotomies were associated with a slowing of motor function. Our studies, in contrast, found a non-significant post-operative improvement in dominant (and non-dominant) dexterity on a timed peg-moving task, and on a test designed to assess writing speed over one minute.

Although abolishing tremor unilaterally restores function in such patients, residual contralateral tremor remains distressing to many patients. A bilateral thalamotomy has unacceptable potential side effects, and we are increasingly performing bilateral thalamic stimulator implants in cases of tremor-predominant parkinsonian patients.

PALLIDOTOMY

Patient Selection

Patients with PD were selected from those referred to movement disorder clinics staffed by a neurologist and neurosurgeon in each center. All patients satisfied United Kingdom PD Society brain bank diagnostic criteria for idiopathic PD (Gibb and Lees, 1988), and most had intractable drug-induced dyskinesia and had failed optimum medical therapy. Any general medical, surgical, or neuropsychological contra-indications were identified, including, for example, significant dysarthria, dysphagia, excessive salivation/drooling, gait freezing "on" medication, dementia or major psychiatric disorder. Patients had to be L-dopa responsive and show that even for a brief period during the day they had windows of quality time. On average, approximately one in three of all patients seen in clinic were considered appropriate for a PVP.

Outcome was assessed in a total of 75 Parkinsonian patients who had unilateral (left n=23, right n=12) or bilateral (n=40) pallidotomies (UPVP and BPVP, respectively). Sequential BPVP was reserved for patients in whom UPVP resulted in inadequate symptomatic relief (N=5). Simultaneous BPVP (N=35) was performed in cases where there were essentially symmetrical bilateral symptoms. All patients underwent identical pre-operative and 3-6 month follow-up neurological assessments, and further follow-up has continued annually thereafter (average follow-up 6.5 months, range 3-22). Changes in medication were minimized during this period.

Patient Assessment

Patients completed a multidisciplinary assessment protocol in the four weeks preceding surgery and 3-6 months post-operatively. The latter time frame was considered long enough for post-operative recovery and patient re-engagement in everyday routines, but short enough to minimize the likelihood of interference effects from disease progression. Neurological assessments were completed independently of the surgeon. The assessment protocol comprised the Unified Parkinson's Disease Rating Scale (UPDRS), Hoehn and Yahr grade and Schwab and England scores. Confrontational visual fields were plotted using a Humphrey field analyser. The last recorded assessments are reported in this series. For patients undergoing staged bilateral procedures only the final assessment is included.

Neuropsychosocial and Functional Assessments

Patients were evaluated in the "on" state in the four weeks preceding surgery and 3-6 months postoperatively. The assessment protocol typically included a semi-structured interview to screen for any Axis I DSM IV diagnoses, questionnaire measures of functional disability, quality of life and psychological symptomatology, and a comprehensive battery of cognitive tests (Scott et al., 1998). The current protocol has, however, evolved with the benefit of clinical experience since 1994; therefore, not all data sets are available for all patients in the neuropsychological series.

Neurological Outcome

Median and mean scores with 95% confidence intervals were calculated for subscores of the UPDRS – namely Part II, Activities of Daily Living (ADL's), Part III (Motor), Gait, Tremor, Rigidity and Bradykinesia. Medications were converted to a total equivalent L-dopa dose (100mg L-dopa = 1mg Pergolide, 10mg Bromocriptine, 1mg Pramipexole, 1mg Cabergoline, 6mg Ropinirole, 1mg Lysuride, 20mg Apomorphine). Pre- and post-operative scores were analysed with Student's t test. Individual UPDRS items such as dyskinesia severity/duration and "off" time were compared with the Wilcoxon ranked sum test and median results are reported. In view of the large number of analyses undertaken, significance was assumed at $P < 0.001$. However, P values between 0.05 and 0.01 were treated with caution and reported if relevant.

The results reported in parentheses refer to the: (percentage of patients improved%, overall reduction in group score%).

Unilateral Pallidotomy

Motor scores "off" medication showed 90 % of patients experienced a reduction in the Part III and II subscore, averaging 27% for the whole group. Bradykinesia (74%, 20%), tremor (68%, 37%), rigidity (69%, 29%), and gait (84%, 32%) scores were significantly improved.

In the "on" medication state, similar tremor and gait changes were seen, but a smaller reduction in bradykinesia and rigidity scores failed to reach significance. Interestingly, a 32% reduction in Part II was observed despite less impressive changes in Part III. This may represent the effect of dyskinesia/off time reduction on ADL's.

In medication, no significant changes in overall L-dopa or LEU were observed.

For dyskinesia, duration was improved in 64% (median pre-post, 2 – 1) and disability in 73% (median pre-post, 2 – 0). Although 82% of patients experiencing pain improved this just failed to achieve significance (p=0.0011) probably because only 40% scored greater than zero pre-operatively and the median pre-op score was low.

"Off" times were unchanged in the majority of patients (51%); however, 42% improved, resulting in an overall median group reduction from 2 to 1, which approached significance (p=0.0082). 70% of patients stopped having sudden "off" switching.

Bilateral Pallidotomy

Part III and Part II motor scores "off" medication were improved in 90% of patients, with an overall reduction of 31% and 35% respectively. Bradykinesia (79%, 26%), tremor (60%, 53%), rigidity (81%, 39%) and gait (87%, 30%) were improved.

In the "on" medication state, all scores improved to a lesser degree, although bradykinesia, tremor and rigidity only approached significance.

No significant changes in medication were observed.

Dyskinesia duration, disability, and pain were improved in 85%, 93%, and 100% of patients to a median post-operative score of zero.

Although "off" time was only improved in 33% of patients, this was significant. Sudden "off" periods were abolished in 88% of patients.

Neuropsychological and Functional Outcome

Patient characteristics. The sample of patients on whom neuropsychological and functional outcome can be reliably reported differs from those with neurological outcomes. It includes 37 PD patients who had UPVP (left UPVP = 21; right PVP 16) and 37 who had BPVP. All but two of the patients studied were right handed. There were 42 men and 32 women.

Patients acted as their own controls, and a paired sample t-test was employed to test the null hypothesis of no difference between baseline follow-up scores on questionnaire measures and psychometric tests. Given the large number of t-tests applied in these analyses, an alpha of 0.01 was chosen to reduce the chances of a type 1 error. For non-parametric data the Wilcoxon signed rank test was employed.

UPVP outcomes. Total UPDRS scores fell by 28%, from 75 to 54 (p<0.001). Although scores on almost all questionnaire subscales moved towards improved 'quality of life' (and reduced functional disability/psychological symptomatology), only the physical functioning, energy-vitality, and bodily pain subscales of the SF-36, and the mobility and ADL subscales of the PDQ-39, improved significantly (p<0.001).

Following a left UPVP there has been some evidence to suggest 'ipsilateral' cognitive deficits that mirror other large, well studied series (Trepanier et al 1998). There was a trend toward improved (10%) speech articulation rates, but some language functions deteriorated. Following a right UPVP there was a similar (10%) trend toward improved speech articulation, but no significant changes in language function. Rather, a deterioration of scores in a visuo-spatial cognitive domain was observed. There were no other significant cognitive changes following UPVP, and no subjective reports of any cognitive or emotional difficulties.

BPVP outcomes. Total UPDRS fell significantly by 35%, from a mean of 94 to 61 (p<0.001). Similarly there was a 31% significant fall in Total % disability scores on the Functional Limitations Profile, from 29% to 20% (p<0.001). In addition to those subscale changes indicating improved quality of life following a UPVP, significant improvement was also observed in the anxiety subscale of the Hospital Anxiety and Deperrssion Scale (p<0.001), the bodily discomfort, stigma, and emotional well-being subscales of the Parkinsons Disease Questionnaire-39, and the general mental health and social functioning subscales of the SF-36.

Cognitive outcome following BPVP included the language deficits observed following a left UPVP, but not those visuo-spatial deficits observed following a right UPVP. This was not an unexpected finding given that lesions in BPVP were intentionally asymmetrical (and typically smaller on the non-dominant side). However, there were additional changes, including a 26% fall in phonemic verbal fluency and a 30% increase in visual scanning time on a test of visual neglect (p<0.001). There was also a trend toward post-operative deterioration in performance on various selective tests of attention and organization. These changes are suggestive of an emerging and relatively selective constellation of executive (i.e., 'frontal-striatal') deficits. Interestingly, Trepanier et al. (1998) have also reported a deterioration in 'frontal-executive' functioning in 25%-30% of their series of 42 patients following *unilateral* PVP, and deteriorating performance on similar cognitive measures, although the frequency and extent of these changes was perhaps greater than we have observed following either UPVP or BPVP.

There were 7 subjective reports of cognitive or emotional changes following BPVP. In one of these cases there was some evidence to suggest relatively global cognitive difficulties and possibly lowered arousal, which were later found to be associated with a misplaced lesion. In the remaining 6 cases (16%), there were reports of some flattening of emotionality and/or initiation.

There were no new individual instances of HADS caseness or neuropsychiatric sequelae following either UPVP or BPVP.

Neurological Complications Following Pallidotomy

There were no visual field deficits and no deaths. One patient developed a frontal hematoma that required surgical evacuation and left her dysphasic. There were three subdural hemorages, of which one required burr hole evacuation at 34 days post-op, and three mild hemipareses, of which one persisted long term and four dysarthrias.

SUBTHALAMIC NUCLEUS SURGERY

Given our experience with thalamotomy and pallidotomy, we felt that subthalamic surgery should be offered to patients with severe axial symptoms of freezing, mid-line dyskinesia, and a rapid history of development of the side effects of L-dopa therapy. "Off" medication they would have marked tremor, rigidity, and bradykinesia. Further optimism was also based on the non human primate and clinical studies published by other groups (Aziz et al., 1992; Hamada and DeLong, 1992; Benazzouz et al., 1993; Guridi et al., 1994). Despite reports of promising results from lesioning, the only published studies were on patients who had stimulators implanted in the STN. It was not clear whether one would prove better than the other.

With these objectives in mind, we lesioned the STN in fourteen patients. Small lesions had a very dramatic effect on tremor, rigidity, bradykinesia, and dyskinesia, but by three months the symptoms returned. Making the lesions larger in subsequent patients had a sustained anti-parkinsonian effect but had a high incidence of hypophonia, mild hemiparesis, and prolonged hemi-chorea such that we abandoned lesioning the STN.

Implantation of 13 bilateral STN stimulators has been more successful with the reservation that such procedures are of marginal benefit in patients with previous thalamotomy or pallidotomy and if they are not responsive to L-dopa. Simultaneous bilateral STN stimulator implantation has resulted in significant post-operative global confusion in two patients, which lasted for nearly one month followed by an excellent clinical response.

A full comparative study of lesioning *vs* stimulation of the STN awaits the completion of follow-up in this cohort of patients, yet we are at this moment unclear as to whether the outcome of STN surgery will exceed that of pallidal surgery. These reservations are upheld by recent studies (Krack et al., 1998; Kumar et al., 1998; Favre et al., 1999), albeit in small groups of patients, showing that pallidal and STN stimulation are not significantly different in their efficacy in alleviating the symptoms of Parkinson's disease, although the latter allows drug reduction.

DISCUSSION

It is becoming clear that there are specific patterns of improvement that can be achieved by each of the available procedures. Deciding on the best procedure would be easy were it not for the complex and highly variable indications for surgery that PD provides. Certainly, in cases of benign tremulous Parkinson's disease, thalamic surgery will always be an effective, simple, and permanent option. Usually, a unilateral thalamotomy will restore the functional capabilities of the majority of patients with an acceptable complication rate in experienced hands. If bilateral procedures are contemplated, then implantation of a neurostimulator is preferable.

For patients with a long history of good response to L-dopa but with disabling limb dyskinesias, pallidotomy is an extremely effective procedure. We have not seen the unacceptable complication rates reported by other groups after bilateral pallidotomy and still offer the procedure to selected patients. There is still little data that would help determine whether bilateral pallidal stimulation is safer surgically and cognitively than lesional procedures (Fields et al., 1999).

We felt that STN surgery should be an option for patients who would not be ideal candidates for thalamic or pallidal surgery, based on our experience to date. These are often younger patients with rapid onset of dyskinesias, freezing, severe tremor, rigidity, and bradykinesia. Our initial experience suggests that, contrary to other reports (Gill and Heywood, 1997; Rodriguez et al., 1998), STN lesioning can have an unacceptable rate of complications despite the fact that small numbers may do extremely well. We have since confined ourselves to neurostimulation in STN surgery, pending decisive data.

Overall, there are still no clear guidelines as to the best site, if it exists, for parkinsonian surgery. Most groups will publish outcomes of such procedures that they do well. Patient-centered decision making and cost benefit analyses will be delayed if studies continue to be performed without 'quality of life' measures.

Finally, it may be that safe, effective surgery at any one of the targets will benefit the majority of patients. Lesioning or stimulation will continue to be guided by such factors as patient choice, economic and geographic considerations, and by the specific procedure any one group does best.

REFERENCES

Alterman, R.L., Kelly, P.J., 1998, Pallidotomy technique and results: the New York University experience, *Neurosurg. Clin. N. Am.* 9:337.

Aziz, T.Z., Peggs, D., Agarwal, E., Sambrook, M.A., Crossman, A.R., 1992, Subthalamic nucleotomy alleviates parkinsonism in the 1-methyl-4-phenyl- 1,2,3,6-tetrahydropyridine (MPTP)-exposed primate, *Br. J. Neurosurg.* 6:575.

Baron, M.S., Vitek, J.L., Bakay, R.A., Green, J., Kaneoke, Y., Hashimoto, T., Turner, R.S., Woodard, J.L., Cole, S.A., McDonald, W.M., DeLong, M.R., 1996, Treatment of advanced Parkinson's disease by posterior GPi pallidotomy: 1-year results of a pilot study, *Ann. Neurol.* 40:355.

Benazzouz, A., Gross, C., Feger, J., Boraud, T., Bioulac, B., 1993, Reversal of rigidity and improvement in motor performance by subthalamic high-frequency stimulation in MPTP-treated monkeys, *Eur. J. Neurosci.* 5:382.

Bronstein, J.M., DeSalles, A., DeLong, M.R., 1999, Stereotactic pallidotomy in the treatment of Parkinson disease: an expert opinion, *Arch. Neurol.* 56:1064.

Dalvi, A., Winfield, L., Yu, Q., Cote, L., Goodman, R.R., Pullman, S.L., 1999, Stereotactic posteroventral pallidotomy: clinical methods and results at 1-year follow up, *Mov. Disord.* 14:256.

de Bie, R.M., de Haan, R.J., Nijssen, P.C., Rutgers, A.W., Beute, G.N., Bosch, D.A., Haaxma, R., Schmand, B., Schuurman, P.R., Staal, M.J., Speelman, J.D., 1999, Unilateral pallidotomy in Parkinson's disease: a randomised, single- blind, multicentre trial, *Lancet,* 354:1665.

Dogali, M., Fazzini, E., Kolodny, E., Eidelberg, D., Sterio, D., Devinsky, O., Beric, A., 1995, Stereotactic ventral pallidotomy for Parkinson's disease, *Neurology,* 45:753.

Duff, J., Sime, E., 1997, Surgical interventions in the treatment of Parkinson's disease (PD) and essential tremor (ET): medial pallidotomy in PD and chronic deep brain stimulation (DBS) in PD and ET, *Axone,* 18:85.

Favre, J., Taha, J.M., Baumann, T., Burchiel, K.J., 1999, Computer analysis of the tonic, phasic, and kinesthetic activity of pallidal discharges in Parkinson patients, *Surg. Neurol.* 51:665.

Fields, J.A., Troster, A.I., Wilkinson, S.B., Pahwa, R., Koller, W.C., 1999, Cognitive outcome following staged bilateral pallidal stimulation for the treatment of Parkinson's disease, *Clin. Neurol. Neurosurg.* 101:182.

Ghika, J., Ghika-Schmid, F., Fankhauser, H., Assal, G., Vingerhoets, F., Albanese, A., Bogousslavsky, J., Favre, J., 1999, Bilateral contemporaneous posteroventral pallidotomy for the treatment of Parkinson's disease: neuropsychological and neurological side effects. Report of four cases and review of the literature, *J. Neurosurg.* 91:313.

Gibb, W.R., Lees, A.J., 1988, The relevance of the Lewy body to the pathogenesis of idiopathic Parkinson's disease, *J. Neurol. Neurosurg. Psychiatry,* 51:745.

Gill, S.S., and Heywood, P., 1997, Bilateral dorsolateral subthalamotomy for advanced Parkinson's disease, *Lancet,* 350:1224.

Giller, C.A., Dewey, R.B., Ginsburg, M.I., Mendelsohn, D.B., Berk, A.M., 1998, Stereotactic pallidotomy and thalamotomy using individual variations of anatomic landmarks for localization, *Neurosurgery,* 42:56.

Guridi, J., Herrero, M.T., Luquin, R., Guillen, J., Obeso, J.A., 1994, Subthalamotomy improves MPTP-induced parkinsonism in monkeys, *Stereotact. Funct. Neurosurg.* 62:98.

Hamada, I., DeLong, M.R., 1992, Excitotoxic acid lesions of the primate subthalamic nucleus result in transient dyskinesias of the contralateral limbs, *J. Neurophysiol.* 68:1850.

Houeto, J.L., Damier, P., Bejjani, P.B., Staedler, C., Bonnet, A.M., Arnulf, I., Pidoux, B., Dormont, D., Cornu, P., Agid, Y., 2000, Subthalamic stimulation in Parkinson disease: a multidisciplinary approach, *Arch. Neurol.* 57:461.

Iacono, R.P., Shima, F., Lonser, R.R., Kuniyoshi, S., Maeda, G., Yamada, S., 1995, The results, indications, and physiology of posteroventral pallidotomy for patients with Parkinson's disease, *Neurosurgery,* 36:1118.

Jankovic, J., Lai, E., Ben-Arie, L., Krauss, J.K., Grossman, R., 1999, Levodopa-induced dyskinesias treated by pallidotomy, *J. Neurol. Sci.* 167:62.

Krack, P., Pollak, P., Limousin, P., Hoffmann, D., Xie, J., Benazzouz, A., Benabid, A.L., 1998, Subthalamic nucleus or internal pallidal stimulation in young onset Parkinson's disease, *Brain,* 121:451.

Kumar, R., Lozano, A.M., Kim, Y.J., Hutchison, W.D., Sime, E., Halket, E., Lang, A.E., 1998, Double-blind evaluation of subthalamic nucleus deep brain stimulation in advanced Parkinson's disease, *Neurology,* 51:850.

Kumar, R., Lozano, A.M., Montgomery, E., Lang, A.E., 1998, Pallidotomy and deep brain stimulation of the pallidum and subthalamic nucleus in advanced Parkinson's disease, *Mov. Disord.* 13 Suppl 1:73.

Laitinen, L.V., Bergenheim, A.T., Hariz, M.I., 1992, Leksell's posteroventral pallidotomy in the treatment of Parkinson's disease, *J. Neurosurg.* 76:53.

Lang, A.E., Lozano, A.M., Montgomery, E., Duff, J., Tasker, R., Hutchinson, W., 1997, Posteroventral medial pallidotomy in advanced Parkinson's disease, *N. Engl. J. Med.* 337:1036.

Limousin-Dowsey, P., Pollak, P., Van Blercom, N., Krack, P., Benazzouz, A., Benabid, A., 1999, Thalamic, subthalamic nucleus and internal pallidum stimulation in Parkinson's disease, *J. Neurol.* 246 Suppl 2:42.

Limousin, P., Krack, P., Pollak, P., Benazzouz, A., Ardouin, C., Hoffmann, D., Benabid, A.L., 1998, Electrical stimulation of the subthalamic nucleus in advanced Parkinson's disease, *N. Engl. J. Med.* 339:1105.

Lozano, A.M., Lang, A.E., Galvez-Jimenez, N., Miyasaki, J., Duff, J., Hutchinson, W.D., Dostrovsky, J.O., 1995, Effect of GPi pallidotomy on motor function in Parkinson's disease, *Lancet,* 346:1383.

Mitchell, I.J., Clarke, C.E., Boyce, S., Robertson, R.G., Peggs, D., Sambrook, M.A., Crossman, A.R., 1989, Neural mechanisms underlying parkinsonian symptoms based upon regional uptake of 2-deoxyglucose in monkeys exposed to 1-methyl-4-phenyl- 1,2,3,6-tetrahydropyridine, *Neuroscience,* 32:213.

Moro, E., Scerrati, M., Romito, L.M., Roselli, R., Tonali, P., Albanese, A., 1999, Chronic subthalamic nucleus stimulation reduces medication requirements in Parkinson's disease, *Neurology,* 53:85.

Ondo, W.G., Jankovic, J., Lai, E.C., Sankhla, C., Khan, M., Ben-Arie, L., Schwartz, K., Grossman, R.G.,

Krauss, J.K., 1998, Assessment of motor function after stereotactic pallidotomy, *Neurology,* 50:266.

Papanastassiou, V., Rowe, J., Scott, R., Silburn, P., Davies, L., and Aziz, T., 1998, Use of the Radionics Image Fusion™ and Stereoplan™ programs for target localisation in functional neurosurgery, *J. Clin. Neurosci.* 5:28.

Rodriguez, M.C., Guridi, O.J., Alvarez, L., Mewes, K., Macias, R., Vitek, J., DeLong, M.R., Obeso, J.A., 1998, The subthalamic nucleus and tremor in Parkinson's disease. *Mov. Disord.* 13 Suppl 3:111.

Samuel, M., Caputo, E., Brooks, D.J., Schrag, A., Scaravilli, T., Branston, N.M., Rothwell, J.C., Marsden, C.D., Thomas, D.G., Lees, A.J., Quinn, N.P., 1998, A study of medial pallidotomy for Parkinson's disease: clinical outcome, MRI location and complications, *Brain,* 121:59.

Scott, R., Gregory, R., Hines, N., Carroll, C., Hyman, N., Papanasstasiou, V., Leather, C., Rowe, J., Silburn, P., Aziz, T., 1998, Neuropsychological, neurological and functional outcome following pallidotomy for Parkinson's disease. A consecutive series of eight simultaneous bilateral and twelve unilateral procedures, *Brain,* 121:659.

Shannon, K.M., Penn, R.D., Kroin, J.S., Adler, C.H., Janko, K.A., York, M., Cox, S.J., 1998, Stereotactic pallidotomy for the treatment of Parkinson's disease. Efficacy and adverse effects at 6 months in 26 patients, *Neurology,* 50:434.

Shima F., 1994, Posteroventral pallidotomy for Parkinson's disease: renewal of pallidotomy, *No Shinkei Geka,* 22:103.

Subramanian, T., Emerich, D.F., Bakay, R.A., Hoffman, J.M., Goodman, M.M., Shoup, T.M., Miller, G.W., Levey, A.I., Hubert, G.W., Batchelor, S., Winn, S.R., Saydoff, J.A., Watts, R.L., 1997, Polymer-encapsulated PC-12 cells demonstrate high-affinity uptake of dopamine in vitro and 18F-Dopa uptake and metabolism after intracerebral implantation in nonhuman primates, *Cell Transplant,* 6:469.

Trepanier, L.L., Saint-Cyr, J.A., Lozano, A.M., Lang, A.E., 1998, Neuropsychological consequences of posteroventral pallidotomy for the treatment of Parkinson's disease. *Neurology,* 51:207.

Uitti, R.J., Wharen, R.E. Jr., Turk, M.F., 1998, Efficacy of levodopa therapy on motor function after posteroventral pallidotomy for Parkinson's disease, *Neurology,* 51:1755.

Vitek, J.L., Bakay, R.A., DeLong, M.R., 1997, Microelectrode-guided pallidotomy for medically intractable Parkinson's disease, *Adv. Neurol.* 74:183.

Volkmann, J., Sturm, V., Weiss, P., Kappler, J., Voges, J., Koulousakis, A., Lehrke, R., Hefter, H., Freund, H.J., 1998, Bilateral high-frequency stimulation of the internal globus pallidus in advanced Parkinson's disease, *Ann. Neurol.* 44:953.

THALAMIC SURGERY FOR TREMOR

ANDRES LOZANO, VLADIMIR SHABALOV, CHIHIRO OHYE,
AND FREDERICK LENZ[*]

INTRODUCTION

The ventral intermediate nucleus of the thalamus (Vim) or its afferent axonal projections is currently the most widely accepted surgical target for tremor (Laitinen, 1985). Vim is populated by neurons that fire in synchronous bursts whose timing is similar to peripheral tremor as originally described by Guiot et al. (1962). The close temporal relation between bursting thalamic neuronal activity and tremor has led to the suggestion that these thalamic "tremor cells" could act as tremorogenic pacemakers. In some cases, the mere introduction of an electrode into their midst may suffice to arrest tremor in the contralateral hemibody for several weeks (Tasker et. al., 1982). Intraoperative electrical stimulation in areas of the thalamus populated by tremor-synchronous cells momentarily arrests tremor (reviewed by Tasker et. al., 1982), an observation that is important in the selection of the most effective target for tremor control.

Vim thalamotomy is highly effective but can be associated with motor, sensory, cerebellar, speech and other complications (reviewed by Ojemann et al., 1990). These difficulties have been ascribed to injury to adjacent axonal projections and nuclear structures. The rate of complication is particularly high with thalamotomies that are bilateral. For example, the incidence of memory and speech disturbances with bilateral thalamotomy can be from 30% to 60% (Ojemann et al., 1990; Burchiel, 1995). There is mounting evidence that the incidence of these complications can be reduced by careful mapping of the target as made possible with microelectrode recordings. Nevertheless, because of the potential serious complications and the irreversible nature of lesioning of the nervous system, neurosurgeons have sought alternate procedures to achieve the effectiveness of thalamotomy while reducing the risk of unwanted effects. This is a principal reason for the increasing use of chronic deep brain stimulation (DBS) as an alternative to lesioning. While the present review is restricted to the thalamus, it is important to realize that other procedures, including transplantation (Ugryumov et al., 1996, see also chapter by Ugrumov in this volume), and lesions or deep brain stimulation

[*] ANDRES LOZANO • Division of Neurosurgery, University of Toronto, The Toronto Hospital, Western Division, Toronto, Ontario M5T 2S8. VLADIMIR SHABALOV • Burdenko Institute of Neurosurgery, 4th Tverskaya-Yamskaya str., 16, Moscow, Russia. CHIHIRO OHYE • Functional and Gamma Knife Surgery Center, Hidaka Hospital, Takasaki, Gunma, Japan. FREDERICK LENZ • Department of Neurosurgery, Johns Hopkins University, Baltimore, Maryland 21287.

(DBS) at other targets, such as the internal segment of the globus pallidus (GPi) and the subthalamic nucleus (STN) (Lang and Lozano, 1998a, 1998b), also have a role in controlling motor disturbances including tremor.

VIM THALAMUS LESIONS AND DBS FOR TREMOR

Finding the Target

Thalamic procedures for tremor are divided into three stages: (1) stereotactic imaging, (2) thalamic mapping, and (3) lesioning or electrode implantation. In the case of thalamic DBS there is, in addition, the need for receiver or pulse generator implantation and programming. These procedures are covered in detail elsewhere (Ohye et al., 1989, 1990, 1993; Lozano, 1998).

Physiological mapping is carried out principally to obtain three pieces of information. First, it is important to localize the Vim nucleus of the thalamus by the characteristic neurophysiological features of Vim neurons, including their spontaneous high amplitude activity, their response to kinesthetic inputs and their tremor-synchronous pattern of discharge. Second, the goal of surgery is to observe the effects of electrical microstimulation or macrostimulation on the patient's tremor. Sites within Vim where tremor is arrested with stimulation are potential surgical targets. Lastly, identification of the somatosensory relay, the ventral caudal nucleus (VC) of the thalamus is needed to help in establishing the posterior boundary of Vim and to minimize adverse effects. VC, which lies posterior and inferior to Vim, is populated by neurons that respond to tactile stimulation in small receptive fields. Stimulation in the somatotopically organized VC produces paraesthesias in the corresponding part of the body.

What is the Best Target?

Cells in the ventral tier motor thalamus display functional properties defined by activity related to somatosensory stimuli and to active movements. Four types of cells have been described: (1) cells with activity related to somatosensory stimulation (sensory cells), (2) cells with activity related to active movement (voluntary cells), (3) cells with activity related to both somatosensory stimulation and active movement (combined cells), and (4) cells with activity related to neither somatosensory stimulation nor active movement (no-response cells). Certain combined, voluntary, and no-response cells have a peak of activity at tremor frequency, which is significantly correlated with the electromyogram (EMG).

Figure 1. Effects of electrical stimulation and intrathalamic muscimol microinjections in essential tremor patients. A - Reconstruction of the trajectories of three microelectrode tracks (S2, S3, S4) in a patient with essential tremor on a sagittal map 14.5 mm lateral to the midline. The site of muscimol injection is represented by a dashed ellipse in S3 (i.e., the targeted injection site was Vim). The regions where single-unit responses to joint movement (kinesthetic), tremor, and tactile stimuli were recorded are indicated by the patterned bars (see box for code). Regions where microstimulation at 100 μA induced tremor reduction or arrest are indicated by the filled bars. B - Baseline tremor (control) at one minute before muscimol microinjection as recorded by accelerometer (A) and wrist flexor EMG (F) over 10 seconds of time (T) during postural tremor. C - The inhibitory effect of macrostimulation (0.5 mm exposed tip, 25 gauge tubing; track S3, 400 μA, 0.1 ms pulses, 300 Hz, 3 s train) performed at the microinjection site. Traces as in B; period of stimulation (s) indicated by the horizontal bar. D - The effect of a 5 μL microinjection of muscimol (same site as for macrostimulation) on tremor, 8 minutes after injection. Traces as in B. E - Changes in tremor frequency spectrum. The smoothed power spectrum for the baseline tremor (dark-lined sharp peak) and the tremor present following muscimol administration (light-line broadened peak). F - Plot of the increase in tremor bandwidth (TB) of tremor versus % tremor reduction as determined from accelerometer amplitude after muscimol microinjections. Each dot represents data from one patient. Abbreviations: Vop, nucleus ventralis oralis posterior; Vim, nucleus ventralis intermedius; Vc, nucleus ventralis caudalis. From Pahapil et al., 1999, with permission.

These cells are anterior to the presumed somatosensory relay nucleus VC and are located in the region of thalamus where a lesion stops tremor (Lenz et al., 1987,1994). Analysis of the phase of thalamic activity relative to EMG activity has shown that voluntary and combined cell activity usually precedes the EMG during tremor, while the activity of sensory cells frequently lags behind tremor. The optimal target site for tremor alleviation is believed to contain cells that both show activity at tremor frequency and receive sensory input (Lenz et al., 1995).

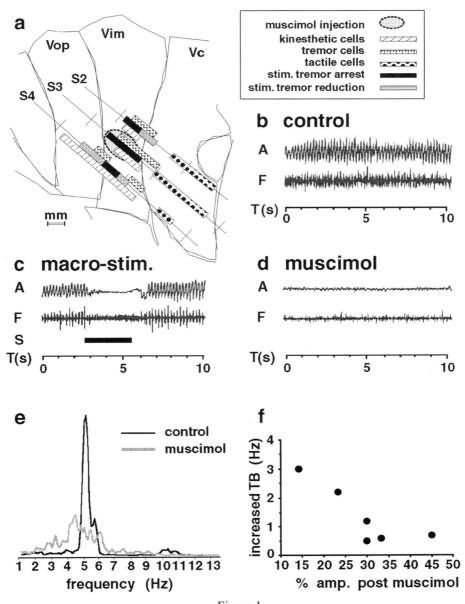

Figure 1

What is the clinical effectiveness of VIM DBS?

Vim thalamic DBS is highly beneficial for tremor control but ineffective for the other disabling features of Parkinson's disease (PD), including bradykinesia, rigidity, and gait and postural disturbances. The work of Benabid and his colleagues (Benabid et. al, 1996) in 117 patients (80 PD, 20 essential tremor (ET), and 17 other) indicates that chronic Vim stimulation is highly effective for tremor, with over 85% of patients having a very good or excellent response with little or no tremor evident in the contralateral arm. In a multi-center trial, Koller et al. (1997) have shown an 80% reduction in contralateral arm tremor in 24 patients with PD tremor and 29 patients with ET with Vim DBS at 1 year follow-up. Vim DBS is associated with a greater improvement in disability for ET patients than for PD patients who have associated non-tremor motor dysfunction (Koller et al., 1997; Ondo et al., 1998). The lack of benefit to the activities of daily living of patients with PD treated with thalamic DBS highlights the ineffectiveness of this therapy for bradykinesia, rigidity, and gait and postural disturbances and emphasizes the need to consider alternate therapies and brain targets in patients with these clinical features. Indeed, it is becoming clear that the majority of patients with PD are more effectively treated with surgical interventions at more versatile targets, including GPi or STN, which offer the possibility of amelioration of all cardinal features of the illness.

Different types of tremor may vary in their response to thalamic DBS. Certain studies suggest that intention tremor responds less than rest and positional tremor (Benabid et al., 1996). Others suggest that they respond approximately equally (Hubble et al., 1997). There is also the suggestion that the target for more proximal arm tremor may lie anterior and dorsal to the basal Vim target most commonly used for distal arm tremor (Nguyen and Degos, 1993). Features other than extremity tremor that have been reported to improve with Vim stimulation include voice tremor (Carpenter et al., 1998) and L-dopa-induced dyskinesias (Caparros-Lefebvre et al., 1993). As with thalamotomy (Shahzadi et al., 1995), thalamic DBS is less effective for tremors associated with cerebellar disease or multiple sclerosis (Geny et al., 1995, 1996) than for the tremors associated with PD and ET. In a recent randomized prospective trial of Vim DBS versus thalamotomy, both therapies were found to be effective, but the DBS group had a lesser number of permanent surgery-related adverse effects and showed a higher level of functioning after surgery (Schuurman et al., 2000).

Mechanism of Action of Thalamic DBS

The similarities in the effectiveness of thalamic DBS and thalamotomy have led investigators to suggest that DBS acts as a reversible lesion of the thalamus, but the mechanism of action of thalamic DBS is far from known. Neuronal blocking and jamming are possibilities (Benabid et al., 1996), but the activation of inhibitory mechanisms is also to be considered (Ashby et al., 1995; Strafella et al., 1997). Pahapill et al. (1999) tested the hypothesis that increasing GABAergic transmission in the motor thalamus may diminish tremor. Six patients undergoing surgery for essential tremor had microinjections of muscimol (a $GABA_A$ receptor agonist) into the Vim nucleus of the thalamus in areas where tremor-synchronous cells were identified electrophysiologically with microelectrode recordings and where tremor reduction occurred with electrical microstimulation. Injections of muscimol, but not saline, reduced tremor in each patient (Figure 1). The effect had a mean latency of 7 minutes and lasted an average of 9 minutes. On the basis of these findings, the authors proposed that inhibition of thalamic neuronal cell bodies may underlie the effectiveness of surgery for tremor and that doing so with GABA analogues could potentially be used therapeutically. These results suggest the possibility that thalamic DBS may work by activating inhibitory mechanisms, causing the release of GABA from the terminals of local circuit or thalamic reticular nucleus neurons that are pre-synaptic to thalamocortical relay cells.

CONCLUSIONS

The stimulation-induced side effects of Vim DBS are reversible and usually mild and acceptable provided the intensity of stimulation produced significant benefits on tremor. Adverse stimulation-related effects include paraesthesia, dystonia (mainly in the foot), dysarthria, and dysequilibrium. Little data are available on the safety and efficacy of bilateral thalamic DBS for tremor. Because of the reversible nature of DBS, this therapy may be safer than lesioning in patients who require bilateral procedures or who require a procedure on the side opposite to a previous thalamotomy. Neuropsychological studies suggest that Vim stimulation does not produce significant adverse cognitive effects (Caparros-Lefebvre et al., 1992). The nature of the fundamental neurophysiological disturbances that produce tremor is still largely unknown, as is the mechanism by which the application of chronic electrical stimulation produces such striking therapeutic effects.

In summary, lesioning or chronic stimulation of the ventral intermediate nucleus (Vim) of the thalamus are effective for the treatment of tremor. Patients with tremor associated with Parkinson's disease and essential tremor respond best, while those with tremor secondary to cerebellar or brainstem pathology respond to a lesser extent. There remains a great deal of controversy regarding the pathogenesis of tremor and how to select the best therapeutic target within the thalamus. While chronic stimulation of the thalamus is producing important benefits, its mechanism of action is poorly understood and is currently a research priority.

REFERENCES

Ashby, P., Lang, A.E., Lozano, A.M., and Dostrovsky, J.O., 1995, Motor effects of stimulating the human cerebellar thalamus, *J. Physiol.* 489:287.

Benabid, A.L., Pollak, P., Gao, D.M., Hoffmann, D., Limousin, P., Gay, E., Payen, I., and Benazzouz, A., 1996, Chronic electrical stimulation of the ventralis intermedius nucleus of the thalamus as a treatment of movement disorders, *J. Neurosurg.* 84:203.

Burchiel, K.J., 1995, Thalamotomy for movement disorders, *Neurosurg. Clin. N. Am.* 6:55.

Caparros-Lefebvre, D., Blond, S., Pecheux, N., Pasquier, F., and Petit, H., 1992, Neuropsychological evaluation before and after thalamic stimulation in 9 patients with Parkinson disease, *Rev. Neurol. (Paris)* 148:117.

Caparros-Lefebvre, D., Blond, S., Vermersch, P., Pecheux, N., Guieu, J.D., and Petit, H., 1993, Chronic thalamic stimulation improves tremor and levodopa induced dyskinesias in Parkinson's disease, *J. Neurol. Neurosurg. Psychiatr.* 56:268.

Carpenter, M.A., Pahwa, R., Miyawaki, K.L., Wilkinson, S.B., Searl, J.P., and Koller, W.C., 1998, Reduction in voice tremor under thalamic stimulation, *Neurology,* 50:796.

Geny, C., N'Guyen, J.P., Cesaro, P., Goujon, C., Brugieres, P., and Degos, J.D., 1995, Thalamic stimulation for severe action tremor after lesion of the superior cerebellar peduncle, *J. Neurol. Neurosurg. Psychiatr.* 59:641.

Geny, C., Nguyen, J.P., Pollin, B., Feve, A., Ricolfi, F., Cesaro, P., and Degos, J.D., 1996, Improvement of severe postural cerebellar tremor in multiple sclerosis by chronic thalamic stimulation, *Mov. Disord.* 11:489.

Guiot, G., Hardy, J., and Albe-Fessard, D., 1962, Delimination precise des structures sous-corticales et identification de noyaux thalamiques chez l'homme par l'electrophysiologie stereotaxique, *Neurochirurgia,* 15:1.

Hubble, J.P., Busenbark, K.L., Wilkinson, S., Pahwa, R., Paulson, G.W., Lyons, K., and Koller, W.C., 1997, Effects of thalamic deep brain stimulation based on tremor type and diagnosis, *Mov. Disord.* 12:337

Koller, W., Pahwa, R., Busenbark, K., Hubble, J., Wilkinson, S., Lang, A., Tuite, P., Sime, E., Lozano, A.M., Hauser, R., Malapira, T., Smith, D., Tarsy, D., Miyawaki, E., Norregaard, T., Kormos, T., and Olanow, C.W., 1997, High-frequency unilateral thalamic stimulation in the treatment of essential and parkinsonian tremor, *Ann. Neurol.* 42:292.

Laitinen, L.V., 1985, Brain targets in surgery for Parkinson's disease. Results of a survey of neurosurgeons, *J. Neurosurg.* 62:349.

Lang, A.E., and Lozano, A.M., 1998a, Parkinson's disease. First of two parts, *N. Engl. J. Med.* 339:1044.

Lang, A.E., and Lozano, A.M., 1998b, Parkinson's disease. Second of two parts, *N. Engl. J. Med.* 339:1130.

Lenz, F.A., Kwan, H.C., Martin, R.L., Tasker, R.R., Dostrovsky J.O., and Lenz Y.E., 1994, Single unit analysis of the human ventral thalamic nuclear group. Tremor-related activity in functionally identified cells, *Brain,* 117:531-543.

Lenz, F.A., Normand, S.L., Kwan, H.C., Andrews, D., Rowland, L.H., Jones, M.W., Seike, M., Lin, Y.C., Tasker, R.R., and Dostrovsky, J.O., 1995, Statistical prediction of the optimal site for thalamotomy in parkinsonian tremor, *Mov. Disord.* 10:318.

Lenz, F.A., Tasker, R.R., Kwan, H.C., Schider, S., Kwong, R., Dostrovsky, J.O., and Murphy, J.T., 1987, Selection of the optimal lesion site for the relief of parkinsonian tremor on the basis of spectral analysis of neuronal firing patterns, *Appl. Neurophysiol.* 50:338.

Lozano, A., 1998, Thalamic deep brain stimulation for the control of tremor, in: *Neurosurgical Operative Atlas*, S. Rengachary and R. Wilkins, eds., American Association of Neurological Surgeons, Parke Ridge, Illinois.

Nguyen, J.-P., and Degos, J.-D., 1993, Thalamic stimulation and proximal tremor: A specific target in the nucleus ventrointermedius thalami, *Arch. Neurol.* 50:498.

Ohye, C., Shibazaki,T., Hirai, T., Wada, H., Hirato, M., and Kawashima, Y., 1989, Further physiological observations on the ventralis intermedius neurons in the human thalamus, *J. Neurophysiol.* 61:488.

Ohye, C., Shibazaki, T., Hirai, T., Kawashima, Y., Hirato, M., and Matsumura, M., 1993, Tremor-mediating thalamic zone studied in humans and in monkeys, *Stereotact. Funct. Neurosurg.* 60:136.

Ohye, C., Shibazaki, T., Hirato, M., Kawashima, Y., and Matsumura, M., 1990, Strategy of selective VIM thalamotomy guided by microrecording, *Stereotact. Funct. Neurosurg.* 54-55:186.

Ojemann, G., and Ward, A., 1990, Abnormal movement disorders, in: *Neurological Surgery*, J. Youmans, ed., WB Saunders, Philadelphia.

Ondo, W., Jankovic, J., Schwartz, K., Almaguer, M., and Simpson, R.K., 1998, Unilateral thalamic deep brain stimulation for refractory essential tremor and Parkinson's disease tremor, *Neurology,* 51:1063.

Pahapill, P.A., Levy, R., Dostrovsky, J.O., Davis, K.D., Rezai, A.R., Tasker, R.R., and Lozano, A.M., 1999, Tremor arrest with thalamic microinjections of muscimol in patients with essential tremor, *Ann. Neurol.* 46:249.

Schuurman, P.R., Bosch, D.A., Bossuyt, P.M., Bonsel, G.J., van Someren, E.J., de Bie, R.M., Merkus, M.P., and Speelman, J.D., 2000, A comparison of continuous thalamic stimulation and thalamotomy for suppression of severe tremor, *N. Engl. J. Med.* 342:461.

Shahzadi, S., Tasker, R.R., and Lozano, A., 1995, Thalamotomy for essential and cerebellar tremor, *Stereotact. Funct. Neurosurg.* 65:11.

Strafella, A., Ashby, P., Munz, M., Dostrovsky, J.O., Lozano, A.M., and Lang, A.E., 1997, Inhibition of voluntary activity by thalamic stimulation in humans: relevance for the control of tremor, *Mov. Disord.* 12:727.

Tasker, R.R., Organ, L.W., and Hawrylyshyn, P., 1982. *The Thalamus and Midbrain of Man: A Physiological Atlas Using Electrical Stimulation*, Charles C. Thomas, Springfield, Ill.

Ugryumov, M.V., Shabalov, V.A., Fedorova, N.V., Popov, A.P., Shtok, V.N., Sotnikova, E.I., Melikian, A.G., Gatina, T.A., Fetisov, S.O., Arkhipova, N.A., Buklina, S.B., and Lutsenko, G.V., 1996, Use of neuro-transplantation in the treatment of Parkinson's disease, *Vestn. Ross. Akad. Med. Nauk.* 8:40. (In Russian)

WHAT IS THE INFLUENCE OF SUBTHALAMIC NUCLEUS STIMULATION ON THE LIMBIC LOOP?

PAUL KRACK, CLAIRE ARDOUIN, AURÉLIE FUNKIEWIEZ,
ELENA CAPUTO, ABDELHAMID BENAZZOUZ, ALIM-LOUIS
BENABID, AND PIERRE POLLAK[*]

INTRODUCTION

High frequency electrical stimulation of the subthalamic nucleus (STN) improves the motor signs and symptoms of Parkinson's disease (PD) (Limousin et al., 1995). In addition to motor disability, cognitive impairment (Dubois and Pillon, 1997) and a wide range of neuropsychiatric disturbances such as depression and apathy are frequently associated with PD (Aarsland et al., 1999). In animals, the three main components (motor, associative, and limbic) of the cortical-basal ganglia-cortical circuits pass through the STN (Parent and Hazrati, 1995). The STN can be divided into a large dorsolateral sensorimotor territory and a smaller ventromedial associative territory. The medial tip of the nucleus, with projections from limbic cortices and ventral (subcommissural) pallidum, represents the limbic territory (Alexander et al., 1986; Groenewegen and Berendse, 1990; Parent and Hazrati, 1995). Given the small size of the STN, current spread from an electrode located within the sensorimotor territory is likely to simultaneously affect the motor and non-motor areas of the STN. However, it is not known whether the limbic part of the STN is involved in the psychiatric problems in PD. The present review analyzes the possibility that high frequency STN stimulation might influence the limbic loop of the basal ganglia.

CLINICAL LINES OF EVIDENCE INDICATE AN INFLUENCE OF STN STIMULATION ON THE LIMBIC LOOP

So far, reports on STN stimulation mainly have focused on the motor (Limousin et al., 1998) or cognitive (Ardouin et al., 1999; Pillon et al., 2000) symptoms of PD. Spontaneous lesions of the STN have been reported to induce euphoria and manic behavior together with dyskinesias (Barraquer-Bordas and Peres-Serra, 1965; Trillet et al., 1995). It is possible that a similar biochemical mechanism can underlie the pathogenesis of behavior and motor

[*] PAUL KRACK • Department of Neurosciences, University of Grenoble, France; and Neurology Department, University of Kiel, 24105 Kiel, Germany. CLAIRE ARDOUIN, AURÉLIE FUNKIEWIEZ, ELENA CAPUTO, ABDELHAMID BENAZZOUZ, ALIM-LOUIS BENABID, AND PIERRE POLLAK • Department of Neurosciences, University of Grenoble, France.

disorders. This is supported by the fact that the dopaminergic striatal activation induces mania and dyskinesias, and the dopaminergic striatal inhibition induces depression and parkinsonism. Drug-induced changes in the dopaminergic striatal activity are associated with changes in STN activity in the opposite direction through the indirect output from the striatum. So far, neither emotion nor behavior has been systematically analyzed in patients treated with high frequency STN stimulation. There are, however, several lines of evidence that high frequency STN stimulation might indeed influence limbic functions.

High Frequency STN Stimulation-Induced Changes in Mood and Behavior

In a few patients, acute stimulation with an electrode located in the STN using high stimulation parameters induced funny associations, leading to laughter and hilarity, whereas therapeutic parameters induced hypomanic behavior and marked improvement of akinesia (Kumar et al., 1999; Krack et al., 2000).

Elation in mood accompanied by hypomanic behavior may be observed during the first postoperative weeks with bilateral high frequency STN stimulation. A raised libido and an increase in penile erections in males have been observed as a part of this hypomanic behavior (Caputo et al., 2000). The combination of high frequency STN stimulation and strong pharmacological activation of the dopaminergic mechanisms by medication may account for the fact that this manic behavior occurs mainly in the post-operative period. After surgery the risk of mania decreases with time, in parallel with the dramatic decrease in the dopaminergic drug doses (Caputo et al., 2000). Full-blown mania can occur but seems to be exceedingly rare (Caputo et al., 2000; Ghika et al., 1999).

Comparison between the pre- and post-operative scores of the Beck Depression Inventory in PD patients with bilateral high frequency STN stimulation showed mild but significant improvement of mood (Ardouin et al., 1999).

Subjective Psychotropic Effects of High Frequency STN Stimulation or Levodopa

A systematic analysis of the changes in subjective psychic effects related to either stimulation or levodopa was performed in fifty PD patients with bilateral high frequency STN implantation. The patients completed the Addiction Research Center Inventory (ARCI), a questionnaire designed to assess subjective positive and negative psychic effects of different drugs, at varying follow-up times after surgery (mean = 18.6 ± 12.4 months) (Martin et al., 1971). The ARCI was evaluated off and on stimulation, using the therapeutical electrical parameters, and off and on drug, using a suprathreshold dose of levodopa. In comparison with the off drug/off stimulation condition, both levodopa and STN stimulation improved all the ARCI subscores indicating psychic stimulation, euphoria, increased motivation, decrease in fatigue, anxiety, and tension. However, high frequency STN stimulation was mildly but significantly less effective than levodopa in four of the five subscores (Funkiewiez et al., 2000).

Apathy may occur after bilateral high frequency STN stimulation and can be related to a dramatic postoperative decrease in the use of dopaminergic drugs, because the resumption of this treatment is usually beneficial (Krack et al., 1998; Caputo et al., 2000).

High Frequency STN Stimulation-Induced Changes in Brain Metabolism

PET analysis of patients performing a free-choice joystick movement shows an increase in rCBF not only in the supplementary motor area (motor loop) and the dorsolateral prefrontal cortex (associative loop), but also in the anterior cingulate cortex (Limousin et al., 1997), which is a part of the limbic system (Papez, 1937).

DISCUSSION

Clinical, neuropsychological, and PET data yield congruent evidence that high frequency STN stimulation has not only motor and cognitive effects, but behavioral and emotional effects as well. Behavior and emotions are regulated by the limbic system. Behavioral and affective effects can be related to changes in the activity of the basal ganglia because anatomical and electrophysiological studies in nonhuman primates have shown that parts of the basal ganglia are connected to limbic structures. Changes in dopaminergic activity mainly affect the striatum with its motor, associative, and limbic components. Surgery or spontaneous lesions of basal ganglia, namely the STN or the internal segment of the globus pallidus (GPi), are generally directed to the sensorimotor area in order to improve motor function. However, selective involvement of the sensorimotor area may be difficult to achieve because of the small size of the STN or because the limbic and motor portions are adjacent or partly intermingled. Moreover, lesions or electrode locations can be misplaced. The observed effects on limbic functions may also be related to current diffusion of stimulation to nearby limbic or limbic-related structures.

Possible Involvement of the Limbic Area of the Basal Ganglia

1. Psychotropic Effects of Dopaminergic Drugs. Levodopa administration in patients with PD often results in an improvement in symptoms such as apathy and lack of interest, the so-called "awakening effect" (Cotzias et al., 1969; Barbeau, 1969; Sacks, 1982). Levodopa has a general mood-elevating effect (Barbeau, 1969), and on-period inappropriate laughter (O'Brien et al., 1971) or euphoria (Cotzias, 1967) may occur. Depression is often alleviated after the introduction of levodopa therapy for PD (Barbeau, 1969; Yahr et al., 1969). Along with motor improvement, patients treated with levedopa often experience "mood spells" characterized by the sudden onset of feelings of intense mental and physical well-being; patients also report feeling less anxiety, greater arousal, and an increased desire for movement and human contact, resulting in accomplishment of neglected tasks (Damasio et al., 1971). Patients experience a sense of amazement about their newly acquired capacity or "awakening from torpor" and may mention a state of permanent arousal (Damasio et al., 1971).

Psychosis, mania, euphoria, and hypersexuality have been reported as behavioral consequences of dopaminergic drugs (Cotzias, 1967; Barbeau, 1969; Cotzias et al., 1969; Sacks, 1982; Giovannoni et al., 2000). The proposed mechanism of these side effects is that overstimulation of the mesocorticolimbic dopamine receptors results in dysfunction of limbic structures (Goetz et al., 1982; Wolters, 1999). The dopamine transporter located in the mesolimbic dopaminergic projections to the nucleus accumbens (ventral striatum) is the target for the behavioral and biochemical actions of amphetamine and cocaine (Giros et al., 1996; Jaber et al., 1997). Dopamine replacement therapy has been classified in the stimulant category of substance abuse along with cocaine and amphetamine derivatives (Giovannoni et al., 2000). The findings of Funkiewiez et al. (2000) also indicate similarities between the psychotropic actions of levodopa, cocaine and amphetamines.

2. Psychotropic Effects of High Frequency STN Stimulation. High frequency STN stimulation may lead to subjective psychostimulant effects. Hypomania with euphoria and increased libido and very rare cases of full-blown mania with psychosis have been observed (Ghika et al., 1999; Caputo et al., 2000). Isolated visual hallucinations related to high frequency STN stimulation have also been described (Diederich et al., 2000). The fact that high frequency STN stimulation mimics the psychotropic effects of levodopa, as shown by Funkiewiez et al., 2000, is a strong indication that the limbic STN, under control of the ventral striatum, is hyperactive in PD just as the sensorimotor STN is. This hyperactivity can be suppressed by either levodopa, through the action of dopamine on the ventral striatum, or directly using the supposed inhibitory effect of high frequency STN stimulation. In rats, electrophysiological studies of the limbic loop from the prelimbic medial orbital area showed the existence of an indirect pathway from the ventral striatum to the STN (Maurice et al., 1998a; Maurice et al., 1998b). The psychotropic effects of high frequency STN stimulation are generally less pronounced than those of dopaminergic stimulation. This may be partly related to the continuous effects of STN stimulation as opposed to the pulsatile effect of levodopa, which can lead to a "kick" or "high" similar to that of psychostimulant drugs (Cummings, 2000; Giovannoni et al., 2000). The psychotropic effects of high frequency STN stimulation and levodopa are additive, explaining why a drastic reduction of dopaminergic treatment may lead to apathy since the psychotropic effects of high frequency STN stimulation seem less powerful. This additive effect may also account for the occurrence of mania if dopaminergic drugs are not reduced after high frequency STN stimulation. It is important to be aware of these additive effects when managing patients following STN surgery.

3. Psychotropic Effects of Basal Ganglia Surgery. Improvement of mood, euphoria, mania, hypersexuality, and increase in weight have been described after internal pallidal surgery (Guiot and Brion, 1953; Lang et al., 1997; Narabayashi et al., 1997; Tröster et al., 1997; Masterman et al., 1998). The case of a PD patient with recurrent manic episodes associated with right, left, and bilateral pallidal stimulation was recently described (Miyawaki et al., 2000). The fact that both GPi and STN stimulation may induce euphoria or mania favors the involvement of the limbic loop within the basal ganglia.

On the other hand, large spontaneous pallidal lesions are known to induce behavioral changes of a frontal nature, most specifically a syndrome of abulia (Bhatia and Marsden, 1994), also described as a loss of psychic autoactivation (Laplane, 1990), pure psychic akinesia (Laplane et al., 1984), and an athymhormic syndrome (Habib and Poncet, 1988; Habib and Galaburda, 1999). This syndrome can also occur after pallidal surgery (Hassler and Riechert, 1954; Hartmann von Monakow, 1959; Trépanier et al., 1998) and is likely to be related to lesions located outside the sensorimotor part of GPi (Laitinen, 2000, Lombardi et al., 2000).

Possible Limbic-Related Targets for the Current Diffusion Hypothesis

Current diffusion to limbic-related brain structures outside the STN may also explain the emotional and behavioral effects of high frequency STN stimulation (Benabid et al., 2000). Indeed, the STN is situated close to several limbic structures, two of which are discussed below.

1. The Medial Forebrain Bundle. The reward/reinforcement circuitry of the mammalian brain depends on the mesolimbic dopaminergic system. Dopaminergic A10 cells in the ventral tegmental area project through the medial forebrain bundle mainly to the nucleus accumbens in the ventral striatum. A smaller part of the A10 fibers projects directly to the medial prefrontal cortex. The mesolimbic system controls emotional behavior. For example, food intake, sexual activity and psychostimulant drugs activate this dopaminergic projection. Intracranial self-stimulation (ICSS) (Olds and Milner, 1954) of the medial forebrain bundle leads to positive reinforcement of behavior (Olds, 1977). Animals with electrodes placed in the ventral tegmental area (A10) cell bodies, or in the medial forebrain bundle, rapidly self-stimulate and increase motor activity. Humans report subjective pleasure when these areas are stimulated (Heath, 1963; Heath, 1972; Crichton, 1972). Both destruction of these ascending dopamine projections and blockade of the dopamine transmission with neuroleptics attenuate this self-stimulation behavior, indicating that dopamine is the neurotransmitter largely responsible for the effects of ICSS related to the reward system (Wise, 1980; Gardner, 1997).

The ventral tegmental area is located medial to the STN and the substantia nigra, lying between the red nucleus and the mamillary bodies (Nieuwenhuys et al., 1988). The medial forebrain bundle crosses the lateral hypothalamus medially to the STN. Both the ventral tegmental area and the medial forebrain bundles are close to the STN and could be influenced by current spread. Since dopamine is the neurotransmitter of the mesolimbic system, the similarities between the effects of levodopa and STN stimulation described by Funkiewiez et al. (2000) could also be explained by current diffusion activating the medial forebrain bundle.

2. The Lateral Hypothalamus. The lateral hypothalamus, which is crossed by the medial forebrain bundle, has connections with the limbic system, thalamus, and the limbic areas of the midbrain. Hypothalamic hamartomas typically induce gelastic seizures with a non-infectious, mirthless, mechanical laughter (Berkovic and Andermann, 1999). Recently, however, a "pressure to laugh" associated with mirth has been described in three patients with small hypothalamic hamartomas. The hamartomas were attached to the floor of the third ventricle and the mamillary bodies (Sturm et al., 2000), just medial and in front of the STN. Foerster & Gagel (1934) reported remarkable findings in humans of mechanical stimulation of the floor of the third ventricle under local anesthesia producing laughter. The lateral hypothalamus is a possible candidate for current diffusion from the STN. Therefore, it is difficult to attribute effects observed with high frequency stimulation of this region with certainty to either the limbic part of the STN or the lateral hypothalamus. Anyway, both are intimately connected with limbic parts of the basal ganglia in a way similar and parallel to the connections of the lateral (sensorimotor) STN with motor-related parts of the basal ganglia (Groenewegen and Berendse, 1990).

CONCLUSIONS

Clinical and PET findings suggest that high frequency STN stimulation influences not only the sensorimotor but also the associative and limbic parts of the basal ganglia circuitry passing through the STN. The fact that emotional and behavioral changes related to high frequency STN stimulation are also seen with GPi stimulation, and mimic those of levodopa, favors a specific target-related limbic effect and not an effect due to current diffusion toward neighboring structures. Since high frequency stimulation is supposed to have an inhibitory effect on the STN, we postulate that some of the psychiatric symptoms of PD, such as loss of motivation and depression, are associated with an overactivity of the non-motor parts of this nucleus. Likewise a hypoactivity of the limbic STN, related to either high frequency stimulation of this target or dopaminergic treatment, may play a contributing role in the development of mania or psychosis, which can occur as complications of both therapeutic approaches.

ACKNOWLEDGMENTS: The authors would like to thank Dr. Rajeev Kumar for helpful discussions.

REFERENCES

Aarsland, D., Larsen, J.P., Lim, N.G., Janvin, C., Karlsen, K., Tandberg, E., and Cummings, J.L., 1999, Range of neuropsychiatric disturbances in patients with Parkinson's disease, *J. Neurol. Neurosurg. Psych.* 67:492.

Alexander, G.E., DeLong, M.R., and Strick, P.L., 1986, Parallel organization of functionally segregated circuits linking basal ganglia and cortex, *Annu. Rev. Neurosci.* 9:357.

Ardouin, C., Pillon, B., Peiffer, E., Bejjani, P., Limousin, P., Damier, P., Arnulf, I., Benabid, A.L., Agid, Y., and Pollak, P., 1999, Bilateral subthalamic or pallidal stimulation for Parkinson's disease affects neither memory nor executive functions. A consecutive series of 62 patients, *Ann. Neurol.* 46:217.

Barbeau, A., 1969, L-Dopa therapy in Parkinsons disease: A critical review of nine years experience, *Can. Med. Ass. J.* 101:791.

Barraquer-Bordas, L., and Peres-Serra , J., 1965, Syndrome subtalamico cerebelo-extrapyramidal oculosimpatico acompanado de insomnio y de expansion del humor, *Arch. Neurobiol. (Madr.)* 28:409.

Benabid, A.L., Koudsie, A., Pollak, P., Kahane, P., Chabardes, S., Hirsch, E., Marescaux, C., and Benazzouz, A., 2000, Future prospects of brain stimulation, *Neurol. Res.* 22:237.

Berkovic, S.F., and Andermann, F., 1999, Pathologic laughter, in: *Movement Disorders in Neurology and Neuropsychology*, A.B. Joseph and R.R. Young, eds., Blackwell Science, Malden, Ms.

Bhatia, K.P., and Marsden, C.D., 1994, The behavioural and motor consequences of focal lesions of the basal ganglia in man, *Brain*, 117:859.

Caputo, E., Ardouin, C., Krack, P., Funkiewiez, A., Van Blercom, N., Fraix, V., Limousin Dowsey, P., Xie, J., Moro, E., Koudsié, A., Benazzouz, A., Benabid, A.L., and Pollak, P., 2000, Changes in mood and behaviour with subthalamic nucleus stimulation in Parkinson´s disease, In preparation.

Cotzias, G., 1967, Aromatic amino acids and modification of parkinsonism, *N. Engl. J. Med.* 276:374.

Cotzias, G.C., Papavasiliou, P.S., and Gellene, R., 1969, Modification of parkinsonism: Chronic treatment with L-Dopa, *N. Engl. J. Med.* 280:337.

Crichton, M., 1972, *Terminal Man*, Ballantine Books, New York.

Cummings, J.L., 2000, A window on the role of dopamine in addiction disorders [editorial], *J. Neurol. Neurosurg. Psych.* 68:404.

Damasio, A.R., Lobo-Antunes, J., and Macedo, C., Psychiatric aspects in parkinsonism treated with L-dopa, *J. Neurol. Neurosurg. Psych.* 34:502.

Diederich, N.J., Alesch, F., and Goetz, C.G., 2000, Visual hallucinations in Parkinson´s disease: a case report on exclusive provocation by bilateral stimulation of the subthalamic nucleus, *Mov. Disord.* 15 (Suppl. 3):61.

Dubois, B., and Pillon, B., 1997, Cognitive deficits in Parkinsons disease, *J. Neurol.* 244:2.

Foerster, O., and Gagel, O., 1934, Ein Fall von Ependymcyste des III. Ventrikels. Ein Beitrag zur Frage der Beziehungen psychischer Störungen zum Hirnstamm, *Z. Ges. Neurol. Psychiatr.* 149:312.

Funkiewiez, A., Ardouin, C., Krack, P., Caputo, E., Van Blercom, N., Fraix, V., Xie, J., Moro, E., Benabid, A.L., and Pollak, P., 2000, Acute psychic effects of bilateral subthalamic nucleus stimulation in Parkinson´s disease, *Mov. Disord.* 15 (Suppl. 3):52.

Gardner, E.L., 1997, Brain reward mechanisms, in: *Substance abuse: A comprehensive textbook*, J.H. Lowinson, P. Ruiz, R.B. Millman, and J.G. Langrod, eds., Williams & Wilkins, Baltimore.

Ghika, J., Vingerhoets, F., Albanese, A., and Villmeure, J.G., 1999, Bipolar swings in mood in a patient with bilateral subthalamic deep brain stimulation (DBS) free of antiparkinsonian medication, *Parkinsonism & Related Dis.* 5, (Suppl.) 1:104.

Giovannoni, G., O´Sullivan, J.D., Turner, K., Manson, A.J., and Lees, A.J., 2000, Hedonistic homeostatic dysregulation in patients with Parkinson´s disease on dopamine replacement therapies, *J. Neurol. Neurosurg. Psych.* 68:423.

Giros, B., Jaber, M., Jones, S.R., Wightman, R.M., and Caron, M.G., 1996, Hyperlocomotion and indifference to cocaine and amphetamine in mice lacking the dopamine transporter, *Nature*, 379:606.

Goetz, C.G., Tanner, C.M., and Klawans, H.L., 1982, Pharmacology of hallucinations induced by long-term drug therapy, *Am. J. Psychiatry,* 139:494.

Groenewegen, H.J., and Berendse, H.W., 1990, Connections of the subthalamic nucleus with ventral striatopallidal parts of the basal ganglia in the rat, *J. Comp. Neurol.* 294:607.

Guiot, G., and Brion, S., 1953, Traitement des mouvements anormaux par la coagulation pallidale. Technique et résultats, *Rev. Neurol.* 89:578.

Habib, M., and Poncet, M., 1988, Perte de l' élan vital, de l'intérêt et de laffectivité (syndrome athymhormique) au cours de lésions lacunaires des corps striés, *Rev. Neurol.* 144:571.

Habid, M.H., and Galaburda, S.M., 1999, Motor and motivational disorders in limbic-paralimbic lesions, in: *Movement Disorders in Neurology and Neuropsychiatry*, A.B. Joseph and R.R. Young, eds., Blackwell Science, Malden, Ms.

Hartmann von Monakow, K., 1959, Halluzinosen nach doppelseitiger stereotaktischer Operation bei Parkinson-Kranken, *Arch. Psychiatr.*, 199:477.

Hassler, R., and Riechert, T., 1954, Indikationen und Lokalisationsmethode der gezielten Hirnoperationen, *Nervenarzt*, 25:441.

Heath, R.G., 1963, Electrical self-stimulation of the brain in man, *Am. J. Psychiatry,* 120:571.

Heath, R.G., 1972, Pleasure and brain activity in man, *J. Nerv. Ment. Dis.* 154:3.

Jaber, M., Jones, S., Giros, B., and Caron, M.G., 1997, The dopamine transporter: a crucial component regulating dopamine transmission, *Mov. Disord.* 12:629.

Krack, P., Kumar, R., Ardouin, C., Limousin Dowsey, P., McVicker, J.M., Benabid, A.L., and Pollak, P., 2000, Mirthful laughter induced by subthalamic nucleus stimulation, *Mov. Disord.* submitted.

Krack, P., Pollak, P., Limousin, P., Hoffmann, D., Xie, J., Benazzouz, A., and Benabid, A.L., 1998, Subthalamic nucleus or internal pallidal stimulation in young onset Parkinson's disease, *Brain*, 121:451.

Kumar, R., Krack, P., and Pollak, P., 1999, Transient acute depression induced by high frequency deep-brain stimulation [letter], *N. Engl. J. Med.* 341:1003.

Laitinen, L.V., 2000, Behavioral complications of early pallidotomy, *Brain Cogn.* 42:313.

Lang, A.E., Lozano, A., Tasker, R., Duff, J., Saint-Cyr, J., and Trépanier, L., 1997, Neuropsychological and behavioral changes and weight gain after medial pallidotomy, *Ann. Neurol.* 41:834.

Laplane, D., Baulac, M, Widlocher, D., and Dubois, B., 1984, Pure psychic akinesia with bilateral lesions of basal ganglia, *J. Neurol. Neurosurg. Psychiat.* 47:377.

Laplane, D., 1990, La perte dauto-activation psychique, *Revue Neurologique*, 146:397.

Limousin, P., Greene, J., Pollak, P., Rothwell, J., Benabid, A.L., and Frackowiak, R., 1997, Changes in cerebral activity pattern due to subthalamic nucleus or internal pallidum stimulation in Parkinson's disease, *Ann. Neurol.* 42:283.

Limousin, P., Krack, P., Pollak, P., Benazzouz, A., Ardouin, C., Hoffmann, D., and Benabid, A.L., 1998, Electrical stimulation of the subthalamic nucleus in advanced Parkinson's disease, *N. Engl. J. Med.* 339:1105.

Limousin, P., Pollak, P., Benazzouz, A., Hoffmann, D., Lebas, J.F., Brousolle, E., Perret, J.E., and Benabid, A.L., 1995, Effect on parkinsonian signs and symptoms of bilateral subthalamic nucleus stimulation, *Lancet*, 345:91.

Lombardi, W.J., Gross, R.E., Trépanier, L.L., Lang, A.E., Lozano, A.M., and Saint-Cyr, J.A., 2000, Relationship of lesion location to cognitive outcome following microelectrode-guided pallidotomy for Parkinson's disease: Support for the existence of cognitive circuits in the human pallidum, *Brain*, 123:746.

Martin, W.R., Sloan, J.W., Sapira, J.D., and Jasinski, D.R., 1971, Physiologic, subjective, and behavioral effects of amphetamine, methamphetamine, ephedrine, phenmetrazine, and methylphenidate in man, *Clin. Pharmacol. Ther.* 12:245.

Masterman, D., DeSalles, A., Baloh, R., Frysinger, R., Foti, D., Behnke, E., Cabatan-Awang, C., Hoetzel, A., Intemann, P.M., Fairbanks, L., and Bronstein, J.M., 1998, Motor, cognitive, and behavioral performance following unilateral ventroposterior pallidotomy for Parkinson disease, *Arch. Neurol.* 55:1201.

Maurice, N., Deniau, J.M., Glowinski, J., and Thierry, A.M., 1998a, Relationships between the prefrontal cortex and the basal ganglia in the rat: physiology of the corticosubthalamic circuits, *J. Neurosci.* 18:9539.

Maurice, N., Deniau, J.M., Menetrey, A., Glowinski, J., and Thierry, A.M., 1998b, Prefrontal cortex-basal ganglia circuits in the rat: involvement of ventral pallidum and subthalamic nucleus, *Synapse*, 29:363.

Miyawaki, E., Perlmutter, J.S., Troster, A.I., Videen, T.O., and Koller, W.C., 2000, The behavioral complications of pallidal stimulation: A case report, *Brain Cogn.* 42:417.

Narabayashi, H., Miyashita, N., Hattori, Y., Saito, K., and Endo, K., 1997, Posteroventral pallidotomy: its effect on motor symptoms and scores of MMPI test in patients with Parkinson's disease, *Parkin. & Rel. Dis.* 3:7.

Nieuwenhuys, R., Voogd, J., and VanHuijzen, C., 1988, *The Human Central Nervous System*, Springer, Berlin.

O'Brien, C.P., DiGiacomo, J.N., Fahn, S. et al., 1971, Mental effects of high-dosage levodopa, *Arch. Gen. Psychiatry*, 24:61.

Olds, J., and Milner, P., 1954, Positive reinforcement produced by electrical stimulation of septal areas and other regions of the rat brain, *J. Comp. Physiol. Psychol.* 47:419.

Olds, J., 1977, *Drives and Reinforcements. Behavioral Studies of Hypothalamic Functions*, Raven Press, New York.

Papez, J.W., 1937, A proposed mechanism of emotion, *Arch. Neurol.* 38:725.

Parent, A. and Hazrati, L.N., 1995, Functional anatomy of the basal ganglia. II. The place of subthalamic nucleus and external pallidum in basal ganglia circuitry, *Brain Res. Rev.* 20:128.

Pillon, B., Ardouin, C., Damier, P., Krack, P., Houeto, J.L., Klinger, M.A., Bonnet, A.M., Pollak, P., Benabid, A.L., and Agid, Y., 2000, Neuropsychological changes between "off" and "on" STN or GPi stimulation in Parkinson's disease, *Neurology*, 55:411.

Sacks, O., 1982, *Awakenings*, Pan Books Ltd., London.

Sturm, J.W., Andermann, F., and Berkovic, S.F., 2000, "Pressure to laugh": An unusual epileptic symptom associated with small hypothalamic hamartomas, *Neurology*, 54:971.

Trépanier, L.L., Saint-Cyr, J.A., Lozano, A., and Lang, A.E., 1998, Neuropsychological consequences of posteroventral pallidotomy for the treatment of Parkinson's disease, *Neurology*, 51:207.

Trillet, M., Vighetto, A., Croisile, B., Charles, N., and Aimard, G., 1995, Hémiballisme avec libération thymo-affective et logorrhée par hématome du noyau sous-thalamique gauche, *Rev. Neurol. (Paris)* 151:416.

Tröster, A.I., Fields, J.A., Wilkinson, S.B., Pahwa, R., Miyawaki, E., Lyons, K.E., and Koller, W.C., 1997, Unilateral pallidal stimulation for Parkinson's disease: Neurobehavioral functioning before and 3 months after electrode implantation, *Neurology,* 49:1078.

Wise, R.A., 1980, Actions of drugs abuse on brain reward systems, *Pharmacol. Biochem. Behavi.* 13:213.

Wolters, E.C., 1999, Dopaminomimetic psychosis in Parkinson's disease patients: Diagnosis and treatment, *Neurology* 52, Suppl. 3:10.

Yahr, M.D., Duvoisin, R.C., Schear, M.J., Barrett, E., and Hoehn, M.N., 1969, Treatment of parkinsonism with levodopa, *Arch. Neurol.* 21:343.

MICROELECTRODE RECORDING-GUIDED DEEP BRAIN STIMULATION IN PATIENTS WITH MOVEMENT DISORDERS (FIRST TRIAL IN KOREA)

JIN WOO CHANG, BAE HWAN LEE, MYUNG SIK LEE,
JONG HEE CHANG, YONG GOU PARK, AND
SANG SUP CHUNG *

INTRODUCTION

The modern surgical treatment for movement disorders has been refined by advances in neuroimaging, greater understanding of basic pathophysiology, and the development of new surgical techniques (Arle and Alerman, 1999). In addition, microelectrode recording techniques further enhance the ability of the surgeon to target subcortical areas with precision by correcting for variability among patients (Lozano and Hutchison, 1997).

Microelectrode recording and stimulation during stereotactic procedures in the thalamus, globus pallidus, and subthalamus serve as examples of this enhanced methodology (Lozano, 1998; Lozano and Hutchison, 1997; Tasker, 1999). Appropriate localization within thalamus, or the internal segment of globus pallidus, or subthalamus, respectively, may be made at a level of precision superior to that achieved by imaging techniques alone. Furthermore, microelectrode recording techniques make a new surgical approach to the subthalamic nucleus feasible by precisely and reproducibly localizing the subthalamic nucleus (Benabid, 1999). We have used deep brain stimulation (DBS) for movement disorders since February, 2000. In this report, we analyze the findings of microelectrode recording data and report the preliminary surgical outcomes in Korean movement disorder patients.

---

* JIN WOO CHANG • Department of Neurosurgery, and Brain Research Institute, Yonsei University College of Medicine, Seoul 120-752, Korea. BAE HWAN LEE • Medical Research Center, and Brain Research Institute, Yonsei University College of Medicine, Seoul 120-752, Korea. MYUNG SIK LEE • Department of Neurology, Yonsei University College of Medicine, Seoul 120-752, Korea. JONG HEE CHANG • Department of Neurosurgery, Yonsei University College of Medicine, Seoul 120-752, Korea. YONG GOU PARK • Department of Neurosurgery, Medical Research Center, and Brain Research Institute, Yonsei University College of Medicine, Seoul 120-752, Korea. SANG SUP CHUNG • Department of Neurosurgery, and Brain Research Institute, Yonsei University College of Medicine, Seoul 120-752, Korea.

MATERIAL AND METHODS

Patient Selection

Since February 2000, we have used DBS for treatment of movement disorders (Table 1). One patient suffering from disabling right hand tremor was treated surgically. The age at the time of surgery was 58 years and the duration of the disease was 10 years. The patient had tried propranolol and Cogentin®/Artane® at maximal tolerated dosage without effective clinical benefits on daily activities (ADL). Three mainly akinetic and rigid idiopathic advanced parkinsonian patients were operated on for disabling symptoms. All patients were severely impaired in ADL. All patients were maximally treated with levodopa combined with peripheral Dopa decarboxylase inhibitor associated with a dopaminergic agonist. All patients showed severe on-off fluctuation with these medications.

Microelectrode Recording Technique

The methods were similar to those previously reported by the Toronto Group (Lozano, 1998; Lozano and Hutchison, 1997; Tasker, 1999). The surface EMG was recorded from contralateral wrist & leg flexors. Neuronal impulses were amplified, filtered, and fed to a window discriminator and audio monitor, and signals were displayed on oscilloscopes and stored using a video recorder. Units were selected if they were stable and well isolated. Data was replayed, digitized, and collected with an interface (CED 1401) and saved in computer files for manipulation and display (Spike 2). Single unit histograms were constructed and evaluated for differences from baseline. Units were recorded for at least 60 s (usually 120 s) and the mean firing rates calculated. The location in the trajectory of each recorded cell was noted and the neurons plotted on the appropriate sagittal brain map.

Table 1. Clinical profiles of DBS for movement disorders

Patient	Sex/Age	Diagnosis	Recording sites	DBS insertion
1	M/58	Essential tremor	STN x 3, Vim x 2	Vim, unilateral
2	F/55	Advanced PD	STN x 4	STN, bilateral
3	M/44	Advanced PD	STN x 4	STN, bilateral
4	F/59	Advanced PD	STN x 4	STN, bilateral

Operative Procedure

At our institute, we use the Leksell G frame (Elekta, Inc) and its accessories to localize the target. We utilize custom-made software (K-N Plan version 1.0) to localize the anatomical target with MR imaging from a Signa 1.5-tesla MRI magnet (GE, Milwaukee, WI, USA). Since March, 1999, we have used this microelectrode recording technique for movement disorders. After localization of the target, electrodes were implanted either in the Vim (DBS 3387 electrode; Medtronic, Minneapolis, MN, USA) or the subthalamic nucleus (DBS 3389 electrode; Medtronic, Minneapolis, MN, USA). Programmable stimulators (Itrell II; Medtronic) were subsequently implanted in the subclavicular area.

Parameter of Stimulation

The neurologist selected electrical settings after surgery and at each follow-up visit.

The patients could not modify the stimulation themselves and were also unaware of the conditions of stimulation applied. A constant pulse width of 60 μs and frequency of 130 Hz was used, and the voltage was progressively changed (voltage ranging from 2 to 3.5 volts).

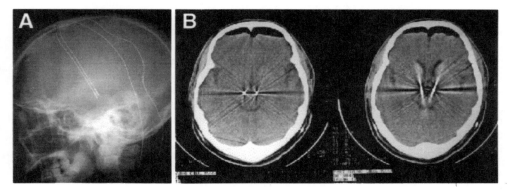

Figure 1. Chronic bilateral subthalamic DBS for idiopathic Parkinson's disease. Skull lateral view (A) and brain CT scan (B) demonstrate the bilateral electrode in the subthalamus of the brain.

Clinical Assessment

Assessments were made of global motor function with the following: Unified Parkinson's Disease Rating Scale (UPDRS); ADL of UPDRS; ADL of Schwab and England; Northwestern disability score; Hoehn & Yahr score; and dyskinesia rating score (Hagel and Widner, 1999). These were carried out in on-phase and off-phase periods (after an overnight withdrawal) before surgery, and then one and two months after surgery.

Table 2. Effect of bilateral subthalamic stimulation in two patients after a follow-up of 2 months

Parameter	Preoperative state		Postoperative state	
	Drug off	Drug on	Drug off/DBS on	Drug on/DBS on
UPDRS motor	51 (61,41)	24.5 (34,15)	36 (40,32)	20 (29,11)
UPDRS ADL	28.5 (37,20)	10 (17,3)	21 (25,17)	8 (13,3)
Schwab & England ADL	30 % (10,50)	75 % (60,90)	55 % (40,70)	90 % (90,90)
Northwestern disability	21 (10,32)	41.5 (34,49)	33 (27,39)	43.5 (40,47)
Hoehn & Yahr score	4 (3,5)	2.5 (3,2)	3.25 (4,2.5)	2.5 (3,2)
Dyskinesia rating scale				
Hyperkinesias	0.5 (0,1)	6 (11,1)	0 (0,0)	2.5 (5,0)
Dystonia	5.5 (8,3)	5 (9,1)	0.5 (0,1)	2.5 (4,1)
Off time				
% of waking time	51.5 % (56,47)		No off time	

UPDRS: Unified Parkinson's Disease Rating Scale, ADL: activities of daily living

RESULTS

Clinical Features

Definite improvement of symptoms occurred from microelectrode recording and DBS electrode insertion with subsequent waning in all patients. At two months follow-up, tremors completely disappeared from the patient with essential tremor. In the two patients with advanced idiopathic Parkinson's disease followed for more than two months, bilateral subthalamic stimulation greatly improved all scores listed in Table 2. Although all features of parkinsonian symptoms improved, the greatest benefit occurred in "off" time and ADL. Interestingly, our two patients did not have "off" time after bilateral stimulation of subthalamic nucleus.

In our preliminary experiences, there were no adverse side effects related to microelectrode recording or DBS procedures in all four patients. Figure 1 shows DBS electrodes which were inserted into the bilateral subthalamic nucleus of advanced idiopathic Parkinson's disease patients.

Microelectrode Recording Findings

Data from microelectrode recording of thalamus for essential tremor were similar to earlier reports (Figure 2). However, in our preliminary findings from microelectrode recording of subthalamic nucleus, we observed several features that were different (Table 3). Most importantly, though, subthalamus showed higher firing rates (55.45±5.1 Hz) than the substantia nigra pars reticularis (29.8±5.2 Hz). Furthermore, mean burst frequency of

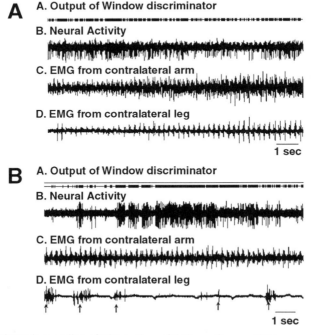

Figure 2. Microelectrode recording of Vim nucleus of thalamus for essential tremor. A: Bursting discharges recorded in Vim of the thalamus correlated with the frequency of patient's tremor, B: Bursting discharges of Vim nucleus were activated by the contralateral leg movements.

Table 3. Findings of microelectrode recording for bilateral subthalamic DBS implantation

Nucleus	Total Neuron	Mean Firing Rate	Number of Bursting Cell	Number of Rhythmic Bursting Cell (%)	Mean Burst Index
VOA	56	24.76±3.4	46	25 (54.3)	2.42±0.19
VOP	16	35.7±5.1	16	8 (50)	2.06±0.17
RN	21	39.3±5.78	18	13 (72.2)	2.42±0.31
ZI	48	28.9±6.2	27	15 (55.5)	2.59±0.16
STN	116	55.45±5.1	58	36 (62.3)	2.85±0.26
SNR	15	29.8±5.2	13	11 (84.6)	1.63±0.28

VOA: ventralis oralis anterior, VOP: ventralis oralis posterior, RN: reticular nucleus, ZI: zona incerta, STN: subthalamic nucleus, SNR: substantia nigra pars reticularis

subthalamic nucleus (2.85±0.26) was much higher than that of substantia nigra pars reticularis (1.63±0.28) (Figure 2). However, we need to collect more data because of the limited number of patients.

DISCUSSION

The modern surgical treatment of movement disorders was introduced more than three decades ago when an unexpected intra-operative complication occurred during pyramidotomy, leading Cooper and colleagues to the discovery that destruction of portions of thalamus suppressed tremor in a patient with Parkinson's disease (Arle, 1999). The surgical options for treating movement disorders are expanding with a variety of new treatment modalities (Benabid, 1999; Lozano, 1998; Limousin, 1998; Renai, 1999; Tronnier, 1999; Tsubokawa and Katayama, 1999). Perhaps the most exciting development in movement disorder surgery is bilateral subthalamic stimulation, which improves the most disabling and medically resistant symptoms of Parkinson's disease. In addition, microelectrode recording techniques further enhance the surgeon's ability to target subcortical areas with precision by correcting for variability among patients (Lozano and Hutchison, 1997; Tasker, 1999).

The authors have used this microelectrode recording for the treatment of movement disorder since March, 1999. Also, since February 2000, we have applied the implantation of DBS for the treatment of movement disorder.

As we expected, all patients showed definite improvements of symptoms after surgery. Interestingly, all patients demonstrated an immediate improvement of symptoms from microelectrode recording and DBS electrode insertion. There was also a subsequent waning of ameliorative effect, and its duration was variable in each patient. Although all features of parkinsonian symptoms improved, the greatest benefit occurred in "off" time and ADL. Our two patients who were followed-up for more than two months did not have "off" time after bilateral stimulation of subthalamic nucleus. We don't know the meaning of this yet. We will collect more data from other patients and observe the changes in this parameter. In our preliminary experiences, we did not observe any adverse side effects related to microelectrode recording and DBS procedures in any of the four patients. This may be an effect of the relatively young age of our patient population.

Firing pattern and mean firing rates of subthalamic nucleus and substantia nigra pars reticularis were reported by several groups (Benabid, 1999; Rezai, 1999). Those reported by Benabid et al. (1999) demonstrated hyperactivity of subthalamic nucleus and low firing rate of substantia nigra pars reticularis. However, the Toronto group reported quite different data, including irregular (mean 36.9±17.0 Hz) and relatively lower firing patterns of the

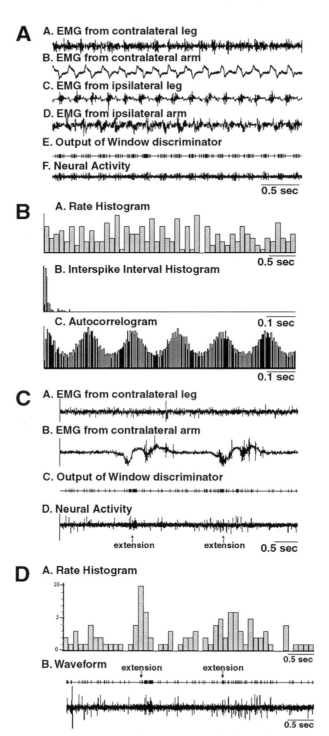

Figure 3. Microelectrode recording of subthalamic nucleus for idiopathic Parkinson's disease. A: Bursting discharges recorded in subthalamic nucleus corresponded with the frequency of patient's tremor, B: Example of autocorrelogram of a tremor correlated cell in subthalamic nucleus, C: Bursting discharges of subthalamic nucleus were activated by the contralateral elbow extension, D: Example of movement-related activity (bottom) and the autocorrelogram of a bursting cell.

subthalamic nucleus and high frequency regular firing rate of substantia nigra pars reticularis (mean 70.6±23.1 Hz). Our data demonstrated several different features. Most importantly, subthalamic nucleus showed a higher firing rate (55.45±5.1 Hz) than that of substantia nigra pars reticularis (29.8±5.2 Hz). Furthermore, mean burst index of subthalamic nucleus (2.85±0.26) was much higher than that of substantia nigra pars reticularis (1.63±0.28) (Figure 2). More data needs to be collected before the significance of these differences becomes clear.

CONCLUSIONS

The first trials of DBS in Korea demonstrated favorable preliminary outcomes for movement disorders. Interestingly, all patients with advanced Parkinson's disease ended up having no "off" time. Also, microlesion effects were remarkable in all patients, with no side effects from DBS. Microelectrode recordings revealed specific activity patterns in subthalamic nucleus and substantia nigra. Following these promising preliminary results, we will continue to find and to evaluate the distinctive features of our Korean patients with movement disorders for DBS.

ACKNOWLEDGMENTS: This work was supported by faculty grants from Yonsei University College of Medicine (1999, S.S. Chung).

REFERENCES

Arle, J.E., and Alerman, R.L., 1999, Surgical options in Parkinson's disease, *Med. Clin. North. Am.* 83:483.

Benabid, A.L. Benazzouz, A., Hoffmann, D., Limousin, P., Gay, E., Payen, I., and Benazzouz, A., 1999, Chronic electrical stimulation of the ventralis intermedius nucleus of the thalamus and of other nuclei as a treatment for Parkinson's disease, *Techn. in Neurosurg.* 5:5.

Hagell, P., and Widner, H., 1999, Clinical rating of dyskinesias in Parkinson's disease: use and reliability of a new rating scale, *Movement Dis.* 14:448.

Limousin, P., Krack, P., Pollak, P., Benazzouz, A., Ardouin, C., Hoffman, D., and Benabid, A.L., 1998, Electrical stimulation of the subthalamic nucleus in advanced Parkinson's disease, *New Eng. J. Med.* 339:1105.

Lozano, A., 1998, Thalamic deep brain stimulation for the control of tremor, *Neurosurg. Op. At.* 7:125.

Lozano, A.M., Hutchison, W.D., 1997, Microelectrode-guided pallidotomy, *Neurosurg. Op. At.* 6:27.

Rezai, A.R., Lozano, A.M., Crawley, A.P., Joy, M.L., Davis, K.D., Kwan, C.L., Dostrovsky J.O., and tasker, R.R., 1999, Chronic subthalamic nucleus stimulation for Parkinson's disease, *Neurosurg. Op. At.* 8:195.

Tasker, R.R., Dostrovsky, J.O., and Hutchison, 1999, Microelectrode recording technology, *Tech. in Neurosurg.* 5:46.

Tronnier, V.M., Krause, M., Heck, A., Kronenburger, M., Bonsanto, M.M., Tronnier, J., and Fogel, W., 1999, Deep brain stimulation for the treatment of movement disorders, *Neurol. Psych. Brain Res.* 6:199.

Tsubokawa, T., and Katayama, Y., 1999, Lesion-making surgery versus brain stimulation for treatment for Parkinson's disease, *Crit. Rev. Neurosurg.* 9:96.

NEUROTRANSPLANTS IN TREATMENT
OF PARKINSON'S DISEASE

MICHAEL V. UGRUMOV[*]

INTRODUCTION

Parkinson's disease is a neurodegenerative disorder characterized mainly by a loss of dopaminergic neurons in the substantia nigra that results in deafferentation of the striatum. In patients, this is manifested in tremor, rigidity, hypokinesia, and postural instability (Agid, 1991; Wichmann and DeLong, 1993; Pogarell and Oertel, 1999). Initial symptoms of Parkinson's disease emerge after a 60-80% loss of the nigral dopaminergic neurons and reduction of striatal dopamine (Berheimer et al., 1973; Kish et al., 1988). The conventional dopamine replacement therapy using L-dopa or dopamine agonists initially provides substantial improvement but finally leads to complications, such as disabling dyskinesias, motor fluctuations, and L-dopa non-responsive features (freezing, postural instability, dementia, and autonomic dysfunction) (King, 1999; Metman and Mouradian, 1999). Such complications prompted the search for alternative treatments to restore impaired functions in patients with Parkinson's disease.

Degeneration of dopaminergic neurons in the substantia nigra has been reproduced in animals by injections of 6-hydroxydopamine (rats, non-human primates) or 1-methyl-4-phenyl-1,2,3,6-tetrahydropyridine (non-human primates) into the nigrostriatal system. The grafting of adrenal medulla or fetal substantia nigra to the denervated striatum of parkinsonian animals improves local dopamine turnover and animal behavior. From a physiological standpoint, the implantation of fetal substantia nigra seemed to be much more efficient than implantation of adrenal medulla (Freed et al., 1981; Brundin et al., 1988; Björklund, 1991; Olanow et al., 1997). More detailed studies showed that the nigral allografts and xenografts (from humans) survive in the brains of animals, and the donor dopaminergic neurons synaptically reinnervate the recipient striatum (Brundin et al., 1988; Björklund, 1991; Lindvall, 1991). These data triggered the development of a new approach to the treatment of neurodegenerative diseases: namely, neurotransplantation, or the

[*] MICHAEL V. UGRUMOV • Institute of Developmental Biology, RAS, and Institute of Normal Physiology, RAMS, Moscow 117808, Russia.

replacement of degenerated neurons in patients with homologous neurons from human fetuses.

Neurotransplantation is based on the idea that implanted cells can function as local "biological minipumps" to supply specific neurotransmitters to neurotransmitter-deficient brain regions (Björklund, 1991). This strategy is particularly promising for treatment of Parkinson's disease for two main reasons: (1) its pathogenesis corresponds to selective degeneration of the well-defined nigrostriatal dopaminergic system, and (2) dopamine replacement therapy provides considerable clinical benefits to Parkinson's patients. Adrenal cells were initially used for transplantation in patients with Parkinson's disease (Backlund et al., 1985; Madrazo et al., 1987), but at present, fetal mesencephalic neurons are used at the majority of transplant centers because of their superior efficiency (Olanow et al., 1997). More than 200 patients world-wide have received unilateral or bilateral grafts and were followed for a maximum of ten years after the operation (Olanow et al., 1996; Lindvall, 1998).

This chapter seeks to summarize the results from implantations of human fetal substantia nigra into the nigrostriatal system of patients with Parkinson's disease accumulated by a number of teams, including the participants of the European program "Network for European CNS Transplantation And Restoration" (NECTAR).

ACQUISITION, CHARACTERIZATION, AND PROCESSING OF HUMAN FETAL MESENCEPHALON

Human embryos (age < 8 weeks) or fetuses (age > 8 weeks) are obtained from women undergoing elective abortions in accordance with federal laws and ethical committee guidelines (Boer, 1994; Olanow et al., 1997). Fetuses resulting from spontaneous abortions have never been used, since they can contain CNS defects as a consequence of genetic or congenital diseases related to the pathological pregnancies. The women donors are carefully examined before abortion in order to avoid transmission of infectious agents, thereby minimizing the risk of recipient contamination and graft rejection. Moreover, the blood of potential donors is screened, usually for the human immunodeficiency virus types 1 and 2, human T-lymphotropic virus, hepatitis A, B, C viruses, cytomegalovirus, toxoplasma, syphilis, and herpes simplex. Donors are excluded when there is evidence of infection, or if there is a history of certain maternal pathologies (Lindvall et al., 1989; Ugryumov et al., 1996; Olanow et al., 1997).

In the course of the abortion, human fetuses are gently aspirated via a plastic curette under ultrasound control. Then, the brain is removed from the skull and the ventral mesencephalon, corresponding to the primordium of the substantia nigra, is dissected out for subsequent processing (Figure 1) (Lindvall et al., 1989; Ugrumov et al., 1996). Clinical benefits to patients with Parkinson's disease strictly depend on the transplant's specific characteristics, such as donor age, number of donors, and type of graft (cell suspension, solid pieces, cell culture). Moreover, the method of preoperative tissue storage and the site of implantation are of particular importance.

The most important variable turned out to be donor age (Freeman and Kordower, 1991), which is estimated by measuring the crown-rump length of the embryos/fetuses, first under ultrasound *in utero* (Lindvall et al., 1989; Ugrumov et al., 1996) and then *ex vivo* after the removal of the embryos/fetuses from the uterus (O'Rahilly and Muller, 1987). Hand and foot morphology can also be useful characteristics for determining the age of the embryos/fetuses if the appendages remain intact during the course of the abortion.

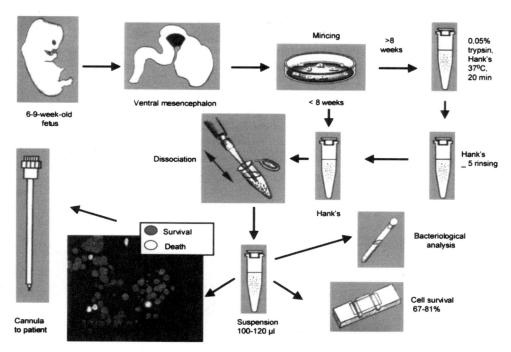

Figure 1. Schematic representation of the processing of the human fetal mesencephalon for grafting to patients with Parkinson's disease (reprinted from Neuroscience Letters, v. 212, Ugrumov et al., Development of the mesencephalic and diencephalic catecholaminergic systems in human fetuses and release of catecholamines *in vitro*, pp. 29-32, 1996, with permission from Elsevier).

The optimal age for graft survival covers the period from the first appearance of dopaminergic neurons in the subventricular zone of the mesencephalon in the embryos at the 6th week, to the initial arrival of their axons in the striatum at the 9th week (Brundin et al., 1988; Freeman et al., 1991, 1995). The earliest dopaminergic neurons were usually detected according to catecholamine histofluorescence or by using immunocytochemistry for tyrosine hydroxylase, the first rate-limiting enzyme of dopamine synthesis (Nobin and Björklund, 1973; Pickel et al., 1980; Freeman et al., 1991; Zecevic and Verney, 1995). Surprisingly, there are no data in the available literature about the ontogenetic schedule of the expression of aromatic L-amino acid decarboxylase, the second enzyme of dopamine synthesis. Indeed, the routine techniques fail to identify dopaminergic neurons with certainty because immunocytochemistry for tyrosine hydroxylase can detect non-dopaminergic neurons with transient expression of this enzyme (Beltramo et al., 1994) or those with no expression of aromatic L-amino acid decarboxylase (Ikemoto et al., 1998; Balan et al., in press). In turn, the histofluorescent technique can detect neurons that contain aromatic L-amino acid decarboxylase but that have no tyrosine hydroxylase (Jaeger, 1986; Balan et al., in press). In these neurons, dopamine is synthetized from extracellular L-dopa (Zoli et al., 1993), but not from L-tyrosine as it occurs in dopaminergic neurons. Therefore, we also applied the "isotopic" biochemical technique to determine whether the neurons/fibers of the ventral mesencephalon of human fetuses express the mechanism for dopamine-specific uptake (Ugryumov et al., 1996). Such neurons were initially detected at the 6th week, the earliest age studied. From the 6th to the 10th week, the dopamine uptake rises gradually (Figure 2), suggesting the continuous increase of dopaminergic neurons/fibers either in number or in size. Surprisingly, the mesencephalic dopaminergic neurons/fibers are not capable of dopamine release in response to membrane depolarization from the 6th to the 12th week, the whole period studied. Still, there was a spontaneous release of dopamine from the 6th week, and its level appeared to decrease with age (Ugryumov et al., 1996).

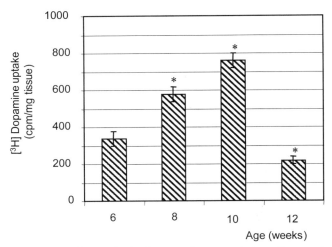

Figure 2. [³H]Dopamine specific uptake in the ventral mesencephalon of human fetuses (modified from Ugryumov et al., 1996). * P < 0.05.

The second most important characteristic of the graft is the number of donors necessary to provide optimal clinical benefits. This value may be approximated from the fact that 60,000 dopaminergic neurons of the human substantia nigra project their axons to a single putamen (Pakkenberg et al., 1991), whereas only 20,000 (5-20%) dopaminergic neurons survive after grafting of mesencephalic tissue from a single human fetus into immunosuppressed rodents (Brundin et al., 1988). This means that the tissues from a minimum of three donors should be implanted into each putamen in order to restore the local dopaminergic activity. It is also preferable to implant an excessive number of dopaminergic neurons in order to compensate for potential neuron degeneration induced by immune rejection, or an ongoing disease process. Moreover, only a part of the grafted dopaminergic neurons project their axons to the recipient putamen. Recent PET scan and clinical observations suggest that larger quantities of donor tissue (3-5 donors per side) provide enhanced functional recovery and increased striatal [¹⁸F]DOPA uptake (Rémy et al., 1995; Lindvall, 1998).

The third donor variable that can greatly influence graft survival, thereby affecting the graft's clinical benefit, is the immunological reactivity of the implanted cells. The human fetal mesencephalic cells express the major histocompatability complex (HLA-DR) class II antigens from at least the 6th week, the first age studied, and can stimulate *in vitro* proliferation of potential recipients' lymphocytes (Ugrumov et al., 2000). Although the brain is generally considered to be an immunologically privileged organ, rejection of the graft cannot be excluded. This in fact does occur when using xenografts in animals (Brundin et al., 1988; Borlongan et al., 1996). In any case, graft antigens are exposed to the immune system because surgical trauma results in the transient disruption of the blood brain barrier.

Therefore, short-term immunosuppression with cyclosporin A appears to be desirable for graft survival. In addition, this treatment can enhance the viability of the differentiating dopaminergic neurons, as recently demonstrated for rat neurons in culture and in the allograft (Castilho et al., 2000). The majority of patients who have shown graft survival under PET examination have received immunosuppression for a few days before transplantation and for at least 6 months thereafter (Borlongan et al., 1996). The subsequent immunosuppression of patients for up to 3 years did not interfere with graft survival (PET, autopsy) and clinical benefits (Kordower et al., 1995; Wenning et al., 1997; Lindvall, 1998).

Just before surgical intervention, the dissected mesencephali of three to five human fetuses at the ages of 6 to 9 weeks are minced and mechanically dissociated for single cells and small cell conglomerates (Figure 1) (Lindvall et al., 1989; Ugrumov et al., 1996). Cell dissociation can be facilitated by the pretreatment of tissue with 0.05-0.1% trypsin, though there is a risk of damaging mesencephalic cells. The meninges are carefully removed beforehand in order to avoid uncontrolled outgrowth of non-neuronal tissues in the recipient brain. Cell suspension of the embryonic mesencephali is recommended for transplantation if the cell viability exceeds 65%. A small part of the suspension is used for the bacteriological analysis (Lindvall et al., 1989; Ugryumov et al., 1996). A number of transplant centers use small solid pieces (~1 mm^3) of the ventral mesencephali instead of cell suspension (Olanow et al., 1997).

The transplantation procedure usually occurs 3-6 hours after the abortions. During this period the fetuses and then the dissected mesencephalic tissues are stored in Hank's solution at 4°C. There are obvious difficulties in accumulating human embryonic material within a short period of time, i.e., just before transplantation. Mesencephalic tissues can be stored for several days in a chemically defined "hybernation medium" with little or no negative effects on cell viability (Sauer and Brundin, 1991; Olanow et al., 1997). Hybernation of the embryonic tissue makes it possible to prolong the accumulating period. In turn, this enables the accumulation of a considerable number of donors, sufficient even for bilateral grafting.

Attempts to use cryopreservation for long-term storage of embryonic nervous tissue were unsuccessful, since they led to a significant decrease in cell viability and the outgrowth of neural processes (Collier et al., 1993). Still, according to Spencer et al. (1992) embryonic mesencephalic tissues stored in liquid nitrogen for up to 10 months could be successfully used for grafting in Parkinson's disease patients. The problem of accumulation of human embryonic material may also be solved by using primary dissociated cell culture. Freed and co-workers (1999) recently reported that the mesencephalic dopaminergic neurons of 7-8 week-old human embryos, cultured for 1 to 4 weeks, survive in the striatum of patients with Parkinson's disease. However, the clinical benefits have been achieved only in patients younger than 60 years.

IMPLANTATION OF HUMAN FETAL MESENCEPHALON INTO PATIENTS WITH PARKINSON'S DISEASE AND EVALUATION OF ITS FUNCTIONAL ACTIVITY

Although several groups all over the world have performed a total of more than 200 transplantations in patients with Parkinson's disease, it is rather difficult to compare the results because of differences in patient selection, transplant variables, drug management, and methods of evaluation. Still, the Core Assessment Protocol for Intracerebral Transplantation (CAPIT), a standardized protocol of clinical evaluation (Langston et al.,

1992), has been used by a number of teams to facilitate comparison of results among transplant centers. The patients with severe idiopathic Parkinson's disease, who cannot be satisfactorily controlled by conventional therapy, are usually selected for transplantation according to the CAPIT protocol. The majority of the patients met the following requirements: (1) age less than 65; (2) Hoehn-Yahr stage 2.5-4.0; (3) rigidity and hypokinesia as major symptoms; (4) no pronounced dementia; (5) a definite response to either L-dopa or bromcriptine (Langston et al., 1992; Olanow et al., 1997; Lindvall, 1998). The L-dopa responsive patient should show no less than a 33% improvement in their worst "off period" according to the UPDRS score. Patients older than 65 are *a priori* less able to respond to the grafting procedure (Freed et al., 1999), apparently because of dramatic irreversible anatomic changes (Olanow et al., 1997; Lindvall, 1998). Selected patients usually receive immunosuppressive therapy with cyclosporin A (daily dose = 2-10 mg/kg b wt) for a couple of days prior to, and for at least 6 months after the surgical procedure (Lindvall et al., 1989; Spencer et al., 1992; Hauser et al., 1999). However, according to Freed et al. (1992), there is practically no difference in clinical improvement in patients with or without immunsuppressive therapy.

In all operated patients with Parkinson's disease, the fetal nigral graft was implanted in the striatum, the target for dopaminergic afferents, but not in the substantia nigra, the homologous area. This is because of the failure of the axons of differentiating dopaminergic neurons, grafted to the substantia nigra, to extend for such a long distance in the adult brain (Björklund et al., 1994). The nigral neurons grafted in the striatum do not receive adequate afferent innervation and do not have the extracellular microenvironment that would contribute to the regulation of neuron differentiation. This might explain, at least in part, the dramatic degeneration of the grafted dopaminergic neurons and associated incomplete functional restoration observed, both in animals with parkinsonism and in patients with Parkinson's disease.

When using a unilateral graft, the site of implantation in patients is the side opposite to that with more severe parkinsonian symptoms. Moreover, a number of bilateral grafts have also been made. The interval between contralateral unilateral transplantations varied from four weeks (Hauser et al., 1999) to 56 months (Hagell et al., 1999). According to the surgical procedure, a burr hole is made under local or general anesthesia, and a stereotactic transplant needle with an inner diameter of about 1.0-1.5 mm is passed to the target site (Lindvall et al., 1989; Ugryumov et al., 1996). Mesencephalic tissues are most often grafted to both putamen and caudate, or only to putamen (Olanow et al., 1997; Lindvall, 1998). Still, a small number of patients have received a graft only in the caudate (Spencer et al., 1992). The cell suspension or small tissue pieces are usually deposited along three different tracks, two in the putamen and one in the head of the caudate. Furthermore, three to four deposits per track are placed at several millimeter intervals along each track. Taking into account the autopsy and PET data on greater dopamine depletion in the postcommissural putamen than in the anterior putamen/caudate nucleus complex in Parkinson's disease (see below), the postcommissural putamen may be considered the preferable target for grafting (Olanow et al., 1997). Therefore, some teams make six to eight needle tracks and four deposits per track throughout the postcommissural putamen (Olanow et al., 1997). Following surgery, patients are monitored in a hospital for approximately 2-3 days. No significant adverse effects from the implantation are usually registered.

The milestones of neurotransplantation appear to be the graft survival, differentiation of grafted dopaminergic neurons, and the dopaminergic reinnervation of the patient's striatum. PET scan is the most adequate noninvasive technique to assess graft results by evaluating the rate of [^{18}F] dopa incorporation to dopamine synthesis in the striatum.

During the postoperative period from 6 to 12 months, PET scan in most patients showed a progressive increase of [^{18}F] dopa uptake in the grafted putamen, a minor increase, if any, in the grafted caudate, and a decrease in the non-grafted striatum (Hoffer et al., 1992; Sawle et al., 1992; Defer et al., 1996; Wenning et al., 1997; Lindvall, 1998; Hauser et al., 1999). According to the results of the Swedish group, [^{18}F] dopa uptake in the grafted putamen increased 68% on average (when compared with preoperative levels), while it decreased in the non-grafted striatum approximately 28% (Wenning et al., 1997; Hagell et al., 1999). In some patients, [^{18}F] dopa uptake in the grafted striatum continued to increase for a postoperative period longer than one year (Lindvall et al., 1994) and was restored to normal levels (Olanow et al., 1997; Piccini et al., 1999). In contrast to the majority of authors, Spencer et al. (1992) have observed a relatively high level of [^{18}F] dopa uptake in the grafted caudate. In summary, PET scan data suggest an increased [^{18}F] dopa uptake by presynaptic dopaminergic terminals in the grafted striatum and the continuous degeneration of the patients' own dopaminergic neurons (Lindvall et al., 1994).

The suspected graft survival has been recently confirmed by histopathologic studies in two patients with Parkinson's disease who received bilateral grafts from fetuses 6.5 to 9 weeks of age (Kordower et al., 1995, 1996, 1998). The patients died 18 and 19 months after neurotransplantation for reasons unrelated to the treatment. In both cases, an increase of [^{18}F] dopa uptake was associated with graft survival, differentiation of dopaminergic neurons, and dopaminergic reinnervation of the host striatum (Kordower et al., 1995, 1996, 1998). Dopaminergic neurons and fibers were identified in this study by using mono-immunolabeling of tyrosine hydroxylase or dopamine transporting protein. The grafts in either putamen contained between 80,000 and 130,000 tyrosine hydroxylase-immuno-reactive neurons corresponding to approximately 5% of the total grafted dopaminergic neurons. However, only a few neurons (cell bodies) immunoreactive for dopamine transporting protein were observed in the same grafts (Kordower et al., 1996). According to the authors' opinion, all tyrosine hydroxylase-immunoreactive neurons are the true dopaminergic neurons, which are characterized by a low level of either expression or accumulation of dopamine transporting protein (Kordower et al., 1996). Still, it might be possibile that tyrosine hydroxylase is expressed in non-catecholaminergic neurons producing L-dopa as the final synthetic product (Meister et al., 1986, 1988). Indeed, the neurons expressing tyrosine hydroxylase but lacking aromatic L-amino acid decarboxylase were found to be widely distributed throughout the brain in many species, including the substantia nigra in humans (Kitahama et al., 1990; Ikemoto et al., 1998; Balan et al., in press). This problem could be elucidated in a future study using the double-immunolabeling of tyrosine hydroxylase and aromatic L-amino acid decarboxylase.

Reinnervation of the patient striatum, provided by the grafted dopaminergic or at least tyrosine hydroxylase-immunoreactive neurons, was restricted. It extended from each implant in the host striatum for no more than 5-7 millimeters. In general, the dopaminergic neurons synaptically reinnervated from 30% to 78% of the grafted postcomissural putamen. Only sparse dopaminergic fibers were observed in the anterior non-grafted putamen. In spite of the HLA-DR expression by grafted cells (Kordower et al., 1997), minimal immunosuppression (cyclosporin A, 6 mg/kg b wt per day for the first 8-11 weeks and 2 mg/kg b wt per day for subsequent 6 months) was sufficient to prevent graft rejection (Kordower et al., 1998). The histopathologic and PET data were in agreement with neurological examinations showing a progressive amelioration of parkinsonian symptoms in the same patients (Kordower et al., 1996, 1998).

Although [^{18}F] dopa PET scan and histopathologic studies definitely showed that the grafted neurons synthesize dopamine, they failed to determine whether dopamine is released from the graft and interacts with the postsynaptic dopaminergic receptors in a controlled manner. Of particular importance, then, is the recent PET scan study evaluating dopamine synthesis as well as dopamine release and binding by postsynaptic dopaminergic D2 receptors in patients who received transplants 10 years earlier (Piccini et al., 1999). In this study, [^{11}C] raclopride was used as a ligand of the D2 receptors of the host striatal neurons. The dramatic decrease in dopaminergic input to the non-grafted striatum in Parkinson's disease causes up-regulation of postsynaptic D2 receptors, and the individual receptors become more sensitive to dopamine (Rinne et al., 1990). After grafting, the number of raclopride-binding sites returns to normal levels along with the restoration of [^{18}F] dopa uptake (Piccini et al., 1999). These data suggest that dopamine was released from the graft-derived terminals in sufficient amounts to suppress the excessive expression of the dopaminergic D2 postsynaptic receptors. This has been further supported by the use of amphetamine as a stimulant of dopamine release from the axonal terminals. In patients with Parkinson's disease, amphetamine does not influence the raclopride binding in the non-grafted striatum because of a considerable loss of endogenous dopaminergic terminals. In contrast, the grafted striatum showed a 27% reduction of raclopride binding after amphetamine administration due to competition between raclopride and dopamine for dopaminergic receptors (Piccini et al., 1999). In other words, dopamine appeared to be released from the graft-derived synaptic terminals to bind the host-derived postsynaptic sites. Taken together, the aforementioned observations show that nigral grafts can restore spontaneous and regulated dopamine release to a normal level (Piccini et al., 1999; Barker and Dunnett, 1999).

In contrast to the cases described above, patients who received the grafts from fetuses older than 9 weeks or younger than 6.5 weeks showed poor neuron survival and sparse dopaminergic reinnervation of the patient striatum (Redmond et al., 1990; Hitchcock et al., 1994; Folkerth and Durso, 1996). Similar results were obtained when the diencephalo-mesencephalic region of human fetuses was grafted to the cerebral ventricles (Folkerth and Durso, 1996).

Thus, the histopathologic and PET data taken together proved that donor age, tissue source, and site of implantation are of particular importance for the survival and functioning of the graft.

FOLLOW UP OF PATIENTS WITH PARKINSON'S DISEASE WHO RECEIVED NEURAL GRAFTS

As mentioned above, CAPIT, a standardized protocol of clinical evaluation, has been used by a number of transplant centers to follow patients with Parkinson's disease after neurotransplantation. The patients treated by NECTAR groups were followed clinically every second week for three to six months preoperatively and every three months for a non-limited time period postoperatively by using a battery of clinical tests, including scoring of rigidity, timed tests of motor performance (pronation-supination test, hand/arm movement between two points, finger dexterity, and stand-walk-sit test), the single dose L-dopa test in defined "off" state, neuropsychological tests, etc. (Defer et al., 1996; Ugryumov et al., 1996; Wenning et al., 1997; Lindvall, 1998).

The latency up to the appearance of the first positive signs varies from immediately after implantation up to 3 to 6 months, and the progressive improvement usually continues up to 5 to 8 months after grafting (Defer et al., 1996; Levivier et al., 1997; Lindvall, 1998).

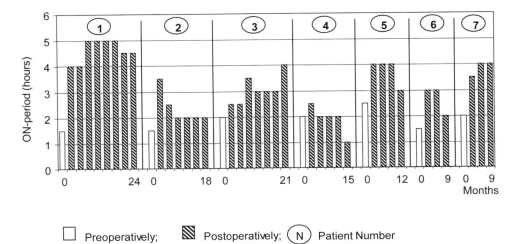

Figure 3. Single-dose L-dopa test in patients with Parkinson's disease before and up to 24 months after neurotransplantation (Modified from Ugryumov et al., 1996).

Although the concrete data from different transplant centers vary slightly, the evolution and patterns of clinical improvement of operated patients with Parkinson's disease exhibit certain common features (Olanow et al., 1997; Lindvall, 1998): (1) a decrease in L-dopa requirements, usually from 10-20%; (2) a prolonged L-dopa-induced "on" period (Figure 3); (3) decrease in both daily time in "off" period and number of daily "off" periods; (4) decreased rigidity; (5) improvement of motor performance (Figure 4); and, finally, (6) general improvement (an average of 32%) according to the Unified Parkinson's Disease Rating Scale (UPDRS) score. However, in almost no cases was there a full recovery (Defer et al., 1996; Ugryumov et al., 1996; Levivier et al., 1997; Wenning et al., 1997; Hauser et al., 1999). It should be mentioned that even after the unilateral grafting, improvement of motor performance is bilateral though it is more pronounced on the side contralateral to the graft implantation (Figure 4) (Defer et al., 1996; Ugryumov et al., 1996; Wenning et al., 1997; Fedorova et al., 2000). In contrast to rigidity and hypokinesia, tremor is not decreased after grafting. Gait seems to be ameliorated in patients with both caudate and putamen implants but not in cases with only putamen grafts (Lindvall, 1997). In spite of certain benefits in most Parkinson's disease patients following neurotransplantation, eventually some patients with particularly severe preoperative disability either revert to the preoperative level or even deteriorate to below their preoperative status (Hoffer et al., 1992; Olanow et al., 1997; Lindvall, 1998; Hagell et al., 1999).

Greater symptomatic relief occurs in patients when the ventral mesencephali of human fetuses at age 6-9 weeks are implanted *ex tempora* to the caudate and/or putamen. This is accompanied by increased [^{18}F] dopa uptake (Olanow et al., 1997; Lindvall, 1998) and the survival of a greater number of dopaminergic neurons in the grafted striatum (Kordower et al., 1998). Clinical trials using cryopreserved fetal materials (Redmond et al., 1990) or fresh materials from fetuses older than 9 weeks were characterized by poor graft survival and almost no clinical benefits (Hitchcock et al., 1994; Folkerth and Durso, 1996). Clinical benefits after grafting to the caudate alone were described in only one paper (Spencer et al., 1992).

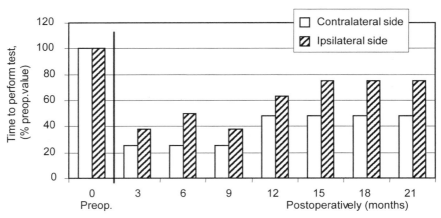

Figure 4. Pronation-supination test in the patient before and up to 21 months after neurotransplantation (modified from Ugryumov et al., 1996).

CONCLUSIONS

1. Human fetal mesencephalic dopaminergic neurons to be grafted express tyrosine hydroxylase, the rate-limiting enzyme of dopamine synthesis, and the mechanism for dopamine-specific uptake, though they do not release dopamine in response to membrane depolarization.
2. Human fetal mesencephalic cells to be grafted express major histocompatibility complex (HLA-DR) class II antigens, making it reasonable to use immunosuppression in patients to be operated.
3. Grafted dopaminergic neurons survive up to 10 years at least and partly reinnervate the striatum in patients with Parkinson's disease.
4. A considerable number of patients with Parkinson's disease display clinical benefits after transplantation, but symptomatic relief is variable, temporary, and incomplete.
5. Neurotransplantation is followed by bilateral improvement of motor performance, although it is more prominent on the side contralateral to the graft.
6. In spite of advances in clinical neurotransplantation, it still remains an experimental procedure.

PERSPECTIVES

It is generally believed that major strategies for improving the functional efficacy of neurotransplantation should include increasing the number of surviving dopaminergic neurons in the graft and promoting dopaminergic reinnervation of the patient striatum. These results could be achieved by an increase in the amount of donor materials and by the administration of neurotrophic/growth factors or the inhibitors of free radical processes, which either protect neurons from death or stimulate the outgrowth of neural processes. The most promising regulators of this kind are brain-derived neurotrophic factor, glial cell line-derived neurotrophic factor, basic fibroblast growth factor, and lazaroid, an inhibitor of free radical processes. Glial cell line-derived neurotrophic factor and lazaroid specifically promote neuron survival and differentiation (Lindvall, 1997; Olanow et al., 1997), whereas brain-derived neurotrophic factor mainly stimulates the sprouting of neural processes (Sauer et al., 1993). Parkinsonian animals that received neurotrophic factors or the inhibitors of free radical processes in addition to nigral grafts (intraventricular or intrastriatal injections) showed an increased number of dopaminergic neurons, an intensified outgrowth of neural processes, and, finally, a more rapid functional recovery (Rosenblad et al., 1996; Lindvall, 1977). Taking into account that the procedure of regular intrastriatal injections of neurotrophic factors cannot be extended to patients, the transfection of the primary mesencephalic cells to be grafted with the neurotrophic factor genes appears to be promising. However, there are certain difficulties in transfection of primary neurons by using traditional molecular biological approaches (Ridet and Privat, 1995).

Although progress has been made in neurotransplantation in Parkinson's disease, the use of human fetal materials will always raise ethical and technical problems (see above). Therefore, the search for or development of alternative cell types for grafting is of particular importance. The ideal hypothetical cell type would: (1) be available in large numbers; (2) proliferate *in vitro*, thus allowing transfection with the genetic materials; (3) be able to survive long-term but cease proliferating when implanted in the recipient brain; (4) be harmless to the functioning of the host target tissue; (5) be non-immunogenic; (6) be able to provide long-lasting functional amelioration after grafting.

At least some of the aforementioned problems may be overcome by using xenografts, though these make it necessary either to mask the major histocompatability complex (HLA-DR) antigens or to use long-term immunosuppression. Encapsulation of dopamine-secreting xenografts may be another way to avoid immunological rejection (Emerich et al., 1992). The histopathologic study showed that porcine dopaminergic neurons grafted to the striatum of an immunosuppressed patient with Parkinson's disease survived, differentiated, and sent their axons to the striatum. However, the number of neurons and the extent of reinnervation of the striatum were significantly less (Deacon et al., 1997) than those observed in the human allografts (Kordower et al., 1996, 1998).

It is believed that in the more distant future, transplantation of genetically modified non-neuronal cells or precursor cells will be used in the treatment of Parkinson's disease (Ridet and Privat, 1995). Different cell types, such as tumor cells (Horellou et a., 1990), primary fibroblasts (Fisher et al., 1991), muscle cells (Jiao et al., 1992), immortalized neural progenitor cell lines (Anton et al., 1994), and astrocytes (Lundberg et al., 1996), have already been used for grafting in parkinsonian animals. Fibroblasts and astrocytes are currently used as carriers of the genes of neurotrophic/growth factors to protect grafted dopaminergic neurons from death (Ridet and Privat, 1995). Moreover, fibroblasts and muscle cells are transfected with the tyrosine hydroxylase gene in order to produce L-dopa. Grafting these cells to the rat striatum results in the reverse of the apomorphine-induced rotation (Wolff et al., 1989; Jiao et al., 1992; Ridet and Privat, 1995). The disadvantages of the use of these cell models are related to failures to secrete L-dopa in a continuous manner, to synthetize dopamine, and to establish specific afferent and efferent connections. In this context, the preparation of the adeno-associated virus vector containing the genes for both tyrosine hydroxylase and aromatic L-amino acid decarboxylase is of particular promise. Indeed, the intrastriatal injections of this vector in parkinsonian non-human primates result in a significant increase of local dopamine synthesis and improvements in parkinsonian behavior (Kaplitt and During, 1995). Over the last few years, perspectives on transplantation of neuroepithelial stem cells, expanded into dopaminergic neurons, have also been widely discussed.

Thus, cell transplantation therapy for treatment of Parkinson's disease could be greatly improved in the future by the use of cells with specifically modified genomes.

ACKNOWLEDGMENTS: This work is supported by the Russian Foundation "Health of the Population of Russia."

REFERENCES

Agid, Y., Parkinson's disease: pathophysiology, *Lancet* 337:1321.

Anton, R., Kordower, J.H., Maidment, N.T., Manaster, J.S., Kane, D.J., Rabizadeh, S., Schueller, S.B., Yang, J., Edwards, R.H., Markham, C.H., and Bredesen, D.E., 1994, Neural-targeted gene therapy for rodent and primate hemiparkinsonism, *Exp. Neurol.* 127:207.

Backlund, E.-O., Granberg, P.-O., Hamberger, B., Knutsson, E., Mårtensson, A., Sedvall, G., Seiger, Å., and Olson, L., 1985, Transplantation of adrenal medullary tissue to striatum in parkinsonism. First clinical trials, *J. Neurosurg.* 62:169.

Balan, I.S., Ugrumov, M.V., Calas, A., Mailly, P., Krieger, M., and Thibault, J., 2000, Tyrosine hydroxylase-expressing and/or aromatic L-amino acid decarboxylase-expressing neurons in the mediobasal hypothalamus of perinatal rats: differentiation and sexual dimorphism, *J. Comp. Neurol.* 424: in press.

Barker, R.A., and Dunnett, S.B., 1999, Functional integration of neural grafts in Parkinson's disease, *Nature Neurosci.* 2:1047.

Beltramo, M., Calas, A., Chernigovskaya, N., Borisova, N., Polenova, O., Tillet, Y., Thibault, J., and Ugrumov, M., 1994, Postnatal development of the suprachiasmatic nucleus in the rat. Morpho-functional characteristics and time course of tyrosine hydroxylase immunopositive fibers, *Neuroscience,* 63:603.

Berheimer, H., Birkmayer, W., Hornykiewicz, O., Jellinger, K., and Seitelberger, F., 1973, Brain dopamine and the syndromes of Parkinson and Huntington: clinical, morphological and neurochemical correlations, *J. Neurol. Sci.* 20:415.

Björklund, A., 1991, Dopaminergic transplants in experimental parkinsonism: cellular mechanisms of graft-induced functional recovery, *Curr. Opin. Neurobiol.* 2:683.

Björklund, A., Dunnett, S.B., and Nikhah, G., 1994, Nigral transplants in the rat parkinson model. Functional limitations and strategies to enhance nigrostriatal reconstruction, in: *Functional Neural Transplantation*, S.B. Dunnett, and A. Björklund, eds., Raven Press, New York.

Boer, G.J., 1994, Ethical guidelines for the use of human embryonic or fetal tissue for experimental and clinical neurotransplantation and research, *J. Neurol.* 242:1.

Borlongan, C.V., Stahl, C.E., Cameron, D.F., Saporta, S., Freeman, T.B., Cahill, D.W., and Sanberg, P.R., 1996, CNS immunological modulation of neural graft rejection and survival, *Neural Res.* 18:297.

Brundin, P., Strecker, R.E., Widner, H., Clarke, D.J., Nilsson, O.G., Åstedt, B., Lindvall, O., and Björklund, A., 1988, Human fetal dopamine neurons grafted in a rat model of Parkinson's disease: immunological aspects, spontaneous and drug-induced behavior, and dopamine release, *Exp. Brain Res.* 70:192.

CAPIT Committee: Langston, J.W., Widner, H., Goetz CG., Brooks, D., Fahn, S., Freeman, T., and Watts, R., 1992, Core assessment program for intracerebral transplantation (CAPIT), *Mov. Disord.* 7:2.

Castilho, R.F., Hansson, O., and Brundin, P., 2000, FK506 and cyclosporin A enhance the survival of cultured and grafted rat embryonic dopamine neurons, *Exp. Neurol.* 164:94.

Collier, T.J., Gallafger, M.J., and Sladek, C.D., 1993, Cryopreservation and storage of embryonic rat mesencephalic dopamine neurons for one year: comparison to fresh tissue in culture and neural grafts, *Brain Res.* 623:249.

Deacon, T., Schumacher, J., Dinsmore, J., Thomas, C., Palmer, P., Kott, S., Edge, D., Penney, D., Kassissieh, S., Dempsey, P., and Isacson, O., 1997, Histological evidence of fetal pig neural cell survival after transplantation into a patient with Parkinson's disease. *Nature Med.* 3:350.

Defer, G.-L., Geny, C., Ricolfi, F., Fenelon, G., Monfort, J.-C., Remy, P., Villafane, G., Jeny, R., Samson, Y., Keravel, Y., Gaston, A., Degos, J.-D., Peschanski, M., Cesaro, P., and Nguyen, J.-P., 1996, Long-term outcome of unilaterally transplanted parkinsonian patients I. Clinical approach, *Brain,* 119:41.

Emerich, D.F., Winn, S.R., Christenson, L., Palmatier, M.A., Gentile, F.T., and Sanberg, P.R., 1992, A novel approach to neural transplantation in Parkinson's disease: use of polymer-encapsulated cell therapy, *Neurosci. Biobehav. Rev.* 16:437.

Fedorova, N., Shabalov, V., Stock, V., Yakovleva, S., Popov, A., Ugrumov, M., 2000, Russian experience in neurotransplantation in Parkinson's disease: surgical procedures and neurological examination, in: *Basal Ganglia and Thalamus in Health and Movement Disorders* (Abstracts), Moscow.

Fisher, L.J., Jinnah, H.A., Kale, L.C., Higgins, G.A., and Gage, F.H., 1991, Survival and function of intrastriatally grafted primary fibroblasts genetically modified to produce L-DOPA, *Neuron,* 6:371.

FQlkerth, R.D., and Durso R., 1996, Survival and proliferation of non-neural tissues with obstruction of cerebral ventricle in a parkinsonian patient treated with fetal allografts, *Neurology,* 46:1219.

Freed, W.J., Morihisa, J.M., Spoor, E., Hoffer, B.J., Olson, L., Seiger, A., and Wyatt, R.J., 1981, Transplanted adrenal chromaffin cells in rat brain reduce lesion-induced rotational behaviour, *Nature,* 292:351.

Freed, C.R., Breeze, R.E., Rosemberg, N.L., Schneck, S.A., Kriek, E., Qi, J.-X., Lone, T., Zhang, Y.-B., Snyder, J.A., Wells, T.H., Ramig, L.O., Thompson, L., Mazziotta, J.C., Huang, S.C., Grafton, S.T., Brooks, D., Sawle, G., Schroter, G., and Ansari, A.A., 1992, Survival of implanted fetal dopamine cells and neurologic improvement 12 to 46 months after transplantation for Parkinson's disease, *N. Engl. J. Med.* 327:1549.

Freed, C.R., Breeze, R.E., Greene, P.E., Eidelberg, D., Tsai, W., Murphy, J., Trojanowski, J.O., Rosenstein, J.M., and Fahn, S., 1999, Double-blind placebo-controlled human fetal dopamine cell transplants in advanced Parkinson's disease, *Soc. Neurosci. Abstr.* 25:212.

Freeman, T.B., and Kordower, J.H., 1991, Human cadaver embryonic substantia nigra grafts: effects of ontogeny, preoperative graft preparation and tissue storage, in: *Intracerebral Transplantation in Movement Disorders,* O. Lindvall, A. Björklund, and H. Widner, eds., Elsevier, Amsterdam.

Freeman, T.B., Spence, M.S., Boss, B.D., Spector, D.H., Strecker, R.E., Olanow, C.W., and Kordower, J.H., 1991, Development of dopaminergic neurons in the human substantia nigra, *Exp. Neurol.* 113:344.

Freeman, T.B., Sanberg, P.R., Nuaret, G.M., Borlongan, C., Liu, E.-Z., Barbara, D., Boss, B.D., Spector, D., Olanow, C.W., and Kordower, J.H., 1995, The influence of donor age on the survival of solid and suspension intraparenchymal human embryonic nigral grafts, *Cell Transplant.* 4:141.

Hagell, P., Schrag, A., Piccini, P., Jahanshahi, M., Brown, R., Rehncrona, S., Widner, H., Brundin, P., Rothwell, J.C., Odin, P., Wenning, G.K., Morrish, P., Gustavii, B., Björklund, A., Brooks, D.J., Marsden, C.D., Quinn, N.P., and Lindvall, O., 1999, Sequential bilateral transplantation in Parkinson's disease. Effects of the second graft, *Brain,* 122:1121.

Hauser, R.A., Freeman, T.B., Snow, B.J., Nauert, M., Gauger, L., Kordower, J.H., and Olanow, C.W., 1999, Long-term evaluation of bilateral fetal nigral transplantation in Parkinson's disease, *Arch. Neurol.* 56:179.

Hitchcock, E.H., Whitwell, H.L., Sofroniew, M., and Bankiewicz, K.S., 1994, Survival of TH-positive and neuromelanin-containing cells in patients with Parkinson's disease after intrastriatal grafting of fetal ventral mesencephalon, *Exp. Neurol.* 129:3.

Hoffer, B.J., Leenders, K.L., Young, D., Gerhardt, G., Zerbe, G.O., Bygdeman, M., Seiger, Å., Olson, L., Strömberg, I., and Freedman, R., 1992, Eighteen-month course of two patients with grafts of fetal dopamine neurons for severe Parkinson's disease, *Exp. Neurol.* 118:243.

Horellou, P., Brundin, P., Kalen, P., Mallet, J., and Björklund, A., 1990, *In vivo* release of DOPA and dopamine from genetically engineered cells grafted to the denervated rat striatum, *Neuron,* 5:393.

Ikemoto, K., Nagatsu, I., Nishimura, A., Nishi, K., and Arai R., 1998, Do all of human midbrain tyrosine hydroxylase neurons synthesize dopamine? *Brain Res.* 255:255.

Jaeger, C.B., 1986, Aromatic L-amino acid decarboxylase in the rat brain: immunocytochemical localization during prenatal development, *Neurosci.* 18:121.

Jiao, S.S., Williams, P., and Wolff, J., 1992, Intracerebral transplantation of genetically engineered muscle cells reduces experimental parkinsonism in rats, *Soc. Neurosci. Abstr.* 18:782.

Kaplitt, M.G., and During, M.J., 1995, Transfer and expression of potentially therapeutic genes into the mammalian central nervous system in vivo using adeno-associated viral vectors, in: *Viral Vectors,* M.G. Kaplitt, and A.D. Loewy, eds., Academic Press, San Diego.

King, D.B., 1999, Parkinson's disease - levodopa complications, *Can. J. Neurol. Sci.* 2, Suppl. 2:13.

Kish, S.J., Shannak, K., and Hornykiewicz, O., 1988, Uneven pattern of dopamine loss in the striatum of patients with idiopathic Parkinson's disease: pathophysiologic and clinical implications, *N. Engl. J. Med.* 318:876.

Kitahama, K., Geffard, M., Okamura, H., Nagatsu, I., Mons, N., and Jouvet, M., 1990, Dopamine and L-DOPA-immunoreactive neurons in the cat forebrain with reference to tyrosine hydroxylase immunohistochemistry, *Brain Res.* 518:83.

Kordower, J.H., Freeman, T.B., Snow, B.J., Vingerhoets, J.G., Mufson, E.J., Sanberg, P.R., Hauser, R.A., Smith, D.A., Nauret, G.M., Perl, D.P., and Olanow, C.W., 1995, Neuropathological evidence of graft survival and striatal reinnervation after the transplantation of fetal mesencephalic tissue in a patient with Parkinson's disease, *N. Engl. J. Med.* 332:1118.

Kordower, J.H., Rosenstein, J.M., Collier, T.J., Burke, M.A., Chen, E.-Y., Li, J.M., Martel, L., Levey, A.E., Mufson, E.J., Freeman, T.B., and Olanow, C.W., 1996, Functional fetal nigral grafts in a patient with Parkinson's disease: chemoanatomic, ultrastructural and metabolic studies, *J. Comp. Neurol.* 370:203.

Kordower, J.H., Styren S., Clarke, M., De Kosky, S.T., Olanow, C.W., and Freeman, T.B., 1997, Fetal grafting for Parkinson's disease: expression of immune markers in two patients with functional fetal nigral grafts, *Cell Transplant.* 6:213.

Kordower, J.H., Freeman, T.B., Chen, E.-Y., Mufson, E.J., Sanberg, P.R., Hauser, R.A., Snow, B., and Olanow, C.W., 1998, Fetal nigral grafts survive and mediate clinical benefit in a patient with Parkinson's disease, *Mov. Disord.* 113:383.

Levivier, M., Dethy, S., Rodesch, F., Peschanski, M., Vandersteene, A., David, P., Wikler, D., Goldman, S., Claes, T., Biver, F., Liesnard, C., Goldmn, M., Hildebrand, J., and Brotchi, J., 1997, Intracerebral transplantation of fetal ventral mesencephalon for patients with advanced Parkinson's disease, *Stereotact. Funct. Neurosurg.* 69:99.

Lindvall, O., 1991, Prospects of transplantation in human neurodegenerative diseases, *Trends Neurosci.* 14:376.

Lindvall, O., 1997, Neural transplantation: a hope for patients with Parkinson's disease, *Neuroreport*, 8:iii.

Lindvall, O., 1998, Update of fetal transplantation: the Swedish experience, *Mov. Disord.* 13, Suppl. 1:83.

Lindvall, O., Rehncrona, S., Brundin, P., Gustavi, B., Åstedt, B., Widner, H., Lindholm, T., Björklund, A., Leenders, K.L., Rothwell, J.C., Frackowiak, R., Marsden, C.D., Johnels, B., Steg, G., Freedman, R., Hoffer, B.J., Seiger, Å., Bygdeman, M., Strömberg, I., and Olson, L., 1989, Human fetal dopamine neurons grafted into the striatum in two patients with severe Parkinson's disease, *Arch. Neurol.* 46:615.

Lindvall, O., Sawle, G., Widner, H., Rothwell, J.C., Björklund, A., Brooks, D., Brundin, P., Frackowiak, R., Marsden, C.D., Odin, P., and Rehncrona, S., 1994, Evidence for long-term survival and function of dopaminergic grafts in progressive Parkinson's disease, *Ann. Neurol.* 35:172.

Lundberg, C., Horellou, P., Mallet, J., and Björklund, A., 1996, Generation of DOPA-producing astrocytes by retroviral transduction of the human tyrosine hydroxylase gene: in vitro characterization and in vivo effects in the rat Parkinson model, *Exp. Neurol.* 139:39.

Madrazo, I., Drücker-Colin, R., Diaz, V., Martinez-Marta, J., Torres, C., and Becerril, J.J., 1987, Open microsurgical autograft of adrenal medulla to the right caudate nucleus in Parkinson's disease: a report of two cases, *N. Engl. J. Med.* 326:831.

Meister, B., Hökfelt, T., Vale, W.W., Sawchenko, P.E., Swanson, L.W., and Goldstein, M., 1986, Coexistence of tyrosine hydroxylase and growth hormone-releasing factor in a subpopulation of tubero-infundibular neurons of the rat, *Neuroendocrinology*, 42:237.

Meister, B., Hökfelt, T., Steinbusch, H.W.M., Skagerberg, G., Lindvall, O., Geffard, M., Joh, T., Cuello, A.C., and Goldstein, M., 1988, Do tyrosine hydroxylase-immunoreactive neurons in the ventro-lateral arcuate nucleus produce dopamine or only L-DOPA? *J. Chem. Neuroanat.* 1:59.

Metman, L.V., and Mouradian, M.M., 1999, Levodopa therapy of Parkinson's disease and associated long-term motor response complications, in: *Parkinson's Disease, The Treatment Options*, P. LeWitt, and W. Oertel, eds., Martin Dunitz Ltd.

Nobin, A., and Björklund, A., 1973, Topography of the monoamine neuron systems in the human brain as revealed in fetuses, *Acta. Physiol. Scand.* 88 Suppl. 388:1.

Olanow, C.W., Freeman, T.B., and Kordower, J.H., 1997, Neural transplantation as a therapy for Parkinson's disease, *Adv. Neurol.* 74:249.

O'Rahilly, R., and Muller, F., 1987, Developmental stages in human embryos. Including a revision of Streeterís "Horizons" and a survey of the Carnegie collection, Carnegie Institution of Washington, Publication 637.

Pakkenberg, B., Moller, A., Gunderson, H.J., Mouritzen-Dam, A., and Pakkenberg, H., 1991, The absolute number of nerve cells in substantia nigra in normal subjects and in patients with Parkinson's disease estimated with an unbiased stereological method, *J. Neurol. Neurosurg. Psychiatry*, 54:30.

Piccini, P., Brooks, D.J., Björklund, A., Gunn, R.N., Grasby, P.M., Rimoldi, O., Brundin, P., Hagell, P., Rehncrona, S., Widner, H., and Lindvall, O., 1999, Dopamine release from nigral transplants visualized *in vivo* in a Parkinson's patient, *Nat. Neurosci.* 2:1137.

Pickel, V.M., Specht, L.A., Sumal, K.K., Joh, T.H., Reis, D.J., and Hervonen, A., 1980, Immunocytochemical localization of tyrosine hydroxylase in the human fetal nervous system, *J. Comp. Neurol.* 194:465.

Pogarell, O., and Oertel, W.H., 1999, Parkinsonian syndromes and Parkinson's disease: diagnosis and differential diagnosis, in: *Parkinson's Disease, The Treatment Options*, P. LeWitt, and W. Oertel, eds., Martin Dunitz Ltd.

Redmond, D.E., Leranth, C., Spencer, D.D., Robbins, R., Vollmer, T., Kim, R.H., Roth, A.J., Dwork, A.J., and Naftolin, F., 1990, Fetal graft survival, *Lancet*, 336:820.

Rémy, P., Samson, Y., Hantraye, Ph., Fontaine, A., Defer, G., Mangin, J.F., Fenelon, G., Ricolfi, F., and Frouin, V., 1995, Neural grafting in five parkinsonian patients: correlations between PET and clinical evaluation, *Ann. Neurol.* 38:580.

Ridet, J.-L., and Privat, A., 1995, Gene therapy in the central nervous system: direct versus indirect gene delivery, *J. Neurosci. Res.* 42:287.

Rinne, U.K., Laihinen, A., Rinne, J.O., Nagren, K., Bergman, J., and Ruotsalainen, U., 1990, Positron emission tomography demonstrates dopamine D2 receptor supersensitivity in the striatum of patients with early Parkinson's disease, *Mov. Disord.* 5:55.

Rosenblad, C., Martinez-Serrano, A., and Björklund, A., 1996, Glial cell line-derived neurotrophic factor increases survival, growth and function of intrastriatal fetal nigral dopaminergic grafts, *Neuroscience*, 75:979.

Sauer H., and Brundin, P., 1991, Effects of cool storage on survival and function of intrastriatal ventral mesencephalic grafts, *Restor. Neurol. Neurosci.* 2:123.

Sauer, H., Fisher, W., Nikkhah, G., Wiegand, S.J., Brundin, P., Lindsay, R.M., and Björklund, A., 1993, Brain-derived neurotrophic factor enhances function rather than survival of intrastriatal dopamine cell-rich grafts, *Brain Res.* 626:37.

Sawle, G.V., Bloomfield, P.M., Björklund, A., Brooks, D.J., Brundin, P., Leenders, K.I., Lindvall, O., Marsden, C.D., Rehncrona, S., Widner, H., and Frackowiak, R.S.J., 1992, Transplantation of fetal dopamine neurons in Parkinson's disease: PET [^{18}F]-6-L-fluorodopa studies in two patients with putaminal implants, *Ann. Neurol.* 31:166.

Spencer, D.D., Robbins, R.J., Naftolin, F., Phil, D., Marek, K.L., Vollmer, T., Leranth, C., Roth, R.H., Proce, L.H., Gjedde, A., Bunney, B.S., Sass, K.J., Elsworth, J.D., Kier, E.L., Makuch, R., Hoffer, P.B., and Redmond, D.E., 1992, Unilateral transplantation of human fetal mesencephalic tissue into the caudate nucleus of patients with Parkinson's disease, *N. Engl. J. Med.* 327:1541.

Ugrumov, M.V., Proshlyakova, E.V., Sapronova, A.Y., and Popov, A.P., 1996, Development of the mesencephalic and diencephalic catecholaminergic systems in human fetuses: uptake and release of catecholamines *in vitro*, *Neurosci. Lett.* 212:29.

Ugrumov, M., Popov, A., Makarenko, I., Zakharova, L., Sapronova, A., Proshlyakova, E., Katunian, P., Baranova, F., Bogdanova, N., Sotnikova, E., and Gatina, T., 2000, Russian experience in neurotransplantation in Parkinson's disease: obtaining, processing and specifying of the donor materials, in: *Basal Ganglia and Thalamus in Health and Movement Disorders* (Abstracts), Moscow.

Ugriumov, M.V., Shabalov, V.A., Fedorova, N.V., Popov, A.P., Shtok, V.N., Sotnikova, Y.I., Melikyan, A.G., Gatina, T.A., Fetisov, S.O., Arkhipova, N.A., Buklina, S.B., and Lutsenko, G.V., 1996, Use of neurotransplantation in the treatment of patients with Parkinson's disease, *Vestnik RAMS,* 8:40.

Wenning, G.K., Odin, P., Morrish, P., Rehncrona, S., Widner, H., Brundin, P., Rothwell, J.C., Brown, R., Gustavi, B., Hagell, P., Jahanshahi, M., Sawle, G., Björklund, A., Brooks, D., Marsden, C.D., Quinn, N.P., and Lindvall, O., 1997, Short- and long-term survival and function of unilateral intrastriatal dopaminergic grafts in Parkinson's disease, *Ann. Neurol.* 42:95.

Wichmann, T., and Delong, M.R., 1993, Pathophysiology of parkinsonian motor abnormalities, *Adv. Neurol.* 60:53.

Wolf, J.A., Fisher, L.J., Xu, L., Jinnah, H.A., Langlais, P.J., Luvone, P.M., O'Malley, K.L., Rosenberg, M.B., Shimohamai, S., Friedmann, T., and Gage, F.H., 1989, Grafting fibroblasts genetically modified to produce L-dopa in a rat model of Parkinson's disease, *Proc. Natl. Acad. Sci.* 86:9011.

Zecevic, N., and Verney, C., 1995, Development of the catecholamine neurons in human embryos and fetuses, with special emphasis on the innervation of the cerebral cortex, *J. Comp. Neurol.* 351:509.

Zoli, M, Agnati LF, Tinner B, Steinbusch HWM, and Fuxe K., 1993, Distribution of dopamine-immunoreactive neurons and their relationships to transmitter and hypothalamic hormone-immunoreactive neuronal systems in the rat mediobasal hypothalamus. A morphometric and microdensitometric analysis, *J. Chem. Neuroanat.* 6:293.

APPENDIX

SUMMARY OF THE TRANSCRIPT OF THE
CLINICAL ROUND TABLE DISCUSSION
ON PALLIDAL AND SUBTHALAMIC SURGERY

SESSION I

The Round Table Discussion consisted of three sessions. The first session, chaired by **Andres Lozano** (Toronto) focused on thalamic surgery and is summarized in the chapter by Lozano et al., in this volume.

SESSION II

Mahlon DeLong (Atlanta) chairs the pallidal session that followed. He opens it with a brief talk on the functional anatomy of the globus pallidus.

Jerrold Vitek (Atlanta) continues the discussion on the microelectrode approach to mapping the basal ganglia and stresses the importance of lesion location in determining clinical outcome. He describes how the different patterns of neuronal activity are used in differentiating the different portions of the basal ganglia and how micro- and macrostimulation is used to identify the location of the optic tract and internal capsule prior to lesioning. The functional heterogeneity of the internal segment of the globus pallidus (GPi) into anatomically segregated motor and nonmotor portions is discussed and the importance in placing lesions to include predominately the motor region stressed. Data from a group of nine patients from an earlier pilot study of pallidotomy for Parkinson's disease conducted in Atlanta is used to support this argument by demonstrating the differential effect of lesioning involving the posterolateral "motor" versus anteromedial "nonmotor" portion of the internal pallidum on clinical outcome. It has been demonstrated that patients with lesions involving the motor portion of the internal pallidum sustained improvement in all the cardinal motor signs of Parkinson's disease for at least two years postoperatively. On the other hand, patients with lesions encroaching on or involving only a small portion of the motor pallidum, although showing similar initial improvement, lost their benefit over the ensuing six to twelve months postoperatively. Several examples of patients who had previously failed pallidotomy at other institutions were found to have lesions outside the posterolateral portion of the GPi. Subsequent mapping and placement of lesions in the posterolateral "motor" portion of the GPi in these patients produced immediate and sustained clinical improvement. These observations stress the importance of lesion placement as a predictor of clinical outcome and re-emphasize the role of microelectrode recording in precisely defining the posterolateral "motor" portion of the pallidum prior to lesion placement.

Jonathan Dostrovsky (Toronto) addresses the basic mechanisms responsible for the presumed inhibitory effect induced by high frequency stimulation (HFS) in current functional targets such as GPi, subthalamic nucleus (STN), and to some extent, the thalamus - Vim. He describes his recent studies in patients on the effects of low and high frequency microstimulation on the spontaneous activity of neurons recorded less than 1 mm away from the stimulating electrode. The findings indicate that in GPi single, low intensity (e.g., 5 μA) stimuli produce short lasting inhibition, and that as the stimulation frequency is increased the overall firing rate decreases to the point where almost complete inhibition is induced at 30 Hz, as a result of the 20 to 40 ms inhibitions that occurr following each pulse. It is suggested that each individual stimulus excites GABAergic terminals, as the current is too low to excite the cells directly, and the released GABA produces IPSPs of about 25 ms duration. Thus, HFS could result in almost complete inhibition of firing, and such inhibition could be one possible mechanism involved in HFS-DBS. However, this hypothesis appears inconsistent with the higher frequencies (over 100 Hz) necessary for obtaining clinically optimal effects suggesting that several mechanisms may be at work here. Furthermore, it is not known what happens after hours of stimulation. Similar microstimulation in STN rarely produces an effect, and in thalamus it frequently results in excitation rather than inhibition. Dostrovsky also presents some preliminary data indicating that at least for some cells in STN and substantia nigra pars reticularis (SNr) marked inhibition can be induced by frequencies of 100 Hz and higher, but not by lower frequencies. In these cases, it is also necessary to use higher currents (200-400 μA, macrostimulation) and the inhibition is observed following termination of the stimulus train. Under such conditions, which resemble the parameters used in deep brain stimulation (DBS) in patients, complete inhibition lasting several hundred milliseconds can be observed. The duration of inhibition tends to increase with increased train frequency, intensity and duration. These currents are probably also exciting the neuron recorded from, but this cannot be documented because of the stimulus artifact. This suggests that there is something specific to HFS that takes place at higher current intensities. Of possible relevance are reports from intracellular recordings in STN slices that high frequency trains can result in prolonged post-inhibition of the STN cells (Bevan and Wilson, *J. Neurosci.* 19:761, 1999). This could be due to an accumulation of calcium causing increased potassium currents. The conclusion is that when stimulation is delivered in any nucleus, there might be a number of possible effects, including presynaptic excitation and GABA or glutamate release. Since the excitability of axons is higher than cell bodies, most of the observed effects are probably due to activation of axons. Since activation of axons leads to both antidromic and orthodromic propagation the effects of stimulation within these nuclei can be complex and diverse. These observations call for more extensive experimental evaluation.

Paul Krack (Kiel and Grenoble) compares pallidal stimulation with pallidotomy and subthalamic stimulation, taking into consideration that the majority of the patient population for these procedures is at advanced disease stages, and therefore, display bilateral signs. In these patients, unilateral surgery would not produce sufficient improvements, particularly in axial symptoms, and therefore one must compare bilateral stimulation or lesioning in STN and in GPi. There are very few recent publications of bilateral pallidotomies, with UPDRS data available. Although the improvement can be quite substantial, serious side effects, mainly deterioration of cognition and speech, have been reported. Therefore, based on the literature of the last ten years, this procedure was not recommended by the Task Force of the American Academy of Neurology. However, Orthner and colleagues have already shown in the early 60's (Orthner et al., 1962, *Acta Neurochir. (Wien)*,10:572-629) in a series of 36 patients with staged bilateral pallidotomy that clinically relevant psychiatric side effects can be absent and speech deterioration rare, if lesions are restricted to the posteroventral and lateral parts of GPi. If this lesson is respected, and with the help of modern imaging, the therapeutic potential of bilateral pallidotomies could be higher than presently thought. There are also few data available on bilateral pallidal stimulation. Improvements in motor UPDRS score vary between 30 and 60% after 3-6 months. Improvements in dyskinesias are very marked, ranging from 60 to 90%. A four year follow-up study by Moro (Grenoble), has shown that these effects of pallidal stimulation are stable in the long-term. A retrospective comparison of STN and GPi HFS by the Grenoble group revealed an advantage of STN stimulation, showing a better improvement in akinesia. Moreover, STN stimulation enabled a decrease in drug dosage, whereas GPi stimulation did not. These findings are confirmed by an unpublished multicenter study. The effects on dyskinesias are equally significant in both methods, but while these effects are a direct result of GPi procedures, they are mostly an indirect effect of the STN procedure, caused by a decrease in drug dosage following clinical improvement. Finally, the consumption of electrical power is about 2 times lower in HFS of STN than in HFS of GPi. This, added to the decrease in medication, makes the STN stimulation more cost-effective than the GPi stimulation.

General Discussion

After these presentations, the discussion deals with several subsequent issues. **Anne Young** (Boston) stresses the point that patients who deteriorate after these treatments might not have been parkinsonian after all, a fact that could have gone unrecognized at the time of patient selection.

The concept based on literature data is brought up in discussion that GPi may be more than one structure, and it could be possible to find areas where a contact would be better for dyskinesias than it would be for akinesia and rigidity.

Mahlon DeLong and Jerrold Vitek stress that the lateral part of the nucleus is a better target than the medial part, and that, when the electrode is deeper, fibers from the pallidal output might also be activated, making the patient akinetic and negating the levodopa effect. **Paul Krack** suggests that stimulating the lower part of pallidum could activate GABAergic pallido-tegmental fibers, which would inhibit PPN.

Marjorie Anderson (Seattle) inquires about which symptoms get worse in patients after STN or GPi procedures, and in particular, whether handwriting and repetitive tasks tend to get worse.

Jerrold Vitek mentions a patient who lost his ability to follow things rhythmically and suggests that studies focusing on these patients' ability to learn new motor tasks may be in order.

Yoland Smith (Atlanta), while recognizing the importance of the procedure location in the sensory motor territory of the targets, asks whether any problems result from lesioning too much.

Andres Lozano (Toronto) comments that if the lesion extends beyond the motor territory in the GPi and encroaches upon the associative or limbic territories, then significant cognitive effects can be observed. He stresses that motor symptoms and other types of effects observed seem to be segregated, which suggests that lesioning in certain areas is better for rigidity, in others for tremor, and in some others for planning. Even in the motor area of GPi, there are subpallidal circuits that interact with the thalamus through different cortical areas, supporting the idea of the existence of an akinesia circuit, a tremor circuit, etc. within the pallidum.

Following **DeLong's** question to **Yoland Smith** about the possible role of GABAergic collaterals and fibers, Smith says that GPi neurons do not have a very dense intrinsic axon collateral organization. So, they play hardly any role in the mechanisms discussed. On the other hand, there are other types of neurons and interneurons in GPi as well as in GPe that are different from the major projection neurons. It would be more important to think along these lines rather than on collaterals.

A final question from the audience about the surgical treatment of post-traumatic parkinsonian patients opens discussion on the correlation between the post-surgical outcome and response to levodopa challenge.

SESSION III

The last session of the round table discussion devoted to the subthalamic nucleus (STN) is chaired by **Alim Benabid** (Grenoble).

The first speaker in this session, **Steven Gill** (Bristol), reports on his two and a half years of experience with STN lesioning procedures. He has performed 17 unilateral lesions and 10 simultaneous bilateral lesions. He targets the dorso-lateral part of the STN that is thought to be the sensory motor area of the nucleus. The ventral part of STN is involved in limbic system-related functions, and the central part with associative functions. According to Gill in addition to interrupting STN, the lesion interrupts the pallidal output fibres to the thalamus running above STN, and also may destroy afferents to the VIM nucleus, which, in turn, could add to an anti-tremor effect. Gill places a 3 mm diameter STN lesion using high resolution MRI imaging for target location, which proved to be more helpful than microrecording. The most effective lesions proved to be in the posterior third of STN, where they induced an initial 70% improvement in UPDRS score. However, these begin to wane about 2 to 3 months after surgery, leaving a sustained improvement of 40%. Making larger bilateral lesions is more likely to produce side effects especially hypophonia, so Gill has developed a method of staged re-lesioning, if required, using an implanted catheter and free hand procedure.

Myung Sik Lee and **Chen Hsu** (Taiwan) report on their short series of 3 patients with STN lesioning, all of whom had post-operative dyskinesias, with two of these long-lasting.

Tipu Aziz (Oxford) reports the experience at Charing Cross based on a large series of pallidotomies and thalamotomies and on their pioneering experimental work on STN lesioning in MPTP monkeys (see chapter by Aziz et al., in this volume).

Paul Krack (Kiel and Grenoble) discusses the HFS effects on mood (see chapter by Krack et al., in this volume).

General Discussion

The discussion that follows focuses on:

1. The mechanism of induction of dyskinesias after STN lesioning. **Jerrold Vitek** calls for comparison of the series of patients of Dr. Gill and Drs. Lee and Chen Hsu. He hypothesizes that the smaller STN lesions made in Taiwan would have spared the adjacent pallidofugal fibers, which are responsible for the induction of dyskinesias through their thalamic projections.

2. The side effects on mood and behavior. As a comment to Paul Krack's presentation, **Jerrold Vitek** reports similar observations of mood changes, specifically sponatneous laughter and euphoria during surgery, and also refers to similar cases of Dr. Kumar (Denver) and of Dr. Doshi (Bombay). **Mahlon DeLong** wonders whether these behaviors are observed mainly in patients who over self-medicated to avoid dysphoric anxious states in the "off" phase. These effects on mood do not seem to be related to a particular side of the brain and do not require bilateral lesions.

3. Comparison of the nuclear volume involved in lesioning and stimulation. **Alim Benabid** points out that in STN, the volume involved in stimulation seems to have the same size as the volume of reported lesions needed to obtain the same effect. GPi stimulation, although carried out with higher parameters, probably does not encompass the entire GPi, but in lesioning procedures it must be almost totally destroyed in order to achieve an effect.

The latter prompts a discussion on the differences between stimulation and lesioning focusing on the involvement of fibers, which are destroyed in lesioning procedures and essentially excited in the stimulation procedures.

Mahlon DeLong argues that for a lesion of a given size, and stimulation involving a similar volume, stimulation will have a much wider effect because of the spread of activation from activated fibers, axon collaterals or fibers of passage. **Alim Benabid** believes that stimulation excites fibers and must therefore spread an excitatory effect at a distance. **Mahlon DeLong** stresses the point that most of the fibers in the area are inhibitory, with very few excitatory fibers present. The experiments based on chronaxy showing that fibers have a lower excitability threshold than cells are mentioned. This tends to support the idea that HFS stimulation involves mainly fibers rather than cells.

This brings the suggestion from **Anne Young** that our surgical methods should be oriented more towards chemical lesioning agents, such as excitotoxins, kainate, AMPA, and ibotenic acid, which specifically target cell bodies and spare the fibers of passage.

Andres Lozano recalls that injections of ibotenic acid into GPi or STN may also result in diffusion of the chemicals to the hippocampus, where neurons are extremely sensitive to exitatory aminoacids and their destruction induces memory deficits.

Kristy Kultas-Ilinsky (Iowa City) comments that one should keep in mind that GPi neurons are rather resistant to the kainic and ibotenic acids due to the paucity of appropriate receptors on them. To sufficiently destroy GPi neurons with these substances one has to use rather high concentrations and large amounts of them. However, kainic acid diffuses readily and if GPi is flooded with it, it can diffuse not only to hippocampus but to the reticular thalamic nucleus whose neurons have large amounts of glutamate receptors on them and will be easily destroyed thus creating a new set of problems.

Anne Young comments that AMPA would be better candidate for a lesioning agent in Gpi since AMPA receptors are abundant in this nucleus.

There is a consensus that some work could be devoted to shape and polish chemical tools that would be hyperselective on specific neuronal populations without affecting the fibers, which are believed to be responsible for most of the side effects observed during surgical and especially lesioning procedures. Gene therapy is mentioned as one of the tools of the future.

Finally, **Alim Benabid** stresses the difficulty in assessing what specific structure is responsible for a set of symptoms or effects observed, either after lesioning or during stimulation.

INDEX

Acetylcholine: *see* Cholinergic
Action potentials, 226–228, 293
Adrenal medulla, 349
Akinesia, 15, 19, 124, 158, 166, 176, 180, 210, 260, 267, 272, 310, 336, 367
   drug-induced, 166
   stimulation-induced, 342
AMPA: *see* Glutamate
Amphetamine, 137, 178, 181, 335, 356
Apomorphine, 137–144, 181, 207–208, 210, 242, 272
Aspiny (nonspiny) neurons, 43–46, 163
Ataxia, 190
Atonia, 181–183, 191
Axo-axonic synapses, 80
Axon collaterals
   intrinsic to globus pallidus and STN, 120–122, 367
   intrinsic to striatum, 32, 33
   of pallidothalamic and nigrothalamic neurons, 175
   of spinocerebellar tract projecting to VL, 110

Baclofen, 129
Ballism, 287, 311; *see also* Hemiballism
Basal Ganglia
   control of behavior, 13, 30, 181–183
   ...and initiation of movements, 14
   inputs to thalamus, 80, 136, 215
   lesions, 14, 16, 119, 137–143, 200, 219–221
   models, 3–4, 6, 9, 11–19, 175, 183, 244, 309
   outputs, 30, 77, 93, 135–136, 140–143, 169, 176–178
   receiving area in thalamus, 77–78, 94, 96, 102, 258, 263
Basal forebrain, 64
Behavior
   cognitive aspects, 73–75
   control by basal ganglia, 13, 30
   emotional, 336–337
   hypomanic, 334
   manic, 333–335, 337
   ...and mood, 334–335
   motor, 67–74, 98, 164–170, 180
   waking motor..., 181
Bicuculine, 180, 197, 244
Blepharospasm, as a model of dystonia, 201–202

Bradykinesia, 15, 19, 102, 141, 151, 180, 200, 260, 262, 267, 272, 320–323, 330
BSIZ (bulbospinal inhibitory zone), 6, 8
Bursting activity, 286–291; 310, 366; *see also* Neurons
   ...Ca^{2+} oscillatory neurons, 275, 277–279
   effect on transmitter release, 252

Calbindin, 32, 55, 120
Calcium binding proteins, 32, 34, 120
Calcium spikes (low threshold), 179, 267, 273, 366
Calretinin, 120
Carbochol, 164–167, 180
Caudate nucleus, 48–50, 163, 190, 357; *see also* Striatum
   moduli in..., 48–50
Center-surround organization, 141, 242
Centrolateral nucleus, 64, 217
Centromedian nucleus, *see also* Intralaminar nuclei
   connections, 12, 64, 123, 178, 217–218
   GABA receptors, 219
   neuronal activity, 277–279
   stimulation, 310
   synapses, 80, 85
Cerebellar
   afferent receiving areas in thalamus, 78, 94, 101, 105, 258–263, 271–272
   axon terminal fields in thalamus, 83
   cortex, electronic model of network, 304
   monosynaptic EPSP's in thalamus, 109
   motor deficits, 114, 273
   multisecond oscillations, 144
   nuclei, 94, 258, 273
   synapses in motor thalamus, 82–83, 106, 215, 217
Cerebral palsy, 287
Cholinergic
   agonists, 164–169, 180
   antagonists, 164–169
   cell groups, 176, 179
   control of movement, 168–170
   interneurons, 32, 137, 164
   mechanisms in behavior, 169–170
   neurons, 32, 177–179, 181
   pathways, 177–179